Remote Sensing Monitoring of Land Surface Temperature (LST)

Remote Sensing Monitoring of Land Surface Temperature (LST)

Editors

Juan Manuel Sánchez
César Coll
Raquel Niclòs

MDPI • Basel • Beijing • Wuhan • Barcelona • Belgrade • Manchester • Tokyo • Cluj • Tianjin

Editors
Juan Manuel Sánchez
Applied Physics Department
University of Castilla-La Mancha
Albacete
Spain

César Coll
Earth Physics and
Thermodynamics Department
University of Valencia
Burjassot
Spain

Raquel Niclòs
Earth Physics and
Thermodynamics Department
University of Valencia
Burjassot
Spain

Editorial Office
MDPI
St. Alban-Anlage 66
4052 Basel, Switzerland

This is a reprint of articles from the Special Issue published online in the open access journal *Remote Sensing* (ISSN 2072-4292) (available at: www.mdpi.com/journal/remotesensing/special_issues/land_tem).

For citation purposes, cite each article independently as indicated on the article page online and as indicated below:

LastName, A.A.; LastName, B.B.; LastName, C.C. Article Title. *Journal Name* **Year**, *Volume Number*, Page Range.

ISBN 978-3-0365-1428-4 (Hbk)
ISBN 978-3-0365-1427-7 (PDF)

Contents

About the Editors

Juan Manuel Sánchez

Juan M. Sánchez is currently an associate professor with the Applied Physics Department and a researcher with the Regional Development Institute at the University of Castilla-La Mancha, Spain. He received his BSc, MSc, and PhD degrees in physics from the University of Valencia, Spain, in 2003, 2005, and 2008, respectively. His research interests focus on thermal infrared remote sensing in general and on surface energy balance in particular. He has published more than 50 papers in international journals, 6 book chapters, and more than 100 conference papers. He has participated in 25 international, national, and regional projects (PI for two of them), and he is a referee in 18 international journals. Dr. Sánchez was awarded the Nobert Gerbier-MUMM Internatinal Award 2010 by the World Meteorological Organization.

César Coll

César Coll is a professor of earth physics with the Department of Earth Physics and Thermodynamics, Faculty of Physics, University of Valencia, since 2010 and was an associate professor at the same institution from 1997. He received his BSc, MSc, and PhD degrees in physics from the University of Valencia, Spain, in 1989, 1992, and 1994, respectively. His research interests focus on the physical processes of thermal-infrared (TIR) remote sensing; atmospheric and emissivity corrections; temperature-emissivity separation; and ground validation of TIR products from AATSR, MODIS, ASTER, and other satellite sensors. He has published more than 60 papers in SCI international journals. Dr. Coll is currently the director of the Remote Sensing Master's degree program at the University of Valencia.

Raquel Niclòs

Raquel Niclos has been an associate professor at the Earth Physics and Thermodynamics Department of the University of Valencia (UV) since 2017. She received her BSc, MSc, and PhD degrees in physics from the UV in 2000, 2002, and 2005, respectively. She has 20 years of research experience in Earth physics and has published more than 40 papers in journals included in the Journal Citation Reports (JCR), with more than 30 within the top quartile. She has also published book chapters and non-JCR papers, and she was an editor of an international book about thermal remote sensing advances. She has been the supervisor to six PhD theses, the principal investigator of three R&D&i projects in competitive calls and of five contracts with outstanding companies, a member of five international scientific committees, a regular reviewer for seven relevant JCR journals, and a member of scientific and organizing committees of symposiums.

Preface to "Remote Sensing Monitoring of Land Surface Temperature (LST)"

The combination of the state-of-the-art in the thermal infrared domain with recent advances in the capabilities provided by already operating and new satellites, or UAV-based or aerial remote sensing is boosting the use of land surface temperature (LST) in a variety of research fields. LST plays a key role in soil–vegetation–atmosphere processes and is crucial in the estimation of surface energy flux exchanges, actual evapotranspiration, or vegetation and soil properties. Additionally, LST is considered one of the Essential Climate Variables (ECV) that critically contributes to the characterization of Earth's climate. The latest advances in data fusion, downscaling, and disaggregation techniques provide a new dimension to LST applications in water resource and agronomic management thanks to improvements in both the temporal and spatial resolutions of thermal products. However, at the same time, continuous research into LST estimation algorithms as well as continuous calibration and validation are still required to improve the accuracy of ground LST data and satellite LST products. Although much research needs to be conducted on the topic of LST monitoring from remote sensing, we truly hope that the selection of papers published in this book can help readers become aware of the potential of orbiting thermal sensors and of the necessity to give them continuity and help them develop and launch higher spatiotemporal resolution platforms.

Juan Manuel Sánchez, César Coll, Raquel Niclòs
Editors

remote sensing

Editorial

Editorial for the Special Issue "Remote Sensing Monitoring of Land Surface Temperature"

Juan M. Sánchez [1,*], César Coll [2] and Raquel Niclòs [2]

1   Applied Physics Department, Regional Development Institute, University of Castilla-La Mancha, Campus Universtiario s/n, 02071 Albacete, Spain
2   Department of Earth Physics and Thermodynamics, Faculty of Physics, University of Valencia, C/Dr. Moliner, 50, 46100 Burjassot, Spain; cesar.coll@uv.es (C.C.); Raquel.Niclos@uv.es (R.N.)
*   Correspondence: juanmanuel.sanchez@uclm.es

check for updates

Citation: Sánchez, J.M.; Coll, C.; Niclòs, R. Editorial for the Special Issue "Remote Sensing Monitoring of Land Surface Temperature". *Remote Sens.* 2021, 13, 1765. https://doi.org/10.3390/rs13091765

Received: 28 April 2021
Accepted: 29 April 2021
Published: 1 May 2021

Publisher's Note: MDPI stays neutral with regard to jurisdictional claims in published maps and institutional affiliations.

The combination of the state-of-the-art in the thermal infrared (TIR) domain [1–3] with the recent advances in the capabilities provided by operating and new satellites [4–10], UAV-based [11] or aerial remote sensing are boosting the use of land surface temperature (LST) in a variety of research fields [5,8,9,11,12]. LST plays a key role in soil–vegetation–atmosphere processes and becomes crucial in the estimation of surface energy flux exchanges, actual evapotranspiration, or vegetation and soil properties [8,9]. The latest advances in data fusion, downscaling and disaggregation techniques provide a new dimension to LST applications in water resource and agronomic management thanks to the improvement in both the temporal and spatial resolution of thermal products [8–10]. However, at the same time, continuous research into LST estimation algorithms, as well as continuous calibration and validation, are still required to improve the accuracy of ground LST data and satellite LST products [1–5,13,14].

Our aim with this Special Issue was to collect recent developments, methodologies, calibration and validation and applications of thermal remote sensing data and derived products, from UAV-based remote sensing, aerial remote sensing and satellite remote sensing. A total of 20 manuscripts were submitted to our Special Issue and after rigorous peer-review process, by around 50 anonymous reviewers, 14 papers were finally selected for publication, by a total of 69 authors. The published papers were those of high-quality content based on their cutting-edge remote sensing techniques. The geographical distribution of the authors´ institutions is global, with the highest number from the USA (18), followed by China (15), Spain (7), UK (6), Sweden and Korea (5 each) and many others, such as The Netherlands, Denmark, Japan, Germany, Italy or Turkey (1 to 3 each).

Published papers cover a wide range of topics, which can be classified in five groups: algorithms, calibration and validation [1–4], improving long-term consistency in satellite LST [5–7], downscaling LST [8–10], LST applications [11,12] and land surface emissivity research [13,14].

In total, three papers have been included dealing with algorithms to retrieve LST from the Landsat series [1–3]. Gerace et al. [1] progressed towards developing an operational split-window algorithm for TIRS on board Landsat 8 and 9, that might improve the accuracy achieved by the current single-channel methodology used to derive LST in the Landsat Collection 2 surface temperature product. The effect of the stray-light correction implemented in Landsat 8 was evaluated by Guo et al. [2] using ground-measured LST from SURFRAD sites. Data from this SURFRAD network, together with ARM, were used by Sekertekin et al. [3] to examine the efficiency of different LST algorithms for daytime and nighttime Landsat 8 images. Despite the feasibility of the assessment results reported, a necessity for more robust and homogeneous validations, using ground-measured datasets, is recognized [2].

Geostationary satellites are also present in this Special Issue. Long-term, consistent LST archives must account for geostationary satellite sensor updates and Pinker et al. [5]

developed a framework to achieve this goal. Choi and Suh [4] developed a nonlinear split-window LST retrieval algorithm for the next-generation geostationary satellite in Korea, GEO-KOMPSAT-2A.

The applicability of remote sensing LSTs is sometimes compromised in areas that are very frequently covered with clouds. Aware of this issue, Zhang et al. [6] and Yoo et al. [7] introduced approaches for the gap-filling of MODIS LST data, by reconstructing 1 km clear-sky LST using Bayesian methods [6] or random forest machine learning [7]. This strategy can improve the applicability of LSTs in a variety of research and practical fields.

As mentioned above, ET modeling from surface energy balance benefits from LST datum as a key input. Field-scale evapotranspiration modelling requires high spatio-temporal resolution in the thermal data. This Special Issue includes recent efforts by [8–10] to fill this gap until next generation of thermal satellites are launched. Sánchez et al. [8] produced LST maps with 10 m spatial resolution from the combination of MODIS/Sentinel-2 images and validated their methodology using a ground-based LST dataset gathered in an agricultural area. Guzinski et al. [9] evaluated several approaches for improving the spatial resolution of the thermal images by merging Sentinel-2 and Sentinel-3 satellite data. The resulting data were used to produce surface energy fluxes that were then validated against flux tower observations in a variety of land covers and climatological conditions. Downscaling approaches also apply to geostationary satellites, increasing the frequency of the LST estimates. Njuki et al. [10], in their work also included in this Special Issue, presented an approach, based on random forest regression, to downscale the coarse-resolution MSG-SEVIRI to 30 m spatial resolution, based on predictor variables derived from Sentinel-2 and the ALOS digital elevation model. Although results reported are promising, particularly for the joint use of the tandem Sentinel-3/Sentinel-2, certain limitations remain that encourage further research.

Urban environments can be explored from a thermal perspective by using high-resolution drone solutions. The Special Issue includes a good example by Naughton and McDonald [11]. Findings shown by these authors elucidate factors that can be applied to develop better temperature mitigation practices to protect human and environmental health. Another potential application of LST data is the use of continuous satellite-derived surface temperatures as input in geophysical models, substituting discrete in situ air temperature registers extrapolated to different elevations using constant lapse rates, then providing more realistic estimates. An example is shown by [12] using MODIS imagery.

Field and laboratory emissivity measurements are essential for improving and validating LST retrievals [13,14]. Temperature and emissivity separation algorithms can be applied when multispectral thermal radiances are available. A manuscript in this Special Issue by [13] explored the influence mechanism of noise on the LST and surface emissivity retrieval errors of the ARTEMISS algorithm. The authors proposed an improved method for thermal data with a high noise level and high spectral resolution, which can reduce LST and emissivity uncertainties. Langsdale et al. [14] made measurements of manmade and natural samples under different environmental conditions, both in situ and at laboratory. Differences between laboratory and field spectral measurements highlighted the importance of field methods for these samples, with the laboratory setup unable to capture sample structure or inhomogeneity. The emissivity box method was faced to FTIR-based approaches, showing significant differences in LST retrieval and then stressing the importance of correct emissivity data specifications.

Although there is much work to be done on the topic of LST monitoring from remote sensing, we truly hope that the selection of papers published in this Special Issue can help research communities to become aware of the potential of the orbiting thermal sensors, the necessity to give them continuity and also to develop and launch higher spatio-temporal resolution platforms.

Author Contributions: The guest editors contributed equally to all aspects of this editorial. All authors have read and agreed to the published version of the manuscript.

Funding: This work has been developed within the framework of the ANIATEL project (SB-PLY/17/180501/000357), financed by the Sport, Culture and Education Council (JCCM, Spain), together with FEDER funds.

Acknowledgments: The guest editors would like to thank the authors who contributed to this Special Issue and to the anonymous reviewers who dedicated their time for providing with valuable and constructive recommendations.

Conflicts of Interest: The guest editors declare no conflict of interest.

References

1. Gerace, A.; Kleynhans, T.; Eon, R.; Montanaro, M. Towards an Operational, Split Window-Derived Surface Temperature Product for the Thermal Infrared Sensors Onboard Landsat 8 and 9. *Remote Sens.* **2020**, *12*, 224. [CrossRef]
2. Guo, J.; Ren, H.; Zheng, Y.; Lu, S.; Dong, J. Evaluation of Land Surface Temperature Retrieval from Landsat 8/TIRS Images before and after Stray Light Correction Using the SURFRAD Dataset. *Remote Sens.* **2020**, *12*, 1023. [CrossRef]
3. Sekertekin, A.; Bonafoni, S. Sensitivity Analysis and Validation of Daytime and Nighttime Land Surface Temperature Retrievals from Landsat 8 Using Different Algorithms and Emissivity Models. *Remote Sens.* **2020**, *12*, 2776. [CrossRef]
4. Choi, Y.; Suh, M. Development of a Land Surface Temperature Retrieval Algorithm from GK2A/AMI. *Remote Sens.* **2020**, *12*, 3050. [CrossRef]
5. Pinker, R.; Ma, Y.; Chen, W.; Hulley, G.; Borbas, E.; Islam, T.; Hain, C.; Cawse-Nicholson, K.; Hook, S.; Basara, J. Towards a Unified and Coherent Land Surface Temperature Earth System Data Record from Geostationary Satellites. *Remote Sens.* **2019**, *11*, 1399. [CrossRef]
6. Zhang, Y.; Chen, Y.; Li, Y.; Xia, H.; Li, J. Reconstructing One Kilometer Resolution Daily Clear-Sky LST for China's Landmass Using the BME Method. *Remote Sens.* **2019**, *11*, 2610. [CrossRef]
7. Yoo, C.; Im, J.; Cho, D.; Yokoya, N.; Xia, J.; Bechtel, B. Estimation of All-Weather 1 km MODIS Land Surface Temperature for Humid Summer Days. *Remote Sens.* **2020**, *12*, 1398. [CrossRef]
8. Sánchez, J.M.; Galve, J.M.; González-Piqueras, J.; López-Urrea, R.; Niclòs, R.; Calera, A. Monitoring 10-m LST from the Combination MODIS/Sentinel-2, Validation in a High Contrast Semi-Arid Agroecosystem. *Remote Sens.* **2020**, *12*, 1453. [CrossRef]
9. Guzinski, R.; Nieto, H.; Sandholt, I.; Karamitilios, G. Modelling High-Resolution Actual Evapotranspiration through Sentinel-2 and Sentinel-3 Data Fusion. *Remote Sens.* **2020**, *12*, 1433. [CrossRef]
10. Njuki, S.; Mannaerts, C.; Su, Z. An Improved Approach for Downscaling Coarse-Resolution Thermal Data by Minimizing the Spatial Averaging Biases in Random Forest. *Remote Sens.* **2020**, *12*, 3507. [CrossRef]
11. Naughton, J.; McDonald, W. Evaluating the Variability of Urban Land Surface Temperatures Using Drone Observations. *Remote Sens.* **2019**, *11*, 1722. [CrossRef]
12. Singh, S.; Bhardwaj, A.; Singh, A.; Sam, L.; Shekhar, M.; Martín-Torres, F.; Zorzano, M. Quantifying the Congruence between Air and Land Surface Temperatures for Various Climatic and Elevation Zones of Western Himalaya. *Remote Sens.* **2019**, *11*, 2889. [CrossRef]
13. Shao, H.; Liu, C.; Xie, F.; Li, C.; Wang, J. Noise-sensitivity Analysis and Improvement of Automatic Retrieval of Temperature and Emissivity Using Spectral Smoothness. *Remote Sens.* **2020**, *12*, 2295. [CrossRef]
14. Langsdale, M.; Dowling, T.; Wooster, M.; Johnson, J.; Grosvenor, M.; de Jong, M.; Johnson, W.; Hook, S.; Rivera, G. Inter-Comparison of Field- and Laboratory-Derived Surface Emissivities of Natural and Manmade Materials in Support of Land Surface Temperature (LST) Remote Sensing. *Remote Sens.* **2020**, *12*, 4127. [CrossRef]

remote sensing

Article

Towards an Operational, Split Window-Derived Surface Temperature Product for the Thermal Infrared Sensors Onboard Landsat 8 and 9

Aaron Gerace *[image_ref id=3], Tania Kleynhans, Rehman Eon and Matthew Montanaro

Rochester Institute of Technology, 54 Lomb Memorial Drive, Rochester, NY 14624, USA; tkpci@rit.edu (T.K.); rse4949@rit.edu (R.E.); montanaro@cis.rit.edu (M.M.)
* Correspondence: gerace@cis.rit.edu; Tel.: +1-585-475-4388

Received: 29 October 2019; Accepted: 6 January 2020; Published: 9 January 2020

Abstract: The split window technique has been used for over thirty years to derive surface temperatures of the Earth with image data collected from spaceborne sensors containing two thermal channels. The latest NASA/USGS Landsat satellites contain the Thermal Infrared Sensor (TIRS) instruments that acquire Earth data in two longwave infrared bands, as opposed to a single band with earlier Landsats. The United States Geological Survey (USGS) will soon begin releasing a surface temperature product for Landsats 4 through 8 based on the single spectral channel methodology. However, progress is being made toward developing and validating a more accurate and less computationally intensive surface temperature product based on the split window method for Landsat 8 and 9 datasets. This work presents the progress made towards developing an operational split window algorithm for TIRS. Specifically, details of how the generalized split window algorithm was tailored for the TIRS sensors are presented, along with geometric considerations that should be addressed to avoid spatial artifacts in the final surface temperature product. Validation studies indicate that the proposed algorithm is accurate to within 2 K when compared to land-based measurements and to within 1 K when compared to water-based measurements, highlighting the improved accuracy that may be achieved over the current single-channel methodology being used to derive surface temperature in the Landsat Collection 2 surface temperature product. Surface temperature products using the split window methodologies described here can be made available upon request for testing purposes.

Keywords: Landsat; land surface temperature; split window algorithm; TIRS; thermal

---

1. Introduction

The United States Geological Survey's (USGS) Earth Resources Observation and Science (EROS) Center will begin distributing higher-level products derived from Landsat image data as part of their Collection 2 release in early 2020. A global surface temperature (ST) product will be included in Collection 2 and will contain over thirty-five years of data collected from the various thermal instruments onboard Landsats 4 through 8. A single-channel algorithm that utilizes the Goddard Earth Observing System, Version 5 (GEOS-5) reanalysis data for atmospheric characterization along with a radiative transfer model (e.g., MODTRAN) will be applied to the existing thermal data archive and to newly collected scenes in a near real-time fashion to produce per-pixel, 30 m surface temperature data [1,2]. On the other hand, the recent Landsat 8 and upcoming Landsat 9 missions both contain the dual-band Thermal Infrared Sensor (TIRS) [3] that will enable the use of the split-window surface temperature algorithm. Recent software updates to the Landsat 8/TIRS image processing flow have mitigated the adverse effects of the stray light issue [4] that precluded the use of the split-window method. Once the radiometric quality of TIRS image data had been brought back to within requirements, work began on developing a split window algorithm tailored to TIRS image data. Several considerations were addressed before using data from these instruments to derive Earth surface temperature with the more accurate and computationally attractive split window algorithm.

This paper presents progress made towards developing and verifying an operational version of the split window algorithm for the TIRS instruments. Specifically, a discussion of the radiometric performance of the split window algorithm versus the single channel methodology implemented in the Collection 2 release is provided with ground-based measurements used as a baseline for reference. Surface measurements from several sites across the continental United States and near-shore and inland buoys were used to demonstrate the improved radiometric performance that may be achieved in future releases of ST products using dual-band Landsat thermal instruments.

An additional discussion regarding the spatial fidelity of split window-derived versus single channel-derived surface temperature products is also provided. Due to the physical layout of the TIRS focal plane, the two thermal channels, Band 10 and Band 11, acquire scene content at slightly different times. As such, an inherent misregistration of image data is evident in the corresponding Level 1 radiance product. Although band-to-band registration is well-within the defined specification [5], applying the difference terms in the split window algorithm (see Equation (1)) leads to undesired artifacts in the resulting surface temperature product. The origin of, and strategies to mitigate, these artifacts are presented along with future considerations necessary to achieve an operational split window product from Landsat 8 and 9 TIRS image data for Collection 3 processing.

2. Methodologies

Leveraging over twenty years of knowledge and refinements, the split window algorithm used in this work was initially proposed by Becker and Li, (1990) for the AVHRR instrument [6]. Wan and Dozier, (1996) generalized the algorithm to enable its utility for other dual-band instruments with a final adjustment made by Wan, (2014) to improve its performance over bare soils for the MODIS instruments [7,8]. The final form of the split window algorithm used here is

$$ST = b_0 + \left(b_1 + b_2 \frac{1-\epsilon}{\epsilon} + b_3 \frac{\Delta\epsilon}{\epsilon^2} \right) \frac{T_i + T_j}{2} + \left(b_4 + b_5 \frac{1-\epsilon}{\epsilon} + b_6 \frac{\Delta\epsilon}{\epsilon^2} \right) \frac{T_i - T_j}{2} + b_7(T_i - T_j)^2, \quad (1)$$

where

- ST is the desired surface temperature $[K]$;
- b_k (for k = 0, 1, ..., 7) are sensor-dependent (and potentially water-vapor-dependent) coefficients that are derived through a training process;
- i and j correspond to the two thermal bands (Bands 10 and 11 for TIRS);
- $\Delta\epsilon = \epsilon_i - \epsilon_j$ or the difference in band-effective emissivities;
- $\epsilon = (\epsilon_i + \epsilon_j)/2$ or the average of the band-effective emissivities;
- T_i, T_j are the apparent temperatures in the two thermal bands.

Once the b-coefficients are derived for a sensor of interest, the ST can be estimated from dual-band thermal image data using Equation (1) if the effective emissivity in each band is known (or can be estimated). This section provides the details of a prototype split window implementation developed for the Landsat 8/TIRS and Landsat 9/TIRS-2 sensors and a corresponding validation effort conducted thus far for Landsat 8/TIRS.

2.1. Derivation of the b-Coefficients

The flowchart in Figure 1 illustrates the training process that was performed to derive the b-coefficients shown in Equation (1). The radiative transfer model, MODTRAN [9], was used to simulate a representative range of environmental acquisition parameters. Atmospheric profiles from the Thermodynamic Initial Guess Retrieval (TIGR) database [10] were used to characterize atmospheric effects; seven surface temperatures bracketing the temperature of the lowest layer of each atmospheric profile were used as input [6,7]; and spectral emissivities of natural materials were obtained from the MODIS UCSB emissivity library [11].

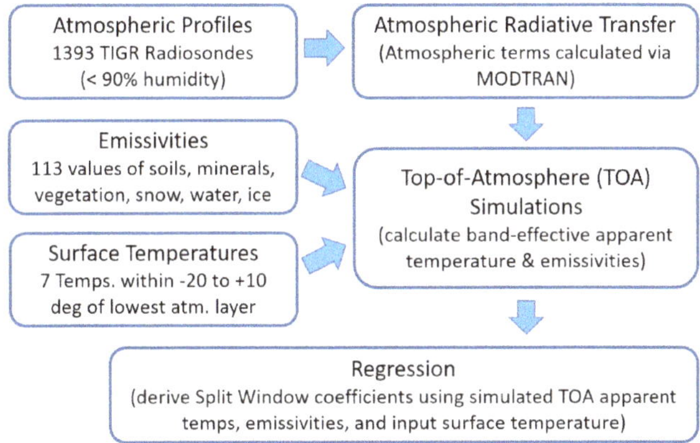

Figure 1. Process flow to derive split window coefficients by propagating modeled surface temperatures, emissivities, and atmospheric conditions to the top of atmosphere and regressing the band-effective apparent temperatures against the known input surface temperatures.

The TIGR database is a climatological library of 2311 unique atmospheric profiles that were categorized from 80,000 radiosondes. The profiles are classified into air masses (i.e., Tropical, Mid-lat1, Mid-lat2, Polar1, Polar2) that are consistent with MODTRAN's default atmospheres but provide a richer and more densified representation of potential atmospheric effects that may be observed from a spaceborne platform. Temperature, water vapor, and ozone data are delivered at 43 predefined pressure levels ranging from 1013 mb (millibars) to 0.0026 mb [10]. Figure 2 shows plots of the 2311 atmospheric profiles provided in the TIGR database categorized by airmass. For comparison, Figure 2 (bottom right) shows the default MODTRAN profiles.

To be consistent with previous efforts and to satisfy the assumption that split window is most appropriate to be used for surface temperature retrieval when the ground temperature is close to the air temperature [6,7], seven surface temperatures bracketing the temperature of the lowest layer of each atmospheric profile were used as input to the forward model. Surface temperatures between $-10\,°C < t_0 < 20\,°C$ in $5\,°C$ steps were used in this study, where t_0 is the temperature of the lowest atmospheric layer. While some specific applications may warrant an extension of the range used to train the model (e.g., studies of urban heat island), the traditional range used here is appropriate for natural materials.

The MODIS UCSB emissivity database was used to provide a representative range of natural materials as input to the forward model [11]. With 74 unique soils and minerals, 28 unique types of vegetation, and 11 forms of water (including snow and ice), this database provides 113 unique spectral emissivities between 8 and 14 μm that can be used to train the model in Equation (1). Note that man-made materials are included in the MODIS UCSB emissivity database but were excluded in this study as the TIRS instrument has a spatial resolution of 100 m and was designed for environmental applications.

Referring again to Figure 1, all parameters described above were provided as input to a forward model that uses MODTRAN for the atmospheric radiative transfer process to generate at-sensor spectral radiance. At-sensor, band-effective radiance was calculated by sampling the simulated top-of-atmosphere spectra with the TIRS spectral response functions, and apparent temperatures (T_i, T_j) were determined by developing and utilizing a predefined look-up table that relates band-effective radiance to blackbody temperature for Bands 10 and 11 of TIRS. Finally, band-effective emissivities were calculated by sampling the 113 spectral emissivities with the TIRS spectral response functions.

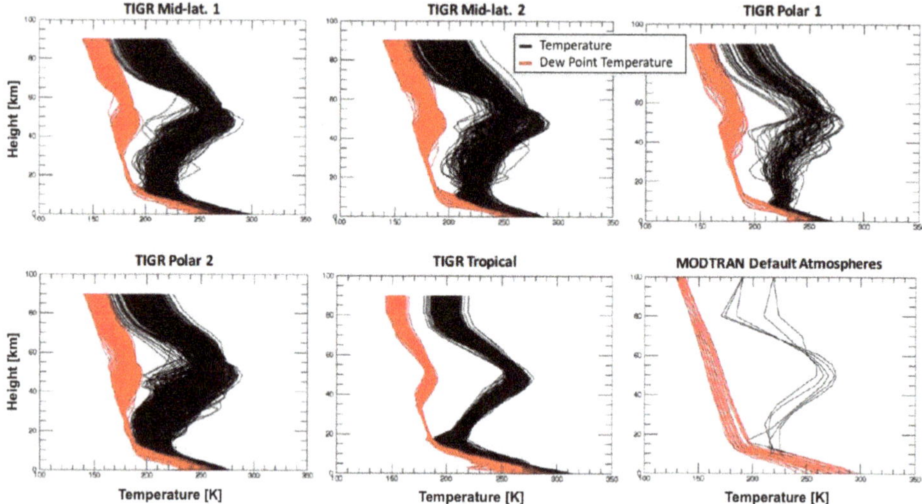

Figure 2. TIGR atmospheric profiles for various atmospheric types compared to MODTRAN's default atmospheres. The black data curves are air temperatures as a function of altitude, while the red data curves are the associated dew point temperatures.

Note from Figure 1 that modeled data were filtered to only include scenarios where the relative humidity is less than 90%. This was performed to remain consistent with previous studies [12] and to eliminate saturated atmospheric conditions, which represents a challenging scenario for ST retrieval. Future work will explore and include higher humidity cases as needed. Nevertheless, with all the components of Equation (1) determined, ST was regressed against the independent variables to determine the b-coefficients that best fit the model in a least-squares sense. Table 1 shows a comparison of the b-coefficients derived in this study versus the b-coefficients derived in Du et al. (2015) [12], along with the residual retrieval error when these coefficients are applied to the simulated data. Note that although the same split window algorithm was used, the derived b-coefficients could be significantly different due to the desired application. For example, Du et al. incorporated man-made materials into their training process to enable the utility of split window applications over urban areas. The impact of this training methodology on environmental applications is discussed in Sections 3 and 4.

Table 1. Split window algorithm b-coefficients derived in this study compared the coefficients derived by Du et al. and the associated root mean square error of the model fits.

	b_0	b_1	b_2	b_3	b_4	b_5	b_6	b_7	RMSE [K]
Du et al. (2015)	−0.4117	0.0052	0.1454	−0.2730	4.0666	−6.9251	−18.2746	0.2447	0.87
Proposed Prototype	2.2925	0.9929	0.1545	−0.3122	3.7186	0.3502	−3.5889	0.1825	0.73

2.2. Emissivity Estimation

Once the b-coefficients are derived, estimation of emissivity remains the one unknown in Equation (1). To be consistent with the single channel methodology used to derive Landsat surface temperature products in the Collection 2 release, the algorithm used to estimate broadband emissivity in the existing single channel workflow was mirrored in this study but extended for the TIRS dual-band instrument. To summarize the existing workflow, ASTER emissivity products that spatially cover the Landsat scene of interest are ingested and a spectral adjustment is made to estimate the equivalent

TIRS emissivities. The spectrally-adjusted emissivities are then modified based on observed in-scene conditions (e.g., emissivities may be modified if snow or vegetation is present in a scene).

The ASTER global emissivity dataset (ASTER-GED) v3 contains worldwide emissivity maps at 100 m spatial resolution. The dataset was compiled using clear-sky scenes acquired between 2000 and 2008. Emissivities were calculated with the temperature emissivity separation algorithm (TES) and the water vapor scaling (WVS) atmospheric correction algorithm, and are available for all five ASTER TIR bands centered at 8.3, 8.6, 9.1, 10.6, and 11.3 μm. The ASTER-GED has been validated to an absolute band error of 1% [13].

To enable an adjustment of the ASTER emissivities to the spectral response of the TIRS bands, a linear relationship that relates ASTER-observed (Bands 13 and 14) to TIRS-estimated (Bands 10 and 11) emissivities was developed. Note that ASTER Bands 13 and 14 were used here as they have the most overlap (spectrally) with the TIRS bands. To develop this relationship, the 113 spectral emissivities from the MODIS UCSB emissivity database described in Section 2.1 were used. Band-effective emissivities for Bands 10 and 11 of TIRS were regressed against the corresponding band-effective emissivities for Bands 13 and 14 of ASTER to derive the coefficients shown in Equations (2) and (3).

$$\epsilon_{10} = c_0 + c_1\epsilon_{13} + c_2\epsilon_{14} \tag{2}$$
$$\epsilon_{11} = c_0 + c_1\epsilon_{13} + c_2\epsilon_{14} \tag{3}$$

where,

$$(c_0, c_1, c_2) = (0.6820, 0.2578, 0.0584) \text{ for TIRS Band 10,}$$
$$(c_0, c_1, c_2) = (-0.5415, 1.4305, 0.1092) \text{ for TIRS Band 11.}$$

Note that an estimation of the residual errors associated with these relationships can be made by applying Equations (2) and (3) to the band-effective ASTER data for the 113 emissivities. The residual errors between the estimated band-effective emissivities can then be compared to the actual band-effective emissivities (as modeled here). The standard deviations of the residual emissivities in this simulated context are 0.001 (0.1%) and 0.005 (0.5%) for Bands 10 and 11, respectively.

Since the ASTER emissivity database represents averages over a nine-year period, modifications were made to the spectrally adjusted emissivities based on observations made by the Operational Land Imager (OLI), the TIRS reflective band counterpart onboard Landsat 8. Specifically, per-pixel normalized difference vegetation indices (NDVI) and normalized difference snow indices (NDSI) were calculated with the OLI. NDSI was computed by dividing the difference in reflectance observed in the Landsat 8 green band (0.53–0.59 μm) and the shortwave infrared band (1.57–1.65 μm) by the sum of the two bands [14]. To make the NDVI adjustment, bare soil locations were estimated when the ASTER NDVI data were less than 0.1, and the Landsat vegetation emissivities adjusted accordingly based on the Landsat calculated NDVI. Snow locations for NDSI were set to 0.9876 and 0.9724 respectively for Band 10 and Band 11, where the calculated NDSI was larger than 0.4. A comprehensive description of the adjustments can be found in Malakar et al. (2018) [1].

2.3. Surface Reference Data Sources

Several sources of surface measurements were used as reference to validate the efficacy of the split window algorithm as trained here for Landsat's TIRS instruments. Several land-based instrumented sites, including three sites from the SURFRAD [15] network and one site from the Ameriflux [16] network, were used in the assessment. Additionally, the National Oceanic and Atmospheric Administration (NOAA) buoy network [17] and the NASA Jet Propulsion Laboratory (JPL) instrumented buoys [18,19] were used to provide reference data over water.

NOAA established the surface radiation budget observing network (SURFRAD) to provide accurate, high-quality broadband solar and thermal upwelled and downwelled irradiance to support climate research, satellite retrieval validation and modeling, and weather forecasting research [15]. The current SURFRAD network consists of seven locations selected to represent diverse climates in the United States [15]. Note that three sites were chosen for this initial analysis due to their high spatial uniformity across an extended region. The three sites consist of agricultural land (Goodwin, Mississippi, US), bare soil (Desert Rock, Nevada, US), and grassland with a high inter-annual variation of snow cover (Fort Peck, Montana, US).

Each SURFRAD site is equipped with two Eppley Precision Infrared Pyrgeometers (model PIR) to collect measurements of broadband (4–50 µm) thermal infrared irradiance. The PIR pyrgeometers have a field-of-view (FOV) of 180 degrees and measure longwave irradiance with an uncertainty of ~1.5% [20], which leads to a reported uncertainty of less than 1 K in the retrieved LST [21]. One pyrgeometer is upward facing and the other is downward facing to measure downwelled atmospheric irradiance and upwelled surface-leaving irradiance, respectively. Data from 1998 to 2009 were collected every three minutes, and every minute thereafter. The data has a quality flag to indicate failed internal quality checks. A detailed description of the SURFRAD instrumentation at each site can be found at: https://www.esrl.noaa.gov/gmd/grad/surfrad/overview.html.

FLUXNET is a vast global network of more than 800 sites for in-situ flux measurement. Regional networks contribute to the FLUXNET data, one of which is a group of sites across the Americas called AmeriFlux. There are hundreds of AmeriFlux sites, with 44 flagged as "core" sites. These core sights deliver timely data, receiving support from the AmeriFlux Management Project (AMP) to ensure high quality data collection at 30 min intervals. Since not all sites measure upwelled and downwelled thermal radiation, the core sights were filtered for spatial uniformity, activity between 2013 to 2018, and having a sufficient number of upwelled and downwelled infrared observations. Only one site passed these criteria; namely, the University of Michigan Biological Station (UMBS) [22]. This site is located within a protected forest of mid-aged northern hardwoods, conifer understory, aspen and old growth hemlock. The UMBS AmeriFlux site is equipped with a CG4 pyrgeometer from Kipp and Zonen to measure broadband (4.5 to 42 µm) thermal irradiance. The CG4 pyrgeometer, similar to the SURFRAD instrumentation, has an FOV of 180 degrees with an instrument uncertainty of less than 3% [20], and temperature uncertainty of ±0.02 K [23].

To estimate the in-situ ST using SURFRAD and AmeriFlux networks, the Stefan–Boltzmann law is manipulated to derive the following relationship [15]:

$$ST_{ground} = \left[\frac{1}{\epsilon\sigma} (E_{upwelled} - (1-\epsilon)E_{downwelled}) \right]^{\frac{1}{4}} [K] \tag{4}$$

where ϵ represents the broadband emissivity, $\sigma = 5.67 \cdot 10^{-8} \left[\frac{W}{m^2K^4} \right]$ is the Stefan–Boltzmann constant, and E is the measured irradiance $\left[\frac{W}{m^2} \right]$. Broadband emissivity can be retrieved from narrowband satellite emissivities via empirical relationships (Wang et al., 2005) [24]. However, this approach uses a combination of broadband emissivity from 8 to 12 µm and 14 to 25 µm. The latter range is from an emissivity library containing only measurements of minerals and does not include data beyond 25 µm because of the strong atmospheric absorption and weak thermal signals. For these reasons, the average emissivity of TIRS Bands 10 and 11 that is estimated from image data (as described in Section 2.2) was used in Equation (4) for this analysis.

When used in conjunction with land-based measurements, water represents a desirable target for surface temperature validation, as its emissivity is spectrally stable and well-defined [25]. NOAA operates a suite of worldwide instrumented buoys that collect, among other variables, water temperature. The data are freely available and delivered through their National Data Buoy Center website [17]. Measurements from thirty-six buoys in the near-shore of the United States coastline were used as reference in this work, with a bulk to surface adjustment, since measurements are obtained at depth [26]. Note that Zeng et al. (1999) estimate the uncertainty in skin temperature estimation to be approximately 0.35 K, which includes measurement uncertainty.

In addition to the NOAA sensor suite, NASA's Jet Propulsion Laboratory's (JPL) instrumented buoys located in Lake Tahoe, California and Salton Sea, California are attractive sources of reference data. Lake Tahoe is approximately 1900 m above sea level, and with average lake temperatures ranging from 5 to 25 °C throughout the year [18], it is an attractive cold water target for surface temperature validation. Alternatively, Salton Sea is located in Southern California and is approximately 70 m below sea level. With lake temperatures exceeding 35 °C, it is an attractive warm water target [19]. The JPL data used for the validation efforts presented here are made freely available by JPL [18,19].

Referring to Table 2, over 1500 Landsat Level-1 Terrain-Corrected (L1T) TIRS scenes acquired between 2013 and 2018 were processed with the split window algorithm and the derived surface temperatures compared to reference measurements acquired from the various sites during the Landsat 8 overpass. For comparison, and to gauge the fidelity of the presented split window implementation, the same L1T scenes were processed to surface temperature using split window with the b-coefficients suggested by Du et al. (2015) [12] and using the single channel methodology [1] that will be delivered to users in Collection 2.

Table 2. A list of the reference data sources along with the number of measurements utilized for this work.

Site Name	Number of Sites	Number of Measurements
SurfRad	3 land sites	727
AmeriFlux	1 land site	186
NOAA Buoys	36 bouys	308
Lake Tahoe (JPL)	4 buoys	234
Salton Sea (JPL)	1 buoy	63
Total		**1518**

2.4. Geometric Considerations

An initial application of Equation (1) to the TIRS (L1T) image data resulted in undesirable artifacts in the final surface temperature product; see an example of Lake Ontario, NY in Figure 3. The ST product derived from the single channel method is shown on the left for visual reference, while the split window-derived surface temperature image is shown on the right. Clearly, ringing artifacts can be observed at sharp transitions (edges) in the data; e.g., along the Lake Ontario shoreline as shown in the zoom windows of Figure 3. Note that the derived surface temperatures in the single channel method are roughly two degrees warmer than the temperature derived from the split window method. This discrepancy will be discussed further in Section 3.

To understand the source of these artifacts, a brief background of the TIRS focal plane is required. Referring to Figure 4, the TIRS focal plane array (FPA) consists of three staggered detector arrays to cover the 185 km cross-track FOV of the instrument at a ground sampling distance (GSD) of approximately 100 m. Spectral filters are placed on the FPA detectors to produce detector rows with the desired spectral band shapes (Band 10 centered at 10.9 μm and Band 11 centered at 12.0 μm). When imaging in the nominal pushbroom mode, image data are recorded from one row of detectors in each filtered region and an image interval of the Earth is assembled as the instrument travels in orbit. Although band-to-band registration is well-within the defined specification for TIRS [5], the physical layout of the detector arrays along with the read-out sample timing in the along-track direction leads to an inherent misregistration between the Band 10 and Band 11 images. This amounts to an along-track offset of the instantaneous fields-of-view (IFOV) of the detectors in the two bands (note that the magnitude of the offset is much less than the size of the pixel). The TIRS 100-m image data is upsampled to 30 m data in the final step of the Landsat product generation process in order to match the spatial resolution of the OLI sensor. The process of upsampling exacerbates the misregistration offsets due to the fact that the along-track band offset is now a significant fraction of the 30 m pixel.

When the differences between the band images are calculated as part of the split window algorithm, the along-track offsets become magnified in the product.

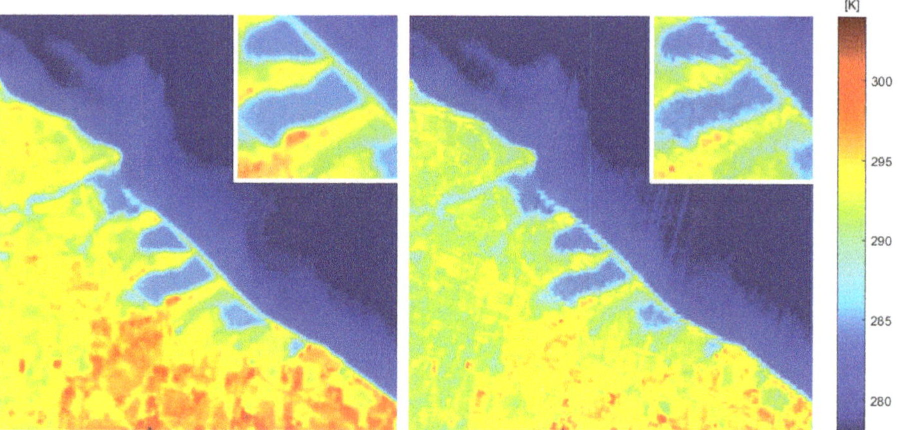

Figure 3. Comparison of the surface temperature product derived from the single channel method (**left**) against the split window method (**right**) for an area around Lake Ontario, NY (Landsat scene ID: LC08_L1TP_016030_20190413_20190422_01_T1). Note the spatial artifacts along edges in the split window product. Zoom windows are shown in the upper right of each image. The image area is roughly 8 by 8 km, and north is up.

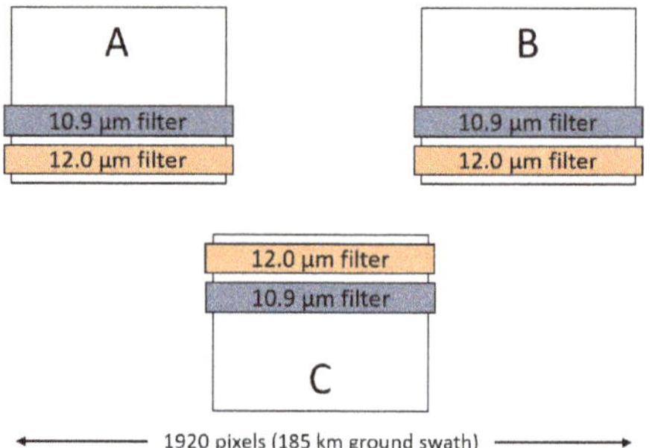

Figure 4. The Thermal Infrared Sensor (TIRS) focal plane array consists of three detector arrays (labeled A, B, and C) arranged to span the cross-track 185 km swath. Spectral filters over the arrays produce the two thermal bands (Bands 10 and 11).

From a technical perspective, applying and delivering a split window-derived ST product at the nominal TIRS resolution (100 m) represents an ideal scenario to avoid artifacts introduced by the algorithm and the upsampling. However, achieving this solution would require a significant deviation from the existing EROS processing pipeline and would result in a product that differs in resolution from the other products (e.g., surface reflectance) being released in Collection 2. Alternative solutions that mitigate the spatial artifacts, yet preserve radiometric fidelity and the 30-meter resolution of the ST product, have been investigated.

To motivate a potentially desirable solution, Figure 5 shows the contributions of each term in Equation (1) to the final surface temperature product for the scene in Figure 3. Columns 3 and 4 of this table were populated by calculating the scene-wide mean and standard deviation, respectively, of each term in Equation (1). Accordingly, column 3 represents the average magnitude of each term's contribution to the final ST product, while column 4 represents the spatial variability introduced by each term to the final ST product. Note from the values in columns 3 and 4 that the additive terms in Equation (1) (highlighted in blue) contribute most of the overall magnitude and variability to the final ST product for the scene in Figure 3. Conversely, note from the values in columns 3 and 4 that the difference terms from Equation (1) (highlighted in gray) contribute significantly less information to the final product. Since the difference terms introduce the artifacts shown in Figure 3, and their radiometric contribution to the final product is relatively small, a proposed solution to mitigate these artifacts is to apply a 5×5 smoothing filter to the Band 10 and 11 apparent temperature images for terms b_4 through b_7 in Equation (1). Recall that the TIRS nominal ground sampling distance is approximately 100 m, but the calculated full width at half maximum (FWHM) of its point-spread function is approximately 200 m (see Wenny et al., 2015) [27]. Therefore, averaging the upsampled 30 m data to 150 m will not significantly alter the image data collected by TIRS. Comparing the nominal standard deviations for the b_4 through b_7 terms (column 4: gray terms) to the 5×5 smoothed standard deviations, as suggested here (zoomed: gray terms), smoothing has a negligible impact (less than 0.1 K) on the scene-wide variability observed in the final proposed ST product for the scene in Figure 3.

Split Window Term	b-value	ST contribution [K]	Std. Dev. [K]	Difference Terms Smoothed (5x5) Std. Dev. [K]
b0	2.2925	2.293	0	
b1	0.9929	283.933	7.607	
b2	0.1545	0.924	0.329	
b3	-0.3122	-0.384	0.317	
b4	3.7186	1.026	0.545	0.462
b5	0.3502	0.002	0.002	0.001
b6	-3.5889	-0.005	0.005	0.005
b7	0.1825	0.071	0.068	0.053

Figure 5. Table illustrating the contribution of each term from Equation (1) to the final surface temperature product shown in Figure 3. Columns 3 and 4 of this table were populated by calculating the scene-wide mean and standard deviation, respectively, of each term in Equation (1). Note that the additive terms (highlighted in blue) contribute most of the overall magnitude (column 3) and variability (column 4) to the final ST product. When compared to the difference terms in column 4 (highlighted in gray), the zoom window suggests that smoothing the difference terms has little impact on scene-wide variability.

While smoothing the difference terms in Equation (1) appears to have negligible impact on radiometric fidelity, its effect on mitigating the geometric artifacts in Figure 3 is dramatic. Figure 6 shows the single channel ST product (left), the nominal split window ST product (middle), and the proposed split window ST product (right). Note that the single channel product is presented here for reference, as it should not exhibit the artifacts described in this section. By visually inspecting the zoom windows in Figure 6, the artifacts present in the nominal split window product (middle) are essentially removed with the proposed methodology (right).

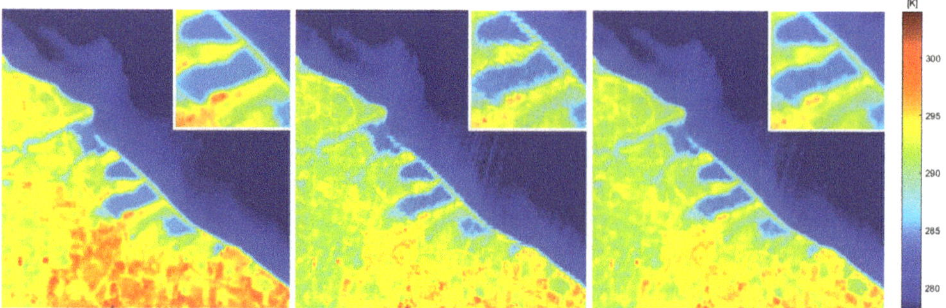

Figure 6. Comparison of surface temperature products: single channel product (**left**), the nominal split window product (**middle**), and the proposed split window product (**right**) (Landsat scene ID: LC08_L1TP_016030_20190413_20190422_01_T1). The scene is roughly 8 by 8 km, and north is up.

3. Results and Validation

The 1518 Landsat 8 scenes corresponding to the ground reference sites listed in Table 2 were processed to surface temperature using the proposed split window method and the coefficients described here (see Table 1). The difference between the derived ST and the measured (reference) ST is shown in Figure 7 for all reference sites. Note that the difference data is displayed as a function of "distance to the nearest cloud (km)" from the pixel where the comparison is made to a reference measurement. As seen in the figure, the temperature error is greatest when the target pixel is in close proximity to a cloud, which adds significant uncertainty to the ST retrieval process. The mean error for the data in Figure 7 is 0.2 K with a standard deviation of 2.73 K. However, ignoring data points within 4 km of a cloud, the mean error becomes 0.02 K with a standard deviation of 1.39 K.

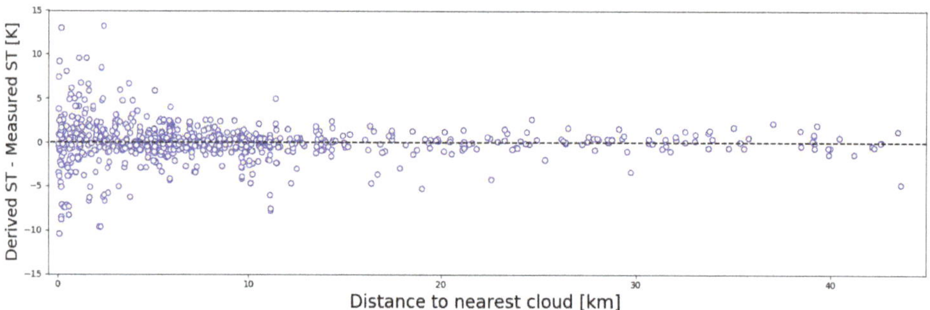

Figure 7. The difference between the reference temperature measurements and the Landsat-derived surface temperatures for the proposed split window methodology as a function of distance to nearest cloud.

Surface temperature products using the single channel methodology and the split window algorithm with the coefficients reported in Du et al. (2015) were also calculated to serve as a comparison to the proposed method. The results from the three methods can be summarized in Figure 8, which shows the average differences between the reference temperature measurements and the Landsat-derived surface temperatures (left), and the corresponding standard deviations of the residuals (right) as functions of distance to cloud.

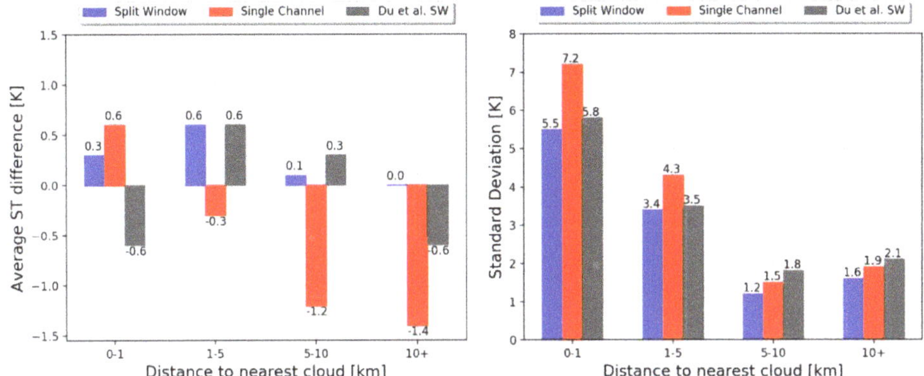

Figure 8. All sites validation: the average differences between the reference temperature measurements and the Landsat-derived surface temperatures for the three retrieval methodologies (**left**) and the corresponding standard deviations of the residuals (**right**).

In general, for the full set of data compiled in this study, the proposed split window implementation (blue bars) has better accuracy and precision than the other two algorithms, as compared to reference data. Figure 8 (left) shows that its retrieved temperatures are, on average, closer to reference measurements with a slight positive bias that diminishes as distance to the nearest cloud increases. Notice that the single channel methodology has a significant bias (compared to reference measurements), which is consistent with the temperature products shown in Figures 3 and 6. Figure 8 (right) indicates that the residuals about the mean are smaller for the proposed split window implementation regardless of cloud proximity for the implementation proposed here.

Two interesting observations can be made when categorizing the data in Figure 8 into "land sites" and "water sites". Figure 9 (left) shows the mean difference between derived and measured surface temperatures for the "land sites" while Figure 9 (right) shows the corresponding standard deviations of the residuals about the mean. The first noteworthy observation from Figure 9 (right) is that for relatively clear scenes (i.e., clouds are over five kilometers from the ground reference measurement), all three methodologies show a standard deviation of approximately 2 K, compared to surface measurements. These values are consistent with those observed in studies using SURFRAD as a reference for validation of other spaceborne thermal instruments [28–30]. Given that this 2 K residual error is consistently observed from several spaceborne platforms and that the pyrgeometers used at the land-based reference sites are sensitive from 4 to 50 μm, this residual error indicates that reflected solar radiation may be contributing to the signal recorded by the pyrgeometers and that broadband emissivity uncertainty is potentially a limiting factor in leveraging these sites to be used as reference sources for applications requiring high accuracy.

A second observation can be made by referring to Figure 10. Figure 10 (left) shows the mean difference between derived and measured surface temperatures for the "water sites", while Figure 10 (right) shows the corresponding standard deviation of the residuals about the mean. The blue bars indicate that the split window algorithm (as presented here) estimates surface temperature more accurately and with less residual error than the single channel method (red bars) and the split window algorithm using the coefficients presented in Du et al. (2015) (gray bars). The under-performance of the Du et al. coefficients for retrieving water temperature is likely due to the inclusion of man-made materials into their training process; i.e., the algorithm coefficients are over-fit to land-based targets. This outcome highlights the potential necessity to develop material-based b-coefficients for an operational split window implementation.

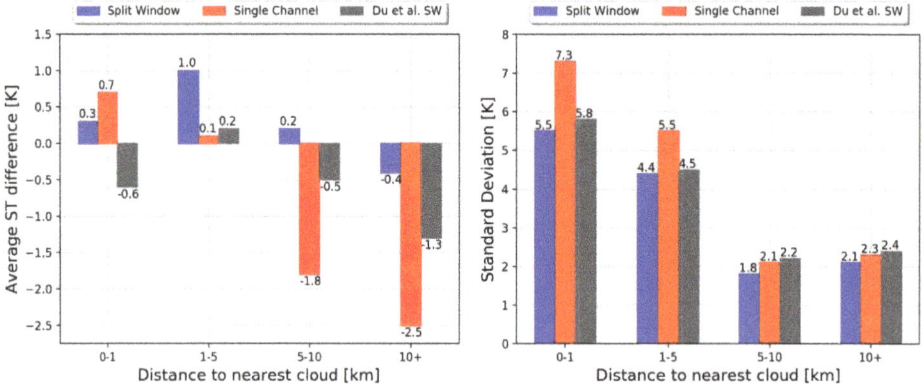

Figure 9. Results over land sites: the average differences between the reference temperature measurements and the Landsat-derived surface temperatures for the three retrieval methodologies (**left**), and the corresponding standard deviations of the residuals (**right**).

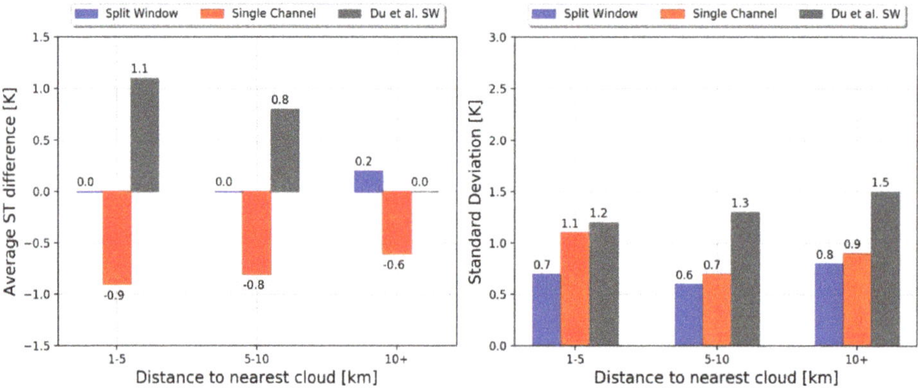

Figure 10. Results over water sites: the average differences between the reference temperature measurements and the Landsat-derived surface temperatures for the three retrieval methodologies (**left**), and the corresponding standard deviations of the residuals (**right**). Note that there is no reference data for clouds within 0–1 km of a water buoy.

4. Conclusions

The TIRS instruments onboard Landsats 8 and 9 contain two thermal channels, enabling the use of the split window methodology to derive Earth surface temperature. This work focused on tailoring the generalized split window algorithm to the specific Landsat bands by deriving appropriate algorithm coefficients and by addressing the inherent aliasing artifacts in the split window temperature product. For the scenes tested here, validation efforts illustrate that the split window ST product is more accurate than the single channel ST product (available in the Landsat Collection 2 release).

The studies presented here demonstrate that smoothing the difference terms in Equation (1) has a dramatic effect on mitigating aliasing artifacts introduced by band-to-band misregistration and upsampling of the nominal TIRS image data. Several comparisons (analogous to Figure 5 in Section 2.4) of the unsmoothed to the smoothed 30 m temperature product indicate that smoothing has little impact on in-scene variability. Considering the fact that the TIRS nominal ground sampling distance is approximately 100 m and the calculated FWHM of its point-spread function is approximately 200 m [27], the solution presented here is believed to be appropriate, although quantifying the impact of smoothing remains an area of ongoing research.

From a radiometric viewpoint, Figure 10 highlights the potential value of deriving per-material split window coefficients for an operational implementation. Considering Du et al. tuned their split window coefficients to support urban heat island applications, the under-performance of their implementation for the water scenes tested here precludes their coefficients from being used for environmental applications requiring less than one Kelvin precision. That being said, the simulated effort conducted by Du et al. highlights potential improvements that can be made to surface temperature retrieval if atmospheric water vapor can be characterized from image data and accounted for in the split window implementation; i.e., the b-coefficients in Equation (1) are categorized as a function of column water vapor. An investigation into the potential improvement of surface temperature estimation using coefficients categorized by material and column water vapor will be conducted.

Other considerations to achieve an operational implementation and validation of a split window algorithm for TIRS in Collection 3 are the development of a quality assurance map and appropriate validation of the product. The single channel algorithm being implemented in Collection 2 to derive surface temperature for the Landsat thermal archive contains a quality assurance (QA) band that was developed based on cloud proximity and atmospheric transmission [2]. While this QA band will provide information regarding the product's accuracy, it is highly dependent on Landsat's cloud mask and was trained using water observations. The investigation and development of an analogous but appropriately-trained quality assurance band to accompany a split window-derived surface temperature product represents an area of ongoing research.

While the efforts reported here represent significant progress toward the development and validation of an operational split window-derived surface temperature product, the considerations described above should be addressed before the final form of the algorithm is achieved. Future validation efforts will include reprocessing of the scenes presented here but with Collection 2 L1T TIRS data, incorporation of additional reference measurements as they become available, and a categorization of the residual errors with a more appropriate metric; i.e., residual errors between retrieved and reference measurements will be categorized as a function of atmospheric column water vapor instead of distance to cloud.

Author Contributions: The authors of this article all had unique contributions that led to the final outcomes presented here. A.G. was responsible for funding acquisition, project administration, initial conceptualization and validation, and draft preparation. T.K. was responsible for data curation and significantly improving the methodologies and validation efforts presented here through software development. M.M. and R.E. provided technical and research support to enable the final form of the TIRS split window algorithm presented here. All authors contributed equally to the review and editing of this article. All authors have read and agreed to the published version of the manuscript.

Funding: This material is based upon work supported by the U.S. Geological Survey under Cooperative Agreement Number G17AC00070. The views and conclusions contained in this document are those of the authors and should not be interpreted as representing the opinions or policies of the U.S. Geological Survey. Mention of trade names or commercial products does not constitute their endorsement by the U.S. Geological Survey. This manuscript is submitted for publication with the understanding that the United States Government is authorized to reproduce and distribute reprints for Governmental purposes.

Acknowledgments: The authors would like to acknowledge James Storey, Michael Choate, and Ray Dittmeier of the USGS. James Storey and Michael Choate provided invaluable insight into the source of the geometric artifacts that can be observed in the nominal split window product and provided suggestions and datasets for testing that facilitated the identification of a potential mitigation solution. Ray Dittmeier provided, and continues to provide, invaluable support in the development of Landsat's Surface Temperature product. He was instrumental in the incorporation and verification of both the single channel and split window methodologies presented here. His tireless efforts made a surface temperature product possible in the Collection 2 timeframe.

Conflicts of Interest: The authors declare no conflict of interest. The funders had no role in the design of the study; in the collection, analyses, or interpretation of data; in the writing of the manuscript, or in the decision to publish the results.

References

1. Malakar, N.; Hulley, G.; Hook, S.; Laraby, K.; Cook, M.; Schott, J. An Operational Land Surface Temperature Product for Landsat Thermal Data: Methodology and Validation. *IEEE Trans. Geosci. Remote Sens.* **2018**, *56*, 5717–5735. [CrossRef]
2. Laraby, K.G.; Schott, J.R. Uncertainty estimation method and Landsat 7 global validation for the Landsat surface temperature product. *Remote Sens. Environ.* **2018**, *216*, 472–481. [CrossRef]
3. Reuter, D.; Richardson, C.; Pellerano, F.; Irons, J.; Allen, R.; Anderson, M.; Jhabvala, M.; Lunsford, A.; Montanaro, M.; Smith, R.; et al. The Thermal Infrared Sensor (TIRS) on Landsat 8: Design Overview and Pre-Launch Characterization. *Remote Sens.* **2015**, *7*, 1135–1153. [CrossRef]
4. Gerace, A.; Montanaro, M. Derivation and validation of the stray light correction algorithm for the Thermal Infrared Sensor onboard Landsat 8. *Remote Sens. Environ.* **2017**, *191*, 246–257. [CrossRef]
5. Storey, J.; Choate, M.; Moe, D. Landsat 8 Thermal Infrared Sensor Geometric Characterization and Calibration. *Remote Sens.* **2014**, *6*, 11153–11181. [CrossRef]
6. Becker, F.; Li, Z.L. Towards a local split window method over land surfaces. *Int. J. Remote Sens.* **1990**, *11*, 369–393. [CrossRef]
7. Wan, Z.; Dozier, J. A generalized split-window algorithm for retrieving land-surface temperature from space. *IEEE Trans. Geosci. Remote Sens.* **1996**, *34*, 892–905. [CrossRef]
8. Wan, Z. New Refinements and Validation of the MODIS Land-Surface Temperature/Emissivity Products. *Remote Sens. Environ.* **2008**, *140*, 59–74. [CrossRef]
9. Berk, A.; Conforti, P.; Kennett, R.; Perkins, T.; Hawes, F.; van den Bosch, J. MODTRAN6: A major upgrade of the MODTRAN radiative transfer code. In Proceedings of the 2014 6th Workshop on Hyperspectral Image and Signal Processing: Evolution in Remote Sensing (WHISPERS), Lausanne, Switzerland, 24–27 June 2014; pp. 113–119. [CrossRef]
10. Thermodynamic Initial Guess Retrieval (TIGR). Available online: http://ara.abct.lmd.polytechnique.fr/index.php?page=tigr (accessed on 22 October 2019).
11. MODIS UCSB Emissivity Library. Available online: https://icess.eri.ucsb.edu/modis/EMIS/html/em.html (accessed on 22 October 2019).
12. Du, C.; Huazhong, R.; Qin, Q.; Meng, J.; Zhao, S. A Practical Split-Window Algorithm for Estimating Land Surface Temperature from Landsat 8 Data. *Remote Sens.* **2015**, *7*, 647–665. [CrossRef]
13. Hulley, G.C.; Hook, S.J.; Abbott, E.; Malakar, N.; Islam, T.; Abrams, M. The ASTER Global Emissivity Dataset (ASTER GED): Mapping Earth's emissivity at 100 meter spatial scale. *Geophys. Res. Lett.* **2015**, *42*, 7966–7976. [CrossRef]
14. Sibandze, P.; Mhangara, P.; Odindi, J.; Kganyago, M. A comparison of Normalised Difference Snow Index (NDSI) and Normalised Difference Principal Component Snow Index (NDPCSI) techniques in distinguishing snow from related land cover types. *S. Afr. J. Geomat.* **2014**, *3*, 197–209. [CrossRef]
15. SURFRAD (Surface Radiation Budget) Network. Available online: https://www.esrl.noaa.gov/gmd/grad/surfrad/ (accessed on 22 October 2019).
16. AmeriFlux Network. Available online: https://ameriflux.lbl.gov/ (accessed on 22 October 2019).
17. National Data Buoy Center. Available online: https://www.ndbc.noaa.gov/ (accessed on 22 October 2019).
18. Lake Tahoe Validation. Available online: https://laketahoe.jpl.nasa.gov/get_met_weather (accessed on 22 October 2019).
19. Salton Sea Validation. Available online: https://saltonsea.jpl.nasa.gov/get_met_weather (accessed on 22 October 2019).
20. Wang, K.; Dickinson, R.E. Global atmospheric downward longwave radiation at the surface from ground-based observations, satellite retrievals, and reanalyses. *Rev. Geophys.* **2013**, *51*, 150–185. [CrossRef]
21. Guillevic, P.C.; Privette, J.L.; Coudert, B.; Palecki, M.A.; Demarty, J.; Ottle, C.; Augustine, J.A. Land Surface Temperature product validation using NOAA's surface climate observation networks—Scaling methodology for the Visible Infrared Imager Radiometer Suite (VIIRS). *Remote Sens. Environ.* **2012**, *124*, 282–298. [CrossRef]
22. University of Michigan Biological Station. Available online: http://flux.org.ohio-state.edu/site-description-umbs/#umbs-ameriflux-towers (accessed on 22 October 2019).
23. Gröbner, J.; Wacker, S. Pyrgeometer Calibration Procedure at the PMOD/WRC-IRS. Davos:[sn]. 2012. Available online: https://library.wmo.int/doc_num.php?explnum_id=7365 (accessed on 8 January 2020).

24. Wang, K.; Wan, Z.; Wang, P.; Sparrow, M.; Liu, J.; Zhou, X.; Haginoya, S. Estimation of surface long wave radiation and broadband emissivity using Moderate Resolution Imaging Spectroradiometer (MODIS) land surface temperature/emissivity products. *J. Geophys. Res. Atmos.* **2005**, *110*. [CrossRef]

25. Masuda, K.; Takashima, T.; Takayama, Y. Emissivity of pure and sea waters for the model sea surface in the infrared window regions. *Remote Sens. Environ.* **1988**, *24*, 313–329. [CrossRef]

26. Zeng, X.; Zhao, M.; Dickinson, R.E.; He, Y. A multiyear hourly sea surface skin temperature data set derived from the TOGA TAO bulk temperature and wind speed over the tropical Pacific. *J. Geophys. Res. Oceans* **1999**, *104*, 1525–1536. [CrossRef]

27. Wenny, B.; Helder, D.; Hong, J.; Leigh, L.; Thome, K.; Reuter, D. Pre- and Post-Launch Spatial Quality of the Landsat 8 Thermal Infrared Sensor. *Remote Sens.* **2015**, *7*, 1962–1980. [CrossRef]

28. Yu, Y.; Tarpley, D.; Privette, J.L.; Flynn, L.E.; Xu, H.; Chen, M.; Vinnikov, K.Y.; Sun, D.; Tian, Y. Validation of GOES-R Satellite Land Surface Temperature Algorithm Using SURFRAD Ground Measurements and Statistical Estimates of Error Properties. *IEEE Trans. Geosci. Remote Sens.* **2012**, *50*, 704–713. [CrossRef]

29. Li, S.; Yu, Y.; Sun, D.; Tarpley, D.; Zhan, X.; Chiu, L. Evaluation of 10 year AQUA/MODIS land surface temperature with SURFRAD observations. *Int. J. Remote Sens.* **2014**, *35*, 830–856. [CrossRef]

30. Liu, Y.; Yu, Y.; Yu, P.; Göttsche, F.M.; Trigo, I. Quality assessment of S-NPP VIIRS land surface temperature product. *Remote Sens.* **2015**, *7*, 12215–12241. [CrossRef]

remote sensing

Article

Evaluation of Land Surface Temperature Retrieval from Landsat 8/TIRS Images before and after Stray Light Correction Using the SURFRAD Dataset

Jinxin Guo [1,2], Huazhong Ren [1,2,*], Yitong Zheng [1,2], Shangzong Lu [1,2] and Jiaji Dong [1,2]

1 Institute of Remote Sensing and Geographical Information System, School of Earth and Space Sciences, Peking University, Beijing 100871, China; guojinxin@pku.edu.cn (J.G.); zhengyitong@pku.edu.cn (Y.Z.); lushangzong@pku.edu.cn (S.L.); dongjiaji@pku.edu.cn (J.D.)

2 Beijing Key Lab of Spatial Information Integration and Its Application, Peking University, Beijing 100871, China

* Correspondence: renhuazhong@pku.edu.cn; Tel.: +86-10-62760375

Received: 10 February 2020; Accepted: 20 March 2020; Published: 22 March 2020

Abstract: Landsat 8/thermal infrared sensor (TIRS) is suffering from the problem of stray light that makes an inaccurate radiance for two thermal infrared (TIR) bands and the latest correction was conducted in 2017. This paper focused on evaluation of land surface temperature (LST) retrieval from Landsat 8 before and after the correction using ground-measured LST from five surface radiation budget network (SURFRAD) sites. Results indicated that the correction increased the band radiance at the top of the atmosphere for low temperature but decreased such radiance for high temperature. The root-mean-square error (RMSE) of LST retrieval decreased by 0.27 K for Band 10 and 0.78 K for Band 11 using the single-channel algorithm. For the site with high temperature, the LST retrieval RMSE of the single-channel algorithm for Band 11 even reduced by 1.4 K. However, the accuracy of two of three split-window algorithms adopted in this paper decreased. After correction, the single-channel algorithm for Band 10 and the linear split-window algorithm had the least RMSE (approximately 2.5 K) among five adopted algorithms. Moreover, besides SURFRAD sites, it is necessary to validate using more robust and homogeneous ground-measured datasets to help solely clarify the effect of the correction on LST retrieval.

Keywords: land surface temperature; Landsat 8; stray light correction; split-window algorithm; single-channel algorithm

---

1. Introduction

Land surface temperature (LST) is a key parameter in the studies of land surface progress, such as the surface energy budget, surface moisture budget, and urban ecology, which is important to nature and human health [1–6]. Remote sensing is an effective way to retrieve LST at the regional and even global scales [7–9]. Many satellites carry thermal infrared (TIR) sensors, such as Terra/advanced spaceborne thermal emission and reflection radiometer (ASTER), NOAA/advanced very high resolution radiometer (AVHRR), Terra and Aqua/moderate resolution imaging spectroradiometer (MODIS), Landsat 5/thematic mapper (TM), Landsat 7/enhanced thematic mapper plus (ETM+), Landsat 8/thermal infrared sensor (TIRS), and Sentinel-3/sea and land surface temperature radiometer (SLSTR). Landsat 8/TIRS is an important part of the Landsat program for monitoring surface energy and temperature. However, it was discovered to have a considerable stray light problem, which resulted in an absolute radiometric calibration error to the TIRS images [10,11]. The inaccurate radiometric calibration of TIRS, especially the excessive error in Band 11, made it difficult to apply the conventional split-window algorithms on retrieving LST from the two TIR bands of TIRS. Therefore, the data provider, United States Geological Survey (USGS), recommends the users to develop a single-channel algorithm for LST retrieval. However, several studies still proposed different split-window algorithms for this sensor [12–14]. In certain validation cases, the LST retrieval accuracy from the split-window algorithms was found to be better than that of the single-channel algorithm [15]. As a result, those reports have made the readers and users puzzled in choosing retrieval algorithms.

A new stray light correction algorithm proposed by Montanaro et al. [16] demonstrates great potential toward removing most stray light effects from TIRS images. The algorithm has also been refined and implemented operationally into the Landsat Product Generation System from early 2017 [17]. Gerace and Montanaro [17] selected 20 scenes (almost offshore water scenes) acquired from TIRS and MODIS onboard the Terra satellite to verify Landsat 8 brightness temperature before and after the stray light correction. They found that the absolute radiometric error of TIRS images was reduced to approximately 0.5% in Bands 10 and 11 on average [17]. This correction should benefit the LST retrieval from TIRS images. Although García-Santos et al. [18] have compared three methods for estimating LST from Landsat 8/TIRS images after the stray light correction with 21 observations, further validation in other regions is necessary for the assessment of LST retrieval accuracy after the correction. Moreover, the comparison of LST retrieval accuracy before and after the correction has not been validated using ground-measured data. Hence, the correction influence on the LST retrieval accuracy in practice remains unknown and thus requires further investigation.

With the above motivations, this study aims to clarify the following two questions: (1) Has the LST retrieval accuracy improved after the stray light correction? (2) Which algorithm (the split-window or the single-channel algorithm) is better for LST retrieval from TIRS images after the stray light correction? To answer the above questions, this paper tries to evaluate the accuracy of the LST retrieved from Landsat 8 TIRS images before and after the stray light correction by using different published two-channel split-window and single-channel algorithms. As a result, the remainder of this paper is organized as follows: Section 2 briefly presents the principles of different split-window algorithms and the single-channel algorithm; Section 3 introduces the involved Landsat 8 images and ground-measured LST and their processing; Section 4 focuses on the band radiance and LST evaluation results to answer the above-mentioned two questions; Sections 5 and 6 provide the discussion and conclusions, respectively.

2. Landsat 8 LST Retrieval Algorithms

Five representative LST retrieval algorithms were selected for evaluation, including three two-channel split-window algorithms and one single-channel algorithm with Bands 10 and 11, respectively. These algorithms have been applied to different TIR sensors and always have been used as the reference algorithms for a new algorithm or as the state-of-art algorithms for LST validation [15,18–20]. Therefore, it is proper and valid to adopt these algorithms for validation in this research. Among the above algorithms, the generalized split-window algorithm was originally developed as the standard algorithm to estimate LST from MODIS TIR images and then refined in 2014 by adding a quadratic term [21]. The new version was confirmed to perform better than its previous version. Du et al. [14] applied this algorithm to conduct the LST retrieval from Landsat 8 TIRS images, and Gerace and Montanaro [17] used their algorithm to check the LST variation before and after the stray light correction. On the basis of the work of Qin et al. [22], a linear split-window algorithm for TIR images was proposed by Rozenstein et al. [13] to retrieve LST from Landsat 8/TIRS and had good performance. The split-window algorithm proposed by Jiménez-Muñoz et al. [12] was inherited from the mathematical form proposed by Sobrino et al. [23] first and modified by Sobrino and Raissouni [24], which has been developed to retrieve LST for several TIR sensors, such as AVHRR and ATSR2 [25]. The single-channel algorithm developed by Jiménez-Muñoz et al. [12] was applied to Landsat 5/TM and Landsat 7/ETM+ with a single TIR band [26,27]. Moreover, USGS recommended users to retrieve LST from Landsat 8 images with single-channel algorithm because of the serious stray light problem in

Band 11. Therefore, this algorithm was also analyzed in this study. The details of the above algorithms are presented in the following part.

2.1. Three Split-Window Algorithms

A. Generalized Split-Window Algorithm by Du et al. [14]

The generalized split-window algorithm Du et al. [14] applied to two Landsat 8 TIR bands expressed as:

$$T_s = b_0 + (b_1 + b_2\frac{1-\varepsilon}{\varepsilon} + b_3\frac{\Delta\varepsilon}{\varepsilon})\frac{T_{10} + T_{11}}{2} + (b_4 + b_5\frac{1-\varepsilon}{\varepsilon} + b_6\frac{\Delta\varepsilon}{\varepsilon})\frac{T_{10} - T_{11}}{2} + b_7(T_{10} - T_{11})^2 \quad (1)$$

where T_{10} and T_{11} are the brightness temperatures at the top of the atmosphere (TOA) from TIRS Bands 10 and 11, respectively; ε is the average emissivity of the two bands; $\Delta\varepsilon$ is the band emissivity difference ($\Delta\varepsilon = \varepsilon_{10} - \varepsilon_{11}$); and b_k ($k = 0, 1, \ldots, 7$) refers to the algorithm coefficients, which are obtained directly from simulation dataset on the basis of the thermodynamic initial guess retrieval (TIGR) [28] atmospheric profile database and the MODTRAN 5.2 [29] atmospheric transmittance/radiance code. Du et al. [14] calculated those coefficients independently on column water vapor (CWV) to reduce the effect of atmospheric water vapor. The CWV was divided into five subranges ($0 < CWV \leq 2.5$ g/cm^2, $2 < CWV \leq 3.5$ g/cm^2, $3 < CWV \leq 4.5$ g/cm^2, $4 < CWV \leq 5.5$ g/cm^2, and $CWV > 5.0$ g/cm^2). This method was also applied on Sentinel-3A [30] and improved by dividing the simulation data into temperature subranges. We refined these algorithm coefficients as Zheng et al. [30] did in two ways to make the evaluation more precise. First, the largest difference between the bottom air temperature (T_{air}) and LST used in the simulation dataset was increased from 20 K in the study of Du et al. [14] to 35 K for a barren or desert surface that probably has high surface temperature. Second, the brightness temperature of Band 10 (T_{10}) was divided into several subranges, which were used to determine the coefficients together with the CWV subranges. T_{10} varies with LST and atmospheric conditions; thus, in accordance with the value ranges of T_{10} under various CWV subranges, T_{10} was divided into four subranges for $0 \leq CWV \leq 2.5$ g/cm^2 as $T_{10} < 270$ K, 270 K $\leq T_{10} < 300$ K, 300 K $\leq T_{10} < 330$ K, and $T_{10} \geq 330$ K. For the other four CWV subranges, T_{10} was divided into $T_{10} < 300$ K and $T_{10} \geq 300$ K. Table 1 lists the new algorithm coefficients of Equation (1) for different combinations of CWV subranges and T_{10} subranges and the root-mean-square error (RMSE) of the predicted temperature compared with the value in the simulation dataset.

B. Linear Split-Window Algorithm by Rozenstein et al. [13]

Based on the linear relationship between band radiance and temperature in specified temperature ranges, the linear split-window algorithm proposed by Rozenstein et al. [13] to estimate LST from Landsat 8 TIRS image is expressed as:

$$T_s = A_0 + A_1 T_{10} - A_2 T_{11} \quad (2)$$

where, A_0, A_1 and A_2 are algorithm coefficients given by following equations derived from thermal radiative transfer equation [31] and linearizing Planck's radiance function:

$$A_0 = E_1 a_{10} + E_2 a_{11} \quad (3a)$$

$$A_1 = 1 + A + E_1 b_{10} \quad (3b)$$

$$A_2 = A + E_2 b_{11} \quad (3c)$$

Table 1. The coefficients b_k in different column water vapor (CWV) and brightness temperature of Band 10 (T_{10}) intervals, and root-mean-square error (RMSE) of the predicted temperature.

			(1) CWV (g/cm^2): [0.0, 2.5]						
T_{10} (K)	b_0	b_1	b_2	b_3	b_4	b_5	b_6	b_7	RMSE (K)
<270	−3.1118	1.0153	0.1658	−0.3046	3.1790	8.7989	34.4917	−0.3746	0.11
[270, 300)	1.6214	0.9968	0.1739	−0.3965	4.3444	5.6164	12.8573	−0.1175	0.30
[300, 330)	7.3937	0.9788	0.1917	−0.3384	3.0247	3.2533	−14.4977	0.1291	0.30
≥330	18.0799	0.9517	0.2043	−0.2870	1.5422	3.1292	−23.0479	0.1694	0.18
			(2) CWV (g/cm^2): [2.0, 3.5]						
T_{10} (K)	b_0	b_1	b_2	b_3	b_4	b_5	b_6	b_7	RMSE (K)
<300	24.9130	0.911	0.174	−0.299	6.351	3.920	−5.582	−0.064	0.54
≥300	27.4670	0.904	0.187	−0.349	5.675	2.842	−7.853	0.023	0.58
			(3) CWV (g/cm^2): [3.0, 4.5]						
T_{10} (K)	b_0	b_1	b_2	b_3	b_4	b_5	b_6	b_7	RMSE (K)
<300	23.7764	0.9123	0.1443	−0.1902	7.1598	5.9811	−11.5454	−0.0597	0.71
≥300	35.3510	0.8780	0.1534	−0.2077	6.0319	5.2617	−14.5807	0.0270	1.01
			(4) CWV (g/cm^2): [4.0, 5.5]						
T_{10} (K)	b_0	b_1	b_2	b_3	b_4	b_5	b_6	b_7	RMSE (K)
<300	9.6135	0.9581	0.1128	−0.1213	7.1210	6.8790	−12.5374	0.0257	0.93
≥300	36.4439	0.8736	0.1160	−0.1181	6.4603	7.0560	−16.3845	0.0305	1.32
			(5) CWV (g/cm^2): [5.0, 6.3]						
T_{10} (K)	b_0	b_1	b_2	b_3	b_4	b_5	b_6	b_7	RMSE (K)
<300	50.7495	0.8021	0.0738	−0.0521	12.3012	9.7371	−15.7669	−0.3001	1.06
≥300	−63.0662	1.2070	0.0466	−0.0323	7.4367	10.3215	−13.6909	−0.0355	1.32

In the above equations, E_1, E_2 and A are coefficients determined by pixel emissivity and atmospheric transmittance, and written as:

$$E_1 = D_{11}(1 - C_{10} - D_{10})/E_0, \quad E_2 = D_{10}(1 - C_{11} - D_{11})/E_0$$

$$A = D_{10}/E_0, \text{ and } E_0 = D_{11}C_{10} - D_{10}C_{11} \tag{4}$$

$$\text{with } C_i = \varepsilon_i \tau_i \text{ and } D_i = (1 - \tau_i)[1 + (1 - \varepsilon_i)\tau_i] \tag{5}$$

where, ε_i and τ_i are the pixel emissivity and atmospheric transmittance of TIRS Band 10 or 11, respectively. The atmospheric transmittance τ_i is obtained from the negative correlation with CWV by the results of MODTRAN 4.0 simulations [13], which is the only empirical fitting step in this algorithm. Therefore, different from the generalized split-window algorithm, this linear split-window algorithm is not dependent on the temperature simulation data, but the direct features of atmospheric profiles. This makes more stable simulation results of this linear split-window algorithm; because the process of temperature simulation needs more other parameters besides the atmospheric profile, such as the input LST and emissivity, while the process of transmittance simulation only needs the atmospheric profile.

C. Split-Window Algorithm by Jiménez-Muñoz et al. [12]

A split-window algorithm was introduced by Jiménez-Muñoz et al. [12] to estimate LST for the TIRS image, that is,

$$T_s = T_{10} + c_1(T_{10} - T_{11}) + c_2(T_{10} - T_{11})^2 + c_0 + (c_3 + c_4 w)(1 - \varepsilon) + (c_5 + c_6 w)\Delta\varepsilon \tag{6}$$

where, ε is the average emissivity of the two bands, and $\Delta\varepsilon$ is the band emissivity difference; they are similar to those in Equation (1). w is the CWV in g/cm^2, and c_0 to c_6 are coefficients. Jiménez-Muñoz et al. [12] regressed those coefficients and calculated the error of the temperature on the basis of simulation data.

Their results are shown in Table 2. Similar to the above generalized split-window, this algorithm also considers the quadratic term of brightness temperature difference of Bands 10 and 11, and obtains the coefficients by fitting the temperature simulation data directly.

Table 2. Coefficients for the split-window algorithm and the land surface temperature (LST) RMSE from the linear regression by Jiménez-Muñoz et al. [12].

c0	c1	c2	c3	c4	c5	c6	LST RMSE (K)
−0.268	1.378	0.183	54.30	−2.238	−129.20	16.40	0.6

2.2. Single-Channel Algorithm

This study used the single-channel algorithm developed by Jiménez-Muñoz et al. [27] to retrieve LST from Landsat 8 TIRS Band 10 or 11 image. This algorithm is expressed as:

$$T_s = \gamma[\frac{1}{\varepsilon}(\psi_1 L_{toa} + \psi_2) + \psi_3] + \delta \tag{7}$$

where L_{toa} is the TOA radiance and calculated from the radiometric calibration on the observation; ε is the emissivity of Band 10 or 11; and γ and δ are two parameters given by

$$\gamma \approx \frac{T_b^2}{b_\gamma L_{toa}}, \text{ and } \delta \approx T_b - \frac{T_b^2}{b_\gamma} \tag{8}$$

In Equation (8), T_b is the brightness temperature of TIRS Band 10 or 11, that is, T_{10} or T_{11}. $b_\gamma = c_2/\lambda$ ($c_2 = 1.43877 \times 10^4$ μm·K; λ is the effective wavelength of TIRS, and b_γ is 1324 K for Band 10 and 1199 K for Band 11, respectively). ψ_1, ψ_2, and ψ_3 are atmospheric terms related to the atmospheric transmittance and downward and upward thermal radiance, and can be nonlinearly related to CWV. Equation (9) shows the coefficients for Band 10 that are estimated by Jimenez-Munoz et al. [12] from the Global Atmospheric Profiles from Reanalysis Information (GAPRI) database [32].

$$\begin{bmatrix} \psi_1 \\ \psi_2 \\ \psi_3 \end{bmatrix} = \begin{bmatrix} c_{11} & c_{12} & c_{13} \\ c_{21} & c_{22} & c_{23} \\ c_{31} & c_{32} & c_{33} \end{bmatrix} \begin{bmatrix} w^2 \\ w \\ 1 \end{bmatrix} = \begin{bmatrix} 0.04019 & 0.02916 & 1.01523 \\ -0.38333 & -1.50294 & 0.20324 \\ 0.00918 & 1.36072 & -0.27514 \end{bmatrix} \begin{bmatrix} w^2 \\ w \\ 1 \end{bmatrix} \tag{9}$$

Jiménez-Muñoz et al. [12] only provided coefficients for Band 10. However, after the stray light correction, Band 11 can also be used in the single-channel algorithm. Therefore, coefficients for this band are required, as well. This study obtained the coefficient matrix C for Band 11 by using the above TIGR atmospheric profiles. The matrix is

$$C_{b11} = \begin{bmatrix} 0.09874 & -0.03212 & 1.06497 \\ -0.81391 & -0.94691 & -0.17172 \\ -0.00676 & 1.40205 & -0.14864 \end{bmatrix} \tag{10}$$

2.3. Determination of Pixel Emissivity and Atmospheric CWV

The above split-window and single-channel algorithms require pixel emissivity and atmospheric CWV as input. The two parameters were both acquired using the same way to maintain constancy among algorithms.

For pixel emissivity, this study adopted the widely used normalized difference vegetation index (NDVI)-threshold method to estimate the pixel emissivity. In this method, land pixels were classified into three types on the basis of their NDVI value, namely, barren soil, fully vegetated, and partly vegetated pixels [24,33]. NDVI was calculated from atmospherically corrected ground red and near-infrared band reflectance. The emissivity of partly vegetated pixel was mainly calculated from the combination of soil and vegetation component emissivities, which are weighted by the fraction of vegetation cover (FVC). For convenience, the emissivities of barren soil and fully vegetated pixels were directly given by soil and vegetation component emissivities, respectively, on the assumption that the component emissivities slightly change in time. Finally, this method is expressed as:

$$\varepsilon_p = \begin{cases} \varepsilon_s & NDVI < NDVI_s (\text{barren soil}) \\ \varepsilon_v f + \varepsilon_s(1-f) + 4 <d\varepsilon> f(1-f), & NDVI_s \leq NDVI \leq NDVI_v (\text{partly vegetated}) \\ \varepsilon_v & NDVI > NDVI_s (\text{fully vegetated}) \end{cases} \quad (11)$$

where ε_p is the pixel emissivity, ε_s is the soil component emissivity, ε_v is the vegetated component emissivity, $<d\varepsilon>$ is the maximum cavity term and is set as 0.01 [34], and f is the FVC. $NDVI_s$ is the NDVI for barren soil pixel, and its value is 0.20; $NDVI_v$ is the NDVI for fully vegetated pixel and valued with 0.86 [35]. Ren et al. [33] improved this method by using flexible component emissivity instead of the fixed component emissivity in the original method on the basis of the different land cover types of the fine-resolution observation and monitoring of global land cover product [36].

For atmospheric CWV, the MODIS/Terra total precipitable water vapor 5-Min L2 Swath 1 km and 5 km (MOD05_L2) product was used in accordance with the location and observation time of Landsat 8. As one of MODIS standard products, the MOD05_L2 product consists of atmospheric CWV amounts estimated from two different algorithms, namely, the near-infrared and infrared algorithms. The spatial resolution of CWV data generated by the near-infrared algorithm is 1 km, whereas that of the infrared algorithm is 5 km [37]. The water vapor at 1 km was used in this study and the temporal variation between MODIS and Landsat 8 overpass time was ignored.

3. Landsat 8 Images and Ground-Measured LST

Landsat 8 images and the corresponding ground-measured data of the surface radiation budget network (SURFRAD) [38] that were usually used for in situ LST validation [15,20,39–47] were obtained to evaluate the different LST retrieval algorithms as mentioned above. The Landsat 8 image pairs before and after the stray light correction were obtained from the USGS website to illustrate the LST retrieval accuracy change before and after the correction. The brightness temperatures T_{10} and T_{11} were consequently calculated from the observation using the radiometric coefficients in the metadata file.

The SURFRAD was established in 1993 through the support of NOAA's Office of Global Programs. The primary objective was to support climate research with accurate, continuous, long-term measurements of the surface radiation budget over the United States. SURFRAD has seven sites, but only four of them (Bondville_IL (BND), Goodwin_Creek_MS (GCM), Penn_State_PA (PSU), and Sioux_Falls_SD (SXF)) are suitable for the validation of moderate-resolution remote sensing images, such as Landsats 5 and 7, owing to the heterogeneity issues [41]. However, given that the error caused by stray light is related to the ground temperature of the surrounding area and the error is greater when the surrounding area is warm [17], the Desert_Rock_NV (DRA) site was also selected for validation to explore the effect of the stray light correction in broader temperature ranges. Since its land cover type is open shrublands, resulting in some high LST of this site. Meanwhile, DRA site has also been used in the LST validation of Landsat 8 [20] and other TIR sensors [46,47] in recent years, which can prove the applicability of the DRA site in some degree. As a result, we finally used the BND, GCM, PSU, SXF, and DRA sites to validate the Landsat 8 LST retrieval results considering heterogeneity issues and sufficient temperature range. Table 3 lists the information of the five sites of the SURFRAD program and the number of clear-sky Landsat 8 images from its launch to August 2017 over each site. We removed the observations that had a standard deviation of temperature exceeding 1 K for 3 pixels × 3 pixels around the site center, in order to reduce the error caused by heterogeneity [41]. Finally, a total of 207 images were obtained for analysis, as shown in Table 3.

Table 3. The information of the surface radiation budget network (SURFRAD) sites and the image numbers under clear-sky conditions.

Site Code	Name	Geolocation	Land Cover	Clear-Sky Image Count
BND	Bondville, Illinois	40.05192°N 88.37309°W	cropland	43
GCM	Goodwin Creek, Mississippi	34.2547°N 89.8729°W	cropland	58
PSU	Penn. State Univ., Pennsylvania	40.72012°N 77.93085°W	cropland	13
SXF	Sioux Falls, South Dakota	43.73403°N 96.62328°W	grassland	31
DRA	Desert Rock, Nevada	36.62373°N 116.01947°W	open shrublands	62

Each site provides a measurement of the upward surface TIR radiance $L^{\uparrow}$ and the downward atmospheric TIR radiance $R^{\downarrow}$ in the wavelength range of 3–50 µm every 1 min [39]. On the basis of the thermal radiative transfer equation of the near surface and the Stefan–Boltzmann law, the ground LST can be calculated as:

$$T_{s\_ground} = \left[\frac{L^{\uparrow} - (1 - \bar{\varepsilon})R^{\downarrow}}{\bar{\varepsilon}\sigma} \right]^{\frac{1}{4}} \tag{12}$$

In Equation (12), σ is the Stefan–Boltzmann constant with the value of 5.67×10^{-8} W/m^2·K^4. $\bar{\varepsilon}$ is the ground broadband emissivity (BBE). BBE in 8–13.5 µm was used, because it is considered as the best wavelength range for estimating the net longwave radiation under clear sky [48,49]. To obtain this BBE, we also used the NDVI-threshold method as stated in Equation (11). However, the broadband component emissivity rather than the band component emissivity was used in Equation (11) for BBE calculation. Some details of the technique can be found in Ren et al. [33].

4. Band Radiance and LST Evaluation Results

The LST in the ground sites was retrieved from Landsat 8 images by using the above five algorithms. For simplification, the generalized split-window algorithm by Du et al. [14], the linear split-window algorithm by Rozenstein et al. [13], and the split-window algorithm by Jiménez-Muñoz et al. [12] were denoted as SW_Du, SW_Rozenstein, and SW_JM, respectively. The single-channel algorithms were denoted as SC_10 for using Band 10 image and as SC_11 for using Band 11 image. This section focuses on the LST evaluation results in two aspects, namely, the band radiance comparison before and after the stray light correction and the LST retrieval result comparison.

4.1. TOA Radiance Comparison Before and After the Stray Light Correction

After the stray light correction, some changes in band TOA radiance should be observed. On the basis of the above clear-sky Landsat 8 images over the ground sites, we investigated the band TOA radiance changes caused by the stray light correction. From the 207 images over the five sites, Figure 1a shows the TOA radiance difference (ΔL) of Band 10 before and after the stray light correction, and Figure 1b is the case of Band 11. The bias of both bands turned out to be slightly positive, which meant that the TIRS data became minimally "hotter" after the stray light correction in general. For Band 10, the histogram of ΔL was mostly concentrated larger than zero, with a bias of 0.08 W·m^{-2}·sr^{-1}·µm^{-1}. A general positive TOA radiance change was observed for this band, although such a change was unremarkable. For Band 11, ΔL showed a broader distribution but had a similar bias with Band 10, indicating that the stray light on Band 11 was corrected in a larger scale and the variance of correction was also greater than that of Band 10.

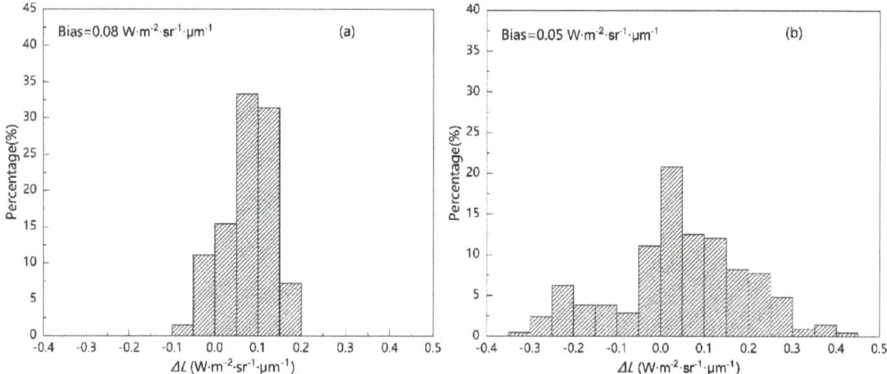

Figure 1. Histograms of the band top of the atmosphere (TOA) radiance change after the stray light correction for Landsat 8 thermal infrared sensor (TIRS) of (**a**) Band 10 and (**b**) Band 11. ΔL is the difference between TOA radiance after and before the stray light correction (after minus before).

The effect of stray light is related to LST [17], hence, the relationship between correction results and LST is necessary to explore. As stated in Section 3, the LST calculated from the SURFRAD data was regarded as the reference LST for the evaluation. Figure 2 shows the scatter plot between ΔL caused by the stray light correction and SURFRAD ground LST (SURFRAD LST). A negative correlation was firstly observed, and the radiance of Band 11 was found to have a more significant negative relationship between ΔL and SURFRAD LST than that of Band 10. In the low LST range of 260–280 K, the stray light correction of Band 11 was 0.2–0.4 $W \cdot m^{-2} \cdot sr^{-1} \cdot \mu m^{-1}$, which was larger than that of Band 10. In the LST range of 290–300 K, the stray light correction of the two bands became nearly the same, with a value of approximately 0.1 $W \cdot m^{-2} \cdot sr^{-1} \cdot \mu m^{-1}$, corresponding to a temperature difference of approximately 0.7–0.8 K. For LST > 310 K, the absolute value of the stray light correction of Band 10 was smaller than that of Band 11. Such modification can result in a remarkable effect on the final LST retrieval, especially for the single-channel method. This study analyzed this effect, as stated in the following Section 4.2, to clarify the change on LST retrieval result before and after the stray light correction.

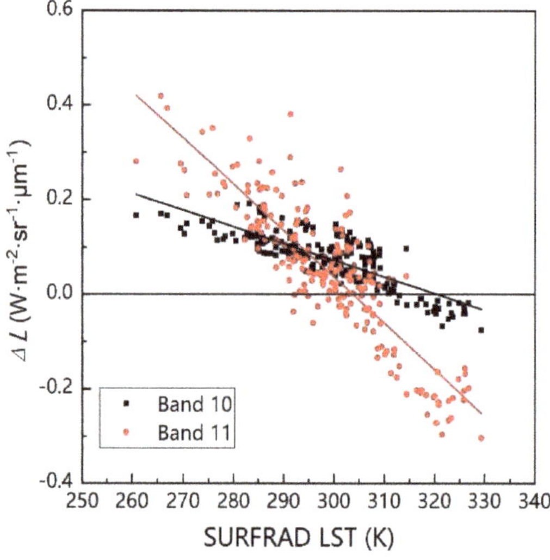

Figure 2. Relationship between ΔL and SURFRAD LST over ground sites.

After considering ΔL of the single band, the relationship between the two bands' brightness temperature difference $(T_{10} - T_{11})$ and the SURFRAD LST before and after the stray light correction was investigated. As illustrated in Figure 3, before the correction, the brightness temperature difference $(T_{10} - T_{11})$ had no evident relationship with the ground LST (see Figure 3a). However, after the stray light correction, the brightness temperature difference $(T_{10} - T_{11})$ had a significant positive correlation with the ground LST (Figure 3b) and the same trend as the simulation data (Figure 3d). This change showed that the stray light correction process not only improved TOA radiance of two TIRS bands respectively as stated in the Introduction section, but also produced a more reasonable correlation on brightness temperature of two Landsat 8 TIRS bands data. Moreover, the change $\Delta(T_{10} - T_{11})$ in the brightness temperature difference $(T_{10} - T_{11})$ before and after the stray light correction was also found to be linearly correlated with SURFRAD LST, as shown in Figure 3c. In high temperature (LST > 320 K), the $(T_{10} - T_{11})$ increased, while in low temperature (LST < 280 K), the $(T_{10} - T_{11})$ decreased. Figure 3d shows that $(T_{10} - T_{11})$ had an obvious relationship with LST in theory, which makes it possible to develop a split-window algorithm with the term $(T_{10} - T_{11})$ to retrieve LST. Therefore, the evident and closer-to-theory relationship between $(T_{10} - T_{11})$ and LST appeared after correction was expected to make the refined TIRS image more suitable for the split-window algorithm than the original TIRS image.

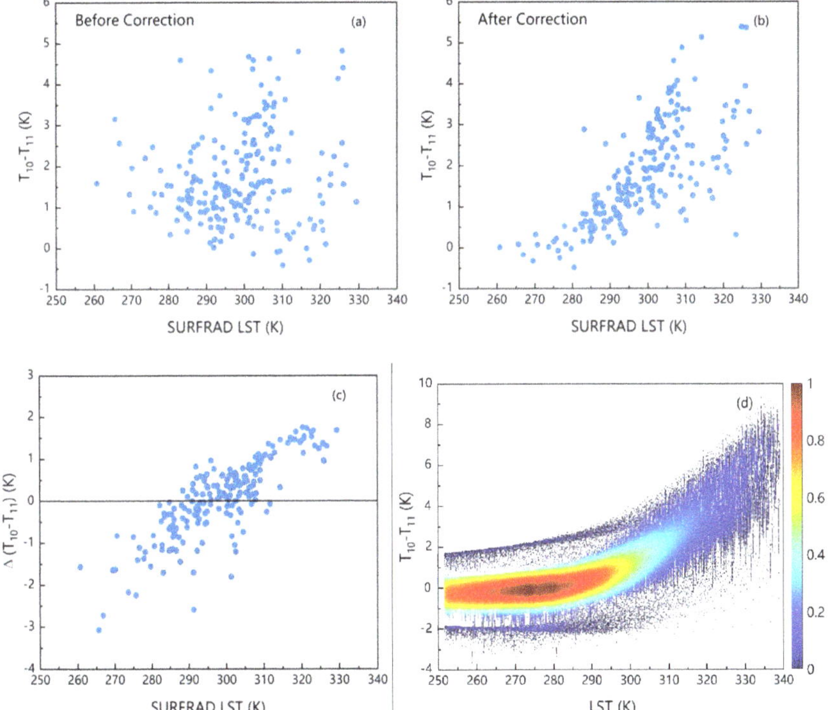

Figure 3. Relationship of brightness temperature difference $(T_{10} - T_{11})$ and ground-measured LST (denoted as SURFRAD LST). (**a**) Before the stray light correction and (**b**) after the stray light correction. (**c**) Change $(\Delta(T_{10} - T_{11}))$ in the brightness temperature difference before and after the stray light correction. (**d**) Density scatter plot of $(T_{10} - T_{11})$ and LST in the simulation dataset.

4.2. LST Retrieval Comparison Before and After the Stray Light Correction

This section focused on the LST retrieval evaluation before and after the stray light correction in two ways. First, the retrieved LST was compared with the ground-measured LST to check whether the

LST retrieval accuracy improved after the stray light correction. Second, the LST retrieval accuracy among different split-window and single-channel algorithms was compared to find the best algorithm for retrieving LST from Landsat 8 TIRS images.

4.2.1. Overall Comparison of LST Retrieval Results

On the basis of the Landsat 8 images before and after the stray light correction, the above split-window and single-channel algorithms were used to estimate the LST over the ground sites. In accordance with the ground reference LST in different sites, Table 4 lists the LST RMSE and bias for the five algorithms before and after the stray light correction over all five sites.

Table 4. The RMSE and bias of Landsat 8 LST retrieval from five algorithms before and after the stray light correction over all five sites.

Algorithm	Before Correction		After Correction	
	RMSE (K)	Bias (K)	RMSE (K)	Bias (K)
SW_Du	2.26	1.14	2.81 (+0.55) [1]	2.05 (+0.91)
SW_Rozenstein	2.76	−1.06	2.46 (−0.30)	−0.09 (+0.97)
SW_JM	2.33	0.58	2.54 (+0.21)	1.37 (+0.79)
SC_10	2.74	−0.17	2.47 (−0.27)	0.54 (+0.71)
SC_11	4.33	0.98	3.55 (−0.78)	1.59 (+0.61)

[1] The value in the bracket is the RMSE or bias change after the stray light correction compared to those of no correction.

Before the stray light correction, three split-window algorithms and SC_10 nearly had RMSEs smaller than 3.0 K, which was obviously smaller than the RMSE (4.33 K) of SC_11. However, the absolute values of bias for SW_Du and SW_Rozenstein were greater than 1.0 K, indicating that the retrieved LST was slightly biased from the ground-measured LST on average. As for the single-channel algorithm, SC_10 performed better than SC_11; the RMSE of SC_10 was 2.74 K with a bias of −0.17 K, whereas SC_11 had a larger RMSE of 4.33 K. From the RMSE and bias, the performance of SW_JM exhibited the best results among the five methods before the stray light correction, with an RMSE of 2.33 K and a bias of 0.58 K. After the stray light correction, the RMSEs of all algorithms were less than 4.0 K. The least RMSE was nearly 2.5 K, which was obtained by SW_Rozenstein, SW_JM, and SC_10. Among the three algorithms, the biases of SW_Rozenstein and SC_10 were within 0.6 K; therefore, these two algorithms may be the best algorithms to retrieve LST after the stray light correction. SC_11 still had the worst performance to retrieve LST from the Landsat TIRS images, with an RMSE of 3.55 K and a bias of 1.59 K.

The comparison of the retrieved results of all sites before and after the stray light correction indicated that the absolute changes in RMSE and bias for the five algorithms were all within 1.0 K, which showed only slight changes in retrieved LSTs after the stray light correction on average. For the split-window algorithms, the RMSE of SW_Rozenstein decreased by 0.3 K, whereas that of SW_Du and SW_JM increased minimally after the stray light correction. For the single-channel algorithm, the RMSE of two bands both decreased, indicating an improved performance of the single-channel algorithm for Landsat 8 TIRS after the correction. The RMSE of SC_11 decreased most among the five algorithms but was still larger than 3.5 K. However, the bias of all algorithms increased greater than 0.6 K, showing that retrieved LSTs were overestimated after the stray light correction on average, and the correction might be biased.

Figure 4 demonstrates the comparison of LST retrieval from the five algorithms before and after the stray light correction. Moreover, because of the particularity of the DRA site mentioned in Section 3, its result is shown separately with different symbols. From Figure 4, it was found that the retrieved LST highly linearly correlated to the ground-measured LST, with coefficients R^2 larger than 0.95 for all cases. For the three split-window algorithms (Figure 4a–c), the retrieved LSTs from the corrected images were generally higher than those from uncorrected images, especially at high ground surface temperature. In the case that the retrieved LST before the correction was higher than the ground LST, the accuracy of LST retrieval naturally decreased if the retrieval LST was higher after the correction. For the SW_Du algorithm (Figure 4a), most retrieved LSTs before correction were higher than ground LST. After correction, the retrieval error was greater in most cases. For the SW_Rozenstein algorithm (Figure 4b), when ground LST was higher than 315 K, the retrieved LST before correction was lower than the ground LST for most cases. Although the retrieved LST after correction was higher than that before the correction (similar to the SW_Du algorithm), the accuracy of LST retrieval increased after the correction, which was different from the SW_Du algorithm. For the SW_JM algorithm (Figure 4c), under low-temperature conditions (LST < 290 K), the retrieved LST became lower after the correction. Consequently, the over-high retrieved LST before the correction was getting closer to the ground LST, leading to the improvement of retrieval accuracy. However, under high-temperature conditions (LST > 300 K), the retrieved LST after the correction became higher in general, resulting in a larger error. With the opposite accuracy change of low and high temperatures, the RMSE change of the SW_JM algorithm finally became very small and was only half of that of the SW_Du algorithm, as presented in Table 4 (+0.55 K for the SW_Du algorithm and +0.21 K for the SW_JM algorithm). For the single-channel algorithm (Figure 4d,e), the retrieved LST after the correction increased at low temperatures but decreased at high temperatures, compared to that of before the correction. The LST change of the SC_11 algorithm was more evident than that of the SC_10 algorithm (−0.27 K for SC_10 and −0.78 K for SC_11), and the scatter plot result of both algorithms were getting closer to the 1:1 line. This finding explained the details of accuracy increase in single-channel algorithms for the two TIRS bands, as shown in Table 4.

The single-channel algorithm is a theoretically derived algorithm; accordingly, the accuracy changes of SC_10 and SC_11 can be used to check whether the TIRS data are better after the stray light correction. Table 4 indicates that the errors of the SC_10 and SC_11 algorithms were reduced, meaning that LST retrieval results got better after the stray light correction and the SC_11 improved more. Combined with the change in the band radiance before and after the correction, the difference of retrieved LST can be analyzed clearly. Figure 2 in Section 4.1 illustrates that the TOA radiance of the two bands after the correction increased at low temperatures and decreased at high temperatures, resulting that the retrieved LST was closer to the ground LST of the single-channel algorithm on two bands. This finding from SURFRAD sites provided the proof supporting that the stray light correction improved not only the TIRS data quality as stated in previous study [17] but also the LST retrieval accuracy in practice. However, the validations over more robust and homogeneous ground-measured datasets, such as desert and water sites, are still needed to clarify the sole effect of the stray light correction on LST retrieval.

Since the DRA site had heterogeneity issues and high ground temperature cases (all from DRA when SURFRAD LST > 320 K) that had distinguishing features in the result of validation in Figure 4, the analysis excluding the DRA site is also necessary to help understand the effect of the stray light correction. Table 5 lists the temperature RMSE and bias of the five algorithms calculated from data of all sites excluding DRA.

Table 5. The RMSE and bias of Landsat 8 LST retrieval from five algorithms before and after the stray light correction over all sites excluding DRA.

Algorithm	Before Correction		After Correction	
	RMSE (K)	Bias (K)	RMSE (K)	Bias (K)
SW_Du	2.40	1.30	2.71 (+0.30) [1]	1.97 (+0.68)
SW_Rozenstein	2.43	−0.42	2.30 (−0.13)	0.32 (+0.74)
SW_JM	2.47	0.82	2.27 (−0.20)	1.22 (+0.40)
SC_10	2.41	−0.83	2.11 (−0.31)	0.10 (+0.93)
SC_11	3.20	−0.17	2.84 (−0.36)	1.02 (+1.19)

[1] The value in the bracket is the RMSE or bias change after the stray light correction compared to those of no correction.

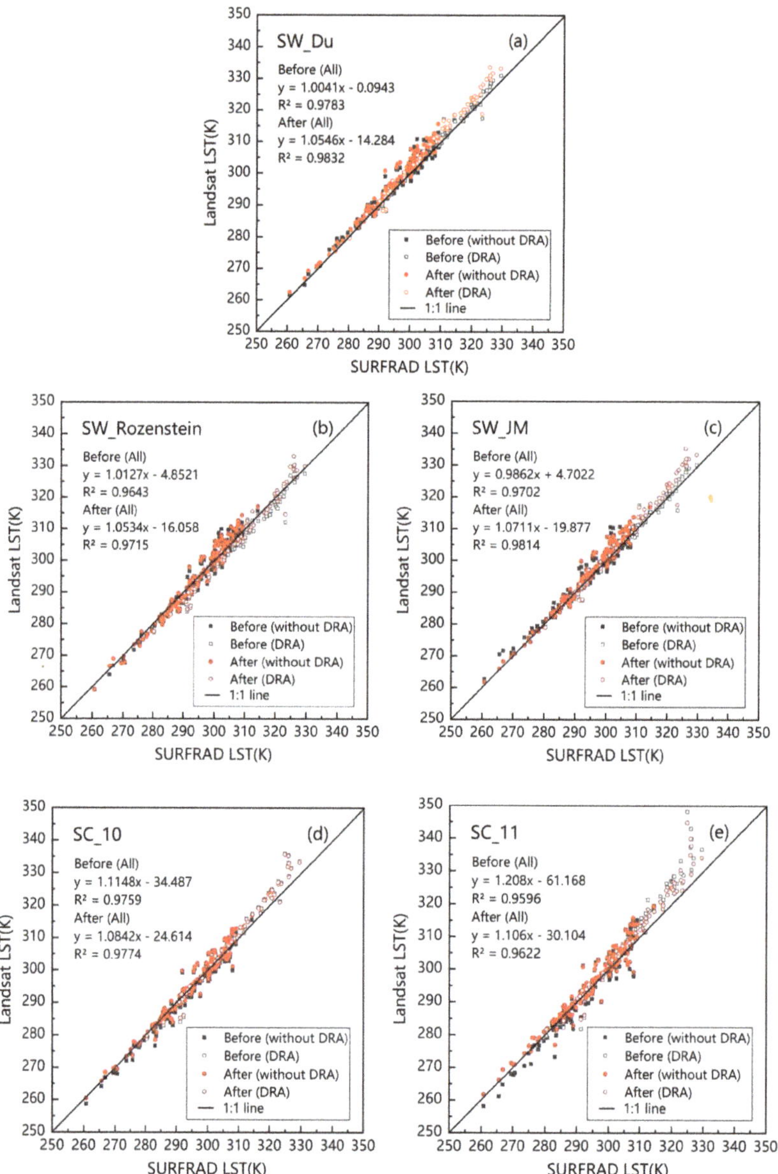

Figure 4. Comparison of retrieved LST before and after the stray light correction for different algorithms. (**a**) for SW_Du, (**b**) for SW_Rozenstein, (**c**) for SW_JM, (**d**) for SC_10, and (**e**) for SC_11. The linear regression coefficients were obtained from data of all five sites. "Before/After (without DRA)" in the figure means the data from the other four SURFRAD sites after excluding DRA before or after the correction; "Before/After (DRA)" in the figure means the data from the DRA site before or after the correction.

From Table 5, when excluding high ground temperature cases of DRA, there were some differences from the results including DRA in Table 4. Except the SW_Du, the RMSE of the other four algorithms all decreased little and RMSEs of five algorithms all were within 3.0 K after the correction. The decreases on RMSE of SC_10 and SC_11 were both around 0.3 K, narrowing the gap of the RMSE change on SC_10 and SC_11 in Table 4. This indicated that the correction and improvement of single-channel algorithm on Band 11 were greater than Band 10 in a high ground temperature range, which was consistent with the result in Figures 2 and 4d,e. However, the RMSE change of split-window algorithms was different from Table 4, especially for SW_JM, which will be analyzed in detail in the Discussion section. Considering the high temperature cases in the practical use, this paper finally kept the DRA site in the statistical analysis to make the conclusion more universal.

4.2.2. Comparison Results over Each SURFRAD Site

On the basis of the comparison results over all ground sites, we can determine the overall change on LST retrieval from different algorithms before and after the stray light correction. However, these sites are different from one another in the land cover type and homogeneity degree, and those differences may cause confusion in understanding the validation results. Therefore, further analysis of the results over each site is necessary. Table 6 lists the temperature RMSE and bias of retrieved LSTs from the five algorithms using the Landsat 8 images before the stray light correction over each site. Table 7 provides the results after the stray light correction. Figure 5 presents the scatter plots between the retrieved LST and ground LST.

Table 6 indicates that before the correction, the four other algorithms performed better than SC_11 over BND, GCM, and PSU sites, and their temperature RMSEs were within the range of 1.5–2.7 K. From Table 7, after the correction, the temperature RMSEs of the single-channel algorithm of two bands decreased over the three sites. However, the temperature RMSE change of the split-window algorithms was not the same. The RMSE of SW_Du for the three sites evidently increased, that is, 0.40 K for BND, 0.66 K for GCM, and 0.75 K for PSU. By contrast, the RMSE of the two split-window algorithms (SW_Rozenstein and SW_JM) over the three sites changed minimally. The temperature bias of all algorithms of the three sites increased, indicating that the retrieved LST was higher after correction on average. Figure 5 depicts that the ground LST of these sites was lower than 310 K; in this temperature range, most of the radiance of Bands 10 and 11 was increased (Figure 2). Therefore, the retrieved LST results were consistent with the change in radiance.

Table 6 presents that the GCM and PSU sites were suitable to retrieve LST using split-window algorithms before the stray light correction. The RMSEs of the three split-window algorithms on the two sites were within the range of 1.5–2.3 K, which were generally smaller than those of the single-channel algorithms of both bands. After the correction, with an increase on RMSEs of the split-window algorithms and a decrease on RMSEs of the single-channel algorithms, the superiority of the split-window algorithms in retrieving LST for the two sites disappeared. SC_10 had the smallest RMSEs with a small bias over the PSU site, similar to the result obtained over the BND site. At the GCM site, SC_10 also showed no weakness compared with some split-window algorithms.

For SXF, before the correction, the performance of the split-window algorithms was worse than that of SC_10 (Table 6), and the SW_Du and SW_JM algorithms even had larger RMSEs than that of SC_11. The combined analysis of Figure 5a–c implied that for the split-window algorithms on SXF, when the ground LST was in the range of 300–310 K, the retrieved LST was obviously higher than the ground LST, corresponding to the poor performance of the split-window algorithms on SXF. After the correction, the obvious error had disappeared, indicating an improvement in the split-window algorithms in retrieving LSTs (the RMSE of three split-window algorithms decreased and the biases of split-window algorithms were closer to 0.0 K). Nevertheless, they still exhibited worse results than SC_10.

Although DRA was unsuitable for accurate validation because of heterogeneity, as mentioned in Section 3, DRA still could reveal important information of the effect of the stray light correction because of its high ground surface temperature, as illustrated in Figure 5. Its RMSE of SC_11 was very large before correction and reduced by 1.4 K after correction. The large RMSE in Figure 5e was caused by the poor performance of SC_11 at high temperature, and the improvement was due to the lower radiance correction at high temperature on Band 11 (Figure 2). This comparison result confirmed the good stray light correction of Band 11 at high-temperature conditions.

In conclusion, the biases of five algorithms increased on most cases after the correction, except for three split-window algorithms on SXF and SC_11 on DRA. However, the RMSE changes were complicated of different split-window algorithms on different sites. SW_Du and SW_JM mostly kept a similar change trend on GCM, PSU, SXF, and DRA, but their RMSEs increased after the stray light correction except on SXF. The RMSEs of SW_Rozenstein decreased on these five sites, except the PSU site, showing the better performance of SW_Rozenstein after the correction on most cases. The RMSEs of the single-channel algorithms decreased over all sites after correction. SC_10 had the smallest RMSEs and good biases on BND, PSU, and SXF after the stray light correction, and its RMSEs were approximately 2.1 K on these sites. The performance of SW_Rozenstein was close to that of SC_10 after the correction in RMSE and bias, but the accuracy of SW_Rozenstein was better than that of SC_10 for the DRA site. Therefore, similar to SC_10, SW_Rozenstein was also a better algorithm than other two split-window algorithms to retrieve LST for Landsat 8 after the stray light correction.

Table 6. The RMSE and bias of the LST retrieved from Landsat 8 TIRS images on five sites before the stray light correction.

Algorithms	BND		GCM		PSU		SXF		DRA	
	RMSE (K)	Bias (K)	RMSE (K)	Bias (K)	RMSE (K)	Bias (K)	RMSE (K)	Bias (K)	RMSE (K)	Bias (K)
SW_Du	2.64	1.76	1.50	0.20	1.90	1.10	3.41	2.80	1.88	0.78
SW_Rozenstein	2.34	−0.04	2.27	−1.37	1.98	−0.50	2.96	0.87	3.41	−2.56
SW_JM	2.61	1.53	1.68	−0.52	2.06	0.61	3.48	2.44	1.97	0.01
SC_10	2.30	−0.42	2.56	−1.49	2.67	−1.54	2.18	0.15	3.37	1.36
SC_11	3.17	0.03	3.27	−0.33	3.34	−1.44	3.05	0.40	6.21	3.67

Table 7. The RMSE and Bias of the LST retrieved from Landsat 8 TIRS images on five sites after the stray light correction.

Algorithms	BND		GCM		PSU		SXF		DRA	
	RMSE (K)	Bias (K)	RMSE (K)	Bias (K)	RMSE (K)	Bias (K)	RMSE (K)	Bias (K)	RMSE (K)	Bias (K)
SW_Du	3.04(+0.40) [1]	2.39(+0.63)	2.16(+0.66)	1.43(+1.23)	2.65(+0.75)	1.83(+0.73)	3.13(−0.28)	2.48(−0.32)	3.05(+1.17)	2.21(+1.43)
SW_Rozenstein	2.33(−0.01)	0.71(+0.75)	2.07(−0.20)	−0.09(+1.28)	2.13(+0.15)	0.24(+0.74)	2.69(−0.27)	0.58(−0.29)	2.79(−0.62)	−1.07(+1.49)
SW_JM	2.52(−0.09)	1.72(+0.19)	1.77(+0.09)	0.67(+1.19)	2.26(+0.20)	1.02(+0.41)	2.70(−0.78)	1.67(−0.77)	3.07(+1.10)	1.70(+1.69)
SC_10	2.19(−0.11)	0.42(+0.84)	2.05(−0.51)	−0.43(+1.06)	2.07(−0.60)	−0.47(+1.07)	2.12(−0.06)	0.88(+0.73)	3.16(−0.21)	1.56(+0.20)
SC_11	2.87(−0.30)	1.22(+1.19)	2.82(−0.45)	0.58(+0.91)	2.48(−0.86)	−0.02(+1.42)	2.98(−0.07)	2.01(+1.61)	4.81(−1.40)	2.91(−0.76)

[1] The value in the bracket means the RMSE and bias change after the stray light correction compared to those of no correction.

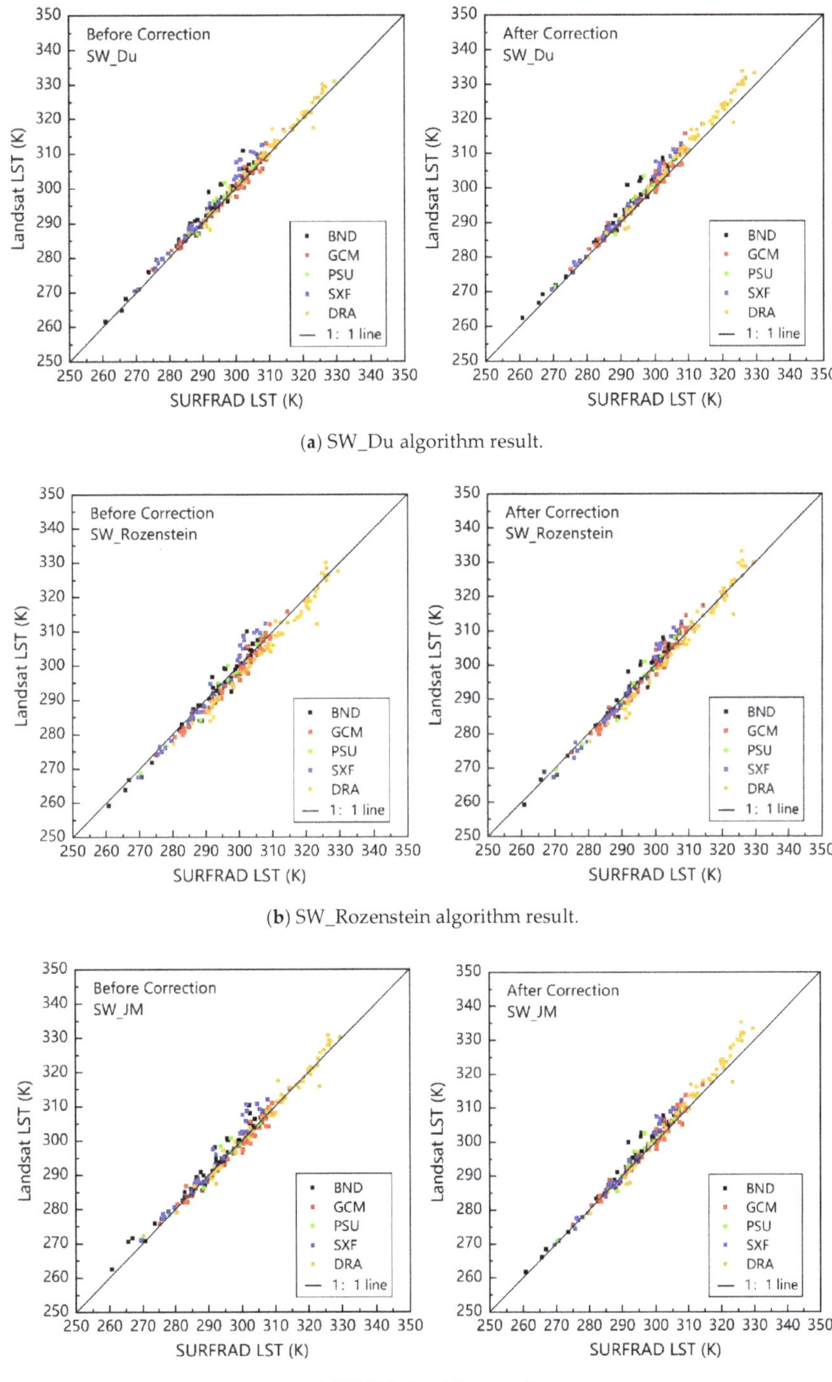

(**a**) SW_Du algorithm result.

(**b**) SW_Rozenstein algorithm result.

(**c**) SW_JM algorithm result.

Figure 5. *Cont.*

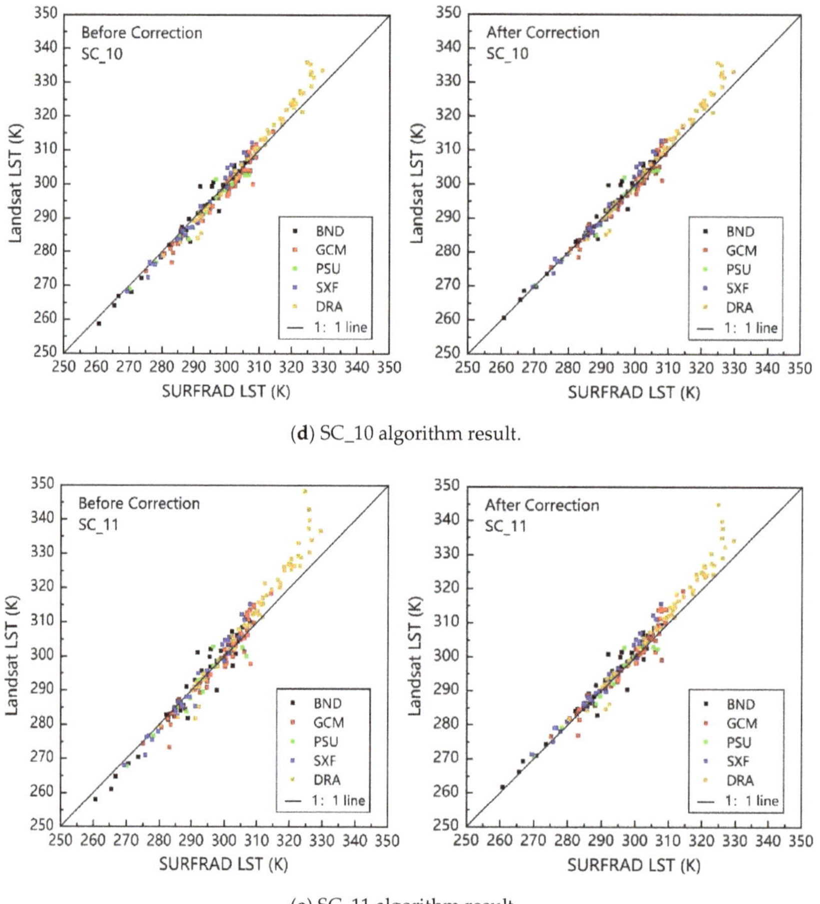

(**d**) SC_10 algorithm result.

(**e**) SC_11 algorithm result.

Figure 5. Comparison of Landsat 8 retrieved LST (Landsat LST) with ground LST (SURFRAD LST) over different sites before and after the stray light correction. (**a**) is for the SW_Du algorithm, (**b**) is for the SW_Rozenstein algorithm, (**c**) is for the SW_JM algorithm, (**d**) is for the SC_10 algorithm, and (**e**) is for the SC_11 algorithm.

5. Discussion

After the stray light correction, the single-channel and split-window algorithms showed different accuracy changes in different directions and ranges. The RMSE of single-channel algorithms of two bands all decreased on each site, but none of the three split-window algorithms showed the same change (all increased or all decreased) on the five sites. Moreover, different split-window algorithms had different accuracy changes on overall sites, and even the same split-window algorithm (SW_JM) had increasing RMSE on all sites when including the DRA site but decreasing RMSE excluding the DRA site. The first confusing point is why split-window algorithms had such complicated accuracy changes that seemed to have no regularity. Secondly, when focusing on some specific temperature ranges, for instance, the high temperature (>320 K), the change of retrieved LSTs by split-window algorithms before and after the stray light correction was unreasonable. Figure 2 in Section 4.1 shows that the TOA radiance of the two bands after the correction was reduced under high-temperature condition (> 320 K). In the case if other factors are regarded as the same, the retrieved LSTs should have reduced in theory. However, as seen in Figure 4, under the high-temperature condition, the retrieved

LSTs of the three split-window algorithms increased after the correction, which was inconsistent with the theoretical expectation. The same unreasonable retrieved LST changes occurred also in some low temperature cases of SW_Du and SW_JM.

To explain the two confusing points, we must first explain the RMSE change of these algorithms was made of two parts: the one was the performance of the algorithm before the correction; the other was the change of retrieved LST before and after the correction. From Figures 4 and 5, it was easy to find that before the correction, the SW_Du overestimated LST for most cases (bias = 1.14 K (with DRA), bias = 1.30 K (without DRA)), the SW_Rozenstein underestimated LST for most cases (bias = −1.06 K (with DRA), bias = −0.42 K (without DRA)), and the SW_JM overestimated LST a little (bias = 0.58 K (with DRA), bias = 0.82 K (without DRA)). Different performance before the correction will certainly result in a different accuracy change if these split-window algorithms have the same retrieved LST change before and after the correction, just as the high temperature cases (>320 K).

Meanwhile, the brightness temperature relationship between Bands 10 and 11, such as the brightness temperature difference of two bands, has an important effect on the performance of split-window algorithms. Taking the SW_Du algorithm as an example, from its structure (see Equation (1)), the coefficients of $\frac{(T_{10}+T_{11})}{2}$ and $\frac{(T_{10}-T_{11})}{2}$ are positive based on the simulation data, and those coefficients of $(T_{10} - T_{11})^2$ are very small. Therefore, the influence of $(T_{10} - T_{11})^2$ can be ignored when analyzing the influence of the brightness temperature change on T_s. Under a high-temperature condition (LST > 320 K), as illustrated in Figure 3, the decreases in T_{10} and T_{11} would tend to cause a low retrieved LST, but the increase in $(T_{10} - T_{11})$ would tend to increase the LST. The result from Figure 4a shows that the retrieved LST of the SW_Du algorithm after the correction was larger than that before the correction for the case LST > 320 K, indicating that the increasing effect on retrieved LST caused by an increase in TOA band radiance difference was greater than the decreasing effect on retrieved LST caused by a decrease in TOA radiance on Bands 10 and 11. The influence of brightness temperature difference finally caused the second confusing point above. Moreover, the different split-window algorithms have different structures and therefore, the brightness temperature relationship (for example the brightness temperature difference) will influence the performance of the algorithm in different degree. This can also cause the first confusing point, or more accurately, it was the different structures that resulted in a different performance of different split-window algorithms before the correction.

In the case that the RMSE change of split-window algorithms was much affected by the brightness temperature difference of two bands, the accuracy of brightness temperature (radiance) difference correction could directly determine the performance of split-window algorithms after the correction. However, the correction did not take into account the relationship between the radiance of two bands. Even though the radiance of Bands 10 and 11 got closer to the real radiance respectively and the relationship between the two bands' brightness temperature difference and LST became closer to the theoretical result mentioned in Section 4.1, the value of the radiance difference between the two bands may still be more biased away from the real value, making the performance of split-window algorithms worse.

Therefore, the structure of split-window algorithms, the performance of split-window algorithms before the correction and the change of brightness temperature difference between Bands 10 and 11 combined together to cause the confusing results of split-window algorithms before and after the correction.

6. Conclusions

This study focused on the evaluation of LST retrieval from Landsat 8/TIRS data before and after the stray light correction on the original observations using ground-measured LST from five SURFARD sites. Three split-window algorithms (SW_Du, SW_JM, and SW_Rozenstein) and two single-channel algorithms (SC_10 and SC_11) were investigated.

The stray light correction increased band TOA radiance for low brightness temperature range (< 305 K for Band 10 and 310 K for Band 11) but decreased such radiance for a high brightness temperature range. The relationship between the two bands' brightness temperature difference and LST became closer to the theoretical result.

The LST retrieval error of the single-channel algorithm was consequently reduced. The RMSE of the single-channel algorithm for Bands 10 and 11 decreased by 0.27 K and 0.78 K, respectively. The improvement in retrieval accuracy for Band 11 at high temperature was obvious. In the high-temperature site (DRA), the decreased RMSE of the single-channel algorithm for Band 11 was 1.40 K. By contrast, the accuracy of split-window algorithm (such as SW_Du) was unexpectedly reduced due to the variation in the brightness temperature difference of the two bands, and was unreasonably inconsistent with the change in radiance. For better use of split-window algorithms, the development of the split-window algorithms specified for Landsat 8, as well as the stray light correction on radiance relationship of Bands 10 and 11 may need more concern.

Among the five algorithms, the best two were SW_Rozenstein and SC_10. Before the stray light correction, the accuracy of the split-window algorithms was generally better than that of the single-channel algorithm. For the corrected images of overall sites, the accuracy of the SC_10 algorithm increased, whereas the accuracy of the SW_Du and SW_JM algorithms decreased, even making the accuracy of the SC_10 algorithm better than that of the SW_Du and SW_JM algorithms. After the correction, the RMSEs of SW_Rozenstein and SC_10 were approximately 2.5 K over all ground sites. In BND, GCM, PSU, and DRA sites, the RMSE of the single-channel algorithm for Band 10 was even within 2.2 K.

The results obtained in this study implied that the accuracy and applicability of the single-channel algorithm on Landsat 8/TIRS LST retrieval improved after the stray light correction. However, it still needs to be careful when using split-window algorithms to retrieve LST from Landsat 8/TIRS images. Meanwhile, it must be noted that the current results were only based on the measurements of five SURFRAD sites. Future evaluation using more ground-measured datasets over more homogeneous sites like desert and water sites remains strongly expected to clarify the quality of Landsat 8/TIRS data and the performance of different LST retrieval algorithms.

Author Contributions: Conceptualization, H.R.; Data curation, J.G., S.L. and J.D.; Formal analysis, J.G.; Funding acquisition, H.R.; Investigation, J.G.; Methodology, J.G., H.R. and Y.Z.; Resources, S.L. and J.D.; Supervision, H.R.; Validation, J.G.; Visualization, J.G. and Y.Z.; Writing—original draft, J.G.; Writing—review and editing, H.R. All authors have read and agreed to the published version of the manuscript.

Funding: This research was funded by the National High-Resolution Earth Observation Project (04-Y30B01-9001-18/20-1-4), National Natural Science Foundation of China (No. 41771369), National key research and development program (2017YFB0503905-05).

Acknowledgments: The Landsat 8 images were downloaded from the USGS datacenter at https://earthexplorer. usgs.gov/; The SURFRAD data was obtained from Earth System Research Laboratory Global Monitoring Division (https://www.esrl.noaa.gov/gmd/grad/surfrad/).

Conflicts of Interest: The authors declare no conflict of interest.

References

1. Mannstein, H. Surface Energy Budget, Surface Temperature and Thermal Inertia. In *Remote Sensing Applications in Meteorology and Climatology*; Springer Science and Business Media LLC: Berlin, Germany, 1987; pp. 391–410.
2. Price, J.C. The potential of remotely sensed thermal infrared data to infer surface soil moisture and evaporation. *Water Resour. Res.* **1980**, *16*, 787–795. [CrossRef]
3. Weng, Q. Thermal infrared remote sensing for urban climate and environmental studies: Methods, applications, and trends. *ISPRS J. Photogramm. Remote Sens.* **2009**, *64*, 335–344. [CrossRef]
4. Weng, Q.; Lu, D.; Schubring, J. Estimation of land surface temperature–vegetation abundance relationship for urban heat island studies. *Remote Sens. Environ.* **2004**, *89*, 467–483. [CrossRef]
5. Yuan, F.; Bauer, M.E. Comparison of impervious surface area and normalized difference vegetation index as indicators of surface urban heat island effects in Landsat imagery. *Remote Sens. Environ.* **2007**, *106*, 375–386. [CrossRef]

6. Koc, C.B.; Osmond, P.; Peters, A. Spatio-temporal patterns in green infrastructure as driver of land surface temperature variability: The case of Sydney. *Int. J. Appl. Earth Obs. Geoinf.* **2019**, *83*, 101903.

7. Mildrexler, D.J.; Zhao, M.; Running, S.W. A global comparison between station air temperatures and MODIS land surface temperatures reveals the cooling role of forests. *J. Geophys. Res. Space Phys.* **2011**, *116*, 15. [CrossRef]

8. Wan, Z.; Wang, P.; Li, X. Using MODIS Land Surface Temperature and Normalized Difference Vegetation Index products for monitoring drought in the southern Great Plains, USA. *Int. J. Remote Sens.* **2004**, *25*, 61–72. [CrossRef]

9. Tran, H.; Uchihama, D.; Ochi, S.; Yasuoka, Y. Assessment with satellite data of the urban heat island effects in Asian mega cities. *Int. J. Appl. Earth Obs. Geoinf.* **2006**, *8*, 34–48. [CrossRef]

10. Montanaro, M.; Barsi, J.; Lunsford, A.; Rohrbach, S.; Markham, B. Performance of the Thermal Infrared Sensor on-board Landsat 8 over the first year on-orbit. *Earth Obs. Syst. XIX* **2014**, *9218*, 921817.

11. Reuter, D.C.; Richardson, C.M.; Pellerano, F.A.; Irons, J.; Allen, R.G.; Anderson, M.C.; Jhabvala, M.D.; Lunsford, A.; Montanaro, M.; Smith, R.L.; et al. The Thermal Infrared Sensor (TIRS) on Landsat 8: Design Overview and Pre-Launch Characterization. *Remote Sens.* **2015**, *7*, 1135–1153. [CrossRef]

12. Jiménez-Muñoz, J.-C.; Sobrino, J.A.; Skoković, D.; Mattar, C.; Cristóbal, J. Land Surface Temperature Retrieval Methods From Landsat-8 Thermal Infrared Sensor Data. *IEEE Geosci. Remote Sens. Lett.* **2014**, *11*, 1840–1843. [CrossRef]

13. Rozenstein, O.; Qin, Z.; Derimian, Y.; Karnieli, A. Derivation of Land Surface Temperature for Landsat-8 TIRS Using a Split Window Algorithm. *Sensors* **2014**, *14*, 5768–5780. [CrossRef] [PubMed]

14. Du, C.; Ren, H.; Qin, Q.; Meng, J.; Zhao, S. A Practical Split-Window Algorithm for Estimating Land Surface Temperature from Landsat 8 Data. *Remote Sens.* **2015**, *7*, 647–665. [CrossRef]

15. Yu, X.; Guo, X.; Wu, Z. Land Surface Temperature Retrieval from Landsat 8 TIRS—Comparison between Radiative Transfer Equation-Based Method, Split Window Algorithm and Single Channel Method. *Remote Sens.* **2014**, *6*, 9829–9852. [CrossRef]

16. Montanaro, M.; Gerace, A.; Rohrbach, S. Toward an operational stray light correction for the Landsat 8 Thermal Infrared Sensor. *Appl. Opt.* **2015**, *54*, 3963. [CrossRef]

17. Gerace, A.; Montanaro, M. Derivation and validation of the stray light correction algorithm for the thermal infrared sensor onboard Landsat 8. *Remote Sens. Environ.* **2017**, *191*, 246–257. [CrossRef]

18. García-Santos, V.; Cuxart, J.; Martínez-Villagrasa, D.; Jiménez, M.A.; Simó, G. Comparison of Three Methods for Estimating Land Surface Temperature from Landsat 8-TIRS Sensor Data. *Remote Sens.* **2018**, *10*, 1450. [CrossRef]

19. Xu, H. Retrieval of the reflectance and land surface temperature of the newly-launched Landsat 8 satellite. *Chin. J. Geophys.* **2015**, *58*, 741–747.

20. Meng, X.; Cheng, J.; Zhao, S.; Liu, S.; Yao, Y. Estimating Land Surface Temperature from Landsat-8 Data using the NOAA JPSS Enterprise Algorithm. *Remote Sens.* **2019**, *11*, 155. [CrossRef]

21. Wan, Z. New refinements and validation of the collection-6 MODIS land-surface temperature/emissivity product. *Remote Sens. Environ.* **2014**, *140*, 36–45. [CrossRef]

22. Qin, Z.; Dall'Olmo, G.; Karnieli, A.; Berliner, P. Derivation of split window algorithm and its sensitivity analysis for retrieving land surface temperature from NOAA-advanced very high resolution radiometer data. *J. Geophys. Res. Space Phys.* **2001**, *106*, 22655–22670. [CrossRef]

23. Sobrino, J.A.; Li, Z.-L.; Stoll, M.P.; Becker, F. Multi-channel and multi-angle algorithms for estimating sea and land surface temperature with ATSR data. *Int. J. Remote Sens.* **1996**, *17*, 2089–2114. [CrossRef]

24. Sobrino, J.A.; Raissouni, N. Toward remote sensing methods for land cover dynamic monitoring: Application to Morocco. *Int. J. Remote Sens.* **2000**, *21*, 353–366. [CrossRef]

25. Jiménez-Muñoz, J.-C.; Sobrino, J.A. Split-Window Coefficients for Land Surface Temperature Retrieval From Low-Resolution Thermal Infrared Sensors. *IEEE Geosci. Remote Sens. Lett.* **2008**, *5*, 806–809. [CrossRef]

26. Sobrino, J.A.; Jiménez-Muñoz, J.-C.; Paolini, L. Land surface temperature retrieval from LANDSAT TM 5. *Remote Sens. Environ.* **2004**, *90*, 434–440. [CrossRef]

27. Jiménez-Muñoz, J.-C.; Cristóbal, J.; Sobrino, J.A.; Soria, G.; Ninyerola, M.; Pons, X. Revision of the Single-Channel Algorithm for Land Surface Temperature Retrieval From Landsat Thermal-Infrared Data. *IEEE Trans. Geosci. Remote Sens.* **2008**, *47*, 339–349. [CrossRef]

28. Chédin, A.P.; Scott, N.A.; Wahiche, C.; Moulinier, P. The Improved Initialization Inversion Method: A High Resolution Physical Method for Temperature Retrievals from Satellites of the TIROS-N Series. *J. Clim. Appl. Meteorol.* **1985**, *24*, 128–143. [CrossRef]

29. Berk, A.; Anderson, G.P.; Acharya, P.K.; Bernstein, L.S.; Muratov, L.; Lee, J.; Fox, M.; Adler-Golden, S.M.; Chetwynd, J.H.; Hoke, M.L.; et al. MODTRAN 5: A reformulated atmospheric band model with auxiliary species and practical multiple scattering options: Update. *Defense Secur.* **2005**, *5806*, 662–667.

30. Zheng, Y.; Ren, H.; Guo, J.; Ghent, D.; Tansey, K.; Hu, X.; Nie, J.; Chen, S. Land Surface Temperature Retrieval from Sentinel-3A Sea and Land Surface Temperature Radiometer, Using a Split-Window Algorithm. *Remote Sens.* **2019**, *11*, 650. [CrossRef]

31. Ottle, C.; Stoll, M. Effect of atmospheric absorption and surface emissivity on the determination of land surface temperature from infrared satellite data. *Int. J. Remote Sens.* **1993**, *14*, 2025–2037. [CrossRef]

32. Mattar, C.; Durán-Alarcón, C.; Jiménez-Muñoz, J.-C.; Artigas, A.S.-; Olivera-Guerra, L.; Sobrino, J.A. Global Atmospheric Profiles from Reanalysis Information (GAPRI): A new database for earth surface temperature retrieval. *Int. J. Remote Sens.* **2015**, *36*, 1–16. [CrossRef]

33. Ren, H.; Liu, R.; Qin, Q.; Fan, W.; Yu, L.; Du, C. Mapping finer-resolution land surface emissivity using Landsat images in China. *J. Geophys. Res. Atmos.* **2017**, *122*, 6764–6781. [CrossRef]

34. Caselles, E.; Valor, E.; Abad, F.; Caselles, V. Automatic classification-based generation of thermal infrared land surface emissivity maps using AATSR data over Europe. *Remote Sens. Environ.* **2012**, *124*, 321–333. [CrossRef]

35. Tang, R.; Li, Z.-L.; Tang, B. An application of the Ts–VI triangle method with enhanced edges determination for evapotranspiration estimation from MODIS data in arid and semi-arid regions: Implementation and validation. *Remote Sens. Environ.* **2010**, *114*, 540–551. [CrossRef]

36. Gong, P.; Wang, J.; Yu, L.; Zhao, Y.; Zhao, Y.; Liang, L.; Niu, Z.; Huang, X.; Fu, H.; Liu, S.; et al. Finer resolution observation and monitoring of global land cover: First mapping results with Landsat TM and ETM+ data. *Int. J. Remote Sens.* **2012**, *34*, 2607–2654. [CrossRef]

37. Gao, B.; Kaufman, Y.J. *MODIS Atmosphere L2 Water Vapor Product*; NASA MODIS Adaptive Processing System, Goddard Space Flight Center: Greenbelt, MD, USA, 2017.

38. Augustine, J.; DeLuisi, J.J.; Long, C.N. SURFRAD—A National Surface Radiation Budget Network for Atmospheric Research. *Bull. Am. Meteorol. Soc.* **2000**, *81*, 2341–2357. [CrossRef]

39. Wang, K.; Liang, S. Evaluation of ASTER and MODIS land surface temperature and emissivity products using long-term surface longwave radiation observations at SURFRAD sites. *Remote Sens. Environ.* **2009**, *113*, 1556–1565. [CrossRef]

40. Sekertekin, A. Validation of Physical Radiative Transfer Equation-Based Land Surface Temperature Using Landsat 8 Satellite Imagery and SURFRAD in-situ Measurements. *J. Atmos. Sol.-Terr. Phys.* **2019**, *196*, 105161. [CrossRef]

41. Malakar, N.; Hulley, G.; Hook, S.; Laraby, K.; Cook, M.; Schott, J.R. An Operational Land Surface Temperature Product for Landsat Thermal Data: Methodology and Validation. *IEEE Trans. Geosci. Remote Sens.* **2018**, *56*, 5717–5735. [CrossRef]

42. Yu, Y.; Tarpley, D.; Privette, J.; Flynn, L.; Vinnikov, K.Y.; Xu, H.; Chen, M.; Sun, D.; Tian, Y. Validation of GOES-R Satellite Land Surface Temperature Algorithm Using SURFRAD Ground Measurements and Statistical Estimates of Error Properties. *IEEE Trans. Geosci. Remote Sens.* **2011**, *50*, 704–713. [CrossRef]

43. Heidinger, A.; Laszlo, I.; Molling, C.C.; Tarpley, D. Using SURFRAD to Verify the NOAA Single-Channel Land Surface Temperature Algorithm. *J. Atmos. Ocean Technol.* **2013**, *30*, 2868–2884. [CrossRef]

44. Guillevic, P.C.; Biard, J.; Hulley, G.; Privette, J.; Hook, S.; Olioso, A.; Göttsche, F.M.; Radocinski, R.; Román, M.; Yu, Y.; et al. Validation of Land Surface Temperature products derived from the Visible Infrared Imaging Radiometer Suite (VIIRS) using ground-based and heritage satellite measurements. *Remote Sens. Environ.* **2014**, *154*, 19–37. [CrossRef]

45. Li, S.; Yu, Y.; Sun, D.; Tarpley, D.; Zhan, X.; Chiu, L. Evaluation of 10 year AQUA/MODIS land surface temperature with SURFRAD observations. *Int. J. Remote Sens.* **2014**, *35*, 830–856. [CrossRef]

46. Duan, S.-B.; Li, Z.-L.; Li, H.; Göttsche, F.-M.; Wu, H.; Zhao, W.; Leng, P.; Zhang, X.; Coll, C. Validation of Collection 6 MODIS land surface temperature product using in situ measurements. *Remote Sens. Environ.* **2019**, *225*, 16–29. [CrossRef]

47. Liu, Y.; Yu, Y.; Yu, P.; Wang, H.; Rao, Y. Enterprise LST Algorithm Development and Its Evaluation with NOAA 20 Data. *Remote Sens.* **2019**, *11*, 2003. [CrossRef]
48. Ogawa, K.; Schmugge, T. Mapping Surface Broadband Emissivity of the Sahara Desert Using ASTER and MODIS Data. *Earth Interact.* **2004**, *8*, 1–14. [CrossRef]
49. Cheng, J.; Liang, S.; Yao, Y.; Zhang, X. Estimating the Optimal Broadband Emissivity Spectral Range for Calculating Surface Longwave Net Radiation. *IEEE Geosci. Remote Sens. Lett.* **2012**, *10*, 401–405. [CrossRef]

remote sensing

Article

Sensitivity Analysis and Validation of Daytime and Nighttime Land Surface Temperature Retrievals from Landsat 8 Using Different Algorithms and Emissivity Models

Aliihsan Sekertekin [1,*] and Stefania Bonafoni [2]

[1]   Department of Geomatics Engineering, Cukurova University, Ceyhan/Adana 01950, Turkey
[2]   Department of Engineering, University of Perugia, via G. Duranti 93, 06125 Perugia, Italy; stefania.bonafoni@unipg.it
*   Correspondence: asekertekin@cu.edu.tr or aliihsan_sekertekin@hotmail.com; Tel.: +90-531-284-6687

Received: 19 July 2020; Accepted: 25 August 2020; Published: 26 August 2020

Abstract: Land Surface Temperature (LST) is a substantial element indicating the relationship between the atmosphere and the land. This study aims to examine the efficiency of different LST algorithms, namely, Single Channel Algorithm (SCA), Mono Window Algorithm (MWA), and Radiative Transfer Equation (RTE), using both daytime and nighttime Landsat 8 data and in-situ measurements. Although many researchers conducted validation studies of daytime LST retrieved from Landsat 8 data, none of them considered nighttime LST retrieval and validation because of the lack of Land Surface Emissivity (LSE) data in the nighttime. Thus, in this paper, we propose using a daytime LSE image, whose acquisition is close to nighttime Thermal Infrared (TIR) data (the difference ranges from one day to four days), as an input in the algorithm for the nighttime LST retrieval. In addition to evaluating the three LST methods, we also investigated the effect of six Normalized Difference Vegetation Index (NDVI)-based LSE models in this study. Furthermore, sensitivity analyses were carried out for both in-situ measurements and LST methods for satellite data. Simultaneous ground-based LST measurements were collected from Atmospheric Radiation Measurement (ARM) and Surface Radiation Budget Network (SURFRAD) stations, located at different rural environments of the United States. Concerning the in-situ sensitivity results, the effect on LST of the uncertainty of the downwelling and upwelling radiance was almost identical in daytime and nighttime. Instead, the uncertainty effect of the broadband emissivity in the nighttime was half of the daytime. Concerning the satellite observations, the sensitivity of the LST methods to LSE proved that the variation of the LST error was smaller than daytime. The accuracy of the LST retrieval methods for daytime Landsat 8 data varied between 2.17 K Root Mean Square Error (RMSE) and 5.47 K RMSE considering all LST methods and LSE models. MWA with two different LSE models presented the best results for the daytime. Concerning the nighttime accuracy of the LST retrieval, the RMSE value ranged from 0.94 K to 3.34 K. SCA showed the best results, but MWA and RTE also provided very high accuracy. Compared to daytime, all LST retrieval methods applied to nighttime data provided highly accurate results with the different LSE models and a lower bias with respect to in-situ measurements.

Keywords: land surface temperature (LST); daytime LST; nighttime LST; validation; land surface emissivity (LSE); single channel algorithm; radiative transfer equation; mono window algorithm; SURFRAD data; Landsat 8

1. Introduction

Land Surface Temperature (LST), also named skin temperature, refers to the surface temperature of the Earth. The International Geosphere and Biosphere Program (IGBP) [1] accepted the LST as one of the high-priority parameters, and the Global Climate Observing System (GCOS) [2] identified it as an Essential Climate Variable (ECV). Considering space-borne, airborne, and ground-based remote sensors, LST represents the accumulative radiometric surface temperature of all materials of the surface cover covering the sensor's field of view in the observation direction [3]. Thus, LST estimation from Thermal Infrared (TIR) images is a complicated procedure since the Earth's surface is composed of dissimilar materials of varying geometry [4–6]. For example, the LST pixel of a densely vegetated area represents the surface temperature of vegetation; however, for a sparsely vegetated area, the surface temperature of vegetation and soil together comprises the LST of the area [5].

LST is a crucial parameter for many fields of interest such as surface energy and water balance, ecology, agriculture, environment, climatology, meteorology, and hydrology [7–9]. Thus, it provides an improved understanding of a wide range of applications involving drought monitoring [10–12], Surface Heat Island (SHI) and urban climate studies [13–17], surface soil moisture and evapotranspiration estimation [18,19], numerical weather prediction and data assimilation [20,21], surface turbulent flux estimation [22], monitoring of heat waves [23], earthquake prediction [24,25], forest fire monitoring [26], and monitoring of geothermal activities [27,28].

The history of satellite-derived LST goes back to TIROS-II satellite, which was launched at the beginning of the 1960s [29,30]. Through meteorological stations, surface temperature estimation from radiance measurements is a classical point-based technique; nevertheless, this technique does not stand for the LST on a large scale. To overcome this drawback, spaceborne TIR remote sensing has been extensively examined for LST retrieval, and regional and global scale monitoring is the main advantage of this technology. However, surface parameters (emissivity and geometry), sensor parameters (spectral range and viewing angle), and atmospheric effects are the major factors that influence the accuracy of the LST retrieval from TIR data of satellites [5,29,31–33]. Thus, accurate estimation of Land Surface Emissivity (LSE) and atmospheric parameters is a crucial procedure to obtain LST from TIR data [34]. Concerning these parameters, various TIR-based multi-channel and single-channel LST retrieval methods have been proposed by the researchers for different sensor types. Namely, these are Temperature-Independent Spectral Indices (TISI) method [35], Split Window Algorithm (SWA) [36–38], Mono Window Algorithm (MWA) [39], Single Channel Algorithm (SCA) [40,41], Radiative Transfer Equation (RTE) [42,43], and Temperature and Emissivity Separation (TES) method [44]. Among the LST retrieval methods above, only SWAs do not need atmospheric parameters such as water vapor profile and/or temperature. The LSE and LST errors arising from the other algorithms largely rely on the input atmospheric profile's uncertainties [45].

There are numerous Earth observation sensors, namely, Geostationary Operational Environmental Satellite (GOES), Moderate Resolution Imaging Spectroradiometer (MODIS), Advanced Along-Track Scanning Radiometer (AATSR), The Spinning Enhanced Visible and Infrared Imager (SEVIRI), The Advanced Very High Resolution Radiometer (AVHRR), The Visible Infrared Imaging Radiometer Suite (VIIRS), and Sentinel-3, providing operational daytime and nighttime LST products with low spatial resolution (from 750 m to 4 km). However, TIR data of Advanced Spaceborne Thermal Emission and Reflection Radiometer (ASTER) and Landsat satellite series have higher spatial resolution but lower temporal resolution than the sensors reported above. Regarding these limitations, LST retrieval having both high temporal and spatial resolution is a challenge for thermal remote sensing studies. However, LST images obtained from Landsat and ASTER TIR data are unique sources to investigate the thermal environment of cities and their surroundings due to the higher spatial resolution in TIR bands. Moreover, Landsat-derived LST is one of the most commonly preferred data for various applications stated above.

The demand for satellite-based LST products has been increasing rapidly. Thus, the quality of the LST data used in the studies should be examined by a validation procedure for accurate and reliable analyses. Validation provides information about the quantitative uncertainty, enabling the proper use and application of the product. Thus, any algorithm or product would not be broadly welcomed without performing thorough calibration and validation [46]. Overall, cross-validation, the Temperature-based method (T-based) and the Radiance-based method (R-based) are three main techniques considered to evaluate space-based LST [31,34]. Many researchers have considered one or two of these methods for satellite-based LST validation derived from Landsat missions [34,47–55], Sentinel-3A [56], GOES [57], SEVIRI [58,59], MODIS [60–62], AATSR [58,62,63], VIIRS [64], ASTER [65,66] and AVHRR [38]. In this work, we utilized the T-based technique for LST validation, and further details about this method are presented in the Methodology Section.

In this study, Landsat 8 data, both daytime and nighttime, were considered for LST retrieval from RTE, SCA, and MWA methods. In the study, SWA was not examined since the USGS do not recommend using Band 11 of Landsat 8 for LST retrieval due to the large calibration uncertainty. Furthermore, we already obtained better results with MWA than with the SWA developed by Mao et al. [36] with coefficients by Yu et al. [51] in our previous research [34]. Considering the literature, in general, researchers have used daytime Landsat data to retrieve LST due to the lack of LSE images in the night. To the best of our knowledge, there is no study published so far that considered nighttime TIR data of Landsat 8 for both retrieval and validation of nighttime LST. Even though the availability of the nighttime Landsat TIR data is limited in time and many researchers are not even aware that Landsat missions acquire nighttime TIR data, it is probable that future Landsat missions may provide much more nighttime TIR data for the sustainability and strength of the scientific studies. As discussed in the previous paper of the authors [34], Normalized Difference Vegetation Index (NDVI)-based LSE retrieval methods are operative and easy to apply for the Landsat data. In this paper, we propose using daytime NDVI-based LSE, whose acquisition is close to nighttime data (the difference ranges from 1 day to 4 days), as an input in the corresponding methods for the nighttime LST retrieval. Besides, the effect of six different NDVI-based LSE models on LST retrieval methods was evaluated for both daytime and nighttime LST analyses. As stated in the day–night algorithm [67], the LSE does not vary dramatically in several days if snow and/or rain does not exist during a short period. Thus, we assumed that the daytime LSE will not change in the night for a few days considering the weather condition of the corresponding time interval. The objectives of this study are to (1) evaluate the efficiency of RTE, MWA, and SCA methods for both daytime and nighttime Landsat 8 data and in-situ measurements, (2) reveal the impact of NDVI-based LSE models on LST retrieval methods for both daytime and nighttime data, (3) encourage the researchers by showing the convenience of the proposed nighttime LST retrieval from Landsat 8 data for the common usage, and (4) provide sensitivity analyses of in-situ measurements and LST retrieval methods for both daytime and nighttime data. Concerning the ground-based LST measurements, upwelling and downwelling thermal radiation measurements were obtained from Atmospheric Radiation Measurement (ARM) and Surface Radiation Budget Network (SURFRAD) stations, established over rural areas, simultaneously with TIR data acquisitions. To carry out the image-processing tasks, we used an automated LST retrieval toolbox, which was provided by the authors for the use of researchers in the previous study [34].

2. Datasets

2.1. In-Situ LST Measurements and Validation Sites

Surface longwave radiation measurements are important sources for the estimation of in-situ LST and emissivity [65,68]. There are some programs, namely, SURFRAD [69], FLUXNET [70], ARM [71], and Baseline Surface Radiation Network (BSRN) [72] that provide long-term and high-quality surface longwave radiation measurements open to the public. In this study, four stations from SURFRAD and five stations from ARM, nine stations in total (Figure 1) over rural areas, were utilized to calculate daytime and nighttime in-situ LST simultaneous with TIR data acquisitions.

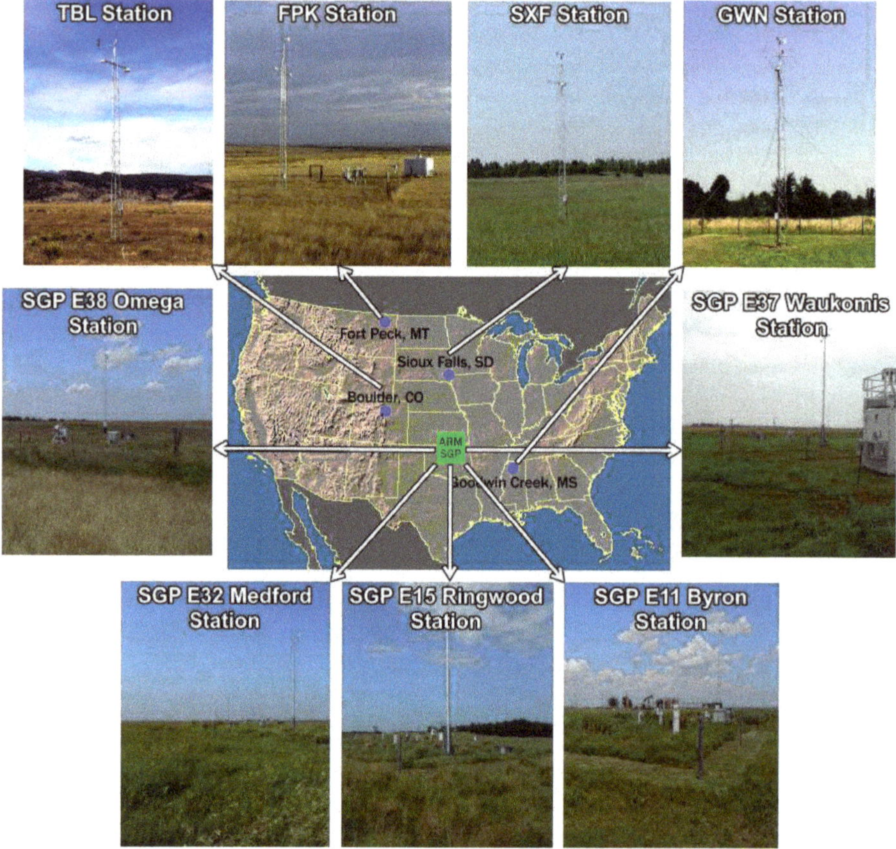

Figure 1. Illustration of the locations and surface covers of the Surface Radiation Budget Network (SURFRAD) and Atmospheric Radiation Measurement (ARM) stations used in this study.

The SURFRAD network was established by National Oceanic and Atmospheric Administration (NOAA) in 1993 to support climate-related research over the United States (US) by providing long-term, continuous, and accurate in-situ surface radiation budget [69]. In 1995, the system started operating with four stations, and now, seven SURFRAD stations have been serving in different climatological regions of the US. The SURFRAD data have been utilized in different studies involving assessment of satellite-based retrievals of surface radiation parameters, climate models, hydrology, and validation of radiation transfer codes and surface physics packages of weather [69]. To calculate in-situ LST, quality-controlled measurements of broadband hemispherical upwelling and downwelling longwave radiation are provided by the SURFRAD stations every 3 min (before 2009) or every minute (after 2009). Many studies have been carried out using SURFRAD measurements to validate LST retrievals from satellites [34,47,54,73–75].

The ARM Program was initially founded in 1989 by the US Department of Energy to examine cloud formation processes. Then, the ARM Climate Research Facility was established in 2003, and this program added further sites and instruments to the available ones as a scientific user facility. All data, providing long-term continuous atmospheric measurements, have been freely available since 2003 (https://www.arm.gov/) [76]. Eastern North Atlantic (ENA), North Slope of Alaska (NSA), and Southern Great Plains (SGP) are three basic ARM sites. In this study, five SGP sites were used for in-situ LST retrieval. As in SURFRAD stations, ARM SGP stations provide quality-controlled measurements of upwelling and downwelling longwave radiation for in-situ LST calculation, and many types of research were carried out using these stations [61,77–79]. Table 1 presents detailed information about both ARM SGP sites and SURFRAD sites considered in this study.

Table 1. Characteristics of the SURFRAD and ARM Southern Great Plains (SGP) validation sites used in the study.

Site Location	Site ID	Latitude	Longitude	Elevation	Land Cover Type
Fort Peck, Montana	FPK	48.308°N	105.102°W	634 m	Grassland
Table Mountain, Boulder, Colorado	TBL	40.125°N	105.237°W	1689 m	Sparse Grassland
Sioux Falls, South Dakota	SXF	43.734°N	96.623°W	473 m	Grassland
Goodwin Creek, Mississippi	GWN	34.255°N	89.873°W	98 m	Cropland/Natural Vegetation Mosaic
Omega, Oklahoma	SGP E38	35.880°N	98.173°W	371 m	Pasture
Waukomis, Oklahoma	SGP E37	36.311°N	97.928°W	379 m	Grassland
Medford, Oklahoma	SGP E32	36.819°N	97.820°W	328 m	Pasture
Ringwood, Oklahoma	SGP E15	36.431°N	98.284°W	418 m	Pasture
Byron, Oklahoma	SGP E11	36.881°N	98.285°W	360 m	Pasture

In the validation sites, two pyrgeometers (Eppley Precision Infrared Radiometer) mounted at 10-m height measure the downwelling and upwelling longwave radiation in the spectral range from 4.0 to 50.0 μm. The instruments are exchanged annually with newly calibrated instruments at each station [69] and world-recognized organizations perform these calibrations [65]. The Eppley pyrgeometer has about 4.2 $W·m^{-2}$ measurement accuracy, and the instrument's precision is around 2 $W·m^{-2}$ for daytime measurements and less than 1 $W·m^{-2}$ for nighttime measurements [80]. Furthermore, Guillevic et al. [75] reported that considering the instrumental error, less than 1 K uncertainty is observed from the retrieved LST. In this study, we also conducted sensitivity/uncertainty analyses for both daytime and nighttime in-situ measurements in Section 4.1. The spatial representativeness of the pyrgeometer is about 70 m × 70 m at the surface [34,65], which is appropriate for the Landsat TIR pixel size (100 m native resampled at 30 m by the US Geological Survey) over homogeneous surfaces. Thus, we selected the validation sites whose footprint on Landsat 8 TIR pixel has homogeneous surface cover. On the other hand, many studies have already considered these ARM SGP and SURFRAD stations to validate low-resolution LST products of MODIS, SEVIRI, GOES, VIIRS, and AATSR [46,58,65,81,82]. Therefore, the use of these stations in the validation of Landsat-derived LST products is highly acceptable.

2.2. Satellite Data

The Landsat mission has been providing moderate-resolution earth observation data from space regularly for almost 50 years. Landsat 8 was launched on 11 February 2013, and it is the recent operational satellite of the Landsat series. Landsat 4 was the first mission providing one thermal band, and the first TIR data of Landsat 4 dates back to 1982, which makes it possible to study long-term LST variations together with all Landsat missions both at a regional and local scale. The Landsat 8 satellite carries two sensors, namely, the Operational Land Imager (OLI) and the TIR sensor (TIRS). The TIRS sensor has two thermal bands (Band 10 and Band 11), while the OLI sensor has nine reflective bands with 30-m spatial resolution. The native spatial resolution of TIR bands is 100-m; however, USGS publishes them at 30-m by resampling.

In this study, 21 pairs of nighttime and daytime Landsat-8 data (Collection 1) from 2013 to 2019 were utilized for the retrieval of daytime and nighttime LST images. Landsat 8 data were freely obtained through the website of the USGS (https://earthexplorer.usgs.gov/). Band 10 of the TIRS sensor, and Band 4 (Red (R)) and Band 5 (Near Infrared (NIR)) of the OLI sensor, for the estimation of NDVI-based LSE, were used in LST retrieval methods. The quality of the used data was checked by the Pixel Quality Assessment (QA) band that provides information for the exclusion of observations affected by sensor factors, clouds, and cloud shadow [83]. The list of the daytime and nighttime Landsat 8 images with corresponding validation site names are reported in Appendix A.

3. Methodologies

3.1. Satellite LST Retrieval Methods

In this study, the following three commonly used methods for LST retrieval are examined: Radiative Transfer Equation (RTE) method, Single Channel Algorithm (SCA) [40], and Mono Window Algorithm (MWA) [39]. The input atmospheric parameters in the methods, such as downwelling radiance ($L_\lambda^\downarrow$), upwelling radiance ($L_\lambda^\uparrow$), and atmospheric transmittance (τ) were calculated using the Atmospheric Correction Parameter Calculator (ACPC) developed by National Aeronautics and Space Administration (NASA) of the US. ACPC uses the atmospheric profiles analyzed by the National Centers for Environmental Prediction (NCEP) as inputs to the radiative transfer codes for a given site and date to calculate the aforementioned atmospheric parameters [84,85].

3.1.1. Brightness Temperature (Tb) Calculation

The brightness temperature of a target refers to the temperature of a blackbody emitting a similar quantity of radiation at a specific wavelength [86], and inverse solution of the Planck function is the way of calculating it. To obtain the brightness temperature image from TIR data, the first step is converting the Digital Number (DN) values to spectral radiance. This radiance conversion for Landsat 8 TIRs can be applied using Equation (1) [87]:

$$L_\lambda^{sen} = M_L \cdot Q_{CAL} + A_L \tag{1}$$

where L_λ^{sen} refers to the TOA spectral radiance in Watts/($m^2 \cdot srad \cdot \mu m$), Q_{CAL} is the calibrated and quantized standard product pixel values (DNs), A_L is the additive rescaling factor of the corresponding band, and M_L is the multiplicative rescaling factor of the corresponding band. A metadata file of the relevant Landsat 8 data contains the values of these parameters. The brightness temperature for Landsat 8 data can be calculated after radiance conversion using Equation (2):

$$Tb = \frac{K_2}{\ln\left(\frac{K_1}{L_\lambda^{sen}} + 1\right)} \tag{2}$$

where Tb is the effective at-satellite brightness temperature in Kelvin, K_1 in Watts/($m^2 \cdot srad \cdot \mu m$) and K_2 in Kelvin refer to the calibration constants. K_1 and K_2 values for the Landsat 8 Band 10 are 774.89 (Watts/($m^2 \cdot srad \cdot \mu m$)) and 1321.08 K, respectively.

3.1.2. Radiative Transfer Equation Method

The inverse solution of the radiative transfer equation (RTE) is a direct method for LST retrieval using a single TIR band. This inverse solution can be given by the following expressions:

$$L_\lambda^{sen} = [\varepsilon B_\lambda(T_s) + (1 - \varepsilon)L_\lambda^\downarrow]\tau + L_\lambda^\uparrow \tag{3}$$

where L_λ^{sen} ($W \cdot m^{-2} \cdot sr^{-1} \cdot \mu m^{-1}$) represents the at-sensor spectral radiance of the corresponding TIR band, ε refers to the LSE, B_λ in $W \cdot m^{-2} \cdot sr^{-1} \cdot \mu m^{-1}$ is the blackbody radiance, T_s is the LST, $L_\lambda^\downarrow$ and $L_\lambda^\uparrow$ represent the downwelling and upwelling radiance, respectively, and τ is the atmospheric transmittance. B_λ at a temperature of T_s is calculated by the inversion of the Equation (3):

$$B_\lambda(T_s) = \frac{L_\lambda^{sen} - L_\lambda^\uparrow - \tau(1 - \varepsilon)L_\lambda^\downarrow}{\tau \varepsilon} \tag{4}$$

and, eventually, T_s (LST) can be obtained from the inversion of Planck's law as in Equation (5):

$$T_s = \frac{K_2}{\ln\left(\dfrac{K_1}{\frac{L_\lambda^{sen} - L_\lambda^\uparrow - \tau(1-\varepsilon)L_\lambda^\downarrow}{\tau\varepsilon}} + 1\right)} \tag{5}$$

where K_1 and K_2 refer to the calibration constants described in the previous section.

3.1.3. Mono Window Algorithm

Qin et al. [39] developed the Mono Window Algorithm (MWA) for the Landsat TM data. Three essential variables, namely, LSE, effective mean atmospheric temperature, and atmospheric transmittance are required for LST retrieval using the MWA method. MWA-based LST can be retrieved by Equation (6):

$$T_s = \{a \cdot (1 - C - D) + [b \cdot (1 - C - D) + C + D] \cdot Tb - D \cdot T_a\} \div C \tag{6}$$

where T_a is the effective mean atmospheric temperature in Kelvin, a (-67.355351) and b (0.458606) are constants of the algorithm, C and D are the parameters of the algorithm calculated as $C = \varepsilon \times \tau$ and $D = (1 - \tau)[1 + (1 - \varepsilon) \times \tau]$. Table 2 provides empirical equations to estimate the T_a through air temperature (T_o), since it is an essential parameter of MWA [39]. In this study, T_a values were computed for the mid-latitude summer region and T_o was obtained from the corresponding validation site.

Table 2. The linear equations for the calculation of the effective mean atmospheric temperature (T_a) from the near-surface air temperature (T_o) [39].

Region	Linear Equations
USA 1976 Region	$T_a = 25.940 + 0.8805 \times T_o$
Tropical Region	$T_a = 17.977 + 0.9172 \times T_o$
Mid-latitude Summer Region	$T_a = 16.011 + 0.9262 \times T_o$
Mid-latitude Winter Region	$T_a = 19.270 + 0.9112 \times T_o$

3.1.4. Single-Channel Algorithm

Jiménez-Muñoz et al. [40] proposed a revised version of SCA for LST retrieval using Landsat TIR data. Concerning the SCA, T_s is obtained from Equation (7):

$$T_s = \gamma\left[\frac{1}{\varepsilon}\left(\psi_1 L_\lambda^{sen} + \psi_2\right) + \psi_3\right] + \delta \tag{7}$$

where ψ_1, ψ_2, and ψ_3 refer to atmospheric functions defined as:

$$\psi_1 = \frac{1}{\tau} \; ; \; \psi_2 = -L_\lambda^\downarrow - \frac{L_\lambda^\uparrow}{\tau} \; ; \; \psi_3 = L_\lambda^\downarrow \tag{8}$$

Concerning the SCA method in this study, $L_\lambda^\uparrow$, $L_\lambda^\downarrow$, and τ obtained from NASA's ACPC were used for the computation of the ψ_1, ψ_2, and ψ_3. On the other hand, the two parameters, γ and δ, are computed by:

$$\gamma \approx \frac{Tb^2}{b_\gamma L_{sen}} \tag{9}$$

$$\delta \approx Tb - \frac{Tb^2}{b_\gamma} \tag{10}$$

where $b_\gamma = c_2/\lambda_i$ and $c_2 = 14{,}387.7$ µm·K, and b_γ is equal to 1320 K for Landsat 8 Band 10. λ_i is the ith band's effective wavelength given by:

$$\lambda_i = \frac{\int_{\lambda_{1,i}}^{\lambda_{2,i}} \lambda f_i(\lambda) d\lambda}{\int_{\lambda_{1,i}}^{\lambda_{2,i}} f_i(\lambda) d\lambda} \tag{11}$$

where $f_i(\lambda)$ is ith band's spectral response function. $\lambda_{1,i}$ and $\lambda_{2,i}$ refer to the lower and upper boundary of $f_i(\lambda)$, respectively.

3.2. NDVI-Based Land Surface Emissivity (LSE) Models

Emissivity of a surface represents the ability of the surface to transform heat energy, relative to a black body, into radiant energy [88]. As presented in the above sections, LSE (ε) is a critical element for accurate TIR-based LST retrieval. Multi-channel Temperature/Emissivity Separation (TES), Physically Based Methods (PBMs), and Semi-Empirical Methods (SEMs) methods are three main types of space-based LSE estimation [31]. The NDVI-Based Emissivity Method (NBEM) [89,90] and Classification Based Emissivity Method (CBEM) [91,92] constitute the SEMs that are convenient for the Landsat-derived LSE. CBEM is not feasible because of the need of a priori information about the test site and in-situ emissivity of each class [93]. NDVI-based LSE models are practical and frequently used methods due to their easy application providing satisfactory results [88,94,95]. Li et al. [31] introduced a comprehensive research revealing limitations, advantages, and disadvantages of LSE models for satellite-derived LST. Moreover, Sekertekin and Bonafoni [34] provided an updated state-of-the-art table from Li et al. [31], presenting the used satellite missions with the corresponding LSE models. In this study, we examined the influence of six NDVI-based LSE models on the performance of three LST algorithms for both daytime and nighttime. To calculate NDVI from Landsat 8 data, firstly, DN values are converted to the TOA reflectance using the Equation (12) [87]. After applying reflectance (ρ_λ) conversion to the R and NIR bands, NDVI is obtained from Equation (13). Specifically:

$$\rho_\lambda = \frac{M_p \cdot Q_{CAL} + A_p}{\sin \theta_{SE}} \tag{12}$$

where Q_{CAL} is the calibrated and quantized standard product pixel values (DNs), A_p is the additive rescaling factor of the corresponding band, M_p is the multiplicative rescaling factor of the corresponding band, and θ_{SE} represents the local sun elevation angle. The values of these parameters are obtained from the Metadata file of the relevant Landsat 8 data.

$$NDVI = \frac{\rho_{NIR} - \rho_R}{\rho_{NIR} + \rho_R} \tag{13}$$

where ρ_{NIR} refers to the reflectance image of the NIR band and ρ_R is the reflectance image of the R band. In addition to NDVI, the Fractional Vegetation Cover (FVC or P_v), i.e., the proportion of vegetation, is another important factor for LSE estimation, and it is calculated from Equation (14) [96] as:

$$P_v = \left[\frac{NDVI - NDVI_{min}}{NDVI_{max} - NDVI_{min}} \right]^2 \tag{14}$$

where $NDVI_{min} = 0.2$ and $NDVI_{max} = 0.5$ in a global context [93]. Table 3 presents the expressions of the six NDVI-based LSE models used in this work (hereafter referred to as LSE1, LSE2, ... , LSE6). More details about these models can be found in the previous paper of the authors [34].

Table 3. The expressions of Normalized Difference Vegetation Index (NDVI)-based Land Surface Emissivity (LSE) models considered in this study.

Sensor	LSE Equations	Model ID	Reference
	$\varepsilon = 1.0094 + 0.047\ln(\text{NDVI})$	LSE1	Van de Griend and Owe [94]
	$\varepsilon = 0.985 P_v + 0.960(1 - P_v) + 0.06 P_v(1 - P_v)$	LSE2	Valor and Caselles [90]
	$\varepsilon = \begin{cases} 0.979 - 0.035\rho_R & \text{NDVI} < 0.2 \\ 0.004 P_v + 0.986 & 0.2 \leq \text{NDVI} \leq 0.5 \\ 0.99 & \text{NDVI} > 0.5 \end{cases}$	LSE3	Sobrino et al. [95]
Landsat 8 (Band 10)	$\varepsilon = \begin{cases} 0.979 - 0.046\rho_R & \text{NDVI} < 0.2 \\ 0.987 P_v + 0.971(1 - P_v) + d\varepsilon & 0.2 \leq \text{NDVI} \leq 0.5 \\ 0.987 + d\varepsilon & \text{NDVI} > 0.5 \end{cases}$	LSE4	Skoković et al. [97]
	$\varepsilon = \begin{cases} 0.973 - 0.047\rho_R & \text{NDVI} < 0.2 \\ 0.9863 P_v + 0.9668(1 - P_v) + d\varepsilon & 0.2 \leq \text{NDVI} \leq 0.5 \\ 0.9863 + d\varepsilon & \text{NDVI} > 0.5 \end{cases}$	LSE5	Yu et al. [51]
	$\varepsilon = \begin{cases} a_{1i} + \sum_{j=2}^{7} a_{ji}\rho_j & \text{NDVI} < 0.2 \\ 0.982 P_v + 0.971(1 - P_v) + d\varepsilon & 0.2 \leq \text{NDVI} \leq 0.5 \\ 0.982 + d\varepsilon & \text{NDVI} > 0.5 \end{cases}$	LSE6	Li and Jiang [98]

$d\varepsilon = (1 - \varepsilon_s)\varepsilon_v F(1 - P_v)$: a term taking the cavity effect into account, which is based on the geometry of the surface. ε_s, ε_v and F refer to soil emissivity, vegetation emissivity and geometrical shape factor (0.55), respectively. ρ_R: the reflectance image of R band; ρ_j: the apparent reflectance in the OLI band j; $a_{1i} - a_{7i}$: the coefficients obtained from [98].

3.3. In-Situ LST Estimation

Station-based (in-situ or ground-based) LST measurements were obtained from four SURFRAD stations and five ARM SGP stations. As stated in Section 2.1, these stations do not measure LST directly; the upwelling and downwelling components of longwave radiation are considered for LST calculation regarding Stefan–Boltzmann law:

$$\text{LST} = \left[\frac{F_\lambda^\uparrow - (1 - \varepsilon_b) \cdot F_\lambda^\downarrow}{\varepsilon_b \cdot \sigma} \right]^{1/4} \tag{15}$$

where $F_\lambda^\downarrow$ and $F_\lambda^\uparrow$ in W/m^2 are the downwelling and upwelling thermal infrared irradiances, respectively, obtained simultaneously with satellite passages. σ is 5.670367×10^{-8} $\text{W·m}^{-2}\text{·K}^{-4}$ that refers to the Stefan–Boltzmann constant. ε_b is the broadband longwave surface emissivity that is not measured by the station instruments, thus [65,68] proposed the computation of the broadband emissivity by regression from narrowband emissivities of MODIS data, and many studies used these regression equations for acquiring the ε_b [52,73,99]. The experimental results in [65,68] revealed that the longwave broadband emissivity can be used as a fixed value of 0.97, which was also considered in the studies of [74,100]. In this study, we assumed the broadband emissivity as 0.97, as well. This phenomenon only affects the accuracy of in-situ LST, not the satellite-based LST accuracy. Heidinger et al. [74] reported that a 0.01 error in broadband emissivity led to 0.25 K LST error in SURFRAD sites. Furthermore, Wang and Liang [65] showed that the LST accuracy of SURFRAD sites ranged from 0.1 K to 0.4 K due to the ±0.01 error in the broadband emissivity. This error is not negligible; however, it is not an overwhelming uncertainty source compared to the magnitude of the other uncertainties in LST retrieval [74]. Concerning this study, we also carried out the uncertainty analysis of broadband longwave surface emissivity and longwave radiation (the downwelling and upwelling components) on ground-based LST measurements in the next section.

3.4. Sensitivity Analysis of In-Situ LST Measurements and LST Retrieval Methods

Sensitivity analysis is an application of how the error of a model output (numerical, statistical, or otherwise) can be divided and allocated to different uncertainty sources in the model inputs [101].

It is difficult to determine the inputs of an algorithm, since these inputs unavoidably have initial errors affecting the accuracy of the LST retrieval methods [34,49]. To investigate the effect of input parameters' errors on LST retrievals from both satellites and stations, the following equation is utilized:

$$\delta T = T_s(x) - T_s(x + \delta x) \tag{16}$$

where δT is the error on the LST; x represents one of the input parameters and δx is the potential error of this parameter; $T_s(x + \delta x)$ and $T_s(x)$ refer to the LST calculated for "$x + \delta x$" and "x", respectively. Some researchers reported the uncertainty of the input parameters on LST retrieval algorithms [49,102,103]. On the other hand, concerning the sensitivity analysis of in-situ LST measurements, [65,74] investigated the sensitivity of SURFRAD LST to broadband emissivity. In the previous paper of the authors [34], we already presented detailed sensitivity analysis for daytime LST retrieval considering MWA, SCA, and RTE. In this study, we mainly focused on the effect of LSE on LST retrieval methods for both daytime and nighttime LST retrievals, since we proposed using the daytime LSE images for nighttime LST retrieval. Furthermore, we also conducted a comprehensive sensitivity analysis for the in-situ LST measurements that is presented in Section 4.1.

3.5. Temperature-Based (T-Based) Validation Method and Performance Metrics

As stated in the introduction, the Radiance-based method (R-based), Temperature-based method (T-based), and cross-validation are the main techniques used to evaluate space-based LST [31,34]. The T-based technique, examined in this research, is a direct way of comparing the satellite-derived LST with in-situ LST simultaneous with satellite pass, and many researchers used this way to validate satellite-derived LSTs [48,52,62,104,105]. The major benefit of the T-based method is that it makes it possible to evaluate satellite sensor's radiometric quality and the efficiency of the LST algorithms based on emissivity and atmospheric parameters. On the other hand, the capability of the T-based technique depends mostly on the accuracy of the in-situ LST measurements and how well they represent the LST at the satellite pixel scale (land cover homogeneity of the study area) [31]. In this study, we considered both issues as we carried out the sensitivity analysis of the SURFRAD LST measurements and selected the validation sites whose footprint on Landsat 8 TIR pixel has homogeneous surface cover.

Satellite-derived LST and Station-based LST were analyzed considering the performance metrics such as Root Mean Square Error (RMSE), Standard Deviation (STD) of Error, and average Bias. The formulas of these metrics are given by:

$$\text{RMSE} = \sqrt{\frac{\sum [T_{L8} - T_{Station}]^2}{n}} \tag{17}$$

$$\text{STD of Error} = \sqrt{\frac{\sum \left[T_{Error} - \overline{T_{Error}}\right]^2}{n}} \tag{18}$$

$$\text{Bias} = \frac{\sum [T_{Station} - T_{L8}]}{n} \tag{19}$$

where T_{L8} and $T_{Station}$ are the Landsat 8-derived LST and Station-based LST, respectively, and n refers to the number of data. T_{Error} refers to the difference between Landsat 8-derived LST and Station-based LST, and $\overline{T_{Error}}$ is the mean value of these differences.

4. Results

To present the results of the LST retrieval methods for daytime and nighttime, 21 pairs of nighttime and daytime Landsat-8 data were utilized to obtain the daytime and nighttime LST images (see Appendix A). Specifically, concerning the daytime LST, 21 Landsat-8 images were used. On the other hand, 21 nighttime images, whose acquisition times are close to daytime data (the difference ranges from one day to four days), were utilized for the nighttime LST retrieval by using the corresponding 21 daytime reflective data for the NDVI-based LSE computation. We verified that rain and/or snow did not occur during these 1–4 days of difference. MWA, RTE, and SCA were performed for both daytime and nighttime LST estimation considering all datasets. The required input atmospheric parameters in the methods (τ, $L_\lambda^\uparrow$, $L_\lambda^\downarrow$) were obtained from ACPC that considers the MODTRAN radiative transfer code, which uses NCEP-based atmospheric profiles as inputs. This section includes two sensitivity analyses: (i) Sensitivity of in-situ LST measurements and (ii) sensitivity of LST retrieval methods to LSE. Lastly, the accuracy assessment of the LST retrieval algorithms and LSE models for both daytime and nighttime at the nine SURFRAD and ARM stations is proposed.

4.1. Sensitivity Results of In-Situ LST Measurements

Concerning the in-situ LST measurements utilized in this work, the average upwelling and downwelling radiances, respectively, were calculated as 482.18 W/m^2 and 331.15 W/m^2 for daytime, and 388.16 W/m^2 and 326.68 W/m^2 for nighttime. In addition, as stated in the previous section, we used a fixed broadband emissivity value as 0.97. Thus, these values were considered in the sensitivity analysis of in-situ LST measurements. To carry out a sensitivity analysis of a method's output to an input parameter, the other input parameters are assumed to be fixed. For instance, to manage the sensitivity analysis of the downwelling radiance in the daytime (Figure 2a), the upwelling radiance and the broadband emissivity was fixed to 482.18 W/m^2 and 0.97, respectively. Then, the sensitivity of the downwelling radiance to in-situ LST accuracy was revealed by changing the downwelling radiance at 5 W/m^2 intervals (Figure 2a). The same procedure was applied to present the sensitivity results of the other parameters. As reported in Section 2.1, the two pyrgeometers have an accuracy of about 4.2 W/m^2 and a precision of around 1–2 W/m^2. Considering the daytime sensitivity results, the following results were obtained: (i) ±5 W/m^2 error in downwelling and upwelling radiance led to ±0.024 K and ±0.8 K error in LST, respectively (Figure 2a,b) and (ii) 0.01 error in the broadband emissivity caused ±0.25 K error in LST (Figure 2c). On the other hand, nighttime sensitivity results showed that (i) ±5 W/m^2 error in downwelling and upwelling radiance led to ±0.029 K and ±0.95 K error in LST, respectively (Figure 2d,e) and (ii) 0.01 error in the broadband emissivity caused ±0.12 K error in LST. It is evident from Figure 2 that the uncertainty of the downwelling and upwelling radiance is almost identical in daytime and nighttime. However, the uncertainty of the broadband emissivity in the nighttime is half of the daytime.

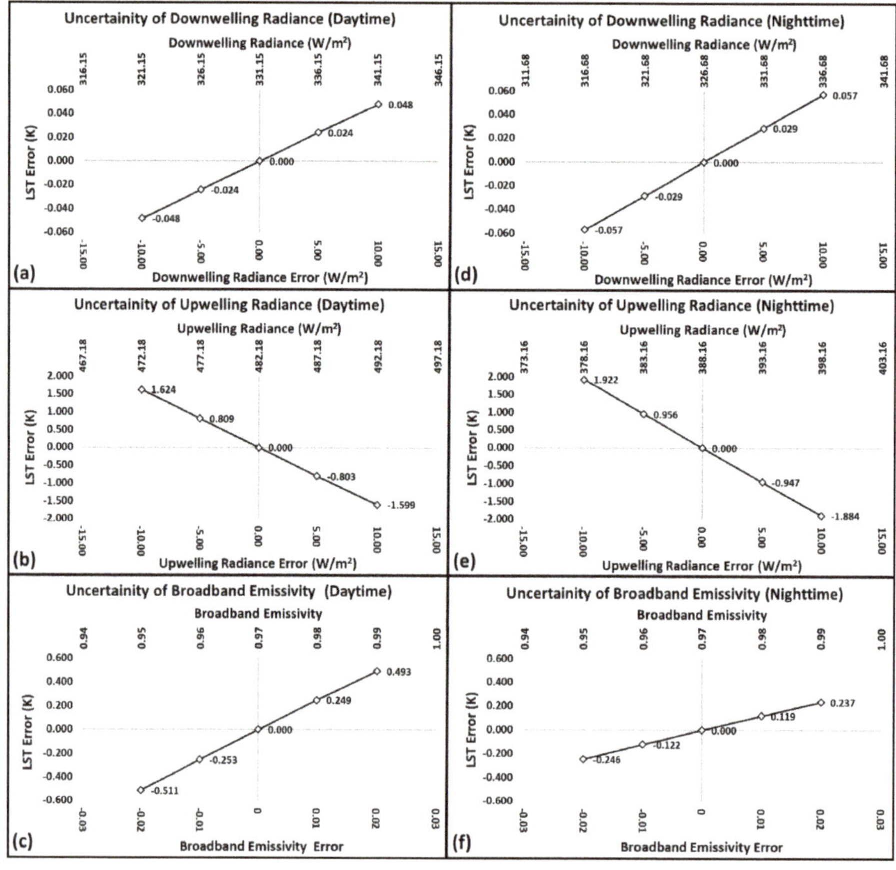

Figure 2. Sensitivity results of in-situ Land Surface Temperature (LST) measurements to downwelling radiance, upwelling radiance, and broadband emissivity, respectively, for both daytime (**a–c**) and nighttime (**d–f**). LST error is computed as in Equation (16).

4.2. Sensitivity Results of LST Retrieval Methods to LSE

In this sensitivity analysis, we mainly focused on the effect of LSE on LST retrieval methods for both daytime and nighttime LST retrievals, since we proposed using the daytime LSE images for nighttime LST calculation. A detailed uncertainty analysis of all parameters on LST retrieval methods (RTE, SCA, and MWA) for daytime can be found in the previous paper of the authors [34]. In the sensitivity analysis of daytime LST images, the following input parameters were utilized based on the current datasets: Air temperature, upwelling and downwelling radiances, atmospheric transmittance, and effective mean atmospheric temperature. Minimum, maximum, and mean near-surface air temperature values from ground stations and simultaneous with the satellite passages were 282.51 K, 302.41 K, and 295.95 K, respectively. Thus, the near-surface air temperature was assumed to be 295.95 K in the sensitivity analysis and, as a consequence, the effective mean atmospheric temperature was computed as 290.12 K. The atmospheric transmittance ranged from 0.63 to 0.94 with a mean value of 0.84, which was used in this analysis. Mean downwelling and upwelling radiances were observed as 2.06 $W·m^{-2}·sr^{-1}·\mu m^{-1}$ and 1.24 $W·m^{-2}·sr^{-1}·\mu m^{-1}$, respectively, and these values were utilized in the sensitivity analyses. The brightness temperature range was assumed between 280 K and 310 K, because the brightness temperature computed from the daytime Landsat scenes ranged from 282.66 K to 314.84 K. The LSE value was fixed as 0.97.

Figure 3 illustrates the sensitivity results of the LST retrieval methods to LSE under a specific brightness temperature range of daytime. Figure 3a,c,e shows the variations in the error of the LST under different brightness temperatures for MWA, RTE, and SCA, respectively, when the LSE error is constant. These figures show that when the LSE error is constant for MWA and SCA, LST error increases with increasing brightness temperature. Instead, when the LSE error is constant for RTE, the LST error is stable with increasing brightness temperature. It is important to note that, since the LST error is computed as in Equation (16), an overestimation (underestimation) of the emissivity produces a positive (negative) value in the LST error. Figure 3b,d,f represents how LSE error impacts the LST error for the MWA, RTE, and SCA, respectively, under different brightness temperature conditions. The findings in these figures support the previous ones (Figure 3a,c,e) by showing that a constant LSE error produces LST error variations under different brightness temperature conditions for MWA and SCA, except for RTE. The intercomparison of the results proves that MWA is more sensitive to LSE error than RTE and SCA under increasing brightness temperatures, while RTE is the least sensitive one.

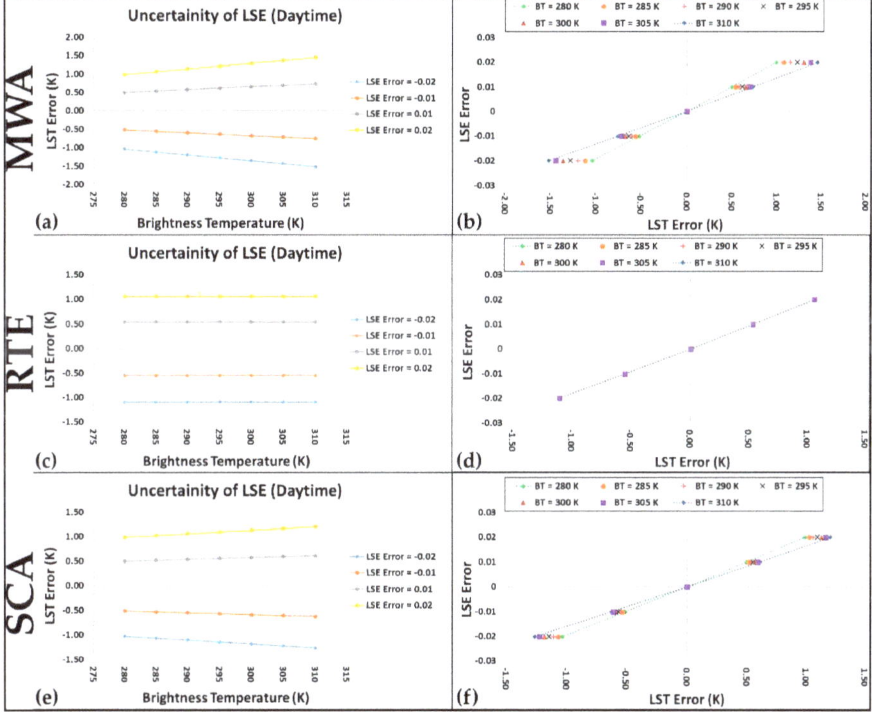

Figure 3. Sensitivity results of Mono Window Algorithm (MWA) (**a,b**), Radiative Transfer Equation (RTE) (**c,d**), and Single Channel Algorithm (SCA) (**e,f**) to LSE for daytime Landsat 8 images. LST error is computed as in Equation (16).

In the sensitivity analysis of the nighttime LST images, minimum, maximum, and mean near-surface air temperature values from the ground stations were 271.65 K, 300.75 K, and 291.07 K, respectively. Considering the mean value (291.07 K), the effective mean atmospheric temperature was 285.60 K. The atmospheric transmittance varied between 0.51 to 0.96 with a mean value of 0.83, while mean upwelling and downwelling radiances were 1.37 $W \cdot m^{-2} \cdot sr^{-1} \cdot \mu m^{-1}$ and 2.20 $W \cdot m^{-2} \cdot sr^{-1} \cdot \mu m^{-1}$. These mean values were utilized in the sensitivity analyses. A brightness temperature range from 270 K to 295 K was investigated since the brightness temperature computed from the nighttime Landsat scenes varied from 267.77 K to 297.22 K. LSE value equal to 0.97 was assumed.

Figure 4 depicts the sensitivity results of the LST retrieval methods to LSE under a specific brightness temperature range of nighttime. Figure 4a,c,e demonstrates the variations in the LST error under different brightness temperatures for MWA, RTE, and SCA, respectively, when the LSE error is constant. Moreover, Figure 4b,d,f represents how LSE error impacts the LST error for the MWA, RTE, and SCA, respectively, varying the brightness temperature values. The sensitivity analysis of nighttime data shows results with a trend similar to the daytime one; however, the variation in the LST error is smaller than daytime, also considering the lower brightness temperature values in the nighttime.

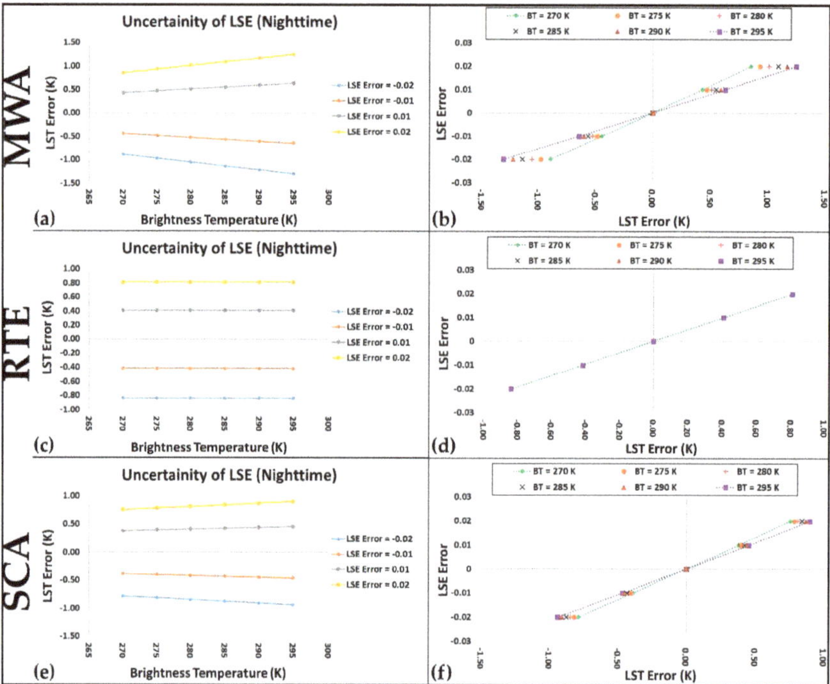

Figure 4. Sensitivity results of MWA (**a,b**), RTE (**c,d**), and SCA (**e,f**) to LSE for nighttime Landsat 8 images. LST error computed as in Equation (16).

4.3. Accuracy of LST Retrieval Algorithms and LSE Models for Daytime

In Figure 5, the accuracy results of the LST retrieval methods for daytime Landsat 8 data are illustrated based on the six NDVI-based LSE models of Table 3. In this validation test at the nine SURFRAD and ARM stations, the Landsat 8 image pixel covering the pyrgeometers was selected, and the estimated LST compared with the corresponding ground LST measurement.

The accuracy varied between 2.17 K RMSE and 5.47 K RMSE considering all LST methods and LSE models. MWA method with LSE4 and LSE6 presented similar and best results for the daytime. Using MWA and LSE4, the RMSE, STD of Error, and Bias were 2.17 K, 1.86 K, and −1.13 K, respectively. Furthermore, the same statistical metrics, in the same order, were 2.17 K, 1.79 K, and −1.24 for MWA with LSE6. In general, the daytime results revealed that for all LSE models, except for LSE2, MWA showed slightly better results than RTE, and RTE demonstrated slightly better results than SCA. LSE1 and LSE2 did not offer satisfying results with any of the LST retrieval algorithms. Apart from that, the other LSE models presented acceptable daytime LST results with MWA, RTE, and SCA. The Bias is always negative regardless of the approach, highlighting a general overestimation of the Landsat 8 retrieval with respect to the in-situ measurements, especially for higher LST values.

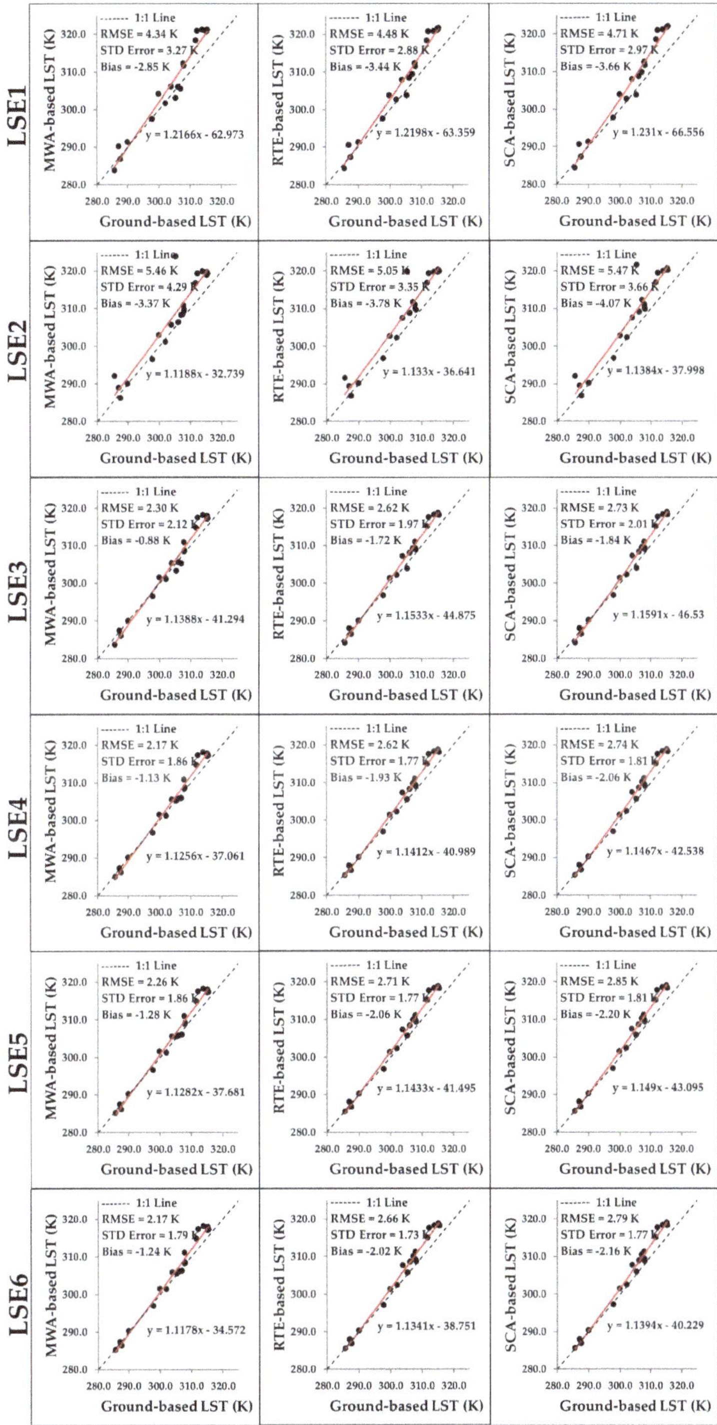

Figure 5. Daytime LST from Landsat 8, period 2013–2019 (see Appendix A): Accuracy assessment of MWA, RTE, and SCA retrieval methods with different LSE models at the nine SURFRAD and ARM stations.

4.4. Accuracy of LST Retrieval Algorithms and LSE Models for Nighttime

In Figure 6, the accuracy assessment of the LST retrieval methods for nighttime Landsat 8 data is reported for the six NDVI-based LSE models. Considering all LST methods and LSE models for the nighttime, the RMSE values ranged from 0.94 K to 3.34 K. In the nighttime LST analysis, the SCA method with LSE5 presented the best results, with RMSE, STD of Error, and Bias equal to 0.94 K, 0.72 K, and 0.60 K, respectively. On the other hand, MWA and RTE also provided very high accuracy with the RMSE equal to 1.01 K and 0.95 K, respectively, when using with LSE5. In general, the nighttime results revealed that for all LSE models, except for LSE2, all LST retrieval methods provided good accuracies with the highest RMSE as 1.51 K. As a summary, Table 4 shows the best LST retrieval methods and LSE models for the proposed daytime and nighttime LST validation test at the nine SURFRAD and ARM stations.

Table 4. Validation test of Landsat 8 LST retrieval at the nine ground stations: The best LST methods and LSE models and accuracy results for daytime and nighttime LST.

Data Type	LSE ID	LST Retrieval Method	RMSE (K)	STD Error (K)	Bias (K)
Daytime LST	LSE4	MWA	2.17	1.86	−1.13
	LSE6	MWA	2.17	1.79	−1.24
Nighttime LST	LSE5	SCA	0.94	0.72	0.60

Compared to the daytime, during nighttime all LST retrieval methods provided highly accurate results with the different LSE models. Moreover, the overestimation of daytime LST retrieval is no longer evident at night, and the bias is clearly reduced. The proposed test with ground measurements as reference suggests that the use of daytime NDVI-based LSE, whose acquisition is close to nighttime data (the difference ranges from one day to four days in this work), is an accurate solution for the nighttime LST retrieval from thermal band observations. We assumed that the LSE does not significantly change in a short time period if rain and/or snow does not occur: This weather condition was verified for the selected images.

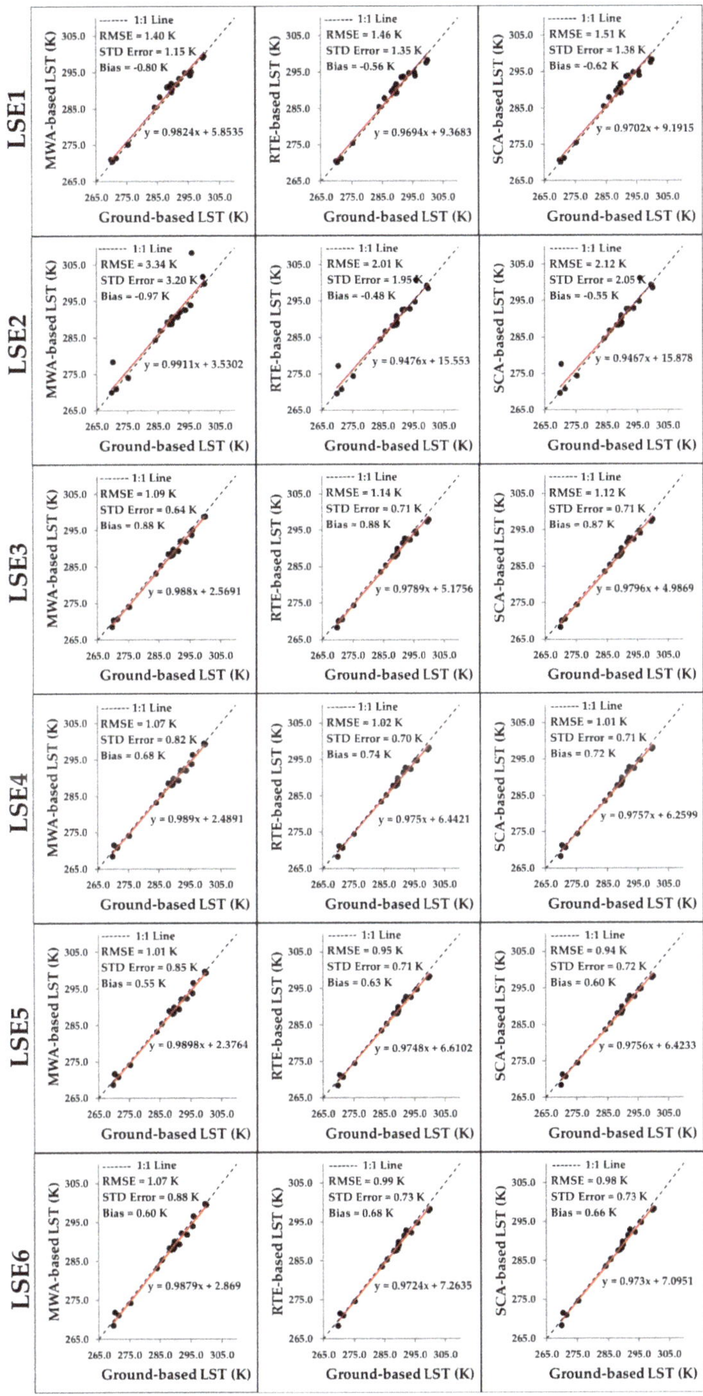

Figure 6. Nighttime LST from Landsat 8, period 2013–2019 (see Appendix A): Accuracy assessment of MWA, RTE, and SCA retrieval methods with different LSE models at the nine SURFRAD and ARM stations.

5. Discussion

Numerous factors affect the accuracy of the LST retrieval from satellite TIR data. Atmospheric profiles, sensor parameters (spectral range and viewing angle), and surface parameters (emissivity and geometry) are amongst the major factors. On the other hand, development of an LST retrieval method has its own error sources due to including some parameterization steps for the retrieval of coefficients and estimation of some initial parameters. Therefore, it is of great importance to conduct sensitivity/uncertainty analyses for a new method by considering all input parameters. Concerning the LST validation procedure in space sciences, two main error sources emerge from both ground-based LST and satellite-based LST. Examining the sensitivity analysis for ground-based LST measurements, it emerges that the reliability of the upwelling radiance measurements is a key factor for the overall accuracy of the LST computation. Then, the effect of LSE on satellite-based LST retrieval methods for both daytime and nighttime were investigated, since we proposed using the daytime LSE images for nighttime LST retrieval. The results showed that the LST sensitivity to LSE error is typically dependent on the brightness temperature values suggesting that areas and study periods with lower Tb could guarantee lower LST errors. Atmospheric parameters needed in the LST retrieval methods were obtained from the NASA's ACPC that is based on MODTRAN radiative transfer code. It is not possible to find in-situ (radiosonde data etc.) atmospheric profiles for any place and any time. Thus, even though this usage (a simulation of profile information on atmosphere with ACPC) affects the accuracy of the methods, it is clear from our results and literature that NASA's ACPC provides satisfactory and effective simulations.

Comparing the results obtained in this research with the ones of other similar studies would be helpful for the readers. The daytime LST results of this study were compatible with the results presented in our previous paper [34]. Yu et al. [51] investigated the daytime LST results from RTE and SCA methods using Landsat 8 data with LSE5. They determined the RMSE values for RTE and SCA as 0.9 K and 1.39 K, respectively. However, we obtained 2.71 K RMSE and 2.85 K RMSE for RTE and SCA, respectively, with the same LSE model. Wang et al. [105] revealed that the generalized SCA and Practical Single-Channel Algorithm (PSCA) presented 2.24 K and 1.77 K, respectively. We obtained 2.73 K RMSE with the SCA and same LSE model (LSE3). Sekertekin [47] obtained 3.12 K RMSE using RTE and LSE4, while it was 2.62 K RMSE in this test. Guo et al. [54] used SCA with daytime Landsat 8 data and obtained 2.74 K and 2.47 K RMSE before and after the stray light correction, respectively. In our study, SCA results ranged from 2.73 K to 2.85 K RMSE under different NDVI threshold-based LSE models. We also observed negative biases for the selected dataset, whereas Guo et al. [54] did not observe biases in their case study. These validation studies of Landsat 8-derived LST refer to the daytime data, and they suggest how the accuracies can differ in similar test sites if the number of scenes and their acquisition time change.

Validation studies were not previously published for nighttime LST from Landsat 8. This test shows that, compared to the daytime, the nighttime accuracy is better, the daytime LST overestimation is no longer present, and the bias is distinctly reduced. It is an interesting and beneficial result for the researchers thinking of using the nighttime LST data from Landsat-8. Further studies can be conducted in different land cover types including also urban areas to confirm the effectiveness of the nighttime LST results. However, it may be difficult to find reliable ground-based LST measurements for accuracy assessments in these different areas.

Satellite-based LST retrieval methods are generally developed considering different conditions and assumptions. Thus, no universal method is yet available to provide accurate LSTs from all satellite TIR data, and it cannot be said that one method is systematically superior to the others. Concerning the stationarity of the methods used in this study, since RTE and SCA are obtained by the radiative transfer equation solution, they are valid for each sensor and atmospheric condition. On the other hand, the MWA is linked to atmospheric parameters and fixed coefficients regardless of the sensor type. However, these coefficients could be refined for different sensors (with different bandwidths), and the results validated.

6. Conclusions

In this study, three LST retrieval algorithms, namely, RTE, SCA, and MWA, were evaluated using daytime and nighttime Landsat 8 OLI/TIRS data. To the best of our knowledge, this is the first study proposing the retrieval and validation of nighttime LST from TIR data of Landsat 8, also with a performance comparison with respect to daytime LST retrieval. Since LSE is one of the most important factors affecting the accuracy of LST retrieval methods, the effects of six NDVI-based LSE models on satellite-based LST accuracy were also investigated.

Concerning nighttime LST retrievals, we proposed the combined use of daytime LSE and nighttime TIR data when the difference in acquisitions of both datasets are close (a few days) and unchanged weather condition is observed. Concerning the evaluation of the LST retrieval methods and LSE models under daytime and nighttime conditions, SURFRAD and ARM SGP sites were used to calculate in-situ LST simultaneous with TIR data acquisitions.

In addition to the accuracy evaluation of the LST methods, we conducted detailed sensitivity/uncertainty analyses for in-situ measurements and sensitivity of LST methods on LSE for both daytime and nighttime. Considering the daytime sensitivity results of in-situ measurements, we proved that ±5 W/m^2 error in downwelling and upwelling radiance led to ±0.024 K and ±0.8 K error in LST, respectively, and 0.01 error in the broadband emissivity caused ±0.25 K error in LST. On the other hand, concerning the nighttime sensitivity results of in-situ measurements, we observed ±5 W/m^2 error in downwelling and upwelling radiance caused ±0.029 K and ±0.95 K error in LST, respectively, and 0.01 error in the broadband emissivity provided ±0.12 K error in LST. The sensitivity results of in-situ LST measurements revealed that the uncertainty of the downwelling and upwelling radiance was almost identical in daytime and nighttime. Nevertheless, the uncertainty of the broadband emissivity in the nighttime was half of that in the daytime.

Then, we investigated the sensitivity of the LST methods to LSE for both daytime and nighttime LST retrievals. The sensitivity results indicated that when the LSE error was constant for MWA and SCA, the LST error increased with increasing brightness temperature. However, when the LSE error was constant for RTE, LST error was stable with increasing brightness temperature. On the other hand, the nighttime sensitivity analysis showed identical trends to daytime ones; however, the variation in the LST error was smaller than daytime mainly due to the lower brightness temperatures.

The accuracy results of the daytime Landsat 8 data at the nine ground stations showed that the MWA method with LSE4 and LSE6 presented the best results for the daytime. In general, for all the LSE models, except for the LSE2, the MWA indicated slightly better results than the RTE, and the RTE demonstrated slightly better results than the SCA for daytime LST retrievals. Considering the nighttime, the SCA method with LSE5 presented the best results. However, MWA and RTE provided very similar results with SCA. Compared to the daytime, all LST retrieval methods provided highly accurate results with the different LSE models in the nighttime. The systematic overestimation of daytime LST retrieval is no longer present at night, with an evident reduced bias. The validation test shows that the use of daytime NDVI-based LSE with reflective data close to nighttime thermal data is a reliable solution for the nighttime LST retrieval.

Author Contributions: Conceptualization, A.S.; methodology, A.S.; software, A.S.; validation, A.S.; writing—review and editing, A.S. and S.B.; supervision, A.S. and S.B. All authors have read and agreed to the published version of the manuscript.

Funding: This research received no external funding.

Acknowledgments: The authors thank USGS for providing Landsat 8 satellite imagery free of charge. In addition, the authors thank NOAA for providing in-situ LST measurements publicly.

Conflicts of Interest: The authors declare no conflict of interest.

Appendix A

Table A1. The list of the daytime and nighttime Landsat 8 images with corresponding validation site IDs.

Acquisition Time	Scene ID	Scene Acquisition Date and Hour (UTC)	Site ID
Daytime	LC80330322013269LGN01	26.09.2013—17:40	TBL
	LC80340322015170LGN01	19.06.2015—17:43	
	LC80340322017239LGN00	27.08.2017—17:44	
	LC80340322017255LGN00	12.09.2017—17:44	
	LC80330322017264LGN00	21.09.2017—17:38	
	LC80340322018258LGN00	15.09.2018—17:44	
	LC80350262017198LGN00	17.07.2017—17:48	FPK
	LC80360262017205LGN00	24.07.2017—17:54	
	LC80350262017230LGN00	18.08.2017—17:48	
	LC80350262017246LGN00	03.09.2017—17:48	
	LC80350262017246LGN00	03.09.2017—17:48	
	LC80230362019296LGN00	23.10.2019—16:38	GWN
	LC80290292019290LGN00	17.10.2019—17:12	SXF
	LC80280352017229LGN00	17.08.2017—17:08	SGP E32 Medford
	LC80280342019203LGN00	22.07.2019—17:08	
	LC80280352017229LGN00	17.08.2017—17:08	SGP E11 Byron
	LC80280342017245LGN00	02.09.2017—17:08	
	LC80280342019203LGN00	22.07.2019—17:08	
	LC80280352019027LGN00	27.01.2019—17:08	SGP E37 Waukomis
	LC80280352019027LGN00	27.01.2019—17:08	SGP E15 Ringwood
	LC80280352019027LGN00	27.01.2019—17:08	SGP E38 Omega
Nighttime	LC81292122013270LGN01	27.09.2013—04:45	TBL
	LC81302122015171LGN01	20.06.2015—04:48	
	LC81302122017240LGN00	28.08.2017—04:49	
	LC81302122017256LGN00	13.09.2017—04:49	
	LC81292122017265LGN00	22.09.2017—04:43	
	LC81302122018259LGN00	16.09.2018—04:49	
	LC81282182017194LGN00	13.07.2017—04:39	FPK
	LC81272182017203LGN00	22.07.2017—04:33	
	LC81282182017226LGN00	14.08.2017—04:39	
	LC81282182017242LGN00	30.08.2017—04:39	
	LC81272182017251LGN00	08.09.2017—04:33	
	LT81212082019295LGN00	22.10.2019—03:52	GWN
	LT81232142019293LGN00	20.10.2019—04:07	SXF
	LC81262102017228LGN00	16.08.2017—04:24	SGP E32 Medford
	LC81262092019202LGN00	21.07.2019—04:23	

Table A1. *Cont.*

Acquisition Time	Scene ID	Scene Acquisition Date and Hour (UTC)	Site ID
	LC81262102017228LGN00	16.08.2017—04:24	
	LC81262102017244LGN00	01.09.2017—04:24	SGP E11 Byron
	LC81262092019202LGN00	21.07.2019—04:23	
	LC81262092019026LGN00	26.01.2019—04:23	SGP E37 Waukomis
	LC81262092019026LGN00	26.01.2019—04:23	SGP E15 Ringwood
	LC81262092019026LGN00	26.01.2019—04:23	SGP E38 Omega

References

1. Townshend, J.R.G.R.; Justice, C.O.O.; Skole, D.; Malingreau, J.-P.P.; Cihlar, J.; Teillet, P.; Sadowski, F.; Ruttenberg, S. The 1 km resolution global data set: Needs of the international geosphere biosphere programme! *Int. J. Remote Sens.* **1994**, *15*, 3417–3441. [CrossRef]
2. GCOS. The Global Observing System for Climate: Implementation Needs. Available online: https://library.wmo.int/doc_num.php?explnum_id=3417 (accessed on 26 May 2020).
3. Yu, Y.; Liu, Y.; Yu, P. Land Surface Temperature Product Development for JPSS and GOES-R Missions. In *Comprehensive Remote Sensing*; Elsevier: Berlin/Heidelberg, Germany, 2018; pp. 284–303.
4. Becker, F.; Li, Z.-L. Surface temperature and emissivity at various scales: Definition, measurement and related problems. *Remote. Sens. Rev.* **1995**, *12*, 225–253. [CrossRef]
5. Dash, P.; Göttsche, F.-M.; Olesen, F.; Fischer, H. Retrieval of land surface temperature and emissivity from satellite data: Physics, theoretical limitations and current methods. *J. Indian Soc. Remote Sens.* **2001**, *29*, 23–30. [CrossRef]
6. Qin, Z.; Karnieli, A. Progress in the remote sensing of land surface temperature and ground emissivity using NOAA-AVHRR data. *Int. J. Remote Sens.* **1999**, *20*, 2367–2393. [CrossRef]
7. Anderson, M.C.; Norman, J.M.; Kustas, W.P.; Houborg, R.; Starks, P.J.; Agam, N. A thermal-based remote sensing technique for routine mapping of land-surface carbon, water and energy fluxes from field to regional scales. *Remote Sens. Environ.* **2008**, *112*, 4227–4241. [CrossRef]
8. Dash, P.; Göttsche, F.-M.; Olesen, F.-S.; Fischer, H. Land surface temperature and emissivity estimation from passive sensor data: Theory and practice-current trends. *Int. J. Remote Sens.* **2002**, *23*, 2563–2594. [CrossRef]
9. Dickinson, R.E. Land Surface Processes and Climate—Surface Albedos and Energy Balance. In *Theory of Climate*; Saltzman, B., Ed.; Elsevier: Berlin/Heidelberg, Germany, 1983; pp. 305–353.
10. Cammalleri, C.; Vogt, J. On the Role of land surface temperature as proxy of soil moisture status for drought monitoring in Europe. *Remote Sens.* **2015**, *7*, 16849–16864. [CrossRef]
11. Wan, Z.; Wang, P.; Li, X. Using MODIS Land Surface Temperature and Normalized Difference Vegetation Index products for monitoring drought in the southern Great Plains, USA. *Int. J. Remote Sens.* **2004**, *25*, 61–72. [CrossRef]
12. Coates, A.; Dennison, P.; Roberts, D.; Roth, K. Monitoring the impacts of severe drought on southern california chaparral species using hyperspectral and thermal infrared imagery. *Remote Sens.* **2015**, *7*, 14276–14291. [CrossRef]
13. Wesley, E.J.; Brunsell, N.A. Greenspace pattern and the surface urban heat island: A biophysically-based approach to investigating the effects of urban landscape configuration. *Remote Sens.* **2019**, *11*, 2322. [CrossRef]
14. Granero-Belinchon, C.; Michel, A.; Lagouarde, J.-P.; Sobrino, J.A.; Briottet, X. Night thermal unmixing for the study of microscale surface urban heat islands with TRISHNA-Like data. *Remote Sens.* **2019**, *11*, 1449. [CrossRef]
15. Zhou, D.; Xiao, J.; Bonafoni, S.; Berger, C.; Deilami, K.; Zhou, Y.; Frolking, S.; Yao, R.; Qiao, Z.; Sobrino, J. Satellite remote sensing of surface urban heat islands: Progress, challenges, and perspectives. *Remote Sens.* **2019**, *11*, 48. [CrossRef]
16. Keeratikasikorn, C.; Bonafoni, S. Urban Heat Island Analysis over the Land Use Zoning Plan of Bangkok by Means of Landsat 8 Imagery. *Remote Sens.* **2018**, *10*, 440. [CrossRef]

17. Sobrino, J.A.; Oltra-Carrió, R.; Sòria, G.; Jiménez-muñoz, J.C.; Franch, B.; Hidalgo, V.; Mattar, C.; Julien, Y.; Cuenca, J.; Romaguera, M.; et al. Evaluation of the surface urban heat island effect in the city of Madrid by thermal remote sensing. *Int. J. Remote Sens.* **2013**, *34*, 3177–3192. [CrossRef]
18. Sun, J.; Salvucci, G.D.; Entekhabi, D. Estimates of evapotranspiration from MODIS and AMSR-E land surface temperature and moisture over the Southern Great Plains. *Remote Sens. Environ.* **2012**, *127*, 44–59. [CrossRef]
19. Galleguillos, M.; Jacob, F.; Prévot, L.; French, A.; Lagacherie, P. Comparison of two temperature differencing methods to estimate daily evapotranspiration over a Mediterranean vineyard watershed from ASTER data. *Remote Sens. Environ.* **2011**, *115*, 1326–1340. [CrossRef]
20. Candy, B.; Saunders, R.W.; Ghent, D.; Bulgin, C.E. The impact of satellite-derived land surface temperatures on numerical weather prediction analyses and forecasts. *J. Geophys. Res. Atmos.* **2017**, *122*, 9783–9802. [CrossRef]
21. Meng, C.L.; Li, Z.-L.; Zhan, X.; Shi, J.C.; Liu, C.Y. Land surface temperature data assimilation and its impact on evapotranspiration estimates from the Common Land Model. *Water Resour. Res.* **2009**, *45*, 2. [CrossRef]
22. Qin, J.; Liang, S.; Liu, R.; Zhang, H.; Hu, B. A Weak-constraint-based data assimilation scheme for estimating surface turbulent fluxes. *IEEE Geosci. Remote Sens. Lett.* **2007**, *4*, 649–653. [CrossRef]
23. Dousset, B.; Gourmelon, F.; Laaidi, K.; Zeghnoun, A.; Giraudet, E.; Bretin, P.; Mauri, E.; Vandentorren, S. Satellite monitoring of summer heat waves in the Paris metropolitan area. *Int. J. Climatol.* **2011**, *31*, 313–323. [CrossRef]
24. Sekertekin, A.; Inyurt, S.; Yaprak, S. Pre-seismic ionospheric anomalies and spatio-temporal analyses of MODIS Land surface temperature and aerosols associated with Sep, 24 2013 Pakistan Earthquake. *J. Atmos. Sol. Terr. Phys.* **2020**, *200*, 105218. [CrossRef]
25. Pavlidou, E.; van der Meijde, M.; van der Werff, H.; Hecker, C. Time series analysis of land surface temperatures in 20 earthquake cases worldwide. *Remote Sens.* **2018**, *11*, 61. [CrossRef]
26. Maffei, C.; Alfieri, S.; Menenti, M. Relating spatiotemporal patterns of forest fires burned area and duration to diurnal land surface temperature anomalies. *Remote Sens.* **2018**, *10*, 1777. [CrossRef]
27. Sekertekin, A.; Arslan, N. Monitoring thermal anomaly and radiative heat flux using thermal infrared satellite imagery –A case study at Tuzla geothermal region. *Geothermics* **2019**, *78*, 243–254. [CrossRef]
28. Mia, M.; Fujimitsu, Y.; Nishijima, J. Monitoring of Thermal Activity at the Hatchobaru–Otake Geothermal Area in Japan Using Multi-Source Satellite Images—With Comparisons of Methods, and Solar and Seasonal Effects. *Remote Sens.* **2018**, *10*, 1430. [CrossRef]
29. Prata, A.J.; Caselles, V.; Coll, C.; Sobrino, J.A.; Ottlé, C. Thermal remote sensing of land surface temperature from satellites: Current status and future prospects. *Remote Sens. Rev.* **1995**, *12*, 175–224. [CrossRef]
30. Wark, D.Q.; Yamamoto, G.; Lienesch, J.H. Methods of Estimating Infrared Flux and Surface Temperature from Meteorological Satellites. *J. Atmos. Sci.* **1962**, *19*, 369–384. [CrossRef]
31. Li, Z.; Tang, B.-H.; Wu, H.; Ren, H.; Yan, G.; Wan, Z.; Trigo, I.F.; Sobrino, J.A. Satellite-derived land surface temperature: Current status and perspectives. *Remote Sens. Environ.* **2013**, *131*, 14–37. [CrossRef]
32. Li, Z.-L.; Becker, F. Feasibility of land surface temerature and emissivity determination from AVHRR data. *Remote Sens. Environ.* **1993**, *43*, 67–85. [CrossRef]
33. Sobrino, J.A.; Caselles, V.; Becker, F. Significance of the remotely sensed thermal infrared measurements obtained over a citrus orchard. *ISPRS J. Photogramm. Remote Sens.* **1990**, *44*, 343–354. [CrossRef]
34. Sekertekin, A.; Bonafoni, S. Land Surface Temperature Retrieval from Landsat 5, 7, and 8 over Rural Areas: Assessment of Different Retrieval Algorithms and Emissivity Models and Toolbox Implementation. *Remote Sens.* **2020**, *12*, 294. [CrossRef]
35. Becker, F.; Li, Z.L. Temperature-independent spectral indices in thermal infrared bands. *Remote Sens. Environ.* **1990**, *32*, 17–33. [CrossRef]
36. Mao, K.; Qin, Z.; Shi, J.; Gong, P. A practical split-window algorithm for retrieving land-surface temperature from MODIS data. *Int. J. Remote Sens.* **2005**, *26*, 3181–3204. [CrossRef]
37. Wan, Z.; Dozier, J. A generalized split-window algorithm for retrieving land-surface temperature from space. *IEEE Trans. Geosci. Remote Sens.* **1996**, *34*, 892–905.
38. Coll, C.; Caselles, V. A split-window algorithm for land surface temperature from advanced very high resolution radiometer data: Validation and algorithm comparison. *J. Geophys. Res. Atmos.* **1997**, *102*, 16697–16713. [CrossRef]

39. Qin, Z.; Karnieli, A.; Berliner, P. A mono-window algorithm for retrieving land surface temperature from Landsat TM data and its application to the Israel-Egypt border region. *Int. J. Remote Sens.* **2001**, *22*, 3719–3746. [CrossRef]

40. Jiménez-muñoz, J.C.; Cristóbal, J.; Sobrino, J.A.; Sòria, G.; Ninyerola, M.; Pons, X. Revision of the single-channel algorithm for land surface temperature retrieval from landsat thermal-infrared data. *IEEE Trans. Geosci. Remote Sens.* **2009**, *47*, 339–349. [CrossRef]

41. Jimenez-Munoz, J.C.; Sobrino, J.A. A Single-Channel Algorithm for Land-Surface Temperature Retrieval From ASTER Data. *IEEE Geosci. Remote Sens. Lett.* **2010**, *7*, 176–179. [CrossRef]

42. Price, J.C. Estimating surface temperatures from satellite thermal infrared data-A simple formulation for the atmospheric effect. *Remote Sens. Environ.* **1983**, *13*, 353–361. [CrossRef]

43. Susskind, J.; Rosenfield, J.; Reuter, D.; Chahine, M.T. Remote sensing of weather and climate parameters from HIRS2/MSU on TIROS-N. *J. Geophys. Res.* **1984**, *89*, 4677. [CrossRef]

44. Gillespie, A.; Rokugawa, S.; Matsunaga, T.; Steven Cothern, J.; Hook, S.; Kahle, A.B. A temperature and emissivity separation algorithm for advanced spaceborne thermal emission and reflection radiometer (ASTER) images. *IEEE Trans. Geosci. Remote Sens.* **1998**, *36*, 1113–1126. [CrossRef]

45. Wan, Z.; Zhang, Y.; Zhang, Q.; Li, Z. liang Validation of the land-surface temperature products retrieved from Terra Moderate Resolution Imaging Spectroradiometer data. *Remote Sens. Environ.* **2002**, *83*, 163–180. [CrossRef]

46. Yu, Y.; Tarpley, D.; Privette, J.L.; Flynn, L.E.; Xu, H.; Chen, M.; Vinnikov, K.Y.; Sun, D.; Tian, Y. Validation of GOES-R Satellite Land Surface Temperature Algorithm Using SURFRAD Ground Measurements and Statistical Estimates of Error Properties. *IEEE Trans. Geosci. Remote Sens.* **2012**, *50*, 704–713. [CrossRef]

47. Sekertekin, A. Validation of Physical Radiative Transfer Equation-Based Land Surface Temperature Using Landsat 8 Satellite Imagery and SURFRAD in-situ Measurements. *J. Atmos. Sol. Terr. Phys.* **2019**, *196*, 105161. [CrossRef]

48. Malakar, N.K.; Hulley, G.C.; Hook, S.J.; Laraby, K.; Cook, M.; Schott, J.R. An Operational Land Surface Temperature Product for Landsat Thermal Data: Methodology and Validation. *IEEE Trans. Geosci. Remote Sens.* **2018**, *56*, 5717–5735. [CrossRef]

49. Wang, L.; Lu, Y.; Yao, Y. Comparison of Three Algorithms for the Retrieval of Land Surface Temperature from Landsat 8 Images. *Sensors* **2019**, *19*, 5049. [CrossRef]

50. Parastatidis, D.; Mitraka, Z.; Chrysoulakis, N.; Abrams, M. Online Global Land Surface Temperature Estimation from Landsat. *Remote Sens.* **2017**, *9*, 1208. [CrossRef]

51. Yu, X.; Guo, X.; Wu, Z. Land surface temperature retrieval from landsat 8 TIRS-comparison between radiative transfer equation-based method, split window algorithm and single channel method. *Remote Sens.* **2014**, *6*, 9829–9852. [CrossRef]

52. Zhang, Z.; He, G.; Wang, M.; Long, T.; Wang, G.; Zhang, X. Validation of the generalized single-channel algorithm using landsat 8 imagery and SURFRAD ground measurements. *Remote Sens. Lett.* **2016**, *7*, 810–816. [CrossRef]

53. Vanhellemont, Q. Combined land surface emissivity and temperature estimation from Landsat 8 OLI and TIRS. *ISPRS J. Photogramm. Remote Sens.* **2020**, *166*, 390–402. [CrossRef]

54. Guo, J.; Ren, H.; Zheng, Y.; Lu, S.; Dong, J. Evaluation of Land Surface Temperature Retrieval from Landsat 8/TIRS Images before and after Stray Light Correction Using the SURFRAD Dataset. *Remote Sens.* **2020**, *12*, 1023. [CrossRef]

55. García-Santos, V.; Cuxart, J.; Martínez-Villagrasa, D.; Jiménez, M.; Simó, G. Comparison of Three Methods for Estimating Land Surface Temperature from Landsat 8-TIRS Sensor Data. *Remote Sens.* **2018**, *10*, 1450. [CrossRef]

56. Zheng, Y.; Ren, H.; Guo, J.; Ghent, D.; Tansey, K.; Hu, X.; Nie, J.; Chen, S. Land Surface Temperature Retrieval from Sentinel-3A Sea and Land Surface Temperature Radiometer, Using a Split-Window Algorithm. *Remote Sens.* **2019**, *11*, 650. [CrossRef]

57. Pinker, R.T.; Ma, Y.; Chen, W.; Hulley, G.; Borbas, E.; Islam, T.; Hain, C.; Cawse-Nicholson, K.; Hook, S.; Basara, J. Towards a Unified and Coherent Land Surface Temperature Earth System Data Record from Geostationary Satellites. *Remote Sens.* **2019**, *11*, 1399. [CrossRef]

58. Martin, M.; Ghent, D.; Pires, A.; Göttsche, F.-M.; Cermak, J.; Remedios, J. Comprehensive In Situ Validation of Five Satellite Land Surface Temperature Data Sets over Multiple Stations and Years. *Remote Sens.* **2019**, *11*, 479. [CrossRef]

59. Niclòs, R.; Galve, J.M.; Valiente, J.A.; Estrela, M.J.; Coll, C. Accuracy assessment of land surface temperature retrievals from MSG2-SEVIRI data. *Remote Sens. Environ.* **2011**, *115*, 2126–2140. [CrossRef]

60. Wan, Z. New refinements and validation of the collection-6 MODIS land-surface temperature/emissivity product. *Remote Sens. Environ.* **2014**, *140*, 36–45. [CrossRef]

61. Wang, W.; Liang, S.; Meyers, T. Validating MODIS land surface temperature products using long-term nighttime ground measurements. *Remote Sens. Environ.* **2008**, *112*, 623–635. [CrossRef]

62. Coll, C.; Caselles, V.; Galve, J.; Valor, E.; Niclos, R.; Sanchez, J.; Rivas, R. Ground measurements for the validation of land surface temperatures derived from AATSR and MODIS data. *Remote Sens. Environ.* **2005**, *97*, 288–300. [CrossRef]

63. Ouyang, X.; Chen, D.; Duan, S.-B.; Lei, Y.; Dou, Y.; Hu, G. Validation and Analysis of Long-Term AATSR Land Surface Temperature Product in the Heihe River Basin, China. *Remote Sens.* **2017**, *9*, 152. [CrossRef]

64. Guillevic, P.C.; Biard, J.C.; Hulley, G.C.; Privette, J.L.; Hook, S.J.; Olioso, A.; Göttsche, F.M.; Radocinski, R.; Román, M.O.; Yu, Y.; et al. Validation of Land Surface Temperature products derived from the Visible Infrared Imaging Radiometer Suite (VIIRS) using ground-based and heritage satellite measurements. *Remote Sens. Environ.* **2014**, *154*, 19–37. [CrossRef]

65. Wang, K.; Liang, S. Evaluation of ASTER and MODIS land surface temperature and emissivity products using long-term surface longwave radiation observations at SURFRAD sites. *Remote Sens. Environ.* **2009**, *113*, 1556–1565. [CrossRef]

66. Sabol, D.E., Jr.; Gillespie, A.R.; Abbott, E.; Yamada, G. Field validation of the ASTER Temperature–Emissivity Separation algorithm. *Remote Sens. Environ.* **2009**, *113*, 2328–2344. [CrossRef]

67. Wan, Z.; Li, Z.-L.; Wan, Z.; Li, Z. A physics-based algorithm for retrieving land-surface emissivity and temperature from EOS/MODIS data. *IEEE Trans. Geosci. Remote Sens.* **1997**, *35*, 980–996.

68. Wang, K.; Wan, Z.; Wang, P.; Sparrow, M.; Liu, J.; Zhou, X.; Haginoya, S. Estimation of surface long wave radiation and broadband emissivity using Moderate Resolution Imaging Spectroradiometer (MODIS) land surface temperature/emissivity products. *J. Geophys. Res.* **2005**, *110*, D11109. [CrossRef]

69. Augustine, J.A.; DeLuisi, J.J.; Long, C.N. SURFRAD—A National Surface Radiation Budget Network for Atmospheric Research. *Bull. Am. Meteorol. Soc.* **2000**, *81*, 2341–2357. [CrossRef]

70. Baldocchi, D.; Falge, E.; Gu, L.; Olson, R.; Hollinger, D.; Running, S.; Anthoni, P.; Bernhofer, C.; Davis, K.; Evans, R.; et al. FLUXNET: A New Tool to Study the Temporal and Spatial Variability of Ecosystem–Scale Carbon Dioxide, Water Vapor, and Energy Flux Densities. *Bull. Am. Meteorol. Soc.* **2001**, *82*, 2415–2434. [CrossRef]

71. Stokes, G.M.; Schwartz, S.E. The Atmospheric Radiation Measurement (ARM) Program: Programmatic Background and Design of the Cloud and Radiation Test Bed. *Bull. Am. Meteorol. Soc.* **1994**, *75*, 1201–1221. [CrossRef]

72. Ohmura, A.; Gilgen, H.; Hegner, H.; Müller, G.; Wild, M.; Dutton, E.G.; Forgan, B.; Fröhlich, C.; Philipona, R.; Heimo, A.; et al. Baseline Surface Radiation Network (BSRN/WCRP): New Precision Radiometry for Climate Research. *Bull. Am. Meteorol. Soc.* **1998**, *79*, 2115–2136. [CrossRef]

73. Li, S.; Yu, Y.; Sun, D.; Tarpley, D.; Zhan, X.; Chiu, L. Evaluation of 10 year AQUA/MODIS land surface temperature with SURFRAD observations. *Int. J. Remote Sens.* **2014**, *35*, 830–856. [CrossRef]

74. Heidinger, A.K.; Laszlo, I.; Molling, C.C.; Tarpley, D. Using SURFRAD to verify the NOAA single-channel land surface temperature algorithm. *J. Atmos. Ocean. Technol.* **2013**, *30*, 2868–2884. [CrossRef]

75. Guillevic, P.C.; Privette, J.L.; Coudert, B.; Palecki, M.A.; Demarty, J.; Ottlé, C.; Augustine, J.A. Land Surface Temperature product validation using NOAA's surface climate observation networks—Scaling methodology for the Visible Infrared Imager Radiometer Suite (VIIRS). *Remote Sens. Environ.* **2012**, *124*, 282–298. [CrossRef]

76. Martin, M.; Göttsche, F.M. *Satellite LST Validation Report*; European Space Agency (ESA): Karlsruhe, Germany, 2016.

77. Faysash, D.A.; Smith, E.A. Simultaneous Retrieval of Diurnal to Seasonal Surface Temperatures and Emissivities over SGP ARM–CART Site Using GOES Split Window. *J. Appl. Meteorol.* **2000**, *39*, 971–982. [CrossRef]

78. Pinker, R.T.; Sun, D.; Hung, M.P.; Li, C.; Basara, J.B. Evaluation of satellite estimates of land surface temperature from GOES over the United States. *J. Appl. Meteorol. Climatol.* **2009**, *48*, 167–180. [CrossRef]

79. Ou, S.C.; Chen, Y.; Liou, K.N.; Cosh, M.; Brutsaert, W. Satellite remote sensing of land surface temperatures: Application of the atmospheric correction method and split-window technique to data of ARM-SGP site. *Int. J. Remote Sens.* **2002**, *23*, 5177–5192. [CrossRef]

80. Philipona, R.; Dutton, E.G.; Stoffel, T.; Michalsky, J.; Reda, I.; Stifter, A.; Wendung, P.; Wood, N.; Clough, S.A.; Mlawer, E.J.; et al. Atmospheric longwave irradiance uncertainty: Pyrgeometers compared to an absolute sky-scanning radiometer, atmospheric emitted radiance interferometer, and radiative transfer model calculations. *J. Geophys. Res. Atmos.* **2001**, *106*, 28129–28141. [CrossRef]

81. Duan, S.-B.; Li, Z.-L.; Li, H.; Göttsche, F.-M.; Wu, H.; Zhao, W.; Leng, P.; Zhang, X.; Coll, C. Validation of Collection 6 MODIS land surface temperature product using in situ measurements. *Remote Sens. Environ.* **2019**, *225*, 16–29. [CrossRef]

82. Xu, H.; Yu, Y.; Tarpley, D.; Gottsche, F.; Olesen, F.-S. Evaluation of GOES-R Land Surface Temperature Algorithm Using SEVIRI Satellite Retrievals With *In Situ* Measurements. *IEEE Trans. Geosci. Remote Sens.* **2014**, *52*, 3812–3822. [CrossRef]

83. Wulder, M.A.; Loveland, T.R.; Roy, D.P.; Crawford, C.J.; Masek, J.G.; Woodcock, C.E.; Allen, R.G.; Anderson, M.C.; Belward, A.S.; Cohen, W.B.; et al. Current status of Landsat program, science, and applications. *Remote Sens. Environ.* **2019**, *225*, 127–147. [CrossRef]

84. Barsi, J.A.; Barker, J.L.; Schott, J.R. An Atmospheric Correction Parameter Calculator for a single thermal band earth-sensing instrument. In Proceedings of the IGARSS 2003. 2003 IEEE International Geoscience and Remote Sensing Symposium. Proceedings (IEEE Cat. No.03CH37477), Toulouse, France, 21–25 July 2003; IEEE: Piscataway, NJ, USA, 2003; Volume 5, pp. 3014–3016.

85. Barsi, J.A.; Schott, J.R.; Palluconi, F.D.; Hook, S.J. Validation of a web-based atmospheric correction tool for single thermal band instruments. In Proceedings of the Earth Observing Systems X, Bellingham, WA, USA, 25 August 2005; SPIE: Bellingham, WA, USA, 2005; Volume 5882, p. 58820.

86. Dozier, J.; Warren, S.G. Effect of viewing angle on the infrared brightness temperature of snow. *Water Resour. Res.* **1982**, *18*, 1424–1434. [CrossRef]

87. Zanter, K. USGS Landsat 8 (L8) Data Users Handbook. Available online: https://prd-wret.s3-us-west-2.amazonaws.com/assets/palladium/production/atoms/files/LSDS-1574_L8_Data_Users_Handbook-v5.0.pdf (accessed on 5 December 2019).

88. Sobrino, J.A.; Raissouni, N.; Li, Z. A Comparative Study of Land Surface Emissivity Retrieval from NOAA Data. *Remote Sens. Environ.* **2001**, *75*, 256–266. [CrossRef]

89. Sobrino, J.A.; Raissouni, N. Toward remote sensing methods for land cover dynamic monitoring: Application to Morocco. *Int. J. Remote Sens.* **2000**, *21*, 353–366. [CrossRef]

90. Valor, E.; Caselles, V. Mapping land surface emissivity from NDVI: Application to European, African, and South American areas. *Remote Sens. Environ.* **1996**, *57*, 167–184. [CrossRef]

91. Peres, L.F.; DaCamara, C.C. Emissivity maps to retrieve land-surface temperature from MSG/SEVIRI. *IEEE Trans. Geosci. Remote Sens.* **2005**, *43*, 1834–1844. [CrossRef]

92. Sun, D.; Pinker, R.T. Estimation of land surface temperature from a Geostationary Operational Environmental Satellite (GOES-8). *J. Geophys. Res.* **2003**, *108*, 4326. [CrossRef]

93. Sobrino, J.a.; Jiménez-Muñoz, J.C.; Paolini, L. Land surface temperature retrieval from LANDSAT TM 5. *Remote Sens. Environ.* **2004**, *90*, 434–440. [CrossRef]

94. Van de Griend, A.A.; Owe, M. On the relationship between thermal emissivity and the normalized difference vegetation index for natural surfaces. *Int. J. Remote Sens.* **1993**, *14*, 1119–1131. [CrossRef]

95. Sobrino, J.A.; Jimenez-Muoz, J.C.; Soria, G.; Romaguera, M.; Guanter, L.; Moreno, J.; Plaza, A.; Martinez, P. Land Surface Emissivity Retrieval From Different VNIR and TIR Sensors. *IEEE Trans. Geosci. Remote Sens.* **2008**, *46*, 316–327. [CrossRef]

96. Carlson, T.N.; Ripley, D.A. On the relation between NDVI, fractional vegetation cover, and leaf area index. *Remote Sens. Environ.* **1997**, *62*, 241–252. [CrossRef]

97. Skokovic, D.; Sobrino, J.a.; Jiménez Muñoz, J.C.; Soria, G.; Julien, Y.; Mattar, C.; Cristóbal, J. Calibration and Validation of land surface temperature for Landsat8- TIRS sensor TIRS LANDSAT-8 CHARACTERISTICS. *Land Prod. Valid. Evol. ESA/ESRIN* **2014**, 1–27.

98. Li, S.; Jiang, G.-M. Land Surface Temperature Retrieval From Landsat-8 Data With the Generalized Split-Window Algorithm. *IEEE Access* **2018**, *6*, 18149–18162. [CrossRef]

99. Wang, S.; He, L.; Hu, W. A Temperature and Emissivity Separation Algorithm for Landsat-8 Thermal Infrared Sensor Data. *Remote Sens.* **2015**, *7*, 9904–9927. [CrossRef]

100. Ndossi, M.; Avdan, U. Inversion of Land Surface Temperature (LST) Using Terra ASTER Data: A Comparison of Three Algorithms. *Remote Sens.* **2016**, *8*, 993. [CrossRef]

101. Saltelli, A. Sensitivity Analysis for Importance Assessment. *Risk Anal.* **2002**, *22*, 579–590. [CrossRef] [PubMed]

102. Wang, F.; Qin, Z.; Song, C.; Tu, L.; Karnieli, A.; Zhao, S. An Improved Mono-Window Algorithm for Land Surface Temperature Retrieval from Landsat 8 Thermal Infrared Sensor Data. *Remote Sens.* **2015**, *7*, 4268–4289. [CrossRef]

103. Wang, H.; Mao, K.; Mu, F.; Shi, J.; Yang, J.; Li, Z.; Qin, Z. A Split Window Algorithm for Retrieving Land Surface Temperature from FY-3D MERSI-2 Data. *Remote Sens.* **2019**, *11*, 2083. [CrossRef]

104. Meng, X.; Cheng, J.; Zhao, S.; Liu, S.; Yao, Y. Estimating Land Surface Temperature from Landsat-8 Data using the NOAA JPSS Enterprise Algorithm. *Remote Sens.* **2019**, *11*, 155. [CrossRef]

105. Wang, M.; Zhang, Z.; Hu, T.; Liu, X. A Practical Single-Channel Algorithm for Land Surface Temperature Retrieval: Application to Landsat series data. *J. Geophys. Res. Atmos.* **2019**, *124*, 2018JD029330. [CrossRef]

 *remote sensing*

Article

Development of a Land Surface Temperature Retrieval Algorithm from GK2A/AMI

Youn-Young Choi and Myoung-Seok Suh *

Department of Atmospheric Science, Kongju National University, 56, Gongjudaehak-ro, Gongju-si, Chungcheongnam-do 32588, Korea; choiyoyo@smail.kongju.ac.kr
* Correspondence: sms416@kongju.ac.kr; Tel.: +82-41-850-8527

Received: 13 August 2020; Accepted: 16 September 2020; Published: 18 September 2020

Abstract: Land surface temperature (LST) is an important geophysical element for understanding Earth systems and land–atmosphere interactions. In this study, we developed a nonlinear split-window LST retrieval algorithm for the observation area of GEO-KOMPSAT-2A (GK2A), the next-generation geostationary satellite in Korea. To develop the GK2A LST retrieval algorithm, radiative transfer model simulation data, considering various impacting factors, were constructed. The LST retrieval algorithm was developed with a total of six equations as per day/night and atmospheric conditions (dry/normal/wet), considering the effects of diurnal variation of LST and atmospheric conditions on LST retrieval. The emissivity of each channel required for LST retrieval was calculated in real-time with the vegetation cover method using the consecutive 8-day cycle vegetation index provided by GK2A. The indirect validation of the results of GK2A LST with Moderate Resolution Imaging Spectroradiometer (MODIS) LST Collection 6 showed a high correlation coefficient (0.969), slightly warm bias (+1.227 K), and root mean square error (RMSE) (2.281 K). Compared to the MODIS LST, the GK2A LST showed a warm bias greater than +1.8 K during the day, but a relatively small bias (<+0.7 K) at night. Based on the results of the validation with in situ measurements from the Tateno station of the Baseline Surface Radiation Network, the correlation coefficient was 0.95, bias was +0.523 K, and RMSE was 2.021 K.

Keywords: land surface temperature; GK2A; split-window method; MODIS; BSRN

1. Introduction

Land surface temperature (LST) can be interpreted as the skin temperature of Earth's land and is derived using the upward longwave radiation measured by a satellite sensor [1]. The LST retrieved from satellites represents the surface temperature of ground pixels that are not contaminated by clouds and is affected by many factors, such as land use/cover, vegetation, soil moisture, snow, etc. [2,3]. Land information is used as input and verification data for numerical/climate models. These are important factors to understand the Earth's system and land–atmosphere interactions [4–6]. LST has various research applications, such as studying land use/cover changes [7–9], drought monitoring [10–12], energy balance estimation [13–15], analyzing urban heat islands [16–18], studying evapotranspiration [19] and so on [20–22]. As various applied studies using LST have been conducted, the demand for long-term LST data with high spatial and temporal resolutions as well as good accuracy is increasing [23,24].

LST is one of the climate elements with very different spatial and temporal variability depending on the ground condition (land cover, vegetation condition, soil moisture, etc.) of the observation point. Due to these characteristics, more detailed spatial and temporal field observations are necessary, but this is not practical in economic and technical terms. So, at present, special observation is only performed in some areas where the ground condition is relatively homogeneous. The LST observed in a field is a value that can represent only a few tens of meters of the radius of the observation point, therefore, there is a limitation of spatial representation. To overcome this limitation, studies on retrieving LST from satellites with spatial resolutions from hundreds of meters to several kilometers have been conducted for more than half a century [25–27]. LSTs retrieved from polar orbit satellites have a relatively high spatial resolution but long revisit cycles; they help to understand a momentary phenomenon in detail for a relatively narrow space, such as urban heat islands [28–32]. Moreover, LSTs retrieved from geostationary satellites have a relatively coarse spatial resolution compared to polar orbit satellites, but they can continuously retrieve the same observation area in a short observation period. Owing to these advantages, LSTs retrieved from geostationary satellites can be used to analyze phenomena occurring in large areas over a long period of time and can also fill in the blank areas of field observation [33–37].

Many studies have been conducted and different methodologies have been developed to retrieve high-quality LST data from satellites. For LST retrieval from the radiation observed by thermal infrared sensors, cloud detection is essential, and it is necessary to know the surface condition and atmospheric effects [38–40]. LST retrieval methods, assuming that the land surface emissivity (LSE) is known a priori, can be roughly divided into three groups: single-channel methods [31,41], multi-channel methods [26,27,33], and multi-angle methods [42,43]. There are also several methods for retrieving LST and LSE at the same time when the LSE is unknown: classification-based emissivity methods [44,45], normalized difference vegetation index (NDVI)-based emissivity methods [46,47], day/night temperature-independent spectral indices-based methods [38,48], two-temperature methods [49,50], and temperature emissivity separation methods [28,51,52]. Among these various LST retrieval methods, a commonly used one for retrieving LST from a geostationary satellite is the split-window (SW) method using two adjacent thermal infrared channels with different absorption capabilities for water vapor and other substances [53–55]. The split-window method is relatively simple, efficient, and convenient to apply to most sensors, but it is assumed that the LSE is accurately known in such cases. In addition, the split-window method has various types of algorithms and thus has different performance characteristics according to the type of algorithm. Another characteristic of the split-window method is that the accuracy is degraded in a specific region where the total column water vapor is high or where the satellite viewing zenith angle (VZA) is large [23].

With improvements in sensor performance on satellites, next-generation geostationary meteorological satellites (Himawari-8, Geostationary Operational Environmental Satellite (GOES)-16/17, GEO-KOMPSAT-2A (GK2A)) have started to conduct their observations with high performances in time and space resolution [56–58]. In addition, the Meteosat Third Generation (MTG) series will be launched in 2021 as the next system after the Meteosat Second Generation (MSG) series of geostationary satellites [59]. With improved sensor performance, the space-time resolution of level 2 products has also been improved. The LST product was designated as the official level 2 product in GOES-16 and GK2A as in previous satellites (GOES-13/15, Communication, Ocean and Meteorological Satellite (COMS)). In addition, the European Space Agency (ESA) has adopted LST as an official product on MTG satellites following MSG satellites, and LST retrieval research is being conducted to prepare for it. In the case of Himawari-8, a study was conducted to retrieve (LST) for research purposes [55,60]. In the Asia-Oceania region, the COMS was retired on March 31, 2020. The COMS' follow-up satellite, GK2A, has been officially operating since 25 July 2019, and replaced COMS' services [61]. Therefore, it is necessary to develop an LST retrieval algorithm using GK2A/AMI (Advanced Meteorological Imager) data.

In this study, we developed an operational LST retrieval algorithm for the GK2A observation area using GK2A/AMI data from next-generation geostationary satellites in Korea. The contents of this paper are as follows. The properties of the data are described in Section 2.1, and the process of developing the simulation dataset using the radiative transfer model (RTM) and LST retrieval process and methodology are described in Section 2.2. The RTM simulation results and the GK2A LST retrieval results are presented in Section 3. In addition, the accuracy, problems, and future improvements of the GK2A LST retrieval algorithm are presented in Section 4.

2. Data and Methods

2.1. Data

2.1.1. GK2A Data

In this study, we used GK2A/AMI data provided by the Korea Meteorological Administration (KMA)/National Meteorological Satellite Center (NMSC). GK2A is located at 128.2°E and started operational services for 16 channels of AMI on 25 July 2019 [58]. To retrieve LST, level-1B of GK2A/AMI, level-2 of GK2A/AMI, auxiliary data of solar zenith angle (SZA), and satellite VZA data were used. The characteristics of the GK2A data used in this study are shown in Table 1. The brightness temperatures of channel 13 (10.35 μm) and channel 15 (12.36 μm) were used for adapting the split-window method. These infrared channels have a spatial resolution of 2 km at the nadir point and a temporal resolution of 10 min. Because LST is retrieved only for clear sky and land pixels, cloud mask and land/sea mask data are necessary. In this algorithm, we used GK2A cloud mask data as developed by NMSC/KMA [62]. In addition, we used the LSE of the two split-window channels derived by the vegetation cover method (VCM) using the land cover map, GK2A VI, and the look-up table of spectral emissivity [63]. The LSE data were produced from the GK2A LSE algorithm, and these data have a spectral resolution of 2 km and a temporal resolution of 1 day. In this study, GK2A LSTs were retrieved for 4 months from July to October 2019.

Table 1. The characteristics of GK2A/AMI data used in this study.

Name	Spatial Resolution (km)	Temporal Resolution	Description
Channel 13	2	10 min	Center wavelength: 10.3539 μm
Channel 15	2	10 min	Center wavelength: 12.3651 μm
GK2A Cloud mask data	2	10 min	Distinguish clear and cloud pixel
GK2A Land surface emissivity of Ch. 13	2	1 day	Input data of GK2A LST
GK2A Land surface emissivity of Ch. 15	2	1 day	Input data of GK2A LST

2.1.2. Validation data

To validate the retrieved LST, the ground measured upward longwave radiation data were used. The Tateno station of the Baseline Surface Radiation Network (BSRN) at 140.126°E and 36.058°N is a ground station in Japan and has been measuring upward longwave radiation every 1 min using the Kipp and Zonen CGR4 pyrgeometer [64–67]. Another type of validation data is high-quality LST data derived from another satellite. The MODIS LST (MOD11_L2, MYD11_L2 swath Collection 6) datasets were used for comparison with GK2A LSTs [32]. The spatial resolution of the MODIS LST was 1 km at nadir. The MODIS LST data are provided every 5 min for the swath region according to the orbit of Terra/Aqua satellites. In this study, GK2A LSTs were validated using the BSRN and MODIS LST data for 4 months, from July to October 2019.

2.2. Methodology

The GK2A LST retrieval algorithm is mainly composed of three parts, as shown in Figure 1. The first part is the development of the LST retrieval algorithm using a simulated pseudo match-up database with the RTM (shown in Figure 1 on the right side). First, we constructed the pseudo match-up database through the RTM simulation called MODTRAN4 under various atmospheric and land surface conditions. In this simulation, to improve the accuracy of the LST retrieval algorithm, we considered various impacting factors, such as the diurnal variations of air temperature and LST, spatial-temporal variations of spectral emissivity, and the non-linear effect of water vapor. To develop LST retrieval algorithms, we used the reference LST and calculated the brightness temperatures of the two SW channels with the inverse Planck function from the simulated radiances. The LST retrieval algorithms were separately developed using a statistical regression method according to the lapse rate (two types: day/night) and atmospheric conditions (three types: wet/normal/dry).

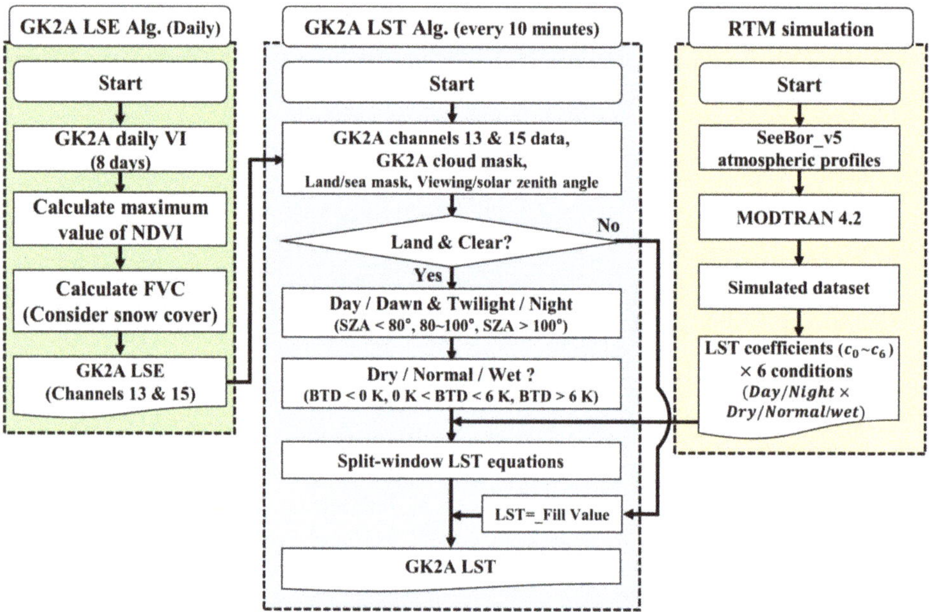

Figure 1. Flow chart of GK2A's land surface temperature retrieval process.

The split-window LST retrieval algorithm assumes that the LSE is prior known. In general, LSE is affected not only by the surface type and status, but also depends on the wavelength. Because the spatial resolution of the GK2A/AMI infrared channel is 2 km, most pixels are composed of a mixture of various vegetation and soil. Moreover, most vegetation not only undergoes seasonal variations, but also, soil conditions are affected by the presence or absence of precipitation (soil moisture and snow cover). The process of retrieving the GK2A LSE is briefly shown on the left side of Figure 1. The GK2A LSE algorithm was developed based on the VCM method and calculates the emissivity of a given pixel as a weighted average according to the fractional coverage of vegetation (FVC) and soil [46,68]. The FVC of a given pixel was derived by the method of Carlson and Ripley (1997) [69] using the 8-days cycle vegetation index data derived from GK2A/AMI. The snow cover fraction was calculated for the pixels covered with snow [70], and this was considered in retrieving the daily GK2A LSE.

The central column of Figure 1 shows the process of calculating LST every 10 min using GK2A/AMI data and pre-calculated emissivity. The GK2A LST was retrieved for clear and land pixels according to day/night and atmospheric conditions. In the process of calculating the GK2A LST, the SZA of each pixel was used for the division of daytime, nighttime, and dawn/twilight, as shown in the center of Figure 1. At this time, the threshold SZA values from the previous study were used [60]. Similarly, the atmospheric conditions were divided into dry/normal/wet according to the brightness temperature difference (BTD) threshold to calculate the GK2A LST.

Unlike the sea surface temperature, the available match-up data (ground-level observational data) for LST over the GK2A/AMI observation region are severely limited [71–73]. Therefore, to develop the GK2A LST algorithm from satellite data, we needed to prepare a similar on-site match-up database. For this purpose, the output from the RTM simulation could be used. As in many previous studies, the pseudo match-up database was generated through simulations using the RTM under various atmospheric and surface conditions. We developed an LST retrieval algorithm using the pseudo match-up database. SeeBor version 5 data have been used as an input profile for RTM simulations in many studies for the generation of a pseudo match-up database. This dataset contains 101 pressure levels of temperature, mixing ratio, and trace gases. For each profile in the dataset, a physically-based characterization of the surface skin temperature and surface emissivity is assigned [74]. The diurnal variation of LST is not provided in atmospheric profiles, so many studies have considered the diverse diurnal variation of LST in the RTM simulations according to the land cover, season, and weather conditions. In the research of [26,34], a large range of temperatures between the near-surface temperature and ground surface, which consists of five surface temperatures, Ta − 5 K, Ta, Ta + 5 K, Ta + 10 K, and Ta + 20 K, was considered for each atmospheric profile. In the research of [27,75], daily variations of LST from Ta − 16 K to Ta + 16 K were considered for each profile. In other studies [29,76], LST was prescribed for each profile in a range as Ta − 15 K < LST < Ta + 15 K with different increments of 1.5 K and 1 K, respectively. In [35], LST was considered asymmetrically from Ta − 6 K to Ta + 14 K in increments of 2 K. In [55], they used six values from Ta − 5 K to Ta + 20 K at 5 K intervals. In this study, the diurnal variation of LST was asymmetric according to the day ($\mathrm{LST_{day}} = \mathrm{Ta} - 2\,\mathrm{K} \sim \mathrm{Ta} + 18\,\mathrm{K}$) and night ($\mathrm{LST_{ngt}} = \mathrm{Ta} - 6\,\mathrm{K} \sim \mathrm{Ta} + 2\,\mathrm{K}$) at 2 K intervals. In addition, we included the diurnal variations of air temperature profiles under the planetary boundary layer using SeeBor data and reference LST, as shown in Figure 2.

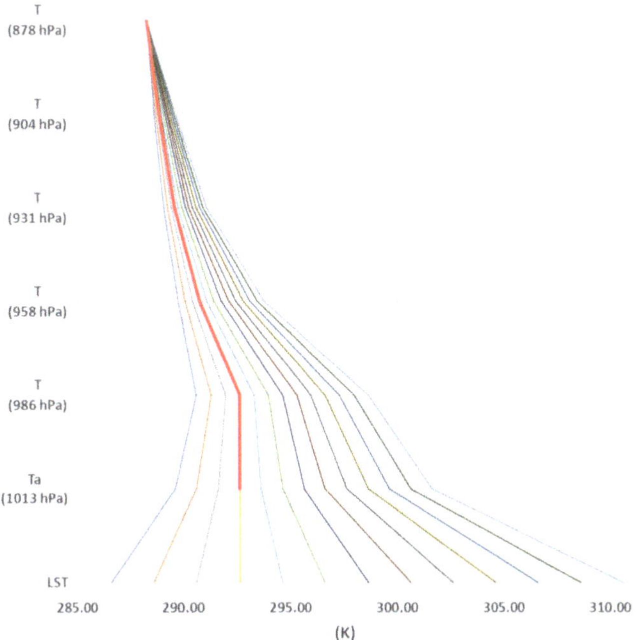

Figure 2. Conceptual atmospheric vertical profiles according to diurnal variation of land surface temperature (LST) and air temperature (bold red line: SeeBor profile; other lines: diurnal variation of air temperature profiles).

To generate the pseudo match-up database of the LST and measured spectral radiance, radiative transfer model simulations under various atmospheric and surface conditions were performed (Table 2). A total of 2694 sets of SeeBor data used were located at a satellite with a VZA of less than 50°. To consider the value of LSE in the RTM simulation, the range of LSE of channels 13 and 15 in the GK2A region was used, as shown in Table 2. Therefore, the total number of simulations for daytime and nighttime were 3,585,714 (2694 (atm) × 13 (LST) × 11 (ε_{ch13}) × 11 ($\Delta\varepsilon$)) and 1,629,870 (2694 (atm) × 5 (LST) × 11 (ε_{ch13}) × 11 ($\Delta\varepsilon$)), respectively.

Table 2. The input conditions of the radiative transfer model simulation according to the various atmospheric and surface conditions.

Impacting Factors	Detailed Conditions		
Atmospheric profiles	2694 SeeBor version 5 database (Viewing zenith angle is lesser than 50°)		
Land surface temperature	Day: Ta − 2 K~Ta + 18 K (a step of 2 K) Night: Ta − 6 K~Ta + 2 K (a step of 2 K)		
Diurnal variation of air temperature		1013 hPa (= Ta) 986 hPa 958 hPa 931 hPa 904 hPa	$\Delta T = LST - Ta$ $T(1013)' = T(1013) + 1/2\,\Delta T$ $T(986)' = T(986) + 1/3\,\Delta T$ $T(958)' = T(958) + 1/6\,\Delta T$ $T(931)' = T(931) + 1/12\,\Delta T$ $T(904)' = T(904) + 1/24\,\Delta T$
Land surface emissivity	ε_{ch13}: 0.9400~0.9900 (a step of 0.005, 11 steps) $-0.02 \leq \Delta\varepsilon \leq 0.01$ (a step of 0.003, 11 steps) if (ε_{ch15}) > 1, $\varepsilon_{ch15} = 0.9999$		

Several LST retrieval algorithms from satellites have been proposed to suit the characteristics of various sensors mounted on each satellite and retrieve LSTs through different approximations and assumptions. Among the various LST retrieval methods, the SW method was used to retrieve LST using different atmospheric absorptions of two adjacent infrared window channels in this study. The SW method assumes that the spectral emissivity of the land surface is a priori known, and the atmospheric effects can be corrected using thermal infrared channels. The LST was estimated through a regression analysis on the relationship between the brightness temperature of two infrared channels and the factors affecting the LST calculation [33,34,36,60,77]. The equation for retrieving LST from GK2A data is given in Equation (1) as follows:

$$LST = c_0 + c_1 BT_{ch13} + c_2(BT_{ch13} - BT_{ch15}) + c_3(BT_{ch13} - BT_{ch15})^2 + c_4(sec\theta - 1) + c_5(1 - \overline{\varepsilon}) + c_6\Delta\varepsilon \quad (1)$$

where BT_{ch13} and BT_{ch15} are the brightness temperatures of GK2A channels 13 and 15, respectively, θ is the satellite VZA, $\overline{\varepsilon}$ is the average LSE of GK2A channels 13 and 15, and $\Delta\varepsilon$ is the LSE difference of GK2A channels 13 and 15. The coefficients of the regression equations (c_0~c_6) were derived by simple regression.

As described in Section 2.1.2, MODIS Collection 6 LST data and BSRN Tateno station data were used for the validation of GK2A/AMI LST. The spatial resolution of the MODIS LST was 1 km. The MODIS LST data were provided at 5 min intervals as Terra/Aqua satellites moved along their orbits. Due to the characteristics of LST with large spatial-temporal variability, strict spatial-temporal collocations are needed. For temporal collocation, the MODIS LST data within ±5 min were used based on the observation time of GK2A, as in previous studies [60]. For the spatial collocation, a simple mean of the nearest 9 MODIS pixels (an average of 3 × 3) surrounding the closest MODIS pixel to the GK2A pixel was used. Validation with MODIS LST data was performed when the pixels had more than five pixels of clear sky and land.

In addition, the in situ measurement data, which were observed every minute from the upward longwave radiation at the Tateno station of the BSRN, were used as the validation data. In the GK2A observation area, there were few upward longwave radiation data or field-observed LST data, so Tateno station data were used as validation data. Validation was performed only when the observation times at GK2A and Tateno station were the same. Spatial collocation was performed by averaging the four GK2A pixels closest to the Tateno station. The measured upward longwave radiation was converted to LST using the Stefan–Boltzmann law. The surface type of Tateno station is grass, so we assumed the average value of the emissivity ($\bar{\varepsilon} = 0.986$) value corresponding to the grassland from the Advanced Spaceborne Thermal Emission and Reflection Radiometer (ASTER) spectral library. The process of converting the observed upward longwave radiation to LST using the Stefan–Boltzmann law is shown in Equation (2):

$$L_{sensor} = \bar{\varepsilon}\sigma T^4$$
$$T = \left(\frac{L_{sensor}}{\bar{\varepsilon}\sigma}\right)^{\frac{1}{4}} \tag{2}$$

where L_{sensor} is the value of the measured upward longwave radiation by the CGR4 pyrgeometer from the Tateno station of the BSRN; $\bar{\varepsilon}$ is the average value of the emissivity corresponding to the grassland; σ is the Stefan-Boltzmann constant ($\sigma = 5.670374 \times 10^{-8}$ Wm^{-2}K^{-4}); T is the value of ground temperature converted from the upward longwave radiation at Tateno station.

3. Results

3.1. Results of the Radiative Transfer Model Simulation

To set the threshold value of the LST algorithm, the distribution of brightness temperature differences (BTD) between channel 13 and channel 15 of GK2A was analyzed. This time, only the land pixels without clouds were analyzed. The analysis dates were selected as the same day each month from July 2019 to October 2019, and the count of BTDs accumulated was observed every 10 min. Figure 3 shows the frequency of BTDs between channel 13 and channel 15 of GK2A. BTD values were distributed from −3 K to 15 K. Most of the BTDs were counted at over 10 million from 1 K to 7 K. The results showed a similar distribution to that of the RTM simulated results, which are shown in Figure 4. Figure 4 shows the BTD distribution between channel 13 and channel 15 of GK2A from the simulated pseudo match-up database with the RTM. The coefficients of the LST equation (Equation (1)) were obtained from the simulated pseudo match-up database by regression analysis. Using these coefficients, the root mean square error (RMSE) values between the reference LST and the estimated LST were calculated. Overall, the distribution of simulated BTDs was similar to that of the actual GK2A BTDs, except for the −3 K to 0 K range. The RMSE value was relatively large when the simulated BTD value was less than −1 K or greater than 6 K. The difference between the actual GK2A BTD and the RTM simulation BTD was less than 0 K. This difference could have been caused by the small amount of aerosols given the atmospheric profile (SeeBor_v5), which was used for the RTM simulation [74]. Using these results, the threshold values of BTD for dry/normal/wet conditions were determined to be 0 K and 6 K, respectively.

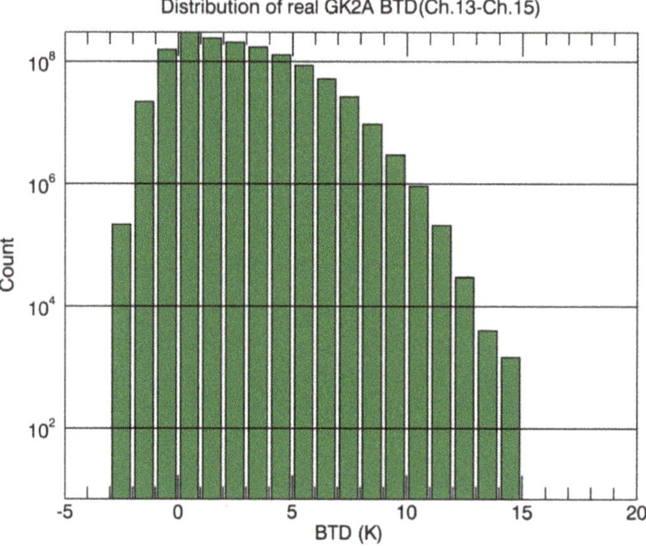

Figure 3. Distribution of brightness temperature difference between channel 13 and channel 15 of GK2A for selected days (25 July 2019; 25 August 2019; 22 September 2019; and 4 October 2019).

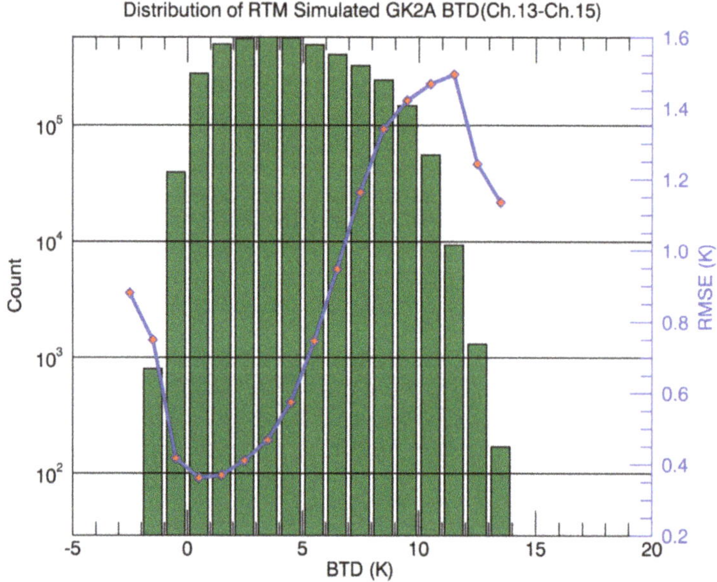

Figure 4. Distribution of radiative transfer model simulated brightness temperature differences between channel 13 and channel 15 of GK2A (green histogram) and their root mean square error (RMSE) values (blue line).

The coefficients of the GK2A/AMI LST equation (Equation (1)) for day/night and dry/normal/wet conditions were obtained from the RTM simulation results through regression analysis, as shown in Table 3.

Table 3. The coefficients of the GK2A/AMI equation (Equation (1)) according to the day/night and dry/normal/wet conditions.

Conditions		c_0	c_1	c_2	c_3	c_4	c_5	c_6
	Dry	−3.7535	1.0146	0.4355	−0.7514	0.5270	46.4021	−76.7542
Day	Normal	−2.5794	1.0094	0.5482	0.1148	1.0890	57.0411	−71.3507
	Wet	44.8058	0.8136	3.3273	−0.0664	2.7271	62.8262	−74.7224
	Dry	2.4418	0.9920	0.7575	−0.3311	0.0106	45.8389	−75.3720
Night	Normal	−4.8096	1.0181	0.2986	0.1573	1.0668	50.1998	−49.2833
	Wet	21.1556	0.8973	3.5049	−0.1219	1.7965	51.9677	−52.6384

To evaluate the RTM simulation results and the GK2A LST retrieval algorithm, we analyzed the relationship between the reference LST value and estimated LST values, as shown in Figure 5.

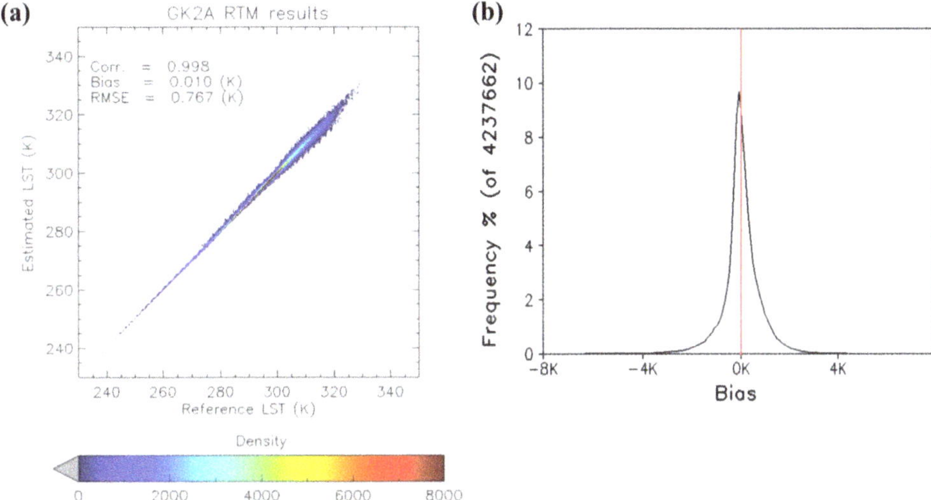

Figure 5. (**a**) Scatter plot (left) and (**b**) bias distribution (right) between the reference LST and estimated LST values using the GK2A/AMI LST retrieval algorithm (gray dotted line in (**a**) represents the 1:1 line).

The estimated LST generally matched well with the reference LST over a wide range from 245 K to 330 K. The comparison of the retrieved LST with the reference LST showed that the LST retrieval algorithm had a high retrieval accuracy in terms of correlation (0.998), bias (0.010 K), and RMSE (0.767 K). The LST retrieval algorithm, most of the reference LSTs, and the estimated LSTs were distributed based on a 1:1 line. In addition, most of the bias was less than ±3 K, with a peak value at 0 K. However, when the LST was greater than 300 K, the LST retrieval algorithm slightly over/underestimated the LSTs compared to the reference LST.

Figure 6 shows the distribution of RMSEs between the estimated LSTs and reference LSTs based on the impact factors (lapse rate, BTD, surface emissivity differences, and satellite VZAs that affect the retrieval accuracy of the LST). Most of the RMSEs are less than 2.1 K, excluding some ranges where the VZA is larger than 40° and the BTD value is larger than 7 K. The reason that the RMSE was large when the VZA was large appears to be due to the fact that the effect of the emissivity becomes significant when the VZA is large, as suggested in many studies [78,79]. One of the most distinctive features of all the cases is that the RMSE was slightly larger when the BTD values were larger than 6 K. The magnitude of BTDs is mainly caused by the different sensitivities of the two window channels to aerosol (channel 13) and water vapor (channel 15). In addition, the GK2A/AMI LST retrieval algorithm is significantly influenced by the VZA and the surface lapse rate among several impacting factors. Compared to previous studies that retrieved LST from Himawari-8, RMSE decreased in all ranges [60]. In particular, when BTD was a positive large value in the previous study, most of the RMSEs were significantly reduced. These results seem to be related to the consideration of the diurnal variation of the boundary layer temperature in the RTM simulations and the consideration of the effect of nonlinearity of water vapor in the LST calculation formula.

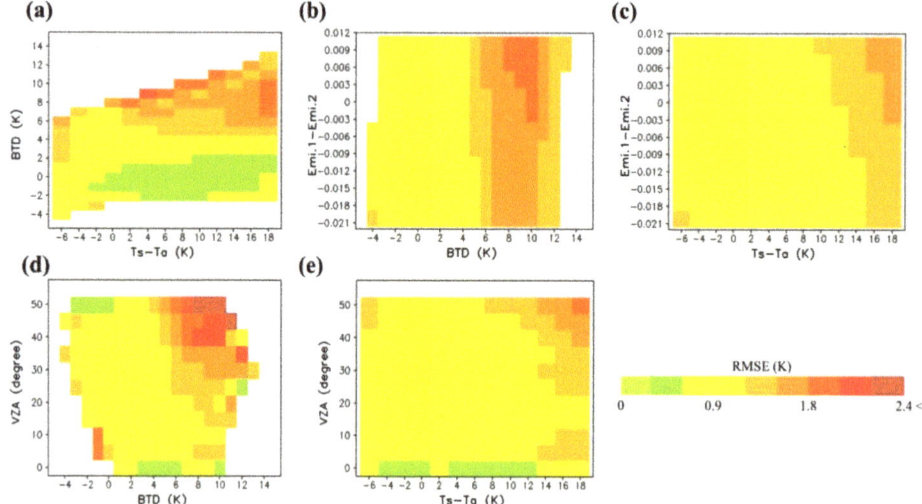

Figure 6. Distribution of RMSEs for the LST retrieval algorithms based on the different impacting factors: (**a**) brightness temperature difference (BTD) and surface lapse rate; (**b**) emissivity difference (Δε) and BTD; (**c**) Δε and surface lapse rate; (**d**) satellite viewing zenith angle (VZA) and BTD; (**e**) VZA and surface lapse rate.

3.2. Comparison of Retrieved GK2A LST and MODIS LST Products

To evaluate the GK2A/AMI LST retrieval algorithm, GK2A LSTs were retrieved for 4 months, from July 2019 to October 2019, when GK2A started operational observation. Unlike the RTM simulations, when retrieving LSTs from GK2A/AMI data, the criterion for dividing day and night was the SZA of each pixel, which is the same as that in previous studies [36,60]. Only clear sky and land pixels that satisfied the strict spatial-temporal matching conditions were compared and validated, as mentioned in Section 2.1.2. Figures 7–10 show the spatial distribution of the LSTs retrieved from GK2A/AMI data and the MODIS LSTs, along with the spatial distribution of their differences between two datasets for the selected days.

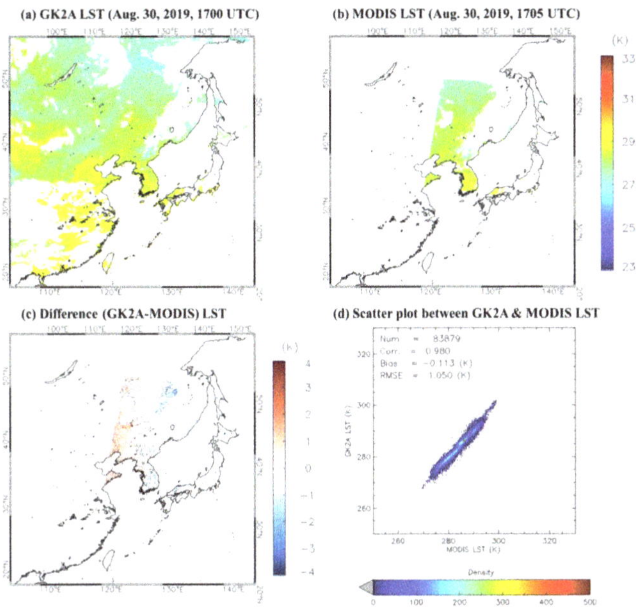

Figure 7. Spatial distribution of (**a**) GK2A LST, (**b**) MODIS (MYD11_L2) LST, (**c**) differences between GK2A and MODIS LSTs, and (**d**) scatter plot between GK2A and MODIS (MYD11_L2) LSTs for 30 August 2019, 1700 and 1705 UTC (gray dotted line in (**d**) represents the 1:1 line).

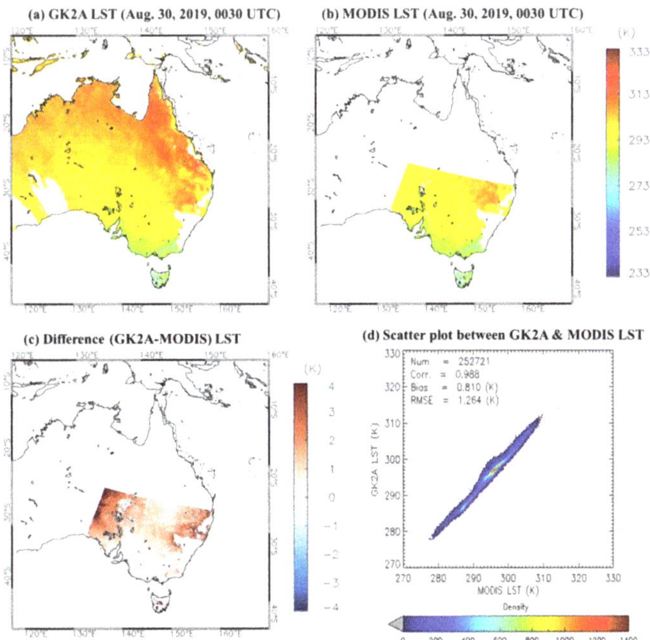

Figure 8. Spatial distribution of (**a**) GK2A LST, (**b**) MODIS (MOD11_L2) LST, (**c**) differences between the GK2A and MODIS LSTs, and (**d**) scatter plot between the GK2A and MODIS (MOD11_L2) LSTs for 30 August 2019, 0030 UTC (gray dotted line in (**d**) represents the 1:1 line).

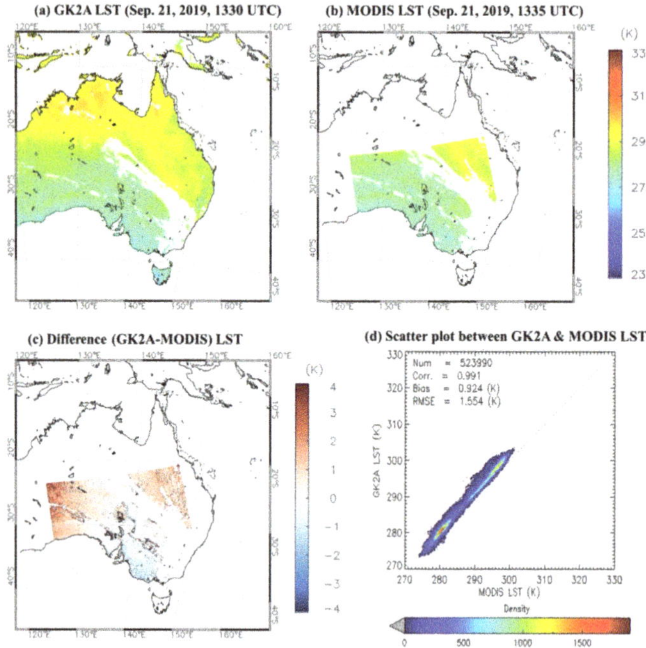

Figure 9. Spatial distribution of (**a**) GK2A LST, (**b**) MODIS (MOD11_L2) LST, (**c**) differences between GK2A and MODIS LSTs, and (**d**) scatter plot between GK2A and MODIS (MOD11_L2) LSTs for 21 September 2019, 1330 and 1335 UTC (gray dotted line in (**d**) represents the 1:1 line).

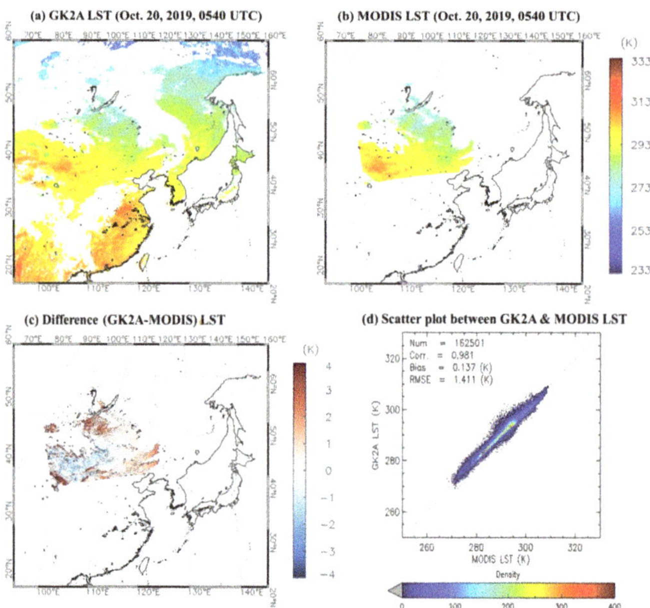

Figure 10. Spatial distribution of (**a**) GK2A LST, (**b**) MODIS (MYD11_L2) LST, (**c**) differences between the GK2A and MODIS LSTs, and (**d**) scatter plot between the GK2A and MODIS (MYD11_L2) LSTs for 20 October 2019, 0540 UTC (gray dotted line in (**d**) represents the 1:1 line).

Figure 7 shows the spatial distribution of the GK2A LST and MODIS LST and their differences at 1700 and 1705 UTC on 30 August 2019. The two LSTs show similar spatial distributions in large areas of the Korean Peninsula and northeast China. The difference between the two LSTs was within ±2 K. In the scatter plot of the two LSTs, the distribution spread slightly to both sides around the 1:1 line from 270 K to 305 K. The bias and RMSE of this scene are −0.113 K and 1.05 K, respectively.

Figure 8 shows the spatial distribution of the two LSTs at 0030 UTC on 30 August 2019. The area in which the LST was retrieved is a region between central and southern Australia consisting of desert, bare soil, and grassland. The difference between the two LSTs in the spatial distribution showed, where the GK2A LST was greater than the MODIS LST. In particular, the warm bias of the GK2A LST was more significant in the desert regions of south Australia than in other regions. For this region, even in the scatter plot, the GK2A LST was higher than the MODIS LST over a wide range (275–310 K). In this case, the bias and RMSE are 0.81 K and 1.264 K, respectively.

The spatial distribution of the two LSTs and their differences, along with the scatter plot on 21 September 2019, are shown in Figure 9. The observation time of the two LSTs differs by 5 min. The spatial distribution of the two LSTs is generally similar, but the spatial distribution of the difference between the two LSTs showed a warm bias in the desert region of Australia's inland but a cold bias in south Australia. As shown in the scatter plot, the overall GK2A LST was higher than the MODIS LST in all ranges (270–305 K). As a result, the GK2A LST shows a warm bias of 0.924 K and an RMSE of 1.554 K.

The spatial distribution of the GK2A LST, the MODIS LST, and their differences at 0540 UTC on 20 October 2019, are shown in Figure 10. The two LSTs have similar spatial distributions in large areas of Mongolia and China. The differences between the two LSTs were within ±3 K. In the scatter plot of the two LSTs, the distribution showed a 1:1 line from 270 K to 295 K, but some pixels were spread around the 1:1 line. The bias and RMSE of this scene are 0.137 K and 1.411 K, respectively.

The indirect validation results of the GK2A LST with the MODIS LST for 4 months (July, August, September, and October 2019) are shown in Table 4. The correlation coefficient between GK2A LST and MODIS LST was greater than 0.96 regardless of the satellite (Terra/Aqua) and months. In addition, the bias showed a warm bias of less than 1.6 K every month. The combined results of the two validation satellites show a high correlation coefficient (0.969), slightly warm bias (1.227 K), and RMSE (2.281 K).

Because the diurnal variation pattern of LST and the retrieval algorithm are very different according to day and night, the performance of the GK2A LST algorithm was evaluated accordingly. Table 5 shows the comparison results between the GK2A and MODIS LSTs according to the daytime and nighttime. The performance of the GK2A LST algorithm is dependent on the validation time, day, and night. In general, the skill of the GK2A LST retrieval algorithm for daytime was about two times worse than that of nighttime, in terms of bias and RMSE, irrespective of the months. The relatively strong warm biases of the GK2A LST during daytime can be attributed to the current status of the MODIS LST [80,81]. The biggest feature is that the GK2A LST overestimated more than the MODIS LST during the daytime, showing a large warm bias (above 1.8 K) and RMSE (above 2.8 K). Moreover, for nighttime, the RMSE was small due to a relatively high correlation and a small bias compared to the daytime.

Table 4. Comparison results of the GK2A LST data and the MODIS LST product (Collection 6) for the selected four months (July, August, September, and October 2019).

Month	MOD11_L2				MYD11_L2			
	# of Pixel (×1000)	Corr.	Bias (K)	RMSE (K)	# of Pixel (×1000)	Corr.	Bias (K)	RMSE (K)
2019.07	22,412	0.961	1.051	2.139	20,485	0.965	1.182	2.236
2019.08	7022	0.960	1.049	2.045	7279	0.974	1.117	2.129
2019.09	21,262	0.958	1.406	2.484	18,120	0.977	1.163	2.257
2019.10	10,577	0.980	1.598	2.601	9397	0.988	1.260	2.237
Total and Ave.	61,275	0.963	1.269	2.328	55,282	0.974	1.180	2.229

# of pixel: 116,557,577, Corr. = 0.969, bias = 1.227 K, RMSE = 2.281 K.

Table 5. Comparison results of the GK2A LST data and the MODIS LST data (Collection 6) for the selected four months, according to the daytime and nighttime (July, August, September, and October 2019).

Month	Daytime				Nighttime			
	# of Pixel (×1000)	Corr.	Bias (K)	RMSE (K)	# of Pixel (×1000)	Corr.	Bias (K)	RMSE (K)
2019.07	18,905	0.948	1.890	2.981	23,992	0.976	0.502	1.559
2019.08	6,969	0.953	1.889	2.839	7,332	0.981	0.318	1.375
2019.09	18,076	0.946	2.340	3.486	21,306	0.985	0.407	1.441
2019.10	9,595	0.978	2.269	3.419	10,378	0.990	0.671	1.515
Total and Ave.	53,547	0.953	2.110	3.211	63,010	0.982	0.476	1.490

# of pixel: 116,557,577, Corr. = 0.969, bias = 1.227 K, RMSE = 2.281 K.

3.3. Validation Using In-Situ Observation Data

To evaluate the the GK2A LST algorithm, we conducted a quantitative validation of the GK2A LST using in situ observation data. As mentioned in Section 2.1.2, the upward longwave radiation data observed at the Tateno station of the BSRN were used as validation data. Approximately 718 sets from July 2019 to October 2019 satisfying the spatial-temporal matching conditions with the GK2A LST were used for the validation of the GK2A LST. In addition, the performance of the GK2A LST algorithm was evaluated according to daytime and nighttime. The validation results of the GK2A LST with the data from Tateno station are shown in Figure 11. The GK2A LST is similar to or slightly warmer than the ground observed LST at Tateno station, regardless of the temperature and time (day and night). The total correlation coefficient, bias, and RMSE between the GK2A LST and the Tateno LST were 0.95, 0.523 K, and 2.021 K, respectively. However, the GK2A LST was warmer than the ground observed LST; in particular, the LST was greater than 305 K during the daytime. As shown in Figure 11, the GK2A LST shows a greater warm bias (0.84 K) and RMSE (2.13 K) during the day than at night (bias: 0.32 K, RMSE: 1.948 K).

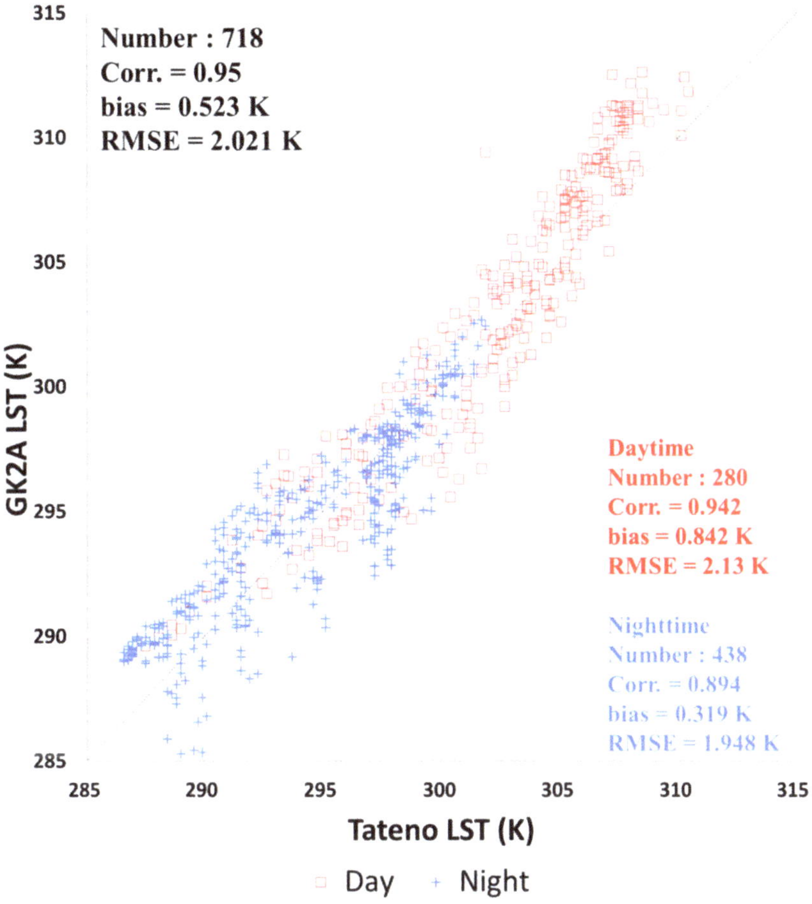

Figure 11. Scatter plot between the GK2A LST and the LST at Tateno station from upward longwave radiation (red square symbol: daytime, blue cross symbol: nighttime, gray dotted line represents the 1:1 line).

4. Discussion

In this study, we developed a GK2A SW LST retrieval algorithm using two adjacent infrared channels in the atmospheric window. GK2A/AMI has three infrared channels (channels 13, 14, and 15) corresponding to the atmospheric window [58]. Channel 15 is more sensitive to water vapor than the other two channels. By comparison, channels 13 and 14 are relatively less sensitive to water vapor, but the sensitivities of the two channels to aerosol and water vapor are slightly different. To select two of the three channels that can be used to retrieve the GK2A LST, RTM simulation results under the same RTM input conditions using channels 13 and 15, as well as channels 14 and 15, were analyzed. The regression coefficients (c_0~c_6) for the LST retrieval Equation (1) using channels 13 and 15 and channels 14 and 15, respectively, were derived from the simulated dataset. The scatter plot results of the estimated LSTs from each dataset are shown in Figure 12.

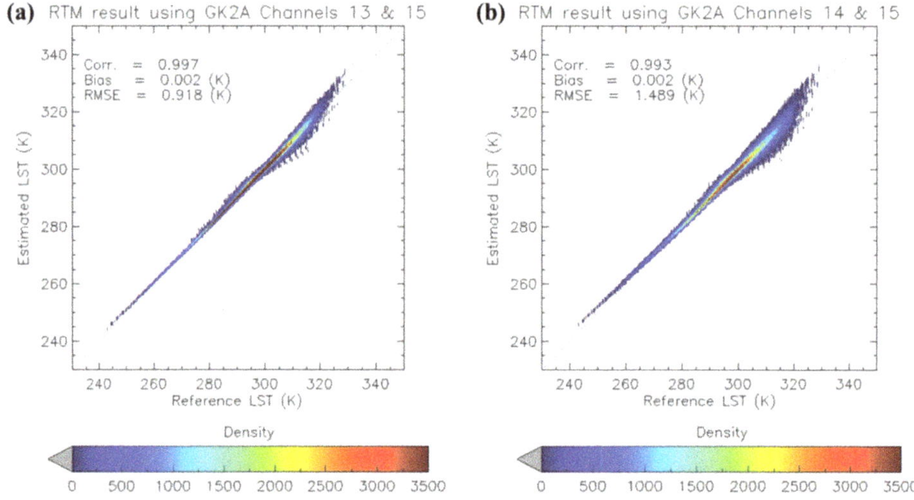

Figure 12. Scatter plots between reference LST and estimated LST from RTM simulation using two channels: (**a**) channels 13 and 15, and (**b**) channels 14 and 15.

The two sets of GK2A LST algorithms estimated LSTs in the range of 240 K to 330 K, but the correlation coefficient and RMSE using channels 13 and 15 showed better results than those using channels 14 and 15. In the scatter plots, the distribution used channels 14 and 15 showed a wider spread than that by channels 13 and 15 at over 300 K. This result is similar to that of a previous study using Himawari-8 [60]. Therefore, channels 13 and 15 were selected for this study.

In addition, several forms of linear and non-linear equations can be selected in the LST retrieval formula of the SW method [26,27]. When retrieving LST, the linear equation of SW LST algorithms showed a large error in wet and hot atmospheric conditions, therefore, non-linear equations of SW LST algorithms have been developed [26,47,77,82,83]. In this study, linear and non-linear equations of the SW LST algorithms were developed and applied to real GK2A data to compare the accuracy of the algorithm. The simulation conditions for the RTM are shown in Table 2, and the coefficient of c_3 in the LST retrieval equation (Equation (1)) is set to zero to represent a linear algorithm. In addition, the coefficients of LST retrieval algorithms were derived by dividing into day/night and dry/normal/wet conditions using the same thresholds of SZA and BTD, as described in Section 3. Table 6 shows the results of quantitatively comparing the GK2A LST and MODIS LST for one month from the linear algorithm and non-linear SW LST algorithm. As a result of verifying the GK2A LST calculated with linear and non-linear algorithms for the September 2019 case with the MODIS LST, the correlation coefficient between the GK2A LST and MODIS LST was very similar in linear and nonlinear algorithms, but the bias and RMSE showed better results in nonlinear algorithms.

Table 6. Comparison results of the GK2A LST data and the MODIS LST product (Collection 6) using a linear algorithm and non-linear algorithm according to the daytime and nighttime in September 2019.

Algorithm	Daytime				Nighttime			
	# of Pixel (×1000)	Corr.	bias (K)	RMSE (K)	# of Pixel (×1000)	Corr.	bias (K)	RMSE (K)
Linear	18,076	0.944	2.504	3.691	21,306	0.985	0.413	1.670
Non-linear	18,076	0.946	2.340	3.486	21,306	0.985	0.407	1.441

Even though the non-linear SW LST algorithm was used, the GK2A LST algorithm showed significant errors during the daytime compared to the MODIS LST. The MODIS Collection 6 LST product, used as validation data, is an improved version of the MODIS Collection 5 LST product by Wan (2014) [32]. One of the improvements in the MODIS Collection 6 LST over the Collection 5 LST is that the MODTRAN simulation is performed by dividing into day and night for the bare soil area and adjusting the emissivity difference in MODIS bands 31 and 32 over bare soil surfaces [32]. In addition, a term including the quadratic difference between the brightness temperatures of bands 31 and 32 was added to the MODIS' generalized SW algorithm. Even though the MODIS Collection 6 LST product had many improvements compared to Collection 5, a cold bias still appeared from -1.4 to -3.7 K during the daytime when compared with in situ measurements [81]. According to the validation study, MODIS Collection 6 LST showed the RMSE of daytime LSTs as 2.59 K, 2.86 K, and 3.05 K for the Gobi area, desert steppe region, and sand desert area, respectively [80]. Considering that the cold bias of the MODIS Collection 6 LST is strong during daytime over bare soil and desert regions, the warm bias of the GK2A LST algorithm can be regarded as normal rather than a serious problem. However, a detailed analysis of bare soil and desert areas using more validation data is needed. In the four-month verification results, the errors in September and October were systematically larger than those in July and August. So, we tried to find the cause but, unfortunately, we could not. In addition, a relatively strong warm bias appeared during the day at Tateno station. This seems to be related to the fact that the land cover at the Tateno point is grass, but most of the area around this point is urban.

To compare the accuracy of the MODIS LST, it was validated using the LST at Tateno station for the same period as the GK2A LST. The validation results of the MODIS (MOD11/MYD11_L2) LST with those from Tateno station are shown in Figure 13. The MODIS LST was slightly colder than Tateno LST, regardless of the daytime and nighttime. The total correlation coefficient, bias, and RMSE between MODIS LST and Tateno LST were 0.925, -1.047 K, and 2.985 K, respectively. Although the number of samples was small, the MODIS LST showed a cold bias compared to the in situ LST in both daytime and nighttime. In particular, the daytime bias (-1.402 K) of the MODIS LST was nearly twice that of the nighttime (-0.767 K). In the comparative validation results of the GK2A LST and the MODIS LST, the reason why the warm bias of GK2A is large during the daytime is also considered to be related to the cold bias of MODIS LST during the daytime.

The number of on-site observation points in the GK2A observation area is not only limited but also the number of data accessible over the internet is small, so we used observation data from the Tateno station. The in situ measured radiation represents a narrow area, whereas the retrieved LST from the satellite is the average temperature corresponding to satellite resolution (2 km × 2 km), so there is a limitation in the spatial representativeness of in situ observation for the satellite. Since the period for retrieving and validating the GK2A LST is as short as four months, there is a limit to evaluating the integrated level of the LST retrieval algorithm.

When retrieving LST from a satellite, the split-window method assumes that the LSE of both channels is known. In this study, the GK2A LSE data derived in real-time using the modified VCM method were used [68]. The fractional vegetation cover of a given pixel was calculated using the GK2A vegetation index (VI) data generated by the maximum value composite with a consecutive 8-day VI [63]. Based on the calculated fractional vegetation cover, the LSEs were retrieved using the look-up table according to the land cover and daily snow cover of each pixel. The VI, land cover classification database, spectral emissivity look-up table, and daily snow cover were calculated from each algorithm, so they contain errors which also affect the accuracy of the retrieved LST. Therefore, to improve the accuracy of the GK2A LST, it is necessary to improve the accuracy of these algorithms.

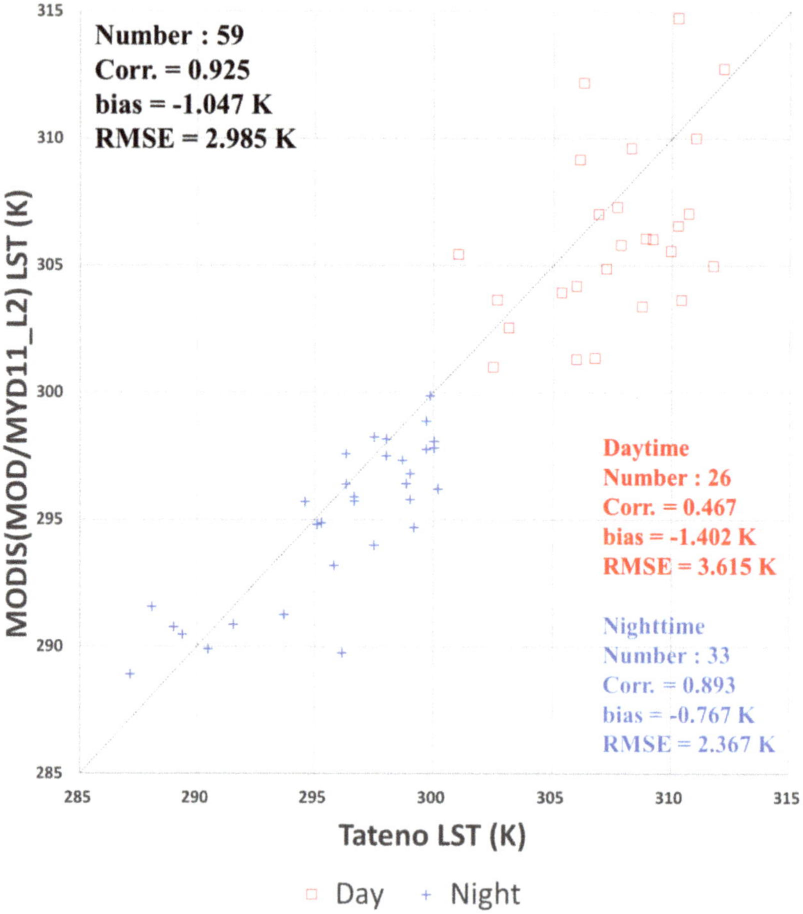

Figure 13. Scatter plot between the MODIS (MOD11_L2/MYD11_L2) LST and the LST at Tateno station from upward longwave radiation (red square symbol: daytime; blue cross symbol: nighttime; gray dotted line represents the 1:1 line).

5. Conclusions

We have developed an operational LST retrieval algorithm for the GK2A viewing area using GK2A/AMI data from Korea's next-generation geostationary satellite. To develop the GK2A LST algorithm, the split-window method was used, and the nonlinearity of water vapor was considered in the algorithm. The RTM simulation data were constructed taking into account various factors affecting the LST calculation in MODTRAN 4 (atmospheric profiles, diurnal variation of LST and air temperature of the boundary layer, and the LSE variations of the two channels). From the RTM simulation data set, regression coefficients were derived according to the actual water vapor (GK2A BTD: dry/normal/wet) during the day and night. The GK2A LST was retrieved using the developed GK2A LST algorithm, and their accuracies were evaluated using MODIS LST and field observations as validation datasets.

As a result of evaluating the output level of the LST calculated by the RTM simulation using the prescribed LST, there was a correlation coefficient of 0.998, a bias of 0.01 K, and an RMSE of 0.767 K. When the BTD value is larger than 6 K and the satellite's VZA is large, the RMSE is large, but the error is relatively smaller than the result of using the linear algorithm of the previous study [61].

Using the GK2A LST algorithm developed in this study, LST was calculated from GK2A data for four months from July 2019 to October 2019. As a result of comparing the GK2A LST with the MODIS LST, the spatial correlation coefficient of the two LSTs was 0.969, the bias was 1.227 K, and the RMSE was 2.281 K. Compared to MODIS LST, GK2A LST shows a warm bias greater than 1.8 K during the day, but a relatively small bias of less than 0. 7 K at night. In particular, the warm bias of the GK2A LST was higher than that of MODIS LST in desert and barren areas during daytime. MODIS LST Collection 6, used as validation data, seems to be influenced by characteristics of cold bias in the desert and on bare soil [80,81]. The results of validation with data from the Tateno station of the BSRN, which were the field observation data, showed that the correlation coefficient is 0.95, the bias is 0.523 K, and the RMSE is 2.021 K. Compared to the Tateno LST, the day bias is +0.5 K greater than the night bias. The reason that the GK2A LST tends to be warmer than the Tateno LST during the day is that the temperature at Tateno station is the temperature of the grass, while the GK2A LST is the temperature of the surrounding urban area.

The GK2A LST algorithm developed in this study uses various outputs of GK2A (cloud detection, VI, snow cover, and LSE) and ancillary data to calculate LST. Therefore, it is possible to improve the accuracy of the GK2A LST by improving the algorithm that produces the basic input data for LST calculation. In addition, it is necessary to evaluate the output level of the GK2A LST algorithm using verification data for a longer time because the surface temperature and atmospheric characteristics differ depending on the season. As other satellite data (Visible Infrared Imaging Radiometer Suite LST, Sentinel-3 Sea and Land Surface Temperature Radiometer LST) and additional field observation data (HiWATER) are starting to be released, it is considered that a cross-comparison study with these data is necessary for globally continuous LST calculation. In addition, if the accuracy of the GK2A LST is improved, it will be able to contribute to the establishment of long-term climatological LST data retrieved from satellites, such as the CCI (Climate Change Initiative) LST project underway by the ESA (European Space Agency) [84].

Author Contributions: Conceptualization, M.-S.S.; Methodology, M.-S.S.; Software, Y.-Y.C.; Validation, Y.-Y.C.; Formal analysis, Y.-Y.C. and M.-S.S.; Investigation, Y.-Y.C.; Data curation, Y.-Y.C.; Writing—original draft preparation, Y.-Y.C.; Writing—review and editing, M.-S.S. and Y.-Y.C.; Visualization, Y.-Y.C.; Supervision, M.-S.S.; Project administration, M.-S.S.; Funding acquisition, M.-S.S.; All authors contributed extensively to the work presented in this paper. All authors have read and agreed to the published version of the manuscript.

Funding: This work was supported by "Development of Scene Analysis and Surface Algorithms" project, funded by ETRI, which is a subproject of the "Development of Geostationary Meteorological Satellite Ground Segment (NMSC-2019-01)" program funded by the NMSC (National Meteorological Satellite Center) of the KMA (Korea Meteorological Administration). This work was also funded by the Korea Meteorological Administration Research and Development Program under Grant KMI2018-06510.

Acknowledgments: We thank the Korean Meteorological Administration's National Meteorological Satellite Center for providing the GK2A/AMI dataset and cloud mask.

Conflicts of Interest: The authors declare no conflict of interest.

References

1. Dash, P.; Göttsche, F.; Olesen, F.-S.; Fischer, H. Land surface temperature and emissivity estimation from passive sensor data: Theory and practice-current trends. *Int. J. Remote Sens.* **2002**, *23*, 2563–2594. [CrossRef]
2. Becker, F.; Li, Z.-L. Surface temperature and emissivity at various scales: Definition, measurement and related problems. *Remote Sens. Rev.* **1995**, *12*, 225–253. [CrossRef]
3. Song, Z.; Li, R.; Qiu, R.; Liu, S.; Tan, C.; Li, Q.; Ge, W.; Han, X.; Tang, X.; Shi, W.-Y.; et al. Global Land Surface Temperature Influenced by Vegetation Cover and PM2.5 from 2001 to 2016. *Remote Sens.* **2018**, *10*, 2034. [CrossRef]
4. Wang, K.; Wan, Z.; Sparrow, M.; Liu, J.; Zhou, X.; Haginoya, S. Estimation of surface long wave radiation and broadband emissivity using Moderate Resolution Imaging Spectroradiometer (MODIS) land surface temperature/emissivity products. *J. Geophys. Res. Space Phys.* **2005**, *110*, D11. [CrossRef]

5. Chen, Y.; Yang, K.; He, J.; Qin, J.; Shi, J.; Du, J.; He, Q. Improving land surface temperature modeling for dry land of China. *J. Geophys. Res. Space Phys.* **2011**, *116*, D20. [CrossRef]

6. Dousset, B.; Gourmelon, F.; Laaidi, K.; Zeghnoun, A.; Giraudet, E.; Bretin, P.; Mauri, E.; Vandentorren, S. Satellite monitoring of summer heat waves in the Paris metropolitan area. *Int. J. Clim.* **2010**, *31*, 313–323. [CrossRef]

7. Julien, Y.; Sobrino, J.A.; Verhoef, W. Changes in land surface temperatures and NDVI values over Europe between 1982 and 1999. *Remote Sens. Environ.* **2006**, *103*, 43–55. [CrossRef]

8. Ibrahim, G.R.F. Urban Land Use Land Cover Changes and Their Effect on Land Surface Temperature: Case Study Using Dohuk City in the Kurdistan Region of Iraq. *Climate* **2017**, *5*, 13. [CrossRef]

9. Hereher, M. Effect of land use/cover change on land surface temperatures—The Nile Delta, Egypt. *J. Afr. Earth Sci.* **2017**, *126*, 75–83. [CrossRef]

10. Zhao, L.; Yang, Z.-L.; Hoar, T.J. Global Soil Moisture Estimation by Assimilating AMSR-E Brightness Temperatures in a Coupled CLM4–RTM–DART System. *J. Hydrometeorol.* **2016**, *17*, 2431–2454. [CrossRef]

11. Marcos, E.; Fernández-García, V.; Fernández-Manso, A.; Quintano, C.; Valbuena, L.; Tárrega, R.; Luis-Calabuig, E.; Calvo, L. Evaluation of Composite Burn Index and Land Surface Temperature for Assessing Soil Burn Severity in Mediterranean Fire-Prone Pine Ecosystems. *Forest* **2018**, *9*, 494. [CrossRef]

12. Maffei, C.; Alfieri, S.; Menenti, M. Relating Spatiotemporal Patterns of Forest Fires Burned Area and Duration to Diurnal Land Surface Temperature Anomalies. *Remote Sens.* **2018**, *10*, 1777. [CrossRef]

13. Elvira, B.G.; Taylor, C.M.; Harris, P.P.; Ghent, D.; Veal, K.L.; Folwell, S.S. Global observational diagnosis of soil moisture control on the land surface energy balance. *Geophys. Res. Lett.* **2016**, *43*, 2623–2631. [CrossRef]

14. Hain, C.; Anderson, M.C. Estimating morning change in land surface temperature from MODIS day/night observations: Applications for surface energy balance modeling. *Geophys. Res. Lett.* **2017**, *44*, 9723–9733. [CrossRef]

15. Jung, H.-S.; Lee, K.-T.; Zo, I.-S. Calculation Algorithm of Upward Longwave Radiation Based on Surface Types. *Asia Pac. J. Atmos. Sci.* **2020**, *56*, 291–306. [CrossRef]

16. Peng, J.; Ma, J.; Liu, Q.; Yanxu, L.; Hu, Y.; Li, Y.; Yue, Y. Spatial-temporal change of land surface temperature across 285 cities in China: An urban-rural contrast perspective. *Sci. Total Environ.* **2018**, *635*, 487–497. [CrossRef] [PubMed]

17. Anastasios, P.; Theleia, M.; Constantinos, C. Quantifying the trends in land surface temperature and surface urban heat island intensity in mediterranean cities in view of smart urbanization. *Urban Sci.* **2018**, *2*, 16. [CrossRef]

18. Zhou, D.; Xiao, J.; Bonafoni, S.; Thau, C.; Deilami, K.; Zhou, Y.; Frolking, S.; Yao, R.; Qiao, Z.; Sobrino, J.A. Satellite Remote Sensing of Surface Urban Heat Islands: Progress, Challenges, and Perspectives. *Remote Sens.* **2018**, *11*, 48. [CrossRef]

19. Song, L.; Liu, S.; Kustas, W.P.; Nieto, H.; Sun, L.; Xu, Z.; Skaggs, T.H.; Yang, Y.; Ma, M.; Xu, T.; et al. Monitoring and validating spatially and temporally continuous daily evaporation and transpiration at river basin scale. *Remote Sens. Environ.* **2018**, *219*, 72–88. [CrossRef]

20. Tomlinson, C.J.; Chapman, L.; Thornes, J.E.; Baker, C. Remote sensing land surface temperature for meteorology and climatology: A review. *Meteorol. Appl.* **2011**, *18*, 296–306. [CrossRef]

21. Zhan, W.; Chen, Y.; Zhou, J.; Wang, J.; Liu, W.; Voogt, J.; Zhu, X.; Quan, J.; Li, J. Disaggregation of remotely sensed land surface temperature: Literature survey, taxonomy, issues, and caveats. *Remote Sens. Environ.* **2013**, *131*, 119–139. [CrossRef]

22. Weng, Q.; Fu, P.; Gao, F. Generating daily land surface temperature at Landsat resolution by fusing Landsat and MODIS data. *Remote Sens. Environ.* **2014**, *145*, 55–67. [CrossRef]

23. Li, Z.-L.; Tang, B.-H.; Wu, H.; Ren, H.; Yan, G.; Wan, Z.; Trigo, I.F.; Sobrino, J.A. Satellite-derived land surface temperature: Current status and perspectives. *Remote Sens. Environ.* **2013**, *131*, 14–37. [CrossRef]

24. Sattari, F.; Hashim, M. A brief review of land surface temperature retrieval methods from thermal satellite sensors. *Middle East J. Sci. Res.* **2014**, *22*, 757–768.

25. Becker, F.; Li, Z.-L. Temperature-independent spectral indices in thermal infrared bands. *Remote Sens. Environ.* **1990**, *32*, 17–33. [CrossRef]

26. Sobrino, J.A.; Li, Z.-L.; Stoll, M.; Becker, F. Improvements in the split-window technique for land surface temperature determination. *IEEE Trans. Geosci. Remote Sens.* **1994**, *32*, 243–253. [CrossRef]

27. Wan, Z.; Dozier, J. A Generalized Split-Window Algorithm for Retrieving Land-Surface Temperature Measurement from Space. *IEEE Trans. Geosci Remote Sens.* **1996**, *34*, 892–905.
28. Gillespie, A.; Rokugawa, S.; Matsunaga, T.; Cothern, J.; Hook, S.; Kahle, A. A temperature and emissivity separation algorithm for Advanced Spaceborne Thermal Emission and Reflection Radiometer (ASTER) images. *IEEE Trans. Geosci. Remote Sens.* **1998**, *36*, 1113–1126. [CrossRef]
29. Yu, Y.; Privette, J.L.; Pinheiro, A. Analysis of the NPOESS VIIRS land surface temperature algorithm using MODIS data. *IEEE Trans. Geosci. Remote Sens.* **2005**, *43*, 2340–2350. [CrossRef]
30. Justice, C.O.; Román, M.O.; Csiszar, I.; Vermote, E.F.; Wolfe, R.E.; Hook, S.J.; Friedl, M.; Wang, Z.; Schaaf, C.; Miura, T.; et al. Land and cryosphere products from Suomi NPP VIIRS: Overview and status. *J. Geophys. Res. Atmos.* **2013**, *118*, 9753–9765. [CrossRef]
31. Yu, X.; Guo, X.; Wu, Z. Land Surface Temperature Retrieval from Landsat 8 TIRS—Comparison between Radiative Transfer Equation-Based Method, Split Window Algorithm and Single Channel Method. *Remote Sens.* **2014**, *6*, 9829–9852. [CrossRef]
32. Wan, Z. New refinements and validation of the collection-6 MODIS land-surface temperature/emissivity product. *Remote Sens. Environ.* **2014**, *140*, 36–45. [CrossRef]
33. Sun, D. Estimation of land surface temperature from a Geostationary Operational Environmental Satellite (GOES-8). *J. Geophys. Res. Space Phys.* **2003**, *108*, 4326. [CrossRef]
34. Sobrino, J.A.; Romaguera, M. Land surface temperature retrieval from MSG1-SEVIRI data. *Remote Sens. Environ.* **2004**, *92*, 247–254. [CrossRef]
35. Hong, K.-O.; Suh, M.-S.; Kang, J.-H. Development of a land surface temperature-retrieval algorithm from MTSAT-1R data. *Asia Pac. J. Atmos. Sci.* **2009**, *45*, 411–421.
36. Cho, A.-R.; Suh, M.-S. Evaluation of Land Surface Temperature Operationally Retrieved from Korean Geostationary Satellite (COMS) Data. *Remote Sens.* **2013**, *5*, 3951–3970. [CrossRef]
37. Göttsche, F.; Olesen, F.-S.; Trigo, I.F.; Bork-Unkelbach, A.; Martin, M. Long Term Validation of Land Surface Temperature Retrieved from MSG/SEVIRI with Continuous in-Situ Measurements in Africa. *Remote Sens.* **2016**, *8*, 410. [CrossRef]
38. Li, Z.-L.; Becker, F. Feasibility of land surface temperature and emissivity determination from AVHRR data. *Remote Sens. Environ.* **1993**, *43*, 67–85. [CrossRef]
39. Ottlé, C.; Stoll, M. Effect of atmospheric absorption and surface emissivity on the determination of land surface temperature from infrared satellite data. *Int. J. Remote Sens.* **1993**, *14*, 2025–2037. [CrossRef]
40. Prata, A.J.; Caselles, V.; Coll, C.; Sobrino, J.A.; Ottlé, C. Thermal remote sensing of land surface temperature from satellites: Current status and future prospects. *Remote Sens. Rev.* **1995**, *12*, 175–224. [CrossRef]
41. Jiménez-Muñoz, J.C.; Sobrino, J.A. A generalized single-channel method for retrieving land surface temperature from remote sensing data. *J. Geophys. Res. Space Phys.* **2003**, *108*, D22. [CrossRef]
42. Prata, A.J. Land surface temperatures derived from the advanced very high resolution radiometer and the along-track scanning radiometer: 1. Theory. *J. Geophys. Res. Space Phys.* **1993**, *98*, 16689–16702. [CrossRef]
43. Sobrino, J.A.; Li, Z.-L.; Stoll, M.P.; Becker, F. Multi-channel and multi-angle algorithms for estimating sea and land surface temperature with ATSR data. *Int. J. Remote Sens.* **1996**, *17*, 2089–2114. [CrossRef]
44. Snyder, W.C.; Wan, Z.; Zhang, Y.; Feng, Y.-Z. Classification-based emissivity for land surface temperature measurement from space. *Int. J. Remote Sens.* **1998**, *19*, 2753–2774. [CrossRef]
45. Peres, L.F.; Dacamara, C.C. Emissivity maps to retrieve land-surface temperature from MSG/SEVIRI. *IEEE Trans. Geosci. Remote Sens.* **2005**, *43*, 1834–1844. [CrossRef]
46. Valor, E. Mapping land surface emissivity from NDVI: Application to European, African, and South American areas. *Remote Sens. Environ.* **1996**, *57*, 167–184. [CrossRef]
47. Sobrino, J.A.; Raissouni, N. Toward remote sensing methods for land cover dynamic monitoring: Application to Morocco. *Int. J. Remote Sens.* **2000**, *21*, 353–366. [CrossRef]
48. Jiang, G.-M.; Li, Z.-L.; Nerry, F. Land surface emissivity retrieval from combined mid-infrared and thermal infrared data of MSG-SEVIRI. *Remote Sens. Environ.* **2006**, *105*, 326–340. [CrossRef]
49. Watson, K. Two-temperature method for measuring emissivity. *Remote Sens. Environ.* **1992**, *42*, 117–121. [CrossRef]
50. Peres, L.F.; Dacamara, C.C.; Trigo, I.F.; Freitas, S.C. Synergistic use of the two-temperature and split-window methods for land-surface temperature retrieval. *Int. J. Remote Sens.* **2010**, *31*, 4387–4409. [CrossRef]

51. Gillespie, A.R.; Abbott, E.A.; Gilson, L.; Hulley, G.; Jiménez-Muñoz, J.C.; Sobrino, J.A. Residual errors in ASTER temperature and emissivity standard products AST08 and AST05. *Remote Sens. Environ.* **2011**, *115*, 3681–3694. [CrossRef]

52. Zhou, S.; Cheng, J. An Improved Temperature and Emissivity Separation Algorithm for the Advanced Himawari Imager. *IEEE Trans. Geosci. Remote Sens.* **2020**, 1–20. [CrossRef]

53. Xu, H.; Yu, Y.; Tarpley, D.; Göttsche, F.; Olesen, F.-S. Evaluation of GOES-R Land Surface Temperature Algorithm Using SEVIRI Satellite Retrievals With In Situ Measurements. *IEEE Trans. Geosci. Remote Sens.* **2013**, *52*, 3812–3822. [CrossRef]

54. Yu, Y.; Liu, Y.; Yu, P.; Wang, H. *Enterprise Algorithm Theoretical Basis Document for Viirs Land Surface Temperature Production*; Version 1.2; NOAA: College Park, MD, USA, 2019. Available online: https://www.star.nesdis.noaa.gov/jpss/documents/ATBD/ATBD_EPS_Land_LST_v1.2.pdf (accessed on 29 June 2020).

55. Yamamoto, Y.; Ishikawa, H.; Oku, Y.; Hu, Z. An Algorithm for Land Surface Temperature Retrieval Using Three Thermal Infrared Bands of Himawari-8. *J. Meteorol. Soc. Jpn.* **2018**, *96*, 59–76. [CrossRef]

56. Bessho, K.; Date, K.; Hayashi, M.; Ikeda, A.; Imai, T.; Inoue, H.; Kumagai, Y.; Miyakawa, T.; Murata, H.; Ohno, T.; et al. An Introduction to Himawari-8/9—Japan's new-generation geostationary meteorological satellites. *J. Meteorol. Soc. Jpn.* **2016**, *94*, 151–183. [CrossRef]

57. Schmit, T.J.; Griffith, P.; Gunshor, M.M.; Daniels, J.; Goodman, S.J.; Lebair, W.J. A Closer Look at the ABI on the GOES-R Series. *Bull. Am. Meteorol. Soc.* **2017**, *98*, 681–698. [CrossRef]

58. Chung, S.-R.; Ahn, M.-H.; Han, K.-S.; Lee, K.-T.; Shin, D.-B. Meteorological Products of Geo-KOMPSAT 2A (GK2A) Satellite. *Asia Pac. J. Atmos. Sci.* **2020**, *56*, 185. [CrossRef]

59. Stuhlmann, R.; Rodriguez, A.; Tjemkes, S.; Grandell, J.; Arriaga, A.; Bezy, J.-L.; Aminou, D.; Bensi, P. Plans for EUMETSAT's Third Generation Meteosat geostationary satellite programme. *Adv. Space Res.* **2005**, *36*, 975–981. [CrossRef]

60. Choi, Y.-Y.; Suh, M.-S. Development of Himawari-8/Advanced Himawari Imager (AHI) land surface temperature retrieval algorithm. *Remote Sens.* **2018**, *10*, 2013. [CrossRef]

61. NMSC. Available online: http://nmsc.kma.go.kr/enhome/html/bbs/selectNews.do?bbsCd=00237&bbsUsq=200676 (accessed on 29 June 2020).

62. Lee, B.-I.; Chung, S.-R.; Baek, S. Development of cloud detection algorithm for GK-2A/AMI. In Proceedings of the 7th Asia-Oceania/2nd AMS-Asia/2nd KMA Meteorological Satellite User's Conference, Songdo City, Korea, 24–27 October 2016; Available online: https://nmsc.kma.go.kr/enhome/html/conference/selectCfrItem.do?cfrUsq=142&cpnUsq=121&cpnDivCd=03 (accessed on 29 June 2020).

63. Seong, N.-H.; Jung, D.; Kim, J.; Han, K.-S. Evaluation of NDVI Estimation Considering Atmospheric and BRDF Correction through Himawari-8/AHI. *Asia Pac. J. Atmos. Sci.* **2020**, *56*, 265–274. [CrossRef]

64. Ijima, O. *Basic and Other Measurements of Radiation at Station Tateno (2019-07)*; Aerological Observatory, Japan Meteorological Agency, PANGAEA: Tokyo, Japan, 2019; Available online: https://doi.pangaea.de/10.1594/PANGAEA.906163 (accessed on 29 June 2020).

65. Ijima, O. *Basic and Other Measurements of Radiation at Station Tateno (2019-08)*; Aerological Observatory, Japan Meteorological Agency, PANGAEA: Tokyo, Japan, 2019; Available online: https://doi.pangaea.de/10.1594/PANGAEA.908094 (accessed on 29 June 2020).

66. Ijima, O. *Basic and Other Measurements of Radiation at Station Tateno (2019-09)*; Aerological Observatory, Japan Meteorological Agency, PANGAEA: Tokyo, Japan, 2019; Available online: https://doi.pangaea.de/10.1594/PANGAEA.909250 (accessed on 29 June 2020).

67. Ijima, O. *Basic and Other Measurements of Radiation at Station Tateno (2019-10)*; Aerological Observatory, Japan Meteorological Agency, PANGAEA: Tokyo, Japan, 2019; Available online: https://doi.pangaea.de/10.1594/PANGAEA.910611 (accessed on 29 June 2020).

68. Caselles, E.; Valor, E.; Abad, F.; Caselles, V. Automatic classification-based generation of thermal infrared land surface emissivity maps using AATSR data over Europe. *Remote Sens. Environ.* **2012**, *124*, 321–333. [CrossRef]

69. Carlson, T.N.; Ripley, D.A. On the relation between NDVI, fractional vegetation cover, and leaf area index. *Remote Sens. Environ.* **1997**, *62*, 241–252. [CrossRef]

70. Lin, J.; Feng, X.; Xiao, P.; Li, H.; Wang, J.; Li, Y. Comparison of Snow Indexes in Estimating Snow Cover Fraction in a Mountainous Area in Northwestern China. *IEEE Geosci. Remote Sens. Lett.* **2012**, *9*, 725–729. [CrossRef]

71. Park, K.-A.; Woo, H.-J.; Chung, S.-R.; Cheong, S.-H. Development of Sea Surface Temperature Retrieval Algorithms for Geostationary Satellite Data (Himawari-8/AHI). *Asia Pac. J. Atmos. Sci.* **2020**, *56*, 1–20. [CrossRef]

72. Korea Meteorological Administration Web Site. Available online: https://data.kma.go.kr/cmmn/main.do (accessed on 10 September 2020).

73. Korea Hydrographic and Oceanographic Agency Web Site. Available online: http://www.khoa.go.kr/oceangrid/khoa/koofs.do (accessed on 10 September 2020).

74. Borbas, E.; Seemann, S.W.; Huang, H.-L.; Li, J.; Menzel, W.P. Global profile training database for satellite regression retrievals with estimates of skin temperature and emissivity. In Proceedings of the 14th International ATOVS Study Conference, Beijing, China, 25–31 May 2005; pp. 763–770.

75. Wan, Z. MODIS Land-Surface Temperature Algorithm Theoretical Basis Document (LST ATBD). Available online: https://modis.gsfc.nasa.gov/data/atbd/atbd_mod11.pdf (accessed on 29 June 2020).

76. Yu, Y.; Tarpley, D.; Privette, J.L.; Goldberg, M.; Raja, M.R.V.; Vinnikov, K.; Xu, H. Developing Algorithm for Operational GOES-R Land Surface Temperature Product. *IEEE Trans. Geosci. Remote Sens.* **2008**, *47*, 936–951. [CrossRef]

77. Coll, C.; Caselles, V. A split-window algorithm for land surface temperature from advanced very high resolution radiometer data: Validation and algorithm comparison. *J. Geophys. Res. Space Phys.* **1997**, *102*, 16697–16713. [CrossRef]

78. Ren, H.; Yan, G.; Chen, L.; Li, Z. Angular effect of MODIS emissivity products and its application to the split-window algorithm. *ISPRS J. Photogramm. Remote Sens.* **2011**, *66*, 498–507. [CrossRef]

79. García-Santos, V.; Coll, C.; Valor, E.; Niclòs, R.; Caselles, V. Analyzing the anisotropy of thermal infrared emissivity over arid regions using a new MODIS land surface temperature and emissivity product (MOD21). *Remote Sens. Environ.* **2015**, *169*, 212–221. [CrossRef]

80. Lu, L.; Zhang, T.; Wang, T.; Zhou, X. Evaluation of Collection-6 MODIS Land Surface Temperature Product Using Multi-Year Ground Measurements in an Arid Area of Northwest China. *Remote Sens.* **2018**, *10*, 1852. [CrossRef]

81. Duan, S.-B.; Li, Z.-L.; Li, H.; Göttsche, F.; Wu, H.; Zhao, W.; Leng, P.; Zhang, X.; Coll, C. Validation of Collection 6 MODIS land surface temperature product using in situ measurements. *Remote Sens. Environ.* **2019**, *225*, 16–29. [CrossRef]

82. François, C.; Ottlé, C. Atmospheric corrections in the thermal infrared: Global and water vapor dependent split-window algorithms-applications to ATSR and AVHRR data. *IEEE Trans. Geosci. Remote Sens.* **1996**, *34*, 457–470. [CrossRef]

83. Atitar, M.; Sobrino, J.A. A split-window algorithm for estimating LST from Meteosat 9 data: Test and comparison with data and MODIS LSTs. *IEEE Geosci. Remote Sens. Lett.* **2008**, *6*, 122–126. [CrossRef]

84. Martin, M.; Ghent, D.; Pires, A.; Göttsche, F.; Cermak, J.; Remedios, J.J. Comprehensive in situ validation of five satellite land surface temperature data sets over multiple stations and years. *Remote Sens.* **2019**, *11*, 479. [CrossRef]

remote sensing

Article

Towards a Unified and Coherent Land Surface Temperature Earth System Data Record from Geostationary Satellites

Rachel T. Pinker [1,*], Yingtao Ma [1], Wen Chen [1], Glynn Hulley [2], Eva Borbas [3], Tanvir Islam [2], Chris Hain [4], Kerry Cawse-Nicholson [2], Simon Hook [2] and Jeff Basara [5]

[1] Department of Atmospheric and Oceanic Science, University of Maryland, College Park, MD 20742, USA; ytma@umd.edu (Y.M.); wchen122@umd.edu (W.C.)

[2] NASA Jet Propulsion Laboratory M/S 183-501, 4800 Oak Grove Drive, Pasadena, CA 91109, USA; glynn.hulley@jpl.nasa.gov (G.H.); Tanvir.Islam@jpl.nasa.gov (T.I.); Kerry-Anne.Cawse-Nicholson@jpl.nasa.gov (K.C.-N.); simon.j.hook@jpl.nasa.gov (S.H.)

[3] University of Wisconsin—Madison, Space Science and Engineering Center (SSEC) Cooperative Institute for Meteorological Satellite Studies (CIMSS), Madison, WI 53706, USA; evab@ssec.wisc.edu

[4] NASA Marshall Space Flight Center, Huntsville, AL 35808, USA; christopher.hain@nasa.gov

[5] School of Meteorology and School of Civil Engineering and Environmental Science, University of Oklahoma, Norman, OK 73019, USA; jbasara@ou.edu

* Correspondence: pinker@atmos.umd.edu; Tel.: +1-301-405-5380

Received: 25 April 2019; Accepted: 4 June 2019; Published: 12 June 2019

Abstract: Our objective is to develop a framework for deriving long term, consistent Land Surface Temperatures (LSTs) from Geostationary (GEO) satellites that is able to account for satellite sensor updates. Specifically, we use the Radiative Transfer for TOVS (RTTOV) model driven with Modern-Era Retrospective Analysis for Research and Applications (MERRA-2) information and Combined ASTER and MODIS Emissivity over Land (CAMEL) products. We discuss the results from our comparison of the Geostationary Operational Environmental Satellite East (GOES-E) with the MODIS Land Surface Temperature and Emissivity (MOD11) products, as well as several independent sources of ground observations, for daytime and nighttime independently. Based on a six-year record at instantaneous time scale (2004–2009), most LST estimates are within one std from the mean observed value and the bias is under 1% of the mean. It was also shown that at several ground sites, the diurnal cycle of LST, as averaged over six years, is consistent with a similar record generated from satellite observations. Since the evaluation of the GOES-E LST estimates occurred at every hour, day and night, the data are well suited to address outstanding issues related to the temporal variability of LST, specifically, the diurnal cycle and the amplitude of the diurnal cycle, which are not well represented in LST retrievals form Low Earth Orbit (LEO) satellites.

Keywords: Land Surface Temperature (LST); satellite retrievals of LST; LST from GOES satellites

---

1. Introduction

Land surface temperature is an important climate parameter due to its control of the components of the surface energy budget, such as turbulent heat and moisture fluxes, and upward terrestrial radiation [1]. For climate applications, information is needed on large scales, and ideally, the diurnal cycle needs to be resolved. In this study, we develop an approach to derive information on LST which is applicable to the GOES satellites across multiple missions and multiple satellite sensors. We report on results obtained during the period (2004–2009) at hourly time intervals, at about 5-km spatial resolution.

Since surface ground observations are limited, shelter temperature has been widely used as a proxy to surface skin temperature to meet large-scale needs. Issues emerging from such an approach have been addressed previously [2]. While observations from satellites are considered useful for inferring LST, only a few satellite sensors observe all the necessary parameters needed to derive LST with high accuracy. Some lack sufficient number of channels to account simultaneously for atmospheric effects (as needed for implementing the "split window" approach) [3–7]. Others do not observe the Earth surface at high frequency to resolve the diurnal cycle, or at high spatial resolution, to minimize the presence of clouds.

Information on surface emissivity is also not readily available at sufficient spectral resolution [8,9]. Moreover, land surface emissivity is generally less than one, and therefore, part of the atmospheric downward radiation is reflected by the surface and has to be accounted for [10] when converting ground observations of radiative flux measurements to LST (which is not always done). The number of successful attempts to derive LST from satellites has been substantial (especially using the well-established split window approach). A full review of what was done is beyond the scope of this paper, but a comprehensive summary can be found in Li et al. [11], and is briefly presented below.

The early effort to retrieve LST from satellites over agricultural land made by Price [3] was done by adopting the Advanced Very High Resolution Radiometer (AVHRR) Sea Surface Temperature (SST) split window algorithm [12,13]. Becker and Li [14] extended the split window method of McMillin [15] for SST to LST and took into account the spectral variability in land surface emissivity. This so called "generalized split window" LST algorithm has been widely used. Additional efforts include the work of Prata, [16], Sobrino et al. [7], Wan and Dozier [5], Francois and Ottle [17], Coll and Caselles [18], Trigo et al. [19], and Wan [20]. The approach for accounting for emissivity has evolved from assignments based on land use to the use of the Normalized Difference Vegetation Index (NDVI) [21–23]; however, the NDVI concept is not applicable for every surface type. Currently, surface emissivity is derived from the Advanced Space Thermal Emission and Reflection Radiometer (ASTER), the Thermal Infrared (TIR) Multispectral Scanner (TIMS), and the Moderate-resolution Imaging Spectroradiometer (MODIS) [24–29] culminating in the Combined ASTER and MODIS Emissivity over Land (CAMEL) product to be used here [30,31].

Most of the above referenced studies focus on polar orbiting satellites such as the National Oceanic and Atmospheric Administration (NOAA)-AVHRR, the Along-Track Scanning Radiometer (ATSR) and the MODIS instrument aboard Terra and Aqua satellites; the temporal measurement frequency of these satellites is approximately twice per day. The Land Surface Diurnal Temperature Range (DTR) is an important element of the climate system and is not captured by the polar orbiting satellites. Geostationary satellites provide diurnal coverage and observe the surface continuously at a nadir pixel resolution of about 4 km [32] which led to the development of several algorithms for GEO satellites [33–35].

While the principles of retrieval methodologies have not changed drastically over time, the development in auxiliary information, quality of such information, and availability of long term records of satellite observations make it feasible to formulate a homogeneous approach across various satellite sensors that can culminate in climatic records of LST. The primary objective of this study is to present a methodology that can be implemented with different GOES observing systems, using consistent auxiliary information of highest possible quality and utilizing radiative transfer models that account for the vertical profiles of atmospheric states for each retrieval. From mid-2004 to 2017, only one thermal channel is available on the GOES series; the focus of this study is on retrievals using such a single channel in order that a consistent, long term record can be generated from all the GOES satellites (including those that allow the implementation of the split window approach). In Section 2, materials used are described; retrieval algorithm development is presented in Section 3; evaluation of GOES-E based LST estimates is presented in Section 4; a discussion is provided in Section 5; and a summary is presented in Section 6.

2. Materials

2.1. GOES Satellite Data

The GOES system is operated by the National Oceanic and Atmospheric Administration, National Environmental Satellite, Data and Information Service (NESDIS). The GOES system is based on the use of satellites designed to operate at an orbit of 35,790 km above the earth, remaining stationary to a given point on the ground. The GOES provides data at high temporal frequency (15 min) with continental-scale coverage (N. and S. America). In this study observations from GOES-12 (4/1/2003–4/14/2010) will be utilized (Table 1). Typically, the GOES imager includes five spectral channels (one visible, four infrared). For GOES 8-10 the channels are located at 3.9, 6.75, 10.7, and 12.2 μm whereas for GOES 11–15, the 6.75 μm channel was moved to 6.5 μm and the 12 μm channel was moved to 13.3 μm. The visible, mid-infrared and 11 μm band are typically used for cloud screening while the two thermal infra-red (TIR) bands (10.2–11.2 μm and 11.5–12.5 μm) are used in what is known as a "split-window" approach to retrieve LST.

Table 1. Summary of GOES-8/GOES-12 channels.

Satellite	Channel	Symbol	Wavelength	Objective	Spatial Resolution (Nadir)
GOES-8 & GOES12	1	R_1	0.67 μm	Cloud	1 km × 1 km
	2	R_2 and T_2	3.9 μm	Cloud and snow	4 km × 4 km
	3	T_3	6.7 μm	Water vapor	4 km × 4 km
	4	T_4	10.7 μm	Surface temperature	4 km × 4 km
GOES-8	5	T_5	12.0 μm	Sea surface temperature and water vapor	4 km × 4 km
GOES12	6	T_6	13.3 μm	… … …	4 km × 4 km

Note: GOES-8 information is provided since cloud algorithm was originally developed for GOES-8. Channel 2 is separated into the reflected solar radiation component (R2) and the emitted infrared radiation component (T2) [52]; Channel 3 is not used in any of the cloud screening tests.

Aspects of GOES satellite systems that need to be addressed before deriving LST include the spectral characteristics of the GOES sensors and their filter functions, calibration of visible and IR channels, cloud screening methodology (that requires snow analysis information) and the selection of cloud screening tests as appropriate for each satellite configuration, with special distinction between night-time and day-time conditions. Over the period of this study, two separate operational GOES Imagers, one located at a longitude of −75° (referred to as GOES-East) and one located at a longitude of −135° (referred to as GOES-West) continuously provided imagery over North and South America. In this study, observation from GOES-East only will be utilized for years 2004–2009. The temporal sampling of the GOES Imager is every 30-min over North America and every 3-h over the full disk. Spectral distribution of the GOES 8–15 series are provided in NOAA NESDIS STAR GOES Imager LST ATBD (Version 3.0).

2.2. Visible and Thermal Channel Calibration

Visible channels are used in the cloud screening part of data processing. Their calibration is done in two stages. First, applied are the pre-launch calibration coefficients to get the first step "nominal reflectance" (*Apre*), and then, the post-lunch calibration coefficients are applied to obtain the final calibrated nominal reflectance (*Apost*).

The nominal reflectance is defined as the ratio of reflected radiance to nominal solar irradiance given as:

$$Apre = \text{pi} * R / F_0 \tag{1}$$

where pi = 3.141593, R is satellite observed upwelling radiance, and F_0 is the solar irradiance at local zenith and mean Sun-Earth distance.

The pre-launch nominal reflectance is:

$$A pre = k \, (X - X space) \tag{2}$$

where k is the pre-launch calibration constant to convert satellite observed digital counts to nominal reflectance. X is the instrument raw digital count, *Xspace* is the raw count of the space scene (has been adjusted to 29 for all GOES imagers at NOAA). The Post-lunch calibration is applied by multiplying the *Apre* by a coefficient:

$$A post = A pre * C \tag{3}$$

The pre-launch coefficient for GOES-12 for radiance calibration was 0.5771 and for nominal reflectance it was 0.001141.

NOAA/STAR monitors the GOES imager and updates the post-launch coefficients every month. Coefficients are considered optimal for the days on or after the second Tuesday of the month following the coefficient generation month. For more details, see: http://www.star.nesdis.noaa.gov/smcd/spb/fwu/homepage/GOES_Imager_Vis_OpCal.php.

Inter-calibration of the infrared channels on the GOES series of satellites has been performed using under-passes of the well calibrated NASA Atmospheric Infrared Sounder (AIRS) sensor on the Aqua platform [36]. Infrared imager data from GOES are stored in GOES Variable Format (GVAR) counts and radiances can be derived from GVAR counts by applying the calibration and scaling coefficients using a procedure described in [37].

Calibration of IR channels of GOES imager data is also done in two stages: (1) converting the imager GVAR raw count to scene radiance; (2) converting the radiance to temperature. To convert a 10-bit GVAR count to scene radiance we use:

$$R = (X - b)/m \tag{4}$$

where X is the GVAR raw count (10-bit, range from 0 to 1023), m and b are calibration coefficients. The values for m and b depend on channel selected, but are constant for a given channel. The obtained radiance can be convert to effective temperature (K) by inverse of the Planck function:

$$T_{eff} = C_2 * n / \ln (1 + C_1 * n^3 / R) \tag{5}$$

$C_1 = 1.191066 \times 10 e^{-5}$ [m·W/(m^2·sr·cm^{-4})], $C_2 = 1.438833$ (K/cm^{-1}), "n" is the central wave-number of the channel and varies from instrument to instrument.

The effective temperature T_{eff} is further converted to actual temperature by:

$$T = a + b * T_{eff} + g * T_{eff}^2 \tag{6}$$

where "a", "b" and "g" are coefficients and their values and central wave-numbers can be found at: https://www.ospo.noaa.gov/Operations/GOES/calibration/gvar-conversion.html#radiance.

Evaluation of calibration was done by comparison of radiances in each channel with NOAA values.

2.3. Emissivity Data

A new land surface emissivity Earth Science Data Systems (ESDS) product has been developed in support of a NASA Making Earth System Data Records for Use in Research Environments (MEaSUREs) project [30,31] MODIS-ASTER Global Infrared Combined Emissivity product produced from the University Wisconsin Global Infrared Land Surface Emissivity (UWIREMIS) and the ASTER Global Emissivity (GED) Database are known as the Combined ASTER and MODIS Emissivity for Land (CAMEL) and represents a combination of MODIS baseline-fit emissivity database (MODBF). The

CAMEL ESDR is produced globally at 5-km resolution at mean monthly time-steps and for 13 bands from 3.6–14.3 micron and extended to 417 bands using a Principal Component (PC) regression approach. This product has been used in our retrievals of LST.

2.4. MERRA-2 Data

The 6 hourly MERRA-2 re-analysis data (MERRA-2 inst6_3d_ana_Np version 5.12.4) (https://disc.gsfc.nasa.gov/datasets/M2I6NPANA_V5.12.4/summary) are used to specify the atmospheric conditions [38]. The re-analysis data are available at [00, 06, 12, 18] hours with a resolution of 0.5° × 0.625°, for 42 pressure levels. Properties include sea-level pressure, surface pressure, geopotential height, air temperature, wind components, and specific humidity. Since the LST is retrieved at hourly time scale, the MERRA-2 data are linearly interpolated in time to give the atmospheric state in between the four analysis times (as needed for input to RTTOV) and linearly interpolated to 0.05° × 0.05° in space to match the formulation of this study.

2.5. MOD11

The GOES based LST estimates will be evaluated against LST retrievals from MOD11 Version 6 Land Surface Temperature and Emissivity product [20] (we use both MOD11_L2 data and MOD11C3 data). The MODIS LST data products are produced as a series of nine products. The sequence begins as a swath at a nominal pixel spatial resolution of 1 km at nadir and a nominal swath coverage of 2030 or 2040 lines along track by 1354 pixels per line in the daily LST product [39]. There are two algorithms used in the daily MODIS LST processing: the generalized split-window LST algorithm [5] and the day/night LST algorithm. New refinements made to these two algorithms are described by *Wan* [20]. The MOD11_L2 version 6 swath product provides per-pixel land surface temperature (LST) and emissivity. It is produced daily in 5-min temporal increments of satellite acquisition and has a pixel size of 1 km. The MOD11C3 Version 6 product provides monthly land surface temperature (LST) and emissivity values in a 0.05 (5600 m × 5600 m) degree latitude/longitude climate modeling grid (CMG), which has a Geographic grid with 7200 columns and 3600 rows representing the entire globe. The MOD11C3 granule consists of day and night LST and their corresponding quality indicator (QC) layers.

2.6. Ground Observations

- SURFRAD/BSRN

NOAA established the Surface Radiation Budget Network (SURFRAD) in 1993 [40] to support climate research by providing accurate, continuous, long-term measurements of the surface radiation budget over the United States. These became the continental U.S. contingent of the International Baseline Surface Radiation Network (BSRN) [41] as described in [42]. The SURFRAD Network uses Eppley Precision Infrared Pyrgeometers (model PIR). The general information about the instrumentation can be found at: https://www.esrl.noaa.gov/gmd/grad/surfrad/overview.html.

Specifically, we used the following sites: Desert Rock, Nevada (DRA: 36.62°N, 116.02°W; 1007 m); Fort Peck, Montana (FPK: 48.32°N, 105.10°W; 634 m); Bondville, Illinois (BON: 40.06°N, 88.37°W; 230 m) and Goodwin Creek, Mississippi (GCM: 34.25°N, 89.87°W; 98 m). BSRN sites that provide data at 1 or 3-min frequency, which makes them suitable for generating information to match the satellite observations. The upwelling and downwelling longwave radiative fluxes are measured with a precision infrared radiometer, which is sensitive in the spectral range from 3000 to 50,000 nm.

- ARM SGP

The Atmospheric Radiation Measurement Program (ARM) Near-Surface Observation Dataset came from the ARM Cloud and Radiation Test Bed site (34°–39°N and 94.5°–100.5°W). The average elevation is 314 m. The surface skin temperature used in this paper is observed at 60-sec intervals at the Central

Facility (36.6°N, 97.48°W) of the Southern Great Plains (SGP) site. The instrument used to observe the skin temperature is the Infrared Thermometer (IRT). It is a ground-based radiation pyrometer that measures the equivalent blackbody brightness temperature of the scene in its field of view. The downwelling version has a narrow field of view for measuring sky temperature and detecting clouds. The upwelling version has a wide field of view for measuring the narrowband radiating temperature of the ground surface (https://www.arm.gov/capabilities/instruments/irt). Time series and scatter plots are produced and inspected to compare surface temperature measured by the IRT and a precision infrared radiometer (PIR). The temperature measuring range is from 173 to 473 K. The accuracy is the greater value of a) ±0.5 K + 0.7% of the temperature difference between the internal reference temperature and the object measured or b) the temperature resolution. The spectral sensitivity is from 9.6 to 11.5 μm [43].

- Oklahoma MESONET

The Oklahoma MESONET is an automated network of over 110 remote, meteorological stations across Oklahoma (http://www.mesonet.org). The surface types are predominantly grassland/wooded, grassland/cropland. In 1999, infrared temperature (IRT) sensors (Apogee Instruments, Inc.) were installed at 89 of the MESONET sites. A combination of automated and manual tests was applied using simultaneous soil and atmospheric measurements to inter-compare observations and ensure that the skin temperature observations are of research quality [44]. The measurements collected by the MESONET provided a unique opportunity to inter-compare observations. Fiebrich et al. [45] provide an evaluation of 5-min-resolution field measurements collected using the sensors. This sensor was chosen for use because it is water resistant and was designed for continuous outdoor use. Sensor accuracy is approximately ± 0.2 °C from 15 °C to 35 °C and 0.3 °C from 35 °C to 45 °C. The sensor is installed at a height of 1.5 m and has a field of view of a diameter circle of 0.5 m. The energy detected by the sensor is converted to a temperature using the Stefan–Boltzmann law and an assumed surface emissivity of 1.0. Slight underestimation is caused because the true emissivity of the land surface is less than 1.0. In addition, slight overestimation is caused by reflected longwave radiation from the target [46]. While surface reflection of downwelling longwave radiation is ignored, Sun et al. [32] discussed the effects of these two factors and found that the total effect may be a slight underestimation of the skin temperature. Generally, estimated impact of uncertainty in relevant parameters on in situ LST are as follows: radiometric calibration uncertainty of ±0.2 to 0.5 K can impact LST as much as 0.2 K; emissivity uncertainty of ±1% can impact LST by as much as 0.3 K; downwelling atmospheric radiance uncertainty of ±10% can impact LST by as much as 0.1 K [47].

2.7. Cloud Detection

For cloud masking, various ancillary data are needed such as surface type, snow-free channel-1 radiance, and a pixel position information, all in the same dimension and location as the satellite images. The land cover data used in this study for cloud screening implementation, are generated at 1-km resolution [48]. This product includes 14 International Geosphere–Biosphere Programme (IGBP) classes and the underlying surface types are aggregated according to the IGBP classification.

A Coupled Cloud and Snow Detection Algorithm (CCSDA) that was developed initially for use with GOES-8 satellite is adjusted as appropriate for each GOES satellite is used. The algorithm is described in [49,50]. Variants of the approach were tested and evaluated in several publications [51]. In the case of the GOES-8 imager four channels were used to detect clouds, snow, and to perform background analysis for each hour of the diurnal cycle. Beginning with GOES-12, Channel 5 is no longer available (Table 1) [52].

The CCSDA algorithm is capable of producing its own snow analysis using an algorithm that applies three tests using three GOES channels. Alternatively, there is a switch to allow the use of a snow analysis from a different source. The advantage of using the snow analysis generated by the CCSDA algorithm is that it is updated hourly, which provides a more accurate analysis of the expected background when applying the cloud tests. If a daily snow analysis is used, the snow conditions cannot change for each hour of the cloud analysis, and this may introduce error.

In Table 2 we present a description of the cloud screening tests used for GOES-8, along with an explanation of how the tests are assembled to determine a clear probability.

Table 2. Cloud Screening Tests for GOES-8.

Test	Apply	Cloud Detection Variable	Cloud That May Be Detected
RGCT	Day	R_1	Highly-reflective cloud
TGCT	Day and Night	T_4	Cold cloud
C2AT	Day	R_2	Weakly Reflective Cloud
TMFT	Day and Night	$T_2 - T_4$	Water cloud + Cirrus + Other Clouds
FMFT	Day and Night	$T_4 - T_5$	Thin Cirrus
ULST	Night	$T_2 - T_4$	Nighttime uniform low stratus
CIRT	Night	$(T_2 - T_4)/T_4$	Nighttime cirrus

Note: Reflectance Gross Contrast Test (RGCT); Channel-2 Albedo Test (C2AT); Thermal Gross Cloud Test (TGCT); Three Minus Five Test (TMFT); Four minus Five Test (FMFT); Uniform Low Stratus Test (ULST); Cirrus IR Test (CIRT).

The ultimate clear probability (P) can be assembled in various ways on the basis of individual test results. In this method:

$$P = \sqrt[n]{\prod_1^n P_i} \tag{7}$$

P_i is the clear probability from each individual cloud screening test and n is the total number of cloud screening tests. This assembly method guarantees that the target pixel is cloudy ($P = 0$) if any individual test identifies it as cloudy ($P_i = 0$). Otherwise, it compiles the confidence levels from all of the individual tests to obtain an overall clear probability.

Referring to Table 2, there was only one cloud screening test that required Channel 5, namely, the FMFT test. In the cloud screening algorithm for GOES-12 and beyond, the FMFT test will not be used, and the test assembly method described for GOES-8 is implemented with one less cloud test.

For each cloud test, threshold levels are used to differentiate between clear and cloudy pixels. For each new satellite, it is necessary to test the thresholds and modify as needed. Here, the cloud mask method applies two spatial tests and one threshold test on an 11–3.7 μm difference image. This fourth test compares the temperature from the 11 μm channel to a 20-day clear-sky composite of 11 μm temperatures, and labels the pixel as cloudy if the difference is greater than the threshold. The pixel level data were gridded to 0.05° and compared to the Pathfinder Atmospheres—Extended (PA TMOS-X) product [34]. Agreement above 95% for various times of the day was found. The only region which showed slight disagreement between the two independent cloud masks is over areas of complex terrain in the Western US, but even over these regions the two cloud masks agreed over 85% of the time.

The major difference between the day time and night time algorithms (Table 2) is that there are no Reflectance Gross Contrast Test (RGCT, when visible channel is missing) and Channel-2 Albedo Test (C2AT) during the night time. This will mostly affect the detection of reflective clouds. For GOES-12, the FMFT test could not be used. We have tested the cloud screening algorithm for use at nighttime. We applied the nighttime algorithm on daytime data and compared to results when the full daytime algorithm is used. Evaluation of LST estimates in each case is presented in Figure 1. As this figure shows, the daytime algorithm provides better agreement with ground observations than the nighttime one yet, the differences are small as illustrated in Figure 1.

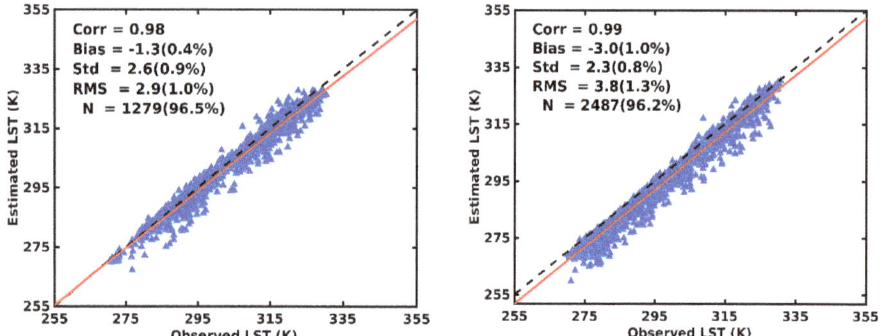

Figure 1. Evaluation of LST for 2004 against a SURFRAD/BSRN station at Desert Rock, NV (DRA) for: Left: daytime; Right: nighttime.

3. LST Retrieval Algorithm Development for GOES Satellites

Satellite observed radiance $R_o^\uparrow$, can be expressed as

$$R_o^\uparrow = \epsilon B(T_s)X + R_a^\uparrow + (1-\epsilon)R_a^\downarrow X \qquad (8)$$

where ϵ is surface emissivity, $B(T_s)$ is blackbody emission at surface temperature T_s, X denotes the atmospheric transmittance, $R_a^\uparrow$ and $R_a^\downarrow$ are atmospheric emission to space and surface, respectively. With known surface emissivity and simulated atmospheric emission and transmittance, the surface temperature can be retrieved

$$T_s = B^{-1}\left[\frac{1}{\epsilon}\left(\frac{1}{X}\left(R_o^\uparrow - R_a^\uparrow\right) - (1-\epsilon)R_a^\downarrow\right)\right] \qquad (9)$$

where B^{-1} denote the inverse of Planck function for GOES-12 channel 4.

Here, the approach is based on the Radiative Transfer for TOVS (RTTOV) model v11.2 [53–55] adjusted for the GEO characteristics and driven with MERRA-2 reanalysis fields. The CAMEL data are also implemented in the method. The advantage of this approach is that it is consistent with the retrieval approach used at JPL to generate the MOD21 product [56]. The processing sequence is described in Figure 2.

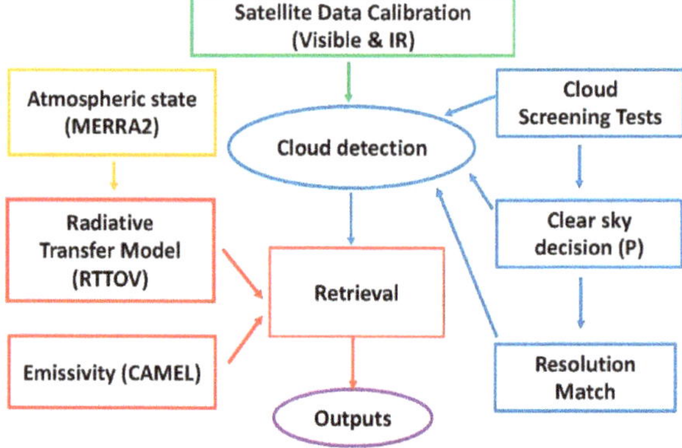

Figure 2. Flow-chart describing the derivation of LST from GOES observations.

Data processing sequence starts from raw digital counts from GOES satellites. Calibration is applied to all channels. Channel 4 (10.2 to 11.2 um) was used for LST retrieval. All channels except channel 3 (6.7 um) were used in the cloud detection algorithm. After cloud screening, GOES observations were resampled to a uniform grid of 0.05° resolution. The atmospheric radiation and transmittance were simulated with the RTTOV model using MERRA-2 fields as input. The MERRA-2 fields were first temporally interpolated to satellite observation time and then collocated to satellite locations. RTTOV calculated upwelling, downwelling radiances and atmospheric transmittance combined with CAMEL surface emissivity to retrieve the LST according to Equation (11).

4. Evaluation of GOES-E Based LST Estimates

We will present results of evaluation for UMD LST retrievals against MOD11 products, the BSRN/SURFRAD network over the USA, the ARM/SGP C1 site over the Southern Great Plains, and the MESONET network over Oklahoma. The issue of evaluation of satellite products of LST against ground measurements is complex, primarily, due to scale issues and known large spatial variability of LST. A comprehensive discussion on all aspects of validation issues are described by Guillevic et al. [43] and Göttsche et al. [57].

4.1. Scale Issues Related to Satellite and Ground Observations

The ground observations are point observations while the satellite LST product is at pixel level gridded to 0.05°. To assess the homogeneity of each site, we use the ASTER Global Emissivity Dataset at 1-km V003 (DOI: 10.5067/Community/ASTER_GED/AG1km.003) available for the period of 2000–2008. It is based on observations from the Advanced Spaceborne Thermal Emission and Reflection Radiometer (ASTER) Global Emissivity Dataset (GED) land surface temperature and emissivity data products using the ASTER Temperature Emissivity Separation (TES) algorithm with a Water Vapor Scaling (WVS) atmospheric correction method with MODIS MOD07 atmospheric profiles and the MODTRAN 5.2 radiative transfer model. The spatial distribution of the emissivity values is illustrated in Figure 3a and their frequency distribution is shown in Figure 3b. As shown, except for the DRA site, the 0.05° boxes show a high degree of homogeneity at the 1-km scale. As seen from Figure 3b, the emissivity values range between 0.965–0.980 with two distinct peaks of 0.965 and 0.975 with some lower values (0.948) at the DRA site. As also shown in the study of Hulley and Hook [58], who compared ASTER emissivity band 11 (8.6 µm) at 90 m spatial resolution to the same at 1 km, the agreement was very good. The spatial matching of ground and satellite observations is done by taking the weighted average of the pixels that fall in the cell box (0.05° × 0.05°) around the target location of the station. The time matching is done by taking the averages of ±15 min around the start scanning time of GOES12 (this interval is selected based on the duration of the satellite scan).

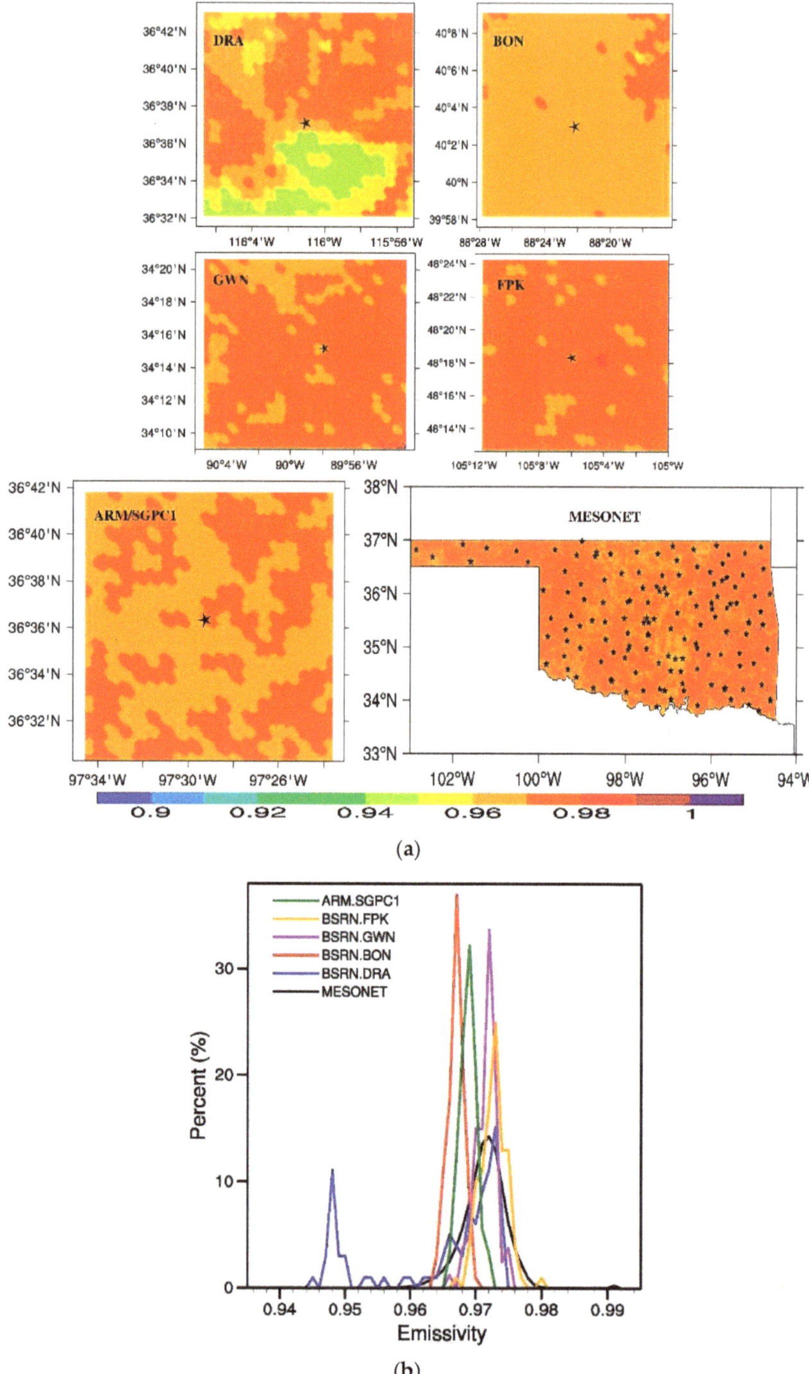

Figure 3. (**a**) Spatial characterization of the sites used in evaluation of the LST products in terms of emissivity as obtained at 1 km spatial resolution using the ASTER Global Emissivity Dataset 1 km V003. (**b**) Frequency distribution of the emissivity values over sites used in evaluation as illustrated in Figure 3a.

4.2. Evaluation against MOD11

We have conducted an inter-comparison between MOD11_L2 and our LST retrievals. Before comparison, the MOD11_L2 data are rescaled to 0.05 × 0.05 degree latitude/longitude grids.

Figure 4 shows an example of comparison between the GOES_RTTOV LST and MOD11_L2 LST on 11 June 2004 UTC 17:15.

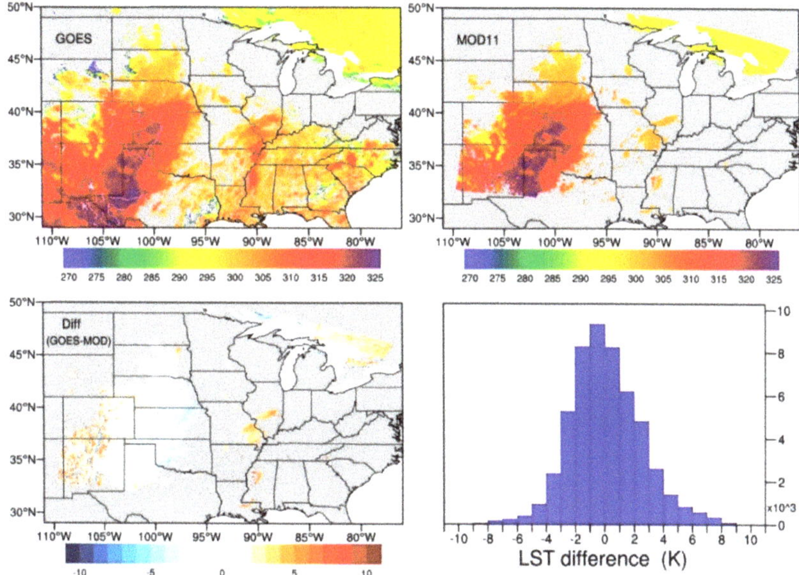

Figure 4. Example case of GOES_RTTOV_LST, MOD11_L2 LST and their difference and its distribution at 11 June 2004 UTC 17:15.

We compared 25 match-up cases for which the two products have overlap and both start scan at 15 min after same hour in 2004. For most cases, the total number of points in the overlap area is more than 40,000. Figure 5 presents the correlation coefficients (*corr*), the mean bias (*bias*), the standard deviation (*std*) and the root mean square error (*rmse*) between the two products. As seen, in most cases, the two products yield close correlation. Only in one case the coefficient is less than 0.8. The averaged *corr* of all cases is 0.91. More than 50% cases have mean bias less than 2 K, and the averaged value is 1.7 K. The averaged *std* and *rmse* are 2.7 and 3.3 K respectively.

4.3. Evaluation against ARM SGP Site at Instantaneous Time Scale

The IRT data are available from two levels of a tower; one instrument was located at 25 m and one at 10 m above ground. The probability distribution of differences between GOES_RTTOV_LST and ARM IRT is shown in Figure 6 using all available retrievals. Obviously, most of the differences fall within the interval of 1 *std*. Less than 20% of the cases are beyond 1 *std*. The correlation between the two data sets is high for both levels, (>than 0.80 for all cases). The mean differences at daytime are smaller than at nighttime at both levels and the same applies to *std* and *rmse*. Numerical values for the cases of Figure 6 are shown in Table 3. Differences due to the height exposure of the instrument can be caused by differences in the field of view of the instrument, and as such, different shading effects.

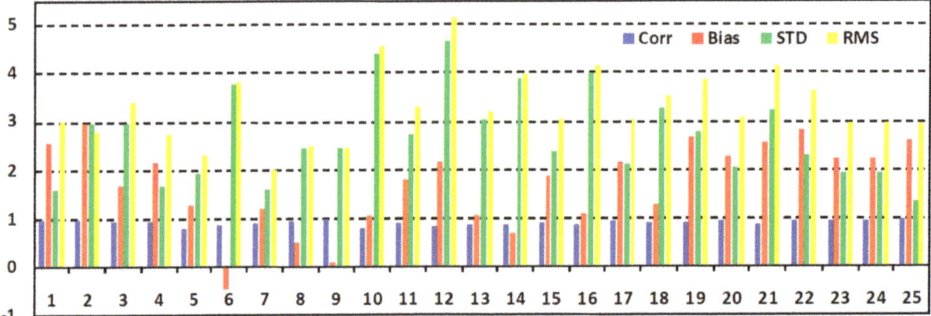

Figure 5. Evaluation of 25 instantaneous match-ups of LST retrievals from GOES observations against MOD11_L2. The *x*-axis provides the numbering of the cases while the *y*-axis indicates the correlation (in blue) and the other variables in W/m^2.

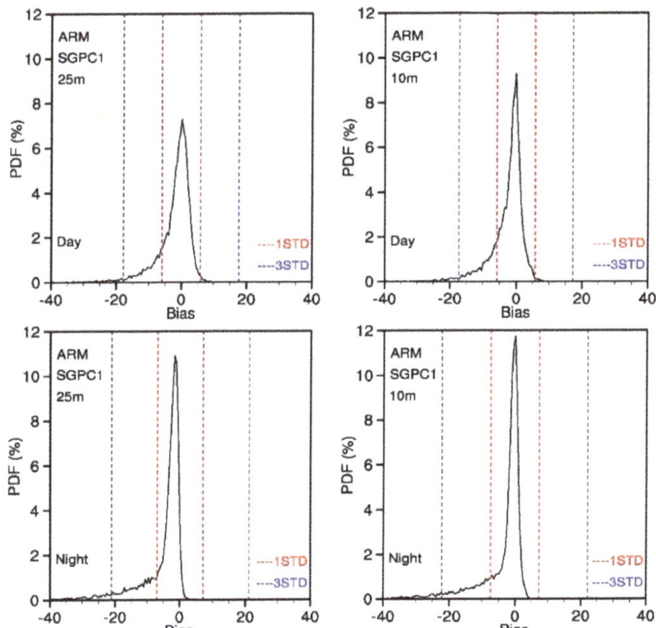

Figure 6. Probability distribution of differences between the GOES_RTTOV and ARM SGP C1 LST. Red dot line: 1 *std*; Blue dot line: 3 *std*.

Table 3. Statistical results for cases illustrated in Figure 6.

	Corr		Mean Bias		Std		RMS		No. Cases	
	Day	Night	Day	Night	Day	Night	Day	Night	Day	Night
25 m	0.89	0.81	−2.09	−5.12	5.92	7.04	6.28	8.7	11,781	12,335
10 m	0.89	0.80	−2.68	−3.64	5.75	7.37	6.34	8.22	11,940	12,639

Figure 7 shows the results of evaluations for 2004–2009 from tower observations: (a) for daytime from 25 m level; (b) same as (a) using observations from 10 m level (year 2006 excluded since this year requires additional quality control); (c) same as (a) for nighttime observations only; (d) same as (b) for nighttime observations only. Only values within 1 *std* are used. The satellite product underestimates the ARM IRT observations, yet, the difference between them is less than 1% and the *std* and *rmse* are also around 1%; the performance at daytime is better than at nighttime, most likely, due to better cloud detection during the daytime when observations from the visible channel are available.

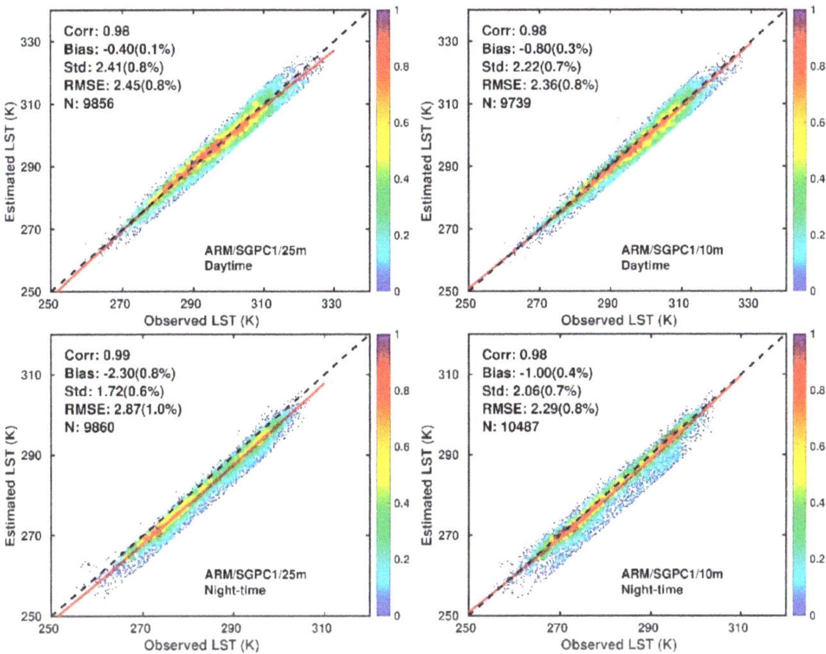

Figure 7. Evaluation of RTTOV based estimates from GOES-E at the SGPC1 ARM test site using observations at hourly intervals during 2004–2009 for daytime and nighttime from 25 m and 10m tower level. Data that have differences of less than one *std* were used.

4.4. Evaluation against SURFRAD/BSRN

The SURFRAD/BSRN network observes upwelling ($F\uparrow$) and down-welling ($F\downarrow$) radiative fluxes which are converted to temperature as follows:

$$F\uparrow = \varepsilon_{IR}\sigma T_S^4 + (1 - \varepsilon_{IR})F\downarrow \tag{10}$$

where ε_{IR} is the surface broadband emissivity assigned by surface type, σ is the Stefan-Boltzmann constant and is equal to 5.669×10^{-8} J m^{-2}s^{-1}K^{-4}. Then

$$T_S = \left[\frac{F\uparrow - (1 - \varepsilon)F\downarrow}{\varepsilon\sigma}\right]^{1/4} \tag{11}$$

The approach we use was also applied by others. The main issue in the conversion is the value of emissivity. Heidinger et al. [34] use a broadband longwave emissivity assumed to be 0.97. They indicate that a 0.1 error in emissivity equates to an error in the SURFRAD LST not exceeding 0.25 K. Yu et al. [59] also used the SURFRAD data to evaluate their LST retrievals following the same procedures. In their approach, the emissivity is estimated by mapping surface type classification of Snyder et al. [60] to emissivity (an approach that was popular for some time when direct information on emissivity was not available). They assume that the mean broadband emissivity of the satellite sensor is applicable. We use the CAMEL emissivity which is derived spectrally and integrated to the window spectral interval of the satellite used, and variable at monthly time scale; namely, for each month and for each location the spectral values are integrated to give a new broadband value. This is the most advanced use of surface emissivity in such retrievals.

The scatter plots of the instantaneous GOES_RTTOV LST against SURFRAD sites for both daytime and nighttime are shown in Figure 8. As seen, the satellite estimates and the ground observations have very high correlation, mostly above 0.98. For daytime (left panel Figure 8) the differences ranged between 0.4 (0.2%) to 1.16 (0.4%) while the *std* ranged between 1.88 (0.6% to 2.53 (0.9%) respectively. For nighttime (right panel Figure 8) the results are comparable to daytime.

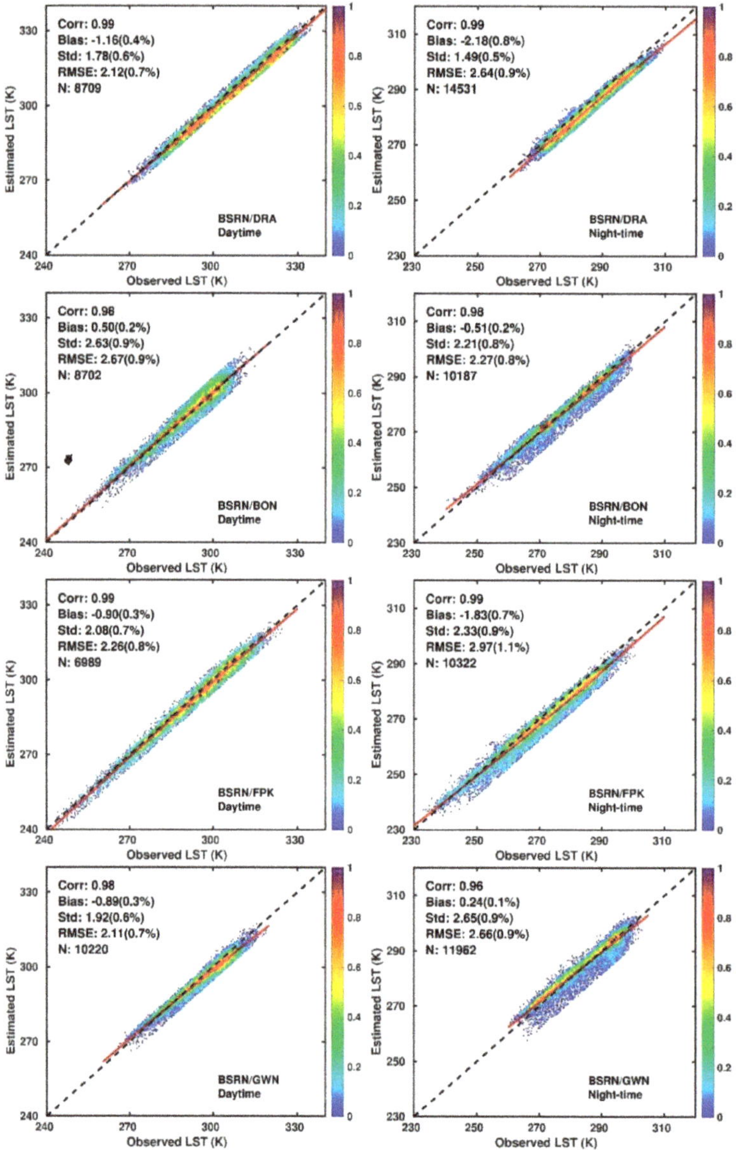

Figure 8. Evaluation of instantaneous GOES based LST estimates at hourly intervals against 4 SURFRAD/BSRN stations, independently for daytime (left panel) and nighttime (right panel) using observations from 2004–2009.

4.5. Evaluation against the Oklahoma MESONET Sites

The distribution of sites used in current evaluation is illustrated in Figure 3a. The evaluations are carried out against all the stations for both daytime and nighttime in January and July during 2004–2007. Results are presented in Figure 9 where outliers outside 1 *std* were removed. Red color designates results that have bias smaller than 1 *std*, ranging from 1.74 to 2.47 K. The retrieved data have high correlation with the in-situ data (>than 0.9). Results of daytime and nighttime for January and July are comparable (unlike the results of Figure 6 where all outliers were used).

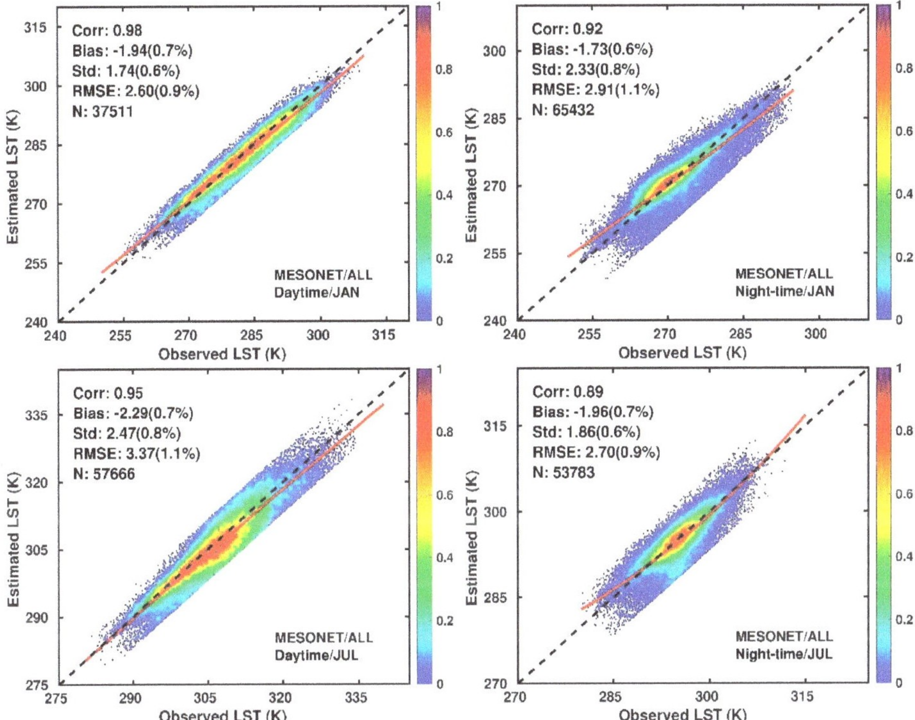

Figure 9. Evaluation of instantaneous GOES based LST estimates at hourly intervals against the MESONET stations, independently for daytime (left panel) and nighttime (right panel) using observations for years 2004–2007.

4.6. Applications

- Seasonal distribution of LST at monthly scale

Since many users of LST data are interested in monthly mean values, we have conducted a comprehensive comparison at such scale. This was possible due to the availability of both ground observations and satellite estimates for a period of six years.

The evaluation was expanded to include ground observations at the USA BSRN/SURFRAD sites and we also used information from MOD11C3, Version: 006 at 0.05° (https://lpdaac.usgs.gov/dataset_discovery/modis/modis_products_table/mod11c3_v006). Derived statistics includes mean values, standard deviation, maximum/minimum, and medium values for each month are shown in Figure 10. It is clearly that the GOES_RTTOV LST has very close distribution pattern as it of SURFRAD/BSRN. It has the ability to describe the annual variability of the LSTs at SURFRAN/BSRN sites. At DRA, MOD11C3v6 LST yields higher estimations against SURFRAD/BSRN for most seasons, the annual mean LST for all study years is 306.6 K, which is 3.1 K higher than the value of SURFRAD/BSRN. While the GOES_RTTOV estimations are much closer to the site value. The annul mean LST of GOES_RTTOV is 301.6 K. For BON, both of the MOD11C3v6 and GOES_RTTOV estimations are close to the site values, except April, May and June. The annul mean LSTs of SUFRAD/BON, GOES_RTTOV and MOD11 are 287.9 K, 288.0 K, 289.8 K respectively. And for GWN, the GOES_RTTOV has relatively lower estimation of annual mean LST which is 293.6 K. The site value is 295.1 K. The MOD11 is 295.7 K. The performance of the satellite estimations at FPK is similar as DRA. The MOD11 annual mean LST is 288.4 K, the GOES_RTTOV is 283.8 K, and the site value is 285.2 K.

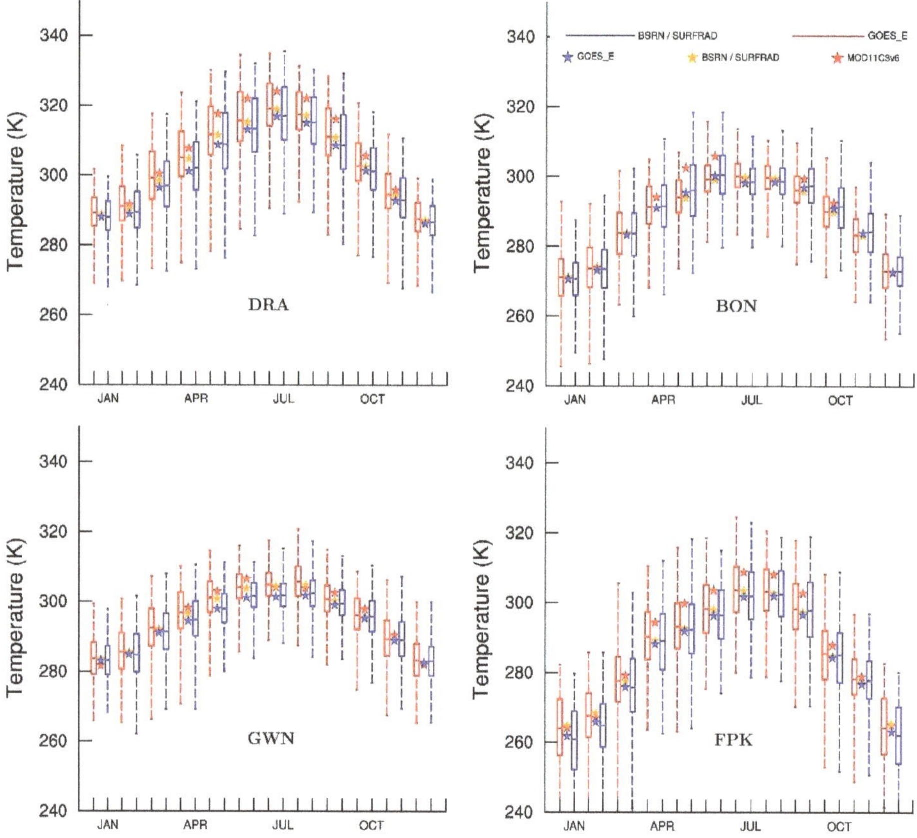

Figure 10. Daytime LST Distribution of GOES_RTTOV, SURFRAD/BSRN sites over 2004–2009 for each month and their monthly mean values compared with MOD11C3v6. Top/Bottom of *dashed line*: maxi/min LST; *Solid "-"*: medium LST: *Solid box:* quartile of LST. Stars are monthly mean LST of SURFRAD/BSRN sites (orange), GOES_RTTOV (blue) and MOD11C3v6 (red).

- A six-year climatology of LST over the US

A six year (2004-2009) monthly means of LST at 0.05° spatial resolution for January and July at UTC 06:15 and at UTC 18:15 are shown in Figure 11 for illustration of the product.

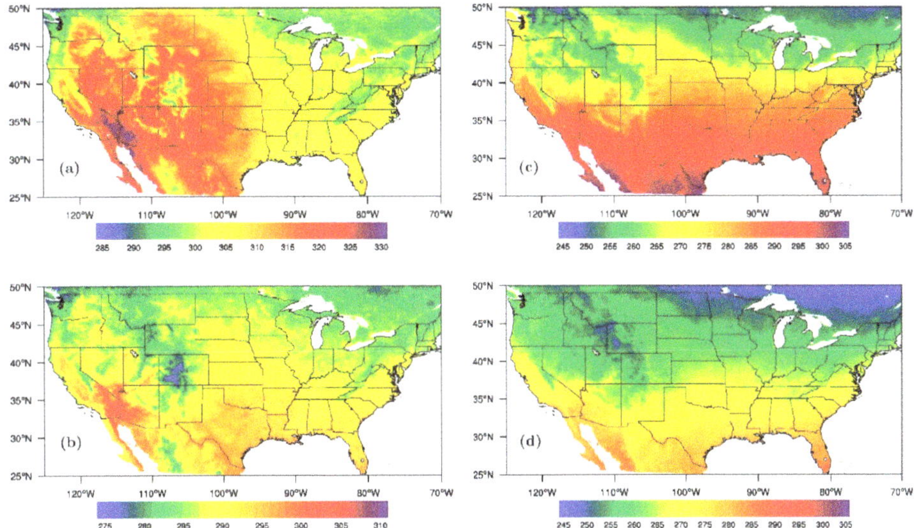

Figure 11. Monthly mean LST at 0.05° spatial resolution averaged over 6 years (2004–2009) for January and July; (**a**) UTC18:15, July; (**b**) UTC06:15:00, July; (**c**) UTC18:15, January; (**d**) UTC06:15, January.

As shown, for July, the differences in surface temperature during these two hours (close to representing daily max and min), are large. During the daytime, the western part of the US is dominated by clear sky conditions (the 100th W longitude is known to separate between the humid and dry parts of the US). During the nighttime, the clear conditions contribute to cooling by emitted LW radiation especially, over high elevations. Noticeable is also the pronounced latitudinal variability in the LST during January, dominated by solar zenith angle dependence of heating by SW radiation. The high spatial and temporal resolution of this product makes it useful for addressing hydrological issues such as modeling of evapotranspiration, snow-melt, or soil moisture estimation (utilizing morning heating rates) [61].

In Figure 12 we depict the diurnal variation of LST as observed at four SURFRAD/BSRN stations and from GOES-12. Notable is the large amplitude at the dry site of Desert Rock (DRA) (characterized as desert, gravel, flat, rural) as compared to the more vegetated regions at the other sites (BON is grass, flat, rural; FPK is grass, flat while GWN is grass, hilly, rural). The effect of latitude is also evident. The amplitude at GWN which is at ~34°N is much smaller than the amplitudes at the higher latitude stations (BON at ~40° and FPK at ~48°). Of interest are the differences between satellite estimates and ground observations which are more noticeable at DRA and FPK than at the other sites. A possible explanation for DRA is the lower homogeneity of the site compared to the others. The FPK is at higher elevation than BND and GCM and also at higher latitude so possibly, the cooling of the ground at the observational site may not represent the grid domain. Additional investigation is needed to better understand the behavior at these four sites during the earlier part of the day. The full agreement between the satellite and ground observations from about noon to late afternoon can possibly be due to more even heating of the ground than at the earlier hours of the day when the higher moisture content can differentially affect the emissivity. While ground observations are very sparse, the findings shown in Figure 12 indicate that satellites alone can be used to characterize the diurnal cycle over the domain of the GOES satellites (a comprehensive analysis over the entire US is needed). This information is of considerable interest since most satellite based estimates of LST use polar orbiters unable to depict the true diurnal cycle.

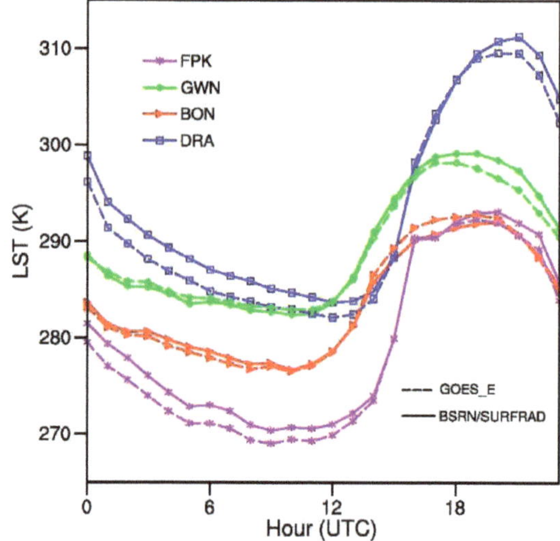

Figure 12. A six-year average (2004–2009) of the hourly LST at four SURFRAD/BSRN stations as observed (solid line) and as derived from the GOES observations (broken line).

5. Discussion

Available information on LST and DTR from remotely sensed data is deficient. Discrepancies and inconsistencies arise due to the quality of the satellite and the ground observations, differences in their spatial, spectral and temporal resolution, as well as differences in the inference methods and auxiliary data used. In principle, the well-established split window approach is known to perform better than the use of a single channel for deriving LST however, the 12 μm channel is not available any more during the operational period of GOES 12-15. To homogenize satellite observations to obtain a consistent long term record requires the utilization of observations from a single channel only.

With the advancement in archiving of satellite data, their maintenance in terms of calibration, geolocation, improved inference schemes and auxiliary information, it is timely to formulate an approach for deriving long-term, consistent, and calibrated data across multiple satellite sensors, as demanded by the user community. Progress has also been made in ground observations in terms of instrument characterization, guidelines for high quality maintenance and calibration. The issue of optimal coupling between satellite and ground observations is still widely debated.

LST is known to have large spatial variability at different temporal scale (diurnal, annual, inter-annual) and this variability has an informative value. For instance, the importance of the diurnal cycle of LST has been widely recognized [62–64] and numerous attempts have been made to estimate it. In an early attempt [65], used were the International Satellite Cloud Climatology Project (ISCCP) data (at 280 km resolution C-2 product) [66] in combination with ground observations to derive the monthly mean diurnal cycle in surface temperature over land (suitable for Global Climate studies). Duan et al. [64] tried to determine it using High Spatial Resolution Clear-Sky MODIS Data while Inamdar et al. [33] dis-aggregated the diurnal cycle of LST at the GOES pixel scale to that of the MODIS pixel scale. Yet, the daytime and nighttime products from polar orbiting satellites (e.g., MODIS) do not fully represent the daily amplitude as feasible from GEO satellites. Our effort represents a contribution to the development of a framework for obtaining long term records of consistent LST and DTR from the entire record of GOES satellites, using a physically based approach and utilizing the best currently available auxiliary information and the best available ground observations to evaluate the proposed approach.

In the evaluation process, factors that play a role include differences in ground instrumentation, their location above the surface, method of estimating LST from the measured outgoing LW radiation, calibration and maintenance of the instruments and scale issues between ground observations and satellite footprints. There is a need to ensure that the satellite observations used represent clear sky condition. Detailed information on each of these factors is needed for a full assessment of errors in the retrieved LST products. While under controlled short term experiments the uncertainty of many of these factors can be minimized, the results obtained are not representative for extended areas and all seasons.

In this paper, the quality of the new product is evaluated against extensive record of best available observations and products that are accepted by the scientific community. Specifically, evaluation of a six-year record of instantaneous LST as well as monthly averages was performed against the DOE Atmospheric Radiation Measurement (ARM) site at the Southern Great Plains central facility, the BSRN/SURFRAD stations, MOD11 products and the Oklahoma MESONET sites.

While the quality of the instrumentation used at each site can be traced to factory specifications, it is not possible to establish how much differences in daily maintenance at each site contribute to the quality of the observations. The hypothesis of our approach is that by using long term observations at numerous sites and seasons, the evaluation results do provide an indication on the robustness of the approach. One of the major factors affecting the evaluation results is related to cloud screening which vary among methodologies as recently discussed in Ermida et al. [67]. Spatial and temporal variability in emissivity are also a contributing factors. As reported in [6] a brightness temperature error due to emissivity error in the 11 µm band is about 3% for a 0.5% error in emissivity and up to 5% for an emissivity error of 2.0%; these estimates are based on global simulations over a wide temperature range. To fully understand discrepancies between products, there is a need in controlled experiments to evaluate independently factors that can cause differences. Till now, available retrievals are based on different satellite observation, different retrieval methodology, atmospheric inputs and time periods. An early attempt to compare the performance of several well-known algorithms was presented in {6]. To make such algorithm comparison consistent the individual methodologies need to be modified; it is necessary to rederive relevant coefficients of the algorithms used in a systematic manner using the same inputs. The need for controlled experiments to facilitate discussion on sources of discrepancies between methods has been recognized by the scientific community and is conducted frequently. Examples are numerical model evaluation as conducted at Lawrence Livermore National Laboratory (https://pcmdi.llnl.gov/?projects/amip/0), while controlled experiments to estimate errors due to aerosols is described in Randles et al. [68]. The objective of the current study is to present a credible methodology to generate long term time series of LST at best available spatial and temporal resolution (that currently are possible with a long term outlook), and evaluate it against best available satellite products and ground observations. Used were long term observations that represent different climatic regions and seasons that provided statistically robust indication on the soundness of the propose approach. The need for further work to investigate the sources of discrepancies is also recognized. Limitations and advantages of each methods and their trade-offs need also to be fully understood.

6. Summary

In principle, the split window approach is known to perform better than a single channel to derive LSTs. But the 12 µm channel is not available any more during the operation periods of GOES 12–15. To homogenize satellite observations to a consistent long term record requires the use of a single channel observation.

We have implemented the RTTOV radiative transfer approach adjusted for GEO channel 4 to derive LST at the high resolution of about 5-km. The model is driven with the MERRA-2 reanalysis profiles for water vapor and temperature and the CAMEL product. A homogeneous six year record of LST at 0.05° spatial resolution at hourly time scale was produced from GOES observations and evaluated for the period of 2004–2009. A six year climatology at monthly time scales was also derived and used to construct representative diurnal cycles for selected surface type.

The results shows that there is a close agreement between the GEO and MOD11 products. The averaged correlation coefficient between them is over 0.9. The averaged difference is less than 2 K and the averaged *rmse* is less than 3.5 K. It was also found that the derived LST has very close correlation with ground-based observations. In most cases, the correlation coefficients are greater than 0.9. The mean differences between the satellite LST and the station LST are less than 1% and over 80% of the differences fall within 1 *std*. The performance of retrieved LST for daytime and nighttime are comparable to each other after elimination of outliers caused by imperfect cloud detection. The estimated quality of the LST information can serve as a guideline for users in a wide range of applications, such as a realistic representation of the diurnal cycle.

Future improvement would be possible by satellite observations of higher spatial resolution, the incorporation of higher temporal resolution of surface emissivity and improved/innovative methodologies to remove cloud contamination [68] and by accounting for anisotropy in emissivity Pinheiro et al. [69], Ermida et al. [70].

Author Contributions: Conceptualization, R.P.; Data curation, G.H., E.B. and J.B.; Formal analysis, W.C.; Funding acquisition, S.H.; Investigation, S.H.; Methodology, R.P. and Y.M.; Project administration, K.C.-N.; Resources, C.H.; Software, Y.M. and T.I.; Validation, W.C.; Writing—original draft, R.P.; Writing—review & editing, K.C.-N.

Funding: This research was funded by the National Aeronautics and Space Administration, grant NNH12ZDA001N-MEASURES to Jet Propulsion Laboratory.

Acknowledgments: We acknowledge the ECMWF for providing the ERA-I data; the MERRA-2 data are provided by the Global Modeling and Assimilation Office (GMAO). Information from the MOD11_L2: MODIS/Terra Land Surface Temperature and Emissivity 5-Minute L2 Swath 1 km V006 data base (Zhengming Wan, PI) (https://search.earthdata.nasa.gov/search/granules/collectiondetails?p=C194001236-LPDAAC_ECS&m=-26.4375!136.6875!0!1!0!0%2C2&tl=1515438649!4!!&q=MOD11_L2%20V006), was provided under the courtesy of the NASA EOSDIS Land Processes Distributed Active Archive Center (LP DAAC), USGS/Earth Resources Observation and Science (EROS) Center, Sioux Falls, South Dakota. GOES data were obtained from the NOAA Comprehensive Large Array data Stewardship system (CLASS) (https://www.bou.class.noaa.gov/saa/products/search?sub_id=0&datatype_family=GVAR_IMG&submit.x=25&submit.y=8). The BSRN/SURFRAD data were provided by the NOAA Earth System Research Laboratory, Global Monitoring Division (https://www.esrl.noaa.gov/gmd/grad/surfrad/). Data were obtained from the Atmospheric Radiation Measurement (ARM) user facility, a U.S. Department of Energy (DOE) office of science user facility managed by the office of Biological and Environmental Research. The team effort in generating and providing all the required data is greatly appreciated. We thank the anonymous Reviewers for constructive comments that helped to improve the manuscript.

Conflicts of Interest: The authors declare no conflict of interest.

References

1. Trenberth, K.E.; Stepaniak, D.P.; Caron, J.M. Accuracy of atmospheric energy budgets from analyses. *J. Clim.* **2002**, *15*, 3343–3360. [CrossRef]

2. Garand, L.; Buehner, M.; Wagneur, N. Background Error Correlation between Surface Skin and Air Temperatures: Estimation and Impact on the Assimilation of Infrared Window Radiances. *J. Appl. Met.* **2004**, *43*, 1853–1863. [CrossRef]

3. Price, J.C. Estimation of surface temperatures from satellite thermal infrared data—A simple formulation for the atmospheric effect. *Remote Sens. Environ.* **1983**, *13*, 353–361. [CrossRef]

4. Becker, F.; Li, Z.L. Toward a local split window method over land surface. *Int. J. Remote Sens.* **1990**, *11*, 369–393. [CrossRef]

5. Wan, Z.M.; Dozier, J. A generalized split-window algorithm for retrieving land-surface temperature from space. *IEEE Trans. Geosci. Remote Sens.* **1996**, *34*, 892–905.

6. Sun, D.; Pinker, R.T. Estimation of land surface temperature from a Geostationary Operational Environmental Satellite (GOES-8). *J. Geophys. Res.* **2003**, *108*, 4326. [CrossRef]

7. Sobrino, J.A.; Li, Z.L.; Stoll, M.P.; Becker, F. Improvements in the split window technique for land surface temperature determination. *IEEE Trans. Geosci. Remote Sens.* **1994**, *32*, 243–253. [CrossRef]

8. Nerry, F.; Labed, J.; Stoll, M.P. Spectral properties of land surfaces in the thermal infrared: 1. Laboratory measurements of absolute spectral emissivity signatures. *J. Geophys. Res.* **1990**, *95*, 7027–7044. [CrossRef]

9. Hulley, G.C.; Hughes, C.G.; Hook, S.J. Quantifying uncertainties in land surface temperature and emissivity retrievals from ASTER and MODIS thermal infrared data. *J. Geophys. Res.* **2012**, *117*, D23113. [CrossRef]
10. Lorenz, D. The effect of the long-wave reflectivity of natural surfaces on surface temperature measurements using radiometers. *J. Appl. Meteorol.* **1986**, *5*, 421–430. [CrossRef]
11. Li, Z.-L.; Tang, B.-H.; Wu, H.; Ren, H.; Yan, G.; Wan, Z.; Trigo, I.F.; Sobrino, J.A. Satellite-derived land surface temperature: Current status and perspectives. *Remote Sens. Environ.* **2013**, *131*, 14–37. [CrossRef]
12. Prabhakara, C.; Dalu, G.; Kunde, V.G. Estimation of sea surface temperature from remote sensing in the 11- and 13-μm window region. *J. Geophys. Res.* **1974**, *79*, 5039–5044. [CrossRef]
13. McClain, E.P.; Pichel, W.G.; Walton, C.C.; Ahmad, Z.; Sutton, J. Multichannel improvements to satellite-derived global sea surface temperatures. *Adv. Space Res.* **1983**, *2*, 43–47. [CrossRef]
14. Becker, F.; Li, Z.-L. Surface temperature and emissivity at various scales: Definition, measurement and related problems. *Remote Sens. Rev.* **1995**, *12*, 225–253. [CrossRef]
15. McMillin, L.M. Estimation of sea surface temperatures from two infrared window measurements with different absorption. *J. Geophys. Res.* **1975**, *80*, 5113–5117. [CrossRef]
16. Prata, A.J. Land surface temperatures derived from the advanced very high resolution radiometer and the along-track scanning radiometer: 2. Experimental results and validation of AVHRR algorithms. *J. Geophys. Res.* **1994**, *99*, 13025–13058. [CrossRef]
17. François, C.; Ottlé, C. Atmospheric corrections in the thermal infrared: Global and water vapor dependent split-window algorithms-applications to ATSR and AVHRR data. *IEEE Trans. Geosci. Remote Sens.* **1996**, *34*, 457–470. [CrossRef]
18. Coll, C.; Caselles, V.; Galve, J.M.; Valor, E.; Niclòs, R.; Sanchez, J.M. Ground measurements for the validation of land surface temperatures derived from AATSR and MODIS data. *Remote Sens. Environ.* **2005**, *97*, 288–300. [CrossRef]
19. Trigo, I.F.; Monteiro, I.T.; Olesen, F.; Kabsch, E. An assessment of remotely sensed land surface temperature. *J. Geophys. Res.* **2008**, *113*, D17108. [CrossRef]
20. Wan, Z. New refinements and validation of the collection-6 MODIS land-surface temperature/emissivity product. *Remote Sens. Environ.* **2014**, *140*, 36–45. [CrossRef]
21. Van de Griend, A.A.; Owe, M. On the relationship between thermal emissivity and the normalized difference vegetation index for natural surfaces. *Int. J. Remote Sens.* **1993**, *14*, 1119–1131. [CrossRef]
22. Valor, E.; Caselles, V. Mapping land surface emissivity from NDVI: Application to European, African, and South American areas. *Remote Sens. Environ.* **1996**, *57*, 167–184. [CrossRef]
23. Karnieli, A.; Agam, N.; Pinker, R.T.; Anderson, M.; Imhoff, M.I.; Gutman, G.G.; Panov, N.; Goldberg, A. Use of NDVI and LST for Assessing Vegetation Health: Merits and Limitations. *J. Clim.* **2010**, *23*, 618–633. [CrossRef]
24. Kealy, P.S.; Hook, S.J. Separating temperature and emissivity in thermal infrared multispectral scanner data: Implications for recovering land surface temperatures. *IEEE Trans. Geosci. Remote Sens.* **1993**, *31*, 1155–1164. [CrossRef]
25. Gillespie, A.; Rokugawa, S.; Matsunaga, T.; Cothern, J.S.; Hook, S.; Kahle, A.B. A temperature and emissivity separation algorithm for Advanced Spaceborne Thermal Emission and Reflection Radiometer (ASTER) images. *IEEE Trans. Geosci. Remote Sens.* **1998**, *36*, 1113–1126. [CrossRef]
26. Wan, Z.; Li, Z.-L. A physics-based algorithm for retrieving land-surface emissivity and temperature from EOS/MODIS data. *IEEE Trans. Geosci. Remote Sens.* **1997**, *35*, 980–996.
27. Liang, S. An optimization algorithm for separating land surface temperature and emissivity from multispectral thermal infrared imagery. *IEEE Trans. Geosci. Remote Sens.* **2001**, *39*, 264–274. [CrossRef]
28. Schmugge, T.; French, A.; Ritchie, J.C.; Rango, A.; Pelgrum, H. Temperature and emissivity separation from multispectral data. *Remote Sens. Environ.* **2001**, *78*, 189–198.
29. Ma, X.L.; Wan, Z.; Moeller, C.C.; Menzel, W.P.; Gumley, L.E. Simultaneous retrieval of atmospheric profiles, land-surface temperature, and surface emissivity from Moderate-Resolution Imaging Spectroradiometer thermal infrared data: Extension of a two-step physical algorithm. *Appl. Opt.* **2002**, *41*, 909–924. [CrossRef] [PubMed]
30. Borbas, E.; Hulley, G.; Feltz, M.; Knuteson, R.; Hook, S. The Combined ASTER MODIS Emissivity over Land (CAMEL) Part 1: Methodology and High Spectral Resolution Application. *Remote Sens.* **2018**, *10*, 643. [CrossRef]

31. Feltz, M.; Borbas, E.; Knuteson, R.; Hulley, G.; Hook, S. The Combined ASTER MODIS Emissivity over Land (CAMEL) Part 2: Uncertainty and Validation. *Remote Sens.* **2018**, *10*, 664. [CrossRef]

32. Sun, D.; Pinker, R.T. Implementation of GOES-based land surface temperature diurnal cycle to AVHRR. *Int. J. Remote Sens.* **2005**, *26*, 3975–3984. [CrossRef]

33. Inamdar, A.K.; French, A.; Hook, S.; Vaughan, G.; Luckett, W. Land surface temperature retrieval at high spatial and temporal resolutions over the southwestern United States. *J. Geophys. Res.* **2008**, *113*, D07107. [CrossRef]

34. Heidinger, A.K.; Evan, A.T.; Foster, M.J.; Walther, A. A naive Bayesian cloud detection scheme derived from CALIPSO and applied with PATMOS-x. *J. Appl. Meteorol. Climatol.* **2012**, *51*, 1129–1144. [CrossRef]

35. Scarino, B.R.; Minnis, P.; Chee, T.; Bedka, K.M.; Yost, C.R.; Palikonda, R. Global clear-sky surface skin temperature from multiple satellites using a single-channel algorithm with angular anisotropy corrections. *Atmos. Meas. Tech.* **2017**, *10*, 351–371. [CrossRef]

36. Gunshor, M.M.; Schmit, T.J.; Menzel, W.P.; Tobin, D.C. Intercalibration of broadband geostationary imagers using AIRS. *J. Atmos. Ocean. Technol.* **2009**, *26*, 746–758. [CrossRef]

37. Weinreb, M.P.; Jamieson, M.; Fulton, N.; Chen, Y.; Johnson, J.X.; Bremer, J.; Smith, C.; Baucom, J. Operational calibration of geostationary operational environmental Satellite-8 and-9 imagers and sounders. *Appl. Opt.* **2007**, *36*, 6895–6904. [CrossRef]

38. Gelaro, R.; McCarty, W.; Suárez, M.J.; Todling, R.; Molod, A.; Takacs, L.; Randles, C.A.; Darmenov, A.; Bosilovich, M.G.; Reichle, R.; et al. The Modern-Era Retrospective Analysis for Research and Applications, Version 2 (MERRA-2). *J. Clim.* **2017**, *30*, 5419–5454. [CrossRef]

39. Wan, Z. *MODIS Land Surface Temperature Products Users' Guide*; Collection-6, ERI; University of California: Santa Barbara, CA, USA, 2013. [CrossRef]

40. Hicks, B.B.; DeLuisi, J.J.; Matt, D.R. The NOAA Integrated Surface Irradiance Study (ISIS)—A new surface radiation monitoring program. *Bull. Am. Meteorol. Soc.* **1996**, *77*, 2857–2864. [CrossRef]

41. Ohmura, A.; Dutton, E.G.; Forgan, B.; Fröhlich, C.; Gilgen, H.; Hegner, H.; Heimo, A.; König-Langlo, G.; McArthur, B.; Müller, G.; et al. Baseline Surface Radiation Network (BSRN/WCRP): New precision radiometry for climate research. *Bull. Am. Meteorol. Soc.* **1998**, *79*, 2115–2136. [CrossRef]

42. Augustine, J.A.; Hodges, G.B.; Cornwall, C.R.; Michalsky, J.J.; Medina, C.I. An update on SURFRAD: The GCOS surface radiation budget network for the continental United States. *J. Atmos. Ocean. Technol.* **2005**, *22*, 1460–1472. [CrossRef]

43. Guillevic, P.; Göttsche, F.; Nickeson, J.; Hulley, G.; Ghent, D.; Yu, Y.; Trigo, I.; Hook, S.; Sobrino, J.A.; Remedios, J.; et al. Land Surface Temperature Product Validation Best Practice Protocol, Version 1.0. In *Best Practice for Satellite-Derived Land Product Validation (p. 60): Land Product Validation Subgroup (WGCV/CEOS)*; Guillevic, P., Göttsche, F., Nickeson, J., Román, M., Eds.; Internal Publication: Brussels, Belgium, 2017. [CrossRef]

44. McPherson, R.A.; Fiebrich, C.A.; Crawford, K.C.; Kilby, J.R.; Grimsley, D.L.; Martinez, J.E.; Basara, J.B.; Illston, B.G.; Morris, D.A.; Kloesel, K.A.; et al. Statewide monitoring of the mesoscale environment: A technical update on the Oklahoma Mesonet. *J. Atmos. Ocean. Technol.* **2007**, *24*, 301–321. [CrossRef]

45. Fiebrich, C.A.; Martinez, J.E.; Brotzge, J.A.; Basara, J.B. The Oklahoma Mesonet's skin temperature network. *J. Atmos. Ocean. Technol.* **2003**, *20*, 1496–1504. [CrossRef]

46. Fuchs, M. Infrared measurement of canopy temperature and detection of plant water stress. *Theor. Appl. Climatol.* **1990**, *42*, 253–261. [CrossRef]

47. Sobrino, J.A.; Skokovic, D. Permanent Stations for Calibration/Validation of Thermal Sensors over Spain. *Data* **2016**, *1*, 10. [CrossRef]

48. Hansen, M.C.; Defries, R.S.; Townshend, J.R.G.; Sohlberg, R. Global land cover classification at 1 km resolution using a classification tree approach. *Int. J. Remote Sens.* **1998**, *21*, 1331–1364. [CrossRef]

49. Li, X.; Pinker, R.T.; Wonsick, M.M.; Ma, Y. Toward improved satellite estimates of short-wave radiative fluxes-Focus on cloud detection over snow: 1. Methodology. *J. Geophys. Res.* **2007**, *112*, D07208. [CrossRef]

50. Pinker, R.T.; Li, X.; Meng, W.; Yegorova, E.A. Toward improved satellite estimates of short-wave radiative fluxes-Focus on cloud detection over snow: 2. Results. *J. Geophys. Res.* **2007**, *112*, D09204. [CrossRef]

51. Wang, H.; Pinker, R.T.; Minnis, P.; Khaiyer, M.M. Experiments with Cloud Properties: Impact on Surface Radiative Fluxes. *J. Atmos. Ocean. Technol.* **2008**, *25*, 1034–1040. [CrossRef]

52. Miller, S.D. Physical decoupling of the GOES daytime 3.9 μm channel thermal emission and solar reflection components using total solar eclipse data. *Int. J. Remote Sens.* **2001**, *22*, 9–34. [CrossRef]

53. Eyre, J.R. *A Fast Radiative Transfer Model. for Satellite Sounding System*; ECMWF Research Department Technical Memorandum 176; European Centre for Medium-Range Weather Forecasts: Reading, UK, 1991.

54. Saunders, R.W.; Matricardi, M.; Brunel, P. An Improved Fast Radiative Transfer Model for Assimilation of Satellite Radiance Observations. *QJRMS* **1991**, *125*, 1407–1425. [CrossRef]

55. Matricardi, M.; Saunders, R. Fast radiative transfer model for simulation of infrared atmospheric sounding interferometer radiances. *Appl. Opt.* **1999**, *38*, 5679–5691. [CrossRef] [PubMed]

56. Hulley, G.; Freepartner, R.; Malakar, N.; Sarkar, S. *Moderate Resolution Imaging Spectroradiometer (MODIS) Land Surface Temperature and Emissivity Product (MOD21) Users' Guide*; Collection-6; Jet Propulsion Laboratory California Institute of Technology: Pasadena, CA, USA, 2016.

57. Göttsche, F.; Olesen, F.S.; Høyer, J.L.; Wimmer, W.; Nightingale, T. *Fiducial Reference Measurements for Validation of Surface Temperature from Satellites (FRM4STS)*; Technical Report 3—A Framework to Verify the Field Performance of TIR FRM; ESA Contract No. 4000113848_15I-LG; Internal Publication: Brussels, Belgium, 2017; pp. 1–75.

58. Hulley, G.C.; Hook, S.J. Generating Consistent Land Surface Temperature and Emissivity Products between ASTER and MODIS Data for Earth Science Research. *IEEE Trans. Geosci. Remote Sens.* **2011**, *49*, 1304–1315. [CrossRef]

59. Yu, Y.; Privette, J.P.; Pinheiro, A.C. Evaluation of split-window land surface temperature algorithms for generating climate data records. *IEEE Trans. Geosci. Remote Sens.* **2008**, *46*, 179–192. [CrossRef]

60. Snyder, W.; Wan, Z.; Zhang, Y.; Feng, Y. Classification based emissivity for land surface temperature measurement from space. *Int. J. Remote Sens.* **1998**, *19*, 2753–2774. [CrossRef]

61. Zhang, D.; Tang, R.; Zhao, W.; Tang, B.; Wu, H.; Shao, K.; Li, Z.-L. Surface Soil Water Content Estimation from Thermal Remote Sensing based on the Temporal Variation of Land Surface Temperature. *Remote Sens.* **2014**, *6*, 3170–3187. [CrossRef]

62. Dai, A.; Trenberth, K.E. The diurnal cycle and its depiction in the Community Climate System Model. *J. Clim.* **2004**, *17*, 930–951. [CrossRef]

63. Aires, F.; Prigent, C.; Rossow, W.B. Temporal interpolation of global surface skin temperature diurnal cycle over land under clear and cloudy conditions. *J. Geophys. Res.* **2004**, *109*, D04313. [CrossRef]

64. Duan, S.-B.; Li, Z.-L.; Tang, B.-H.; Wu, H.; Tang, R.; Bi, Y.; Zhou, G. Estimation of Diurnal Cycle of Land Surface Temperature at High Temporal and Spatial Resolution from Clear-Sky MODIS Data. *Remote Sens.* **2014**, *6*, 3247–3262. [CrossRef]

65. Ignatov, A.; Gutman, G. Monthly mean diurnal cycles in surface temperature over and land for global climate studies. *J. Clim.* **1999**, *12*, 1900–1910. [CrossRef]

66. Rossow, W.B.; Schiffer, R.A. ISCCP Cloud Data Products. *Bull. Am. Meteorol. Soc.* **1991**, *72*, 2–20. [CrossRef]

67. Randles, C.A.; Kinne, S.; Myhre, G.; Schulz, M.; Stier, P.; Fischer, J.; Doppler, L.; Highwood, E.; Ryder, C.; Harris, B.; et al. Intercomparison of shortwave radiative transfer schemes in global aerosol modeling: Results from the AeroCom Radiative Transfer Experiment. *Atmos. Chem. Phys.* **2013**, *13*, 2347–2379. [CrossRef]

68. Ermida, S.L.; Trigo, I.S.; DaCamara, C.C.; Jiménez, C.; Prigent, C. Quantifying the Clear-Sky Bias of Satellite Land SurfaceTemperature Using Microwave-Based Estimates. *J. Geophys. Res.* **2019**, *124*, 844–857. [CrossRef]

69. Pinheiro, A.C.T.; Jeffrey, L. Privette, and Pierre Guillevic, Modeling the Observed Angular Anisotropy of Land Surface Temperature in a Savanna. *IEEE Trans. Geosci. Remote Sens.* **2006**, *44*, 1036–1047. [CrossRef]

70. Ermida, S.L.; Trigo, I.S.; DaCamara, C.C.; Pires, A.C. A Methodology to Simulate LST Directional Effects Based on Parametric Models and Landscape Properties. *Remote Sens.* **2018**, *10*, 1114. [CrossRef]

Article

Reconstructing One Kilometre Resolution Daily Clear-Sky LST for China's Landmass Using the BME Method

Yunfei Zhang [1,2], **Yunhao Chen** [1,2,*] , **Yang Li** [1,2], **Haiping Xia** [1,2] and **Jing Li** [1,2]

1   State Key Laboratory of Remote Sensing Science, Faculty of Geographical Science, Beijing Normal University, Beijing 100875, China; 201431480008@mail.bnu.edu.cn (Y.Z.); leeyang@mail.bnu.edu.cn (Y.L.); xiahp93@mail.bnu.edu.cn (H.X.); lijing@bnu.edu.cn (J.L.)
2   Beijing Key Laboratory of Environmental Remote Sensing and Digital City, Beijing Normal University, Beijing 100875, China
*   Correspondence: cyh@bnu.edu.cn; Tel.: +86-010-5880-4056

Received: 18 September 2019; Accepted: 5 November 2019; Published: 7 November 2019

Abstract: The land surface temperature (LST) is a key parameter used to characterize the interaction between land and the atmosphere. Therefore, obtaining highly accurate, spatially consistent and temporally continuous LSTs in large areas is the basis of many studies. The Moderate Resolution Imaging Spectroradiometer (MODIS) LST product is commonly used to achieve this. However, it has many missing values caused by clouds and other factors. The current gap-filling methods need to be improved when applied to large areas. In this study, we used the Bayesian maximum entropy (BME) method, which considers spatial and temporal correlation, and takes multiple regression results of the Normalized Difference Vegetation Index (NDVI), Digital Elevation Model (DEM), longitude and latitude as soft data to reconstruct space-complete daily clear-sky LSTs with a 1 km resolution for China's landmass in 2015. The average Root Mean Square Error (RMSE) of this method was 1.6 K for the daytime and 1.2 K for the nighttime when we simultaneously covered more than 10,000 verification points, including blocks that were continuous in space, and the average RMSE of a single discrete verification point for 365 days was 0.4 K for the daytime and 0.3 K for the nighttime when we covered four discrete points. Urban and snow land cover types have a higher accuracy than forests and grasslands, and the accuracy is higher in winter than in summer. The high accuracy and great ability of this method to capture extreme values in urban areas can help improve urban heat island research. This method can also be extended to other study areas, other time periods, and the estimation of other geographical attribute values. How to effectively convert clear-sky LST into real LST requires further research.

Keywords: land surface temperature; MODIS; Bayesian Maximum Entropy; interpolation

---

1. Introduction

The land surface temperature (LST), generally defined as the radiative skin temperature of the ground, is closely related to the radiative budget and energy fluxes between the atmosphere and the ground [1–4]. LST plays an important role in the estimation of climate models, environmental models and evapotranspiration models, as well as the calculation of drought indices, soil moisture contents and mortality rates [5–14].

Compared with LST measurements at ground stations, satellite remote sensing observations have the advantages of easy acquisition and complete spatial coverage over large areas. Typical LST products include Advanced Spaceborne Thermal Emission and Reflection Radiometer (ASTER), Moderate Resolution Imaging Spectroradiometer (MODIS) and Meteosat Second Generation Spinning

Enhanced Visible and Infrared Imager (MSG-SEVIRI) datasets, with spatial resolutions of 90 m, 1 km and 3 km, respectively [15–17]. Among them, the MODIS LST product is the most widely used and best suited for our research because of its appropriate spatial resolution (1 km), high temporal resolution (four overpasses per day), wide coverage (globe), and high retrieval accuracy (approximately better than 1 K). The MODIS instruments were launched on the Sun-synchronous satellites of Terra and Aqua in December 1999 and May 2002, respectively [18]. The MODIS LST products are generated with bands 31 and 32 of MODIS's 36 spectral using the split window algorithm [2]. The latest version C6 MODIS LST products have different spatial resolutions of 1 km, 6 km, and 0.05° and different temporal resolutions of daily, eight days and monthly. The MOD11A1 and MYD11A1 are 1 km daily Level 3 products in those MODIS LST products. The transit time of Terra corresponding to MOD11A1 is about 10:30 (22:30), while the transit time of Aqua corresponding to MYD11A1 is about 13:30 (1:30). They are both processed into sinusoidal projection and stored in tiles containing 1200 rows and 1200 columns. The quality of MODIS products is continuing to improve, from more than 2 K in the previous versions to less than 2 K (within ±1 K in most cases) in the C6 version [4]. MODIS LST products have been widely used in LST research [19–21]. However, the MODIS LST product can only provide usable values under clear-sky conditions, and its spatial integrity is thus affected by clouds or other atmospheric disturbances. Taking China as an example, more than half of the pixels per day have no observations on average, and these gaps seriously hinder the application of the MODIS LST product.

Several gap-filling methods have been developed to reconstruct LSTs under cloudy conditions to obtain spatiotemporally-continuous LST products. In general, these methods can be divided into two main groups: clear-sky LST [19–30] and cloudy-sky LST [31–37] methods. Clear-sky LST represents the retrieved LST assuming no cloud effects, whereas cloudy-sky LST represents the actual LST of the reconstruction considering cloud effects. Usually, clear-sky LST is slightly higher than cloudy-sky LST. The methods for reconstructing cloudy-sky LST, mostly based on surface energy balances, often use passive microwave remote sensing data or require ground station measurements or shortwave radiation products. Nonetheless, microwave data have a coarse spatial resolution and an accuracy that needs improvement. Moreover, ground station measurements or shortwave radiation products with a high spatial and temporal resolution are difficult to obtain. This study focused on reconstructing the clear-sky LST, first, because improving the accuracy of clear-sky LST is conducive to further determining the cloudy-sky LST better, and second, because clear-sky LST can be directly applied to research fields such as numerical weather prediction [38], the identification of diurnal patterns of urban heat islands [39], and calculation of the Temperature-Vegetation Dryness Index (TVDI) or Temperature-Vegetation-soil Moisture Dryness Index (TVMDI) [40,41].

The methods for clear-sky LST may be divided into four categories, according to the underlying principles: considering temporal correlation, considering spatial correlation, considering auxiliary information, and the hybrid method. Details of each category are as follows: (1) LST has a temporal correlation because, for the same pixel at different times, the surface properties are the same and only different weather factors, such as solar radiation and wind speed, cause LST differences. Therefore, the first category reconstructs LST based on the temporal correlation using temporal interpolation methods or methods that employ correlations at different times [22,23]; (2) LST also has a spatial correlation because different pixels at the same time have the same weather factors and different surface properties (such as elevation and land cover), but only the surface properties cause LST differences. The second category thus reconstructs LST based on the spatial correlation using spatial interpolation methods [19,24]; (3) in addition, LST is affected by related factors such as elevation and NDVI. The third category thus estimates the missing LST using the empirical relationship between LST and the auxiliary information, which has a similar spatiotemporal resolution to LST and a better spatial coverage integrity than LST [20,21]; (4) finally, the fourth category is hybrid methods that combine two or three of the above methods, such as spatiotemporal gap-filling methods or spatial interpolation methods that consider auxiliary information [25–30]. In general, the hybrid approach is the most promising. Considering only temporal correlation is not suitable for regions with high spatial heterogeneity. If only spatial correlation is taken into account, the results will be inaccurate for areas that have large weather changes in a short period of time. If only auxiliary information is considered, the accuracy of regression and the uncertainty of auxiliary information will affect the final results. In previous studies, there have been relatively few methods suitable for LST reconstruction in large areas. In a region as large as China, where climate change is complex, spatial heterogeneity is high, and auxiliary information has considerable uncertainty, a method is needed that can comprehensively and reasonably consider time correlation, spatial correlation, and auxiliary information, and the uncertainty of auxiliary information should also be considered.

Bayesian maximum entropy (BME) is a spatiotemporal statistical method proposed by Christakos that can provide a systematic and rigorous framework for incorporating hard data, soft data and other sources of information into the estimation of variables [42,43]. BME has several attractive features; it does not need to make any assumptions regarding the linearity of the estimator, the normality of the underlying probability laws, or the homogeneity of the spatial distribution. Moreover, BME is capable of considering uncertainties contained in the data. The method has been successfully applied to numerous areas, such as air pollution, soil properties, water demand and disease [44–61]. It has also achieved good results in the gap-filling of remote sensing data [62–64]. The BME method is suitable for our research because it can not only take advantage of the temporal correlation and spatial correlation of the LST, but can also explicitly consider the uncertainties of the auxiliary information.

In this study, we applied the BME-based interpolation method to reconstruct 1 km resolution daily clear-sky LST for China's landmass considering temporal correlation, spatial correlation and auxiliary information. The goals of this article are to (1) examine the feasibility of the BME method to reconstruct LST for the whole of China, (2) discuss the accuracy of the BME method for different land cover types, and (3) compare the BME method with other commonly used LST reconstruction methods.

2. Materials and Methods

2.1. Study Area

In this study, we selected the land area of 34 provinces in China as our study area. China is located on the eastern side of the Eurasian continent and the western shore of the Pacific Ocean; it spans approximately 5500 km from north to south and 5000 km from west to east. The topography across China is complicated and includes plains, plateaus, mountains, hills and basins. It varies from the Qinghai-Tibet Plateau at more than 4000 m above sea level (peaking at 8848 m) to its eastern coastline on the Pacific Ocean. China's land resources are vast, and its use types are diverse. The cultivated lands are mainly distributed in the eastern region, the forests are distributed in the south and northeast regions, the grasslands are mainly distributed in the central and southwestern regions, and the unused lands are mainly distributed in the northwestern region. China's climate is governed by monsoonal circulations, and winters with low temperatures and little rain significantly differ from summers with high temperatures and abundant rain [65].

2.2. Data Acquisition and Preprocessing

LST can be affected by many factors [20,66,67]. In view of relatively more critical factors across the Chinese scale and the convenience of data acquisition and processing, NDVI, DEM, longitude and latitude were selected as auxiliary data to regress LST. The following specific datasets were used in this study: (1) For LST, we used the MODIS/Aqua LST Daily L3 Global 1 km SIN Grid product (MYD11A1, Collection 6). LSTs observed throughout the year 2015 were used at local 1:30/13:30 overpass times, which approximate daily minimum and maximum LST values; in the later part of the article, we call them daytime LST and nighttime LST; (2) for NDVI, we selected the MODIS/Aqua 16 day 1 km Vegetation Index product (MYD13A2, Collection6), which has great spatial completeness and the spatial resolution we need. There are 23 NDVI data sets of MYD13A2 for the entire year of 2015; (3) for DEM, we used the Shuttle Radar Topography Mission (SRTM) Digital Elevation Data Version 4 at a 90 m spatial resolution produced to provide consistent, high-quality elevation data. The original DEM data were resized using nearest-neighbours from 90 m to 1 km; (4) for land cover data, we used the MODIS Land Cover Type Yearly Global 500 m product (MCD12Q1) from 2015, which was derived using supervised classifications of MODIS Terra and Aqua reflectance data. We combined all the land types into six categories: forests, grasslands, croplands, barren, urban and snow. Evergreen needleleaf forests, evergreen broadleaf forests, deciduous needleleaf forests, deciduous broadleaf forests, mixed forests, closed shrublands and open shrublands were merged into forests, whilst woody savannas, savannas and grasslands were merged into grasslands. Water was not considered in this study. The pixels were resampled to a 1 km resolution in preparation for the subsequent verification phase; (5) for longitude and latitude, we gave each grid one longitude value and one latitude value at a 1 km spatial resolution based on the WGS84 datum.

MYD11A1, MYD13A2 and MCD12Q1 were provided by the Land Processes Distributed Active Archive Center (LP DAAC) site (https://lpdaac.usgs.gov/). DEM datasets are available from the CGIAR-CSI SRTM 90 m Database site (http://srtm.csi.cgiar.org). In this study, all the above data were downloaded, reprojected, stitched and resized with Google Earth Engine (GEE, https://earthengine. google.com). We processed all the data into 3540 rows × 6166 columns to cover the study area.

2.3. Method

As shown in Figure 1, BME was the core method of this study, and data were prepared for adapting the BME procedure. The three aspects of considering auxiliary data, time correlation, and spatial correlation were used to describe how the three most important input parameters of the BME algorithm, hard data, soft data and covariance models, were constructed. Regarding the auxiliary data, for each image to be estimated, the pixel LSTs with MODIS observations were taken as the dependent variable, and the NDVI image that was taken on the date closest to the estimated image and the elevation, longitude and latitude of the corresponding pixels were taken as independent variables to perform multiple linear regression and obtain the regression coefficients. Then, via the four independent variables of the pixels to be estimated and the respective regression coefficients, the regression LSTs of all the pixels to be estimated were calculated. The regression LSTs were used as the mean value and the mean square error (MSE) between the dependent variable and the predicted value as the variance to construct the Gaussian LST distribution as the soft data. The calculation results of the regression coefficients and the regression R^2 of the multiple linear regressions for the daytime and nighttime are shown in Tables A1 and A2. Regarding time correlation, we subtracted the mean LSTs of 15 days (7 days before and after) from both MODIS observed LSTs used as hard data and regression LSTs used as soft data, and the resulting difference values were input into the BME model as real hard data and soft data, respectively. This is equivalent to the 15 day mean LST minus the LST of the estimated day, which can be understood as the average trend after removing the special weather factors of the day; the LST time correlation could then be taken into account by this simple calculation. After such processing, the LSTs did not need to have trends removed before being input into the BME model, and only the spatial correlation needed to be considered, which could be achieved by the covariance function that represents the spatial dependence. The specific cause analysis can be found in lines 4 to 11 of the fourth paragraph of the introduction. Regarding spatial correlation, we calculated the spatial covariance using the real hard data and soft data mentioned above and input the obtained spatial covariance function name and parameters into the BME model. The relevant parameters of the covariance model calculated in this paper are shown in Tables A3 and A4.

From the above, it is worth noting that considering the availability of data and the simplicity of method processing, this study made the following assumptions when using the BME method to reconstruct LST: (1) LST changed linearly in a short time (time correlation was considered by subtracting the 15 day mean LST from the LST of the day to be estimated); (2) for the day to be reconstructed, one omnidirectional covariance model of that day can be used in the whole study area; (3) NDVI of each day can be represented by the NDVI data of its adjacent 7 days (the time resolution of the NDVI data was 15 days, so the time interval between the selected NDVI and the estimated image on any day was 0 to 7 days). The main BME conceptual core and framework and the explanations for its use in this research, are shown below.

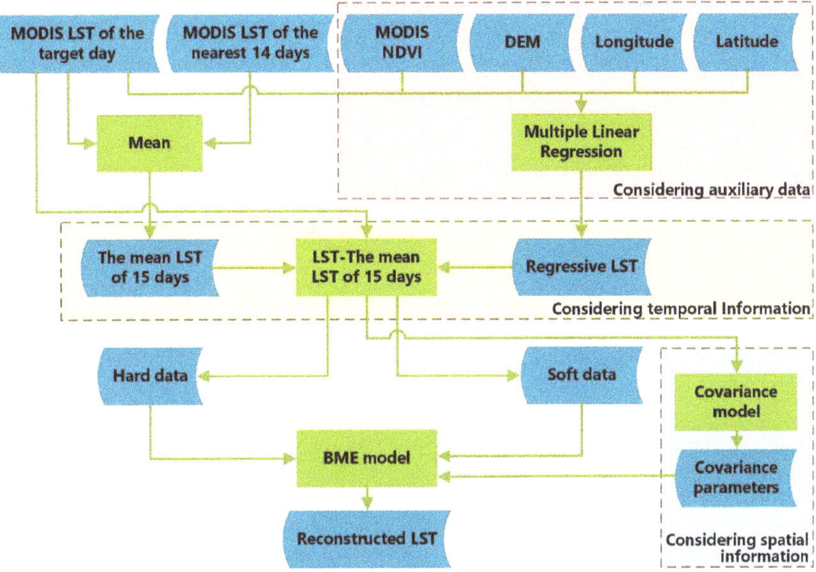

Figure 1. Flowchart describing the land surface temperature (LST) reconstruction model using the Bayesian maximum entropy (BME) method.

2.3.1. BME Epistemic Paradigm and Conceptual Core

The BME approach belongs to modern geostatistics, which provide insights into spatiotemporal variables. The epistemic paradigm of BME distinguishes between three main stages of knowledge acquisition, interpretation, and processing, as follows: (1) The prior stage. Spatiotemporal analysis and mapping always starts with a basic set of assumptions and the general knowledge base G. G refers to the background knowledge and the justified beliefs relative to the overall mapping situation; (2) the meta-prior stage. The specific knowledge base S is considered, including hard and soft data. S refers to a particular occurrence or state of affairs at a particular location and at a particular time. Hard data are considered accurate or have a high degree of confidence. Soft data are uncertain observations expressed in terms of interval values, probability statements, empirical charts, and others. That is to say, soft data can have varying levels of uncertainty and may be derived from the direct calculation of the probabilities or the indirect estimation from accumulated experience; (3) the integration or posterior stage. Information from (1) and (2) is processed by means of logical rules to produce the required spatiotemporal map. Therefore, the conceptual core of the BME method is that it aims at informativeness (in terms of prior information relative to the general knowledge G), as well as cogency (in terms of posterior probability relative to the specific knowledge S). BME combines the maximum entropy theory with operational Bayesian statistics to construct its scientific mathematical framework and to implement the above conceptual heart. In general, BME is used to acquire various knowledge bases and to order these bases in an appropriate manner so that, when taken together, they form a realistic picture of the phenomenon of interest.

2.3.2. BME Framework

BME has a rigorous cognitive system and a mathematical reasoning framework. The complete theoretical basis, mathematical formulas and specific derivation processes can be found in reference [43]. The main BME formulas and steps involved in this study are as follows:

$$f_G(\chi_k, \chi_{data}) = f_G(\chi_{map}) = \frac{\exp\left(\sum_{\alpha=1}^{N} \mu_\alpha(p_{map}) g_\alpha(\chi_{map})\right)}{\int d\chi_{map} \exp\left(\sum_{\alpha=1}^{N} \mu_\alpha(p_{map}) g_\alpha(\chi_{map})\right)}, \tag{1}$$

where χ_k denotes the LST of the estimated pixel, $\chi_{data} = (\chi_{hard}, \chi_{soft})$, χ_{hard} represents hard data, and χ_{soft} represents soft data. In this study, MODIS LST observations were used as hard data and LST Gaussian distributions obtained by multivariate linear regression were used as soft data. The regression process is described above. $f_G(\chi_k, \chi_{data})$ denotes the prior pdf of the map $\chi_{map} = (\chi_k, \chi_{data})$ given the general knowledge base G. $\mu_\alpha(p_{map})$ represents Lagrange multipliers. $g_\alpha(\chi_{map})$ is a set of known functions of χ_{map}. In practical applications, prior knowledge usually includes the first-order statistical moment (mean trend) and second-order statistical moment (covariance). The mean trend was not adopted as the first-order statistical moment in this study; rather, the LST difference on the observation day minus the 15 day mean was used. This was done to take time correlation into account and to remove LST instability caused by different weather factors, which resulted in a more stable LST distribution. The second-order statistical moments employed in this study were the spatial covariance functions derived by the difference values calculated above. Such a priori knowledge in this study could consider both the temporal correlation dominated by weather factors and the spatial correlation dominated by surface properties.

$$f_K(\chi_k) = f_G(\chi_k | \chi_{hard}, \chi_{soft}) = f_G(\chi_k | \chi_{data}) = f_G(\chi_k, \chi_{data}) / f_G(\chi_{data}) \tag{2}$$

In Equation (2), f_K denotes the posterior pdf of the map χ_k, given the total knowledge base K comprised of general knowledge G and specific knowledge S, including hard and soft data. The general knowledge, hard and soft data used in this study were described earlier.

$$\chi_k, \ \text{mean} = \int \chi_k f_K(\chi_k) d\chi_k \tag{3}$$

We used the BME mean value (χ_k, mean) as the final estimated LST. The BME mean value could be calculated from the posterior PDF since we sought to penalize large errors more than smaller ones.

2.3.3. BME Implementation

We used the BMElib algorithm package for BME algorithm implementation in MATLAB [68]. The calculation details for each estimated day are as follows. Firstly, the mean value and mean square error of the soft data of each prediction point were input into the probaGaussian.m function of the software package to obtain the soft data information that meets the requirements of the subsequent input. Secondly, the values and position coordinates of hard and soft data were entered into the covario.m and corefit.m functions of the software package to obtain the covariance function name and parameters. Finally, the main function was used to calculate the final result. Information such as hard data, soft data, and covariance was input into the BMEprobaMoments.m function. In addition, the maximum effective distance was set to 15 km, the maximum hard data point was set to 20 points, and the maximum soft data point was set to 3 points.

3. Results

3.1. Spatial Patterns of the Reconstructed LSTs

We selected the 15th day of each month in 2015 to conduct the method experiment and obtained the spatial distribution results of 12 images for both the daytime and nighttime (Figures 2 and 3). In the daytime, an average of 43% of the pixels of the MODIS LST products in the study area had LST observations, and in the nighttime, the value was 51%. That is, there was a missing rate of nearly one-half before filling gaps. The missing LST could be 100% filled using the BME method to generate a complete spatial distribution (Table 1).

Table 1. Availability of Moderate Resolution Imaging Spectroradiometer (MODIS) observed LST and reconstructed LST for the daytime and nighttime from 15 January to 15 December in 2015.

Date		Observed	Reconstructed		Observed	Reconstructed
15 January, 2015	Daytime	39.8%	100%	Nighttime	54.6%	100%
15 February, 2015		40.6%	100%		53.2%	100%
15 March, 2015		40.1%	100%		45.2%	100%
15 April, 2015		52.8%	100%		58.2%	100%
15 May, 2015		39.6%	100%		46.2%	100%
15 June, 2015		33.6%	100%		30.1%	100%
15 July, 2015		39.1%	100%		48.1%	100%
15 August, 2015		36.7%	100%		42.2%	100%
15 September, 2015		48.6%	100%		56.8%	100%
15 October, 2015		63.0%	100%		74.8%	100%
15 November, 2015		46.2%	100%		48.0%	100%
15 December, 2015		34.8%	100%		52.7%	100%
Average		42.9%	100%		50.9%	100%

In general, the entire study area showed strong spatial heterogeneity that varied in different seasons for both the day and night (Figures 2 and 3).

In winter, the lowest LST for the daytime occurred in northeast China, followed by the Qinghai-Tibet Plateau, whereas the lowest LST for the nighttime appeared in the Qinghai-Tibet Plateau, followed by northeast China. In summer, the lowest LST during the daytime simultaneously occurred in the Qinghai-Tibet Plateau and northeast China, whereas during the nighttime, the LST of the Tibetan Plateau was significantly lower than that in northeast China (Figures 2 and 3). The LST of the Qinghai-Tibet Plateau in southwest China was obviously low due to its high topography, and the LST of northeast China affected by Siberian cold air in winter was also low.

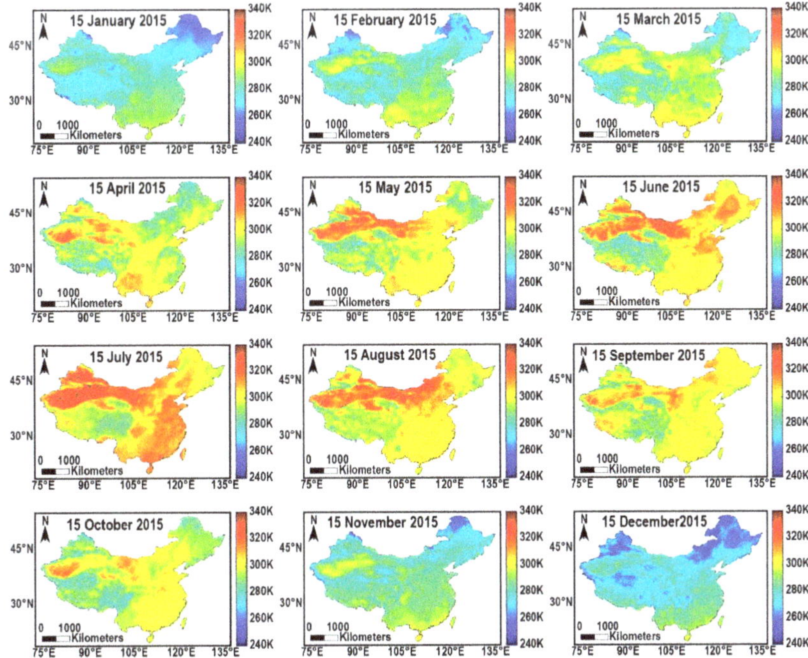

Figure 2. Spatial distribution of reconstructed daytime LST from 15 January to 15 December in 2015.

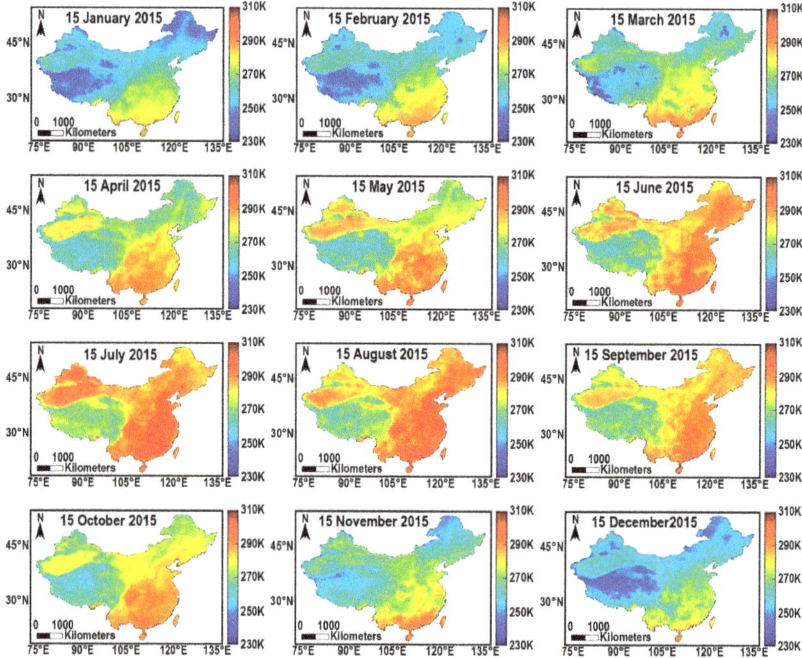

Figure 3. Spatial distribution of reconstructed nighttime LST from 15 January to 15 December in 2015.

In summer, the highest daytime LST was evident in northwest and central Inner Mongolia, whereas the highest nighttime LST was widely distributed in the south-central and southeastern regions, except for the Qinghai-Tibet Plateau. In winter, the highest daytime LST was distributed in the northwest and southeast regions, but the highest nighttime LST was distributed in the southeast coastal areas (Figures 2 and 3). The LST was usually higher in the northwest and Inner Mongolia due to the large number of deserts. LST was also higher in the southern region because of its low latitude and more solar radiation on the ground.

3.2. Accuracy Assessment

Since the study aimed at clear-sky LST reconstruction, it was not necessary to employ ground observation points for verification. First, this is because clear-sky LST is the theoretical value that is assumed not to be affected by clouds, while the ground observed value is the real value that is affected by clouds, so they cannot be directly compared. Secondly, this is because the acquisition time of the ground station is difficult to coincide with that of MODIS products, and there are also scale effects between points and surfaces. The verification method in this paper is for clear-sky LST. We selected some points with MODIS LST observations to cover them, reconstructed the LST of the covered points with the BME method, and then compared the reconstructed LST values with the known observations of MODIS LST. The verification points must have MODIS LST observations as references for the reconstructed LSTs. Therefore, we selected the points where the MODIS LST observations existed on the 15th of each month in 2015 as the verification points, which also helped to show the accuracy change of points with the same positions over time (Figure 4). There were 10,971 test pixels for the daytime (green points in Figure 4), including 330 forest pixels, 2683 grassland pixels, 537 cropland pixels, 7376 barren pixels, 38 urban pixels and 7 snow and ice pixels. There were 14,376 test pixels for the nighttime (blue points in Figure 4), including 218 forest pixels, 6040 grassland pixels, 425 cropland pixels, 7627 barren pixels, 425 urban pixels and 8 snow and ice pixels. Some of the verification points were spatially continuous and they formed regions of various shapes. The maximum diameter of the regions formed by the verification pixels was close to 60 km, and the estimation accuracy was thus also of reference value for the missing LST values caused by large cloud cover.

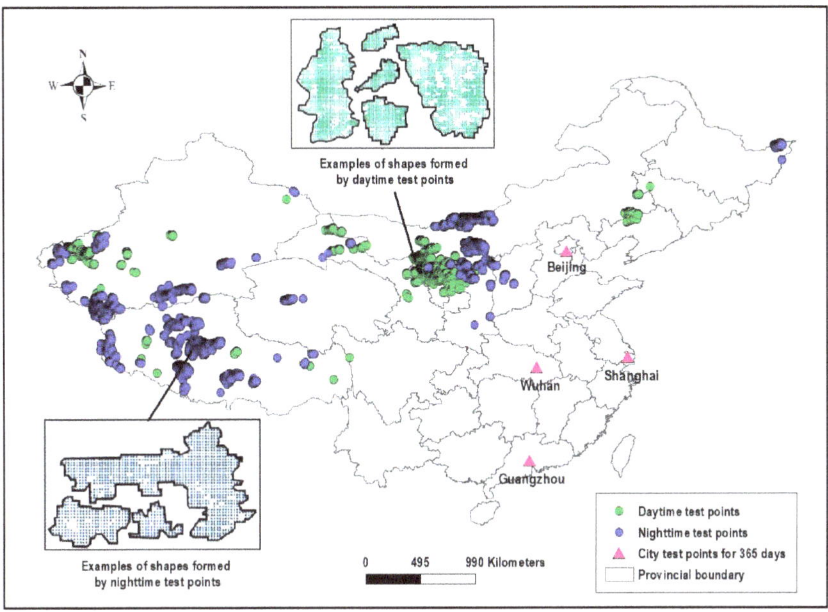

Figure 4. Distribution of 10,971 verification points in the daytime and 14,376 verification points in the nighttime, and four verification points in big cities.

The daytime accuracy was mostly lower than that of the nighttime (Figures 5 and 6). The average mean absolute error (MAE) and RMSE values were 1.1 K and 1.6 K, respectively, in the daytime, whereas the values were 0.8 K and 1.2 K in the nighttime, respectively (Figure 7d–f). The accuracy in summer was generally lower than in winter, which decreased and then increased from January to December (Figures 5, 6 and 7d–f). During the daytime, the maximum RMSE was 3.0 K in July and the minimum was 0.8 K in January. During the nighttime, the maximum RMSE was 1.9 K in June and the minimum was 1.0 K in December (Figures 5 and 6). The higher RMSE in summer than in winter may be due to the higher surface heterogeneity in summer than in winter, and the higher RMSE during the day than during the night may be due to more serious cloud cover and more missing values during the day.

In general, the RMSEs of barren land were the largest, with averages of 1.6 K and 1.4 K during the day and night, respectively, whereas the RMSEs of urban areas were the smallest, with averages of 1.0 K and 0.6 K during the day and night, respectively (Figure 7a,b). During the daytime, RMSEs ranked from high to low for barren, grasslands, forests, croplands, urban and snow and ice. During the nighttime, RMSE was in the order of forests, barren, snow and ice, grasslands, croplands and urban (Figure 7c).

As shown in Figures 5 and 6, the changes between day and night of urban LST were the most obvious, and the seasonal changes of ice and snow LST were most obvious. For the daytime, urban LST was close to the average LST of different land cover types, whereas for the nighttime, urban LST was generally higher than the average LST of different land cover types. This indicates that the urban areas have a relatively strong heat island effect at night. The LST of snow and ice was lower in winter and higher in summer.

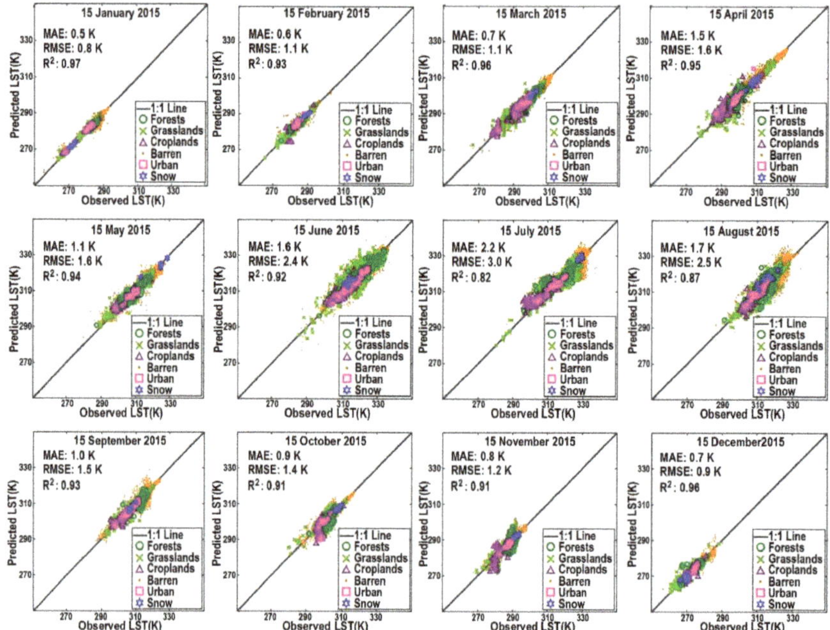

Figure 5. Scatter plots of reconstructed LST versus observed LST for 10,972 pixels for the daytime from 15 January to 15 December in 2015.

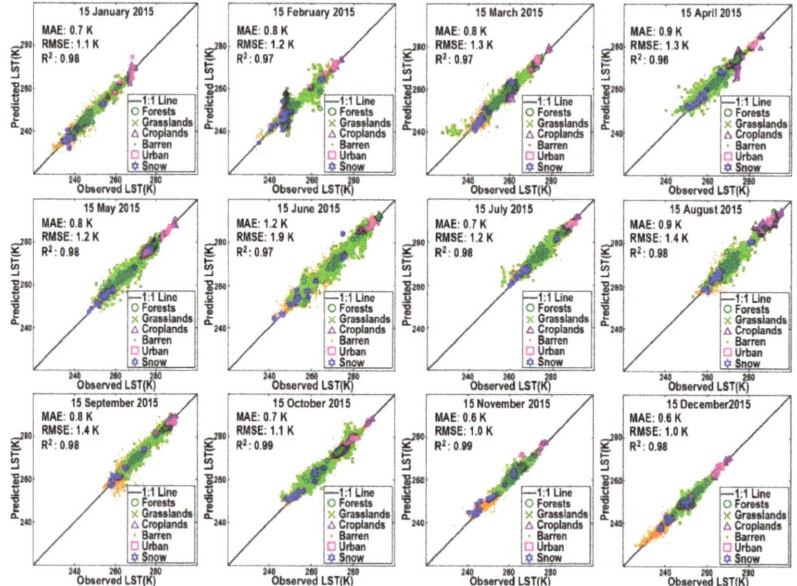

Figure 6. Scatter plots of reconstructed LST versus observed LST for 14,376 pixels for the nighttime from 15 January to 15 December in 2015.

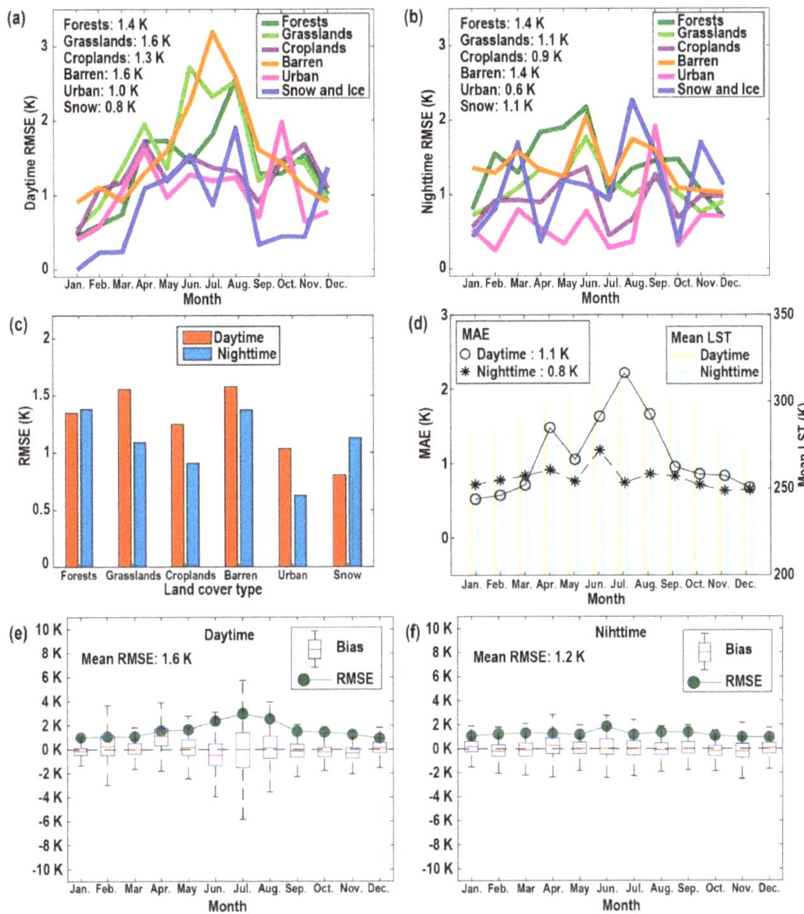

Figure 7. RMSE of (**a**) daytime LST and (**b**) nighttime LST for each land cover type from 15 January to 15 December in 2015; (**c**) overall average RMSE for each land cover type in 2015; (**d**) overall average mean absolute error (MAE) from 15 January to 15 December in 2015; bias and overall average RMSE for the daytime (**e**) and nighttime (**f**) from 15 January to 15 December in 2015.

In addition, we selected one point in Beijing, Wuhan, Shanghai and Guangzhou to estimate the LST for 365 days in 2015 and validated the accuracy with MODIS observations. These four verification points were geographically discrete and located in four major cities of China from north to south (Figure 4). The results are shown in Figure 8. Beijing had fewer than 200 days with LST observations, whereas in the other three cities, the number was less than 100. The maximum number of consecutive missing days was 30 days in Beijing, 40 days in Wuhan, 44 days in Shanghai and 54 days in Guangzhou. BME could fill 100% of the missing LSTs and reconstruct the uninterrupted LST time curve of each pixel for 365 days. As seen from the variation range of the curve, the BME method can describe the change of LST in a relatively fine manner, without smoothing out the maximum and minimum values. The R^2 values of the four urban test sites were all greater than 99%, and, except for the RMSE in Wuhan of 0.6 K, the RMSE in the other cities in the day or night was less than 0.5 K. Therefore, the single point test accuracy of the BME method was very high in large cities. The time distribution of LST demonstrated that there were obvious temperature differences between the day and night in the four cities. The four seasons changed most obviously in Beijing because it is located in a typical north temperate semi-humid continental monsoon climate zone, whereas Guangzhou had the smallest difference between the four seasons because of the Marine subtropical monsoon climate.

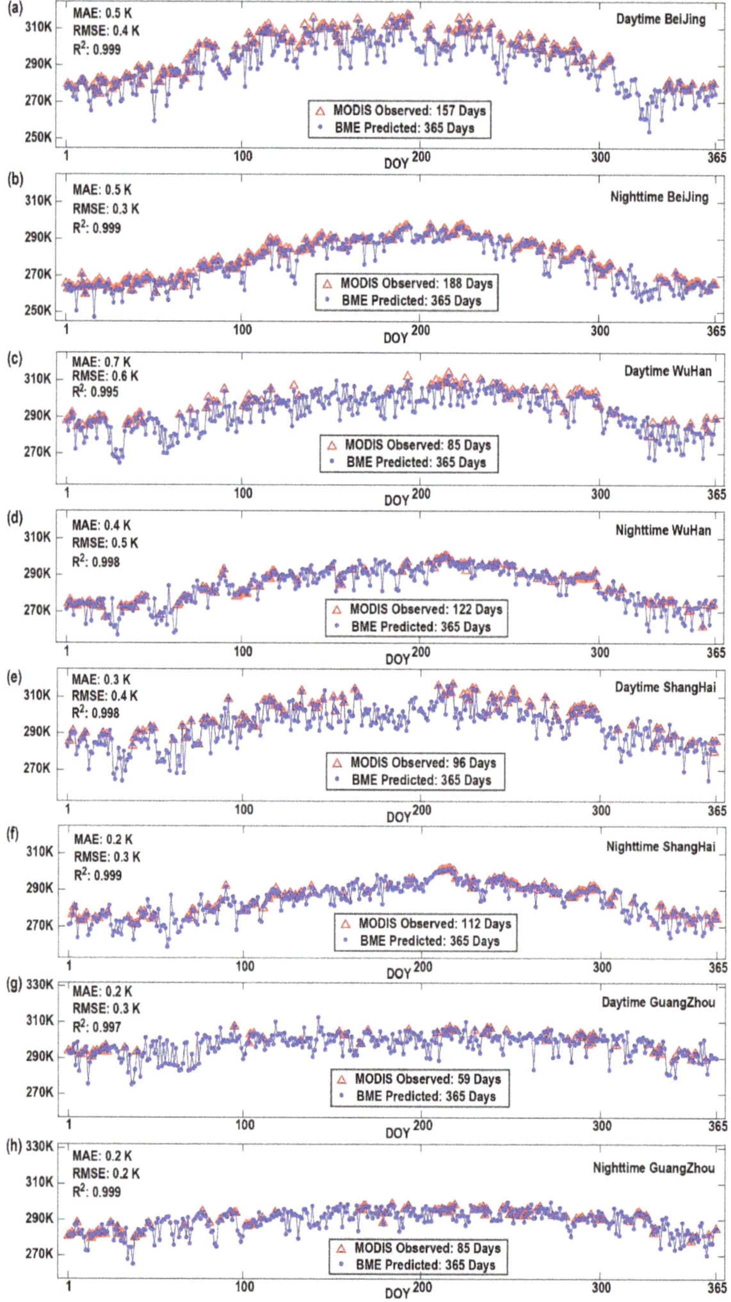

Figure 8. Temporal pattern of observed LST and reconstructed LST for the (**a**) daytime in Beijing, (**b**) nighttime in Beijing, (**c**) daytime in Wuhan, (**d**) nighttime in Wuhan, (**e**) daytime in Shanghai, (**f**) nighttime in Shanghai, (**g**) daytime in Guangzhou, and (**h**) nighttime in Guangzhou.

3.3. Factors that Influence Accuracy

Figures 9 and 10 illustrate the influence of different factors on the LST estimation accuracy represented by RMSE. For the daytime, the Pearson correlation coefficients between the RMSE of the reconstructed LST and the four influencing factors multiple linear regression R^2, average LST, ratio of pixels with LST observations to total pixels (namely, completeness) and average NDVI were 0.32, 0.85, −0.21 and 0.86, respectively; for the nighttime, the corresponding Pearson correlation coefficients were 0.22, 0.55, −0.6 and 0.44, respectively. Therefore, the average temperature and average NDVI were strongly correlated with RMSE in the daytime, and the mean LST and completeness were moderately correlated with RMSE in the nighttime. Three rather interesting aspects emerged from the results: (1) The average LST affected the accuracy of the method, where the higher the average temperature, the larger the RMSE; (2) the completeness, or the number of missing pixels, slightly affected the accuracy of the method; (3) the accuracy R^2 of multiple linear regression did not affect the accuracy of the method. In this study, R^2 varied from 0.39 to 0.90 (from 0.39 to 0.82 during the daytime and from 0.79 to 0.90 during the nighttime). This suggests that the BME method has a great ability to consider the uncertainty of soft data. Since the BME method does not have high requirements for the accuracy of soft data, it can be applied to other large-scale regions, and there is no need to improve the regression accuracy by random forest or other regression methods.

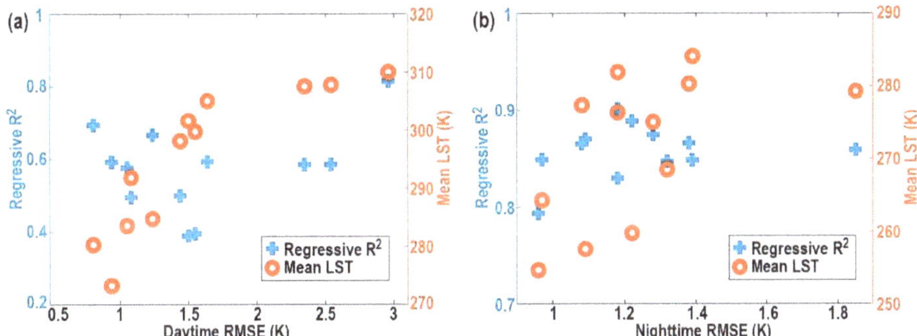

Figure 9. Correlation between multiple regressive R^2, mean LST and RMSE for the (**a**) daytime and (**b**) nighttime.

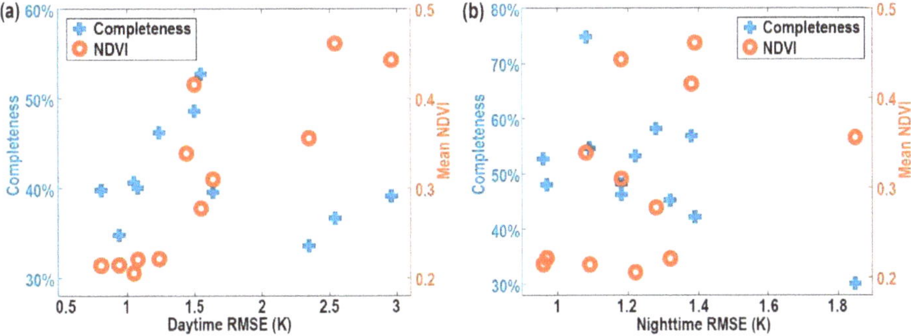

Figure 10. Correlation between the completeness of the observed LST, mean NDVI and RMSE for the (**a**) daytime and (**b**) nighttime.

3.4. Comparisons with Other Methods

We compared the BME method with four other commonly used LST gap-filling methods, including Crosson's method of supplementing MYD data with the same day's MOD data [22], the time interpolation method HANTS [23], the Kriging spatial interpolation method, and the hybrid gap-filling method proposed by Li [30]. It is worth noting that the same hard data and the same spatial covariance model of the BME method were entered into Kriging, and the only difference was that the Kriging method does not consider soft data. The RMSEs and error distributions of each method are shown in Figure 11. In general, the accuracy of each method ranked from high to low, as follows: BME > Kriging > Hybrid > HANTS > Crosson. It appeared that Crosson's method had the lowest accuracy, the BME method had the highest accuracy during the daytime, the Kriging method had the highest accuracy during the nighttime, the Hybrid method had a stable accuracy in the day and night, and HANTS had a significantly higher accuracy in the night than in the day.

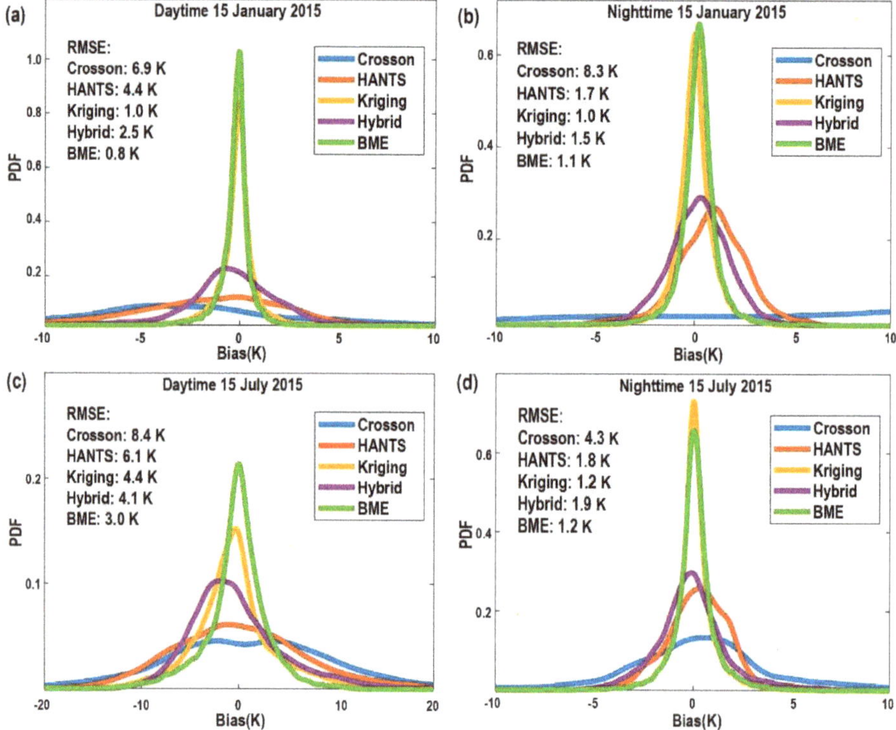

Figure 11. Error distribution of LST using the five methods of Crosson, HANTS, Kriging, Hybrid and BME for the (**a**) daytime on 15 January, 2015; (**b**) nighttime on 15 January, 2015; (**c**) daytime on 15 July, 2015; and (**d**) nighttime On 15 July, 2015.

4. Discussion

4.1. Accuracy Analysis

The mean RMSEs in this study were 1.6 K for the daytime and 1.2 K for the nighttime, which were slightly lower than the RMSEs of 3.3 K for the daytime and 2.7 K for the nighttime in Li's study in a comparably large area [30]. The single point RMSE of approximately 0.5 K is comparable with the RMSE of approximately 2 K under cloud-free conditions in Duan's study, which selected four ground points for validation [31]. The accuracy of this method is acceptable for large areas with complex geographical and climatic conditions. In addition, this method has a high accuracy in estimating urban LST and can be applied to urban heat island research.

There were different accuracies for different land cover types, which indicates that the accuracy is affected by land cover [34]. The accuracy of barren land and forest was lower than that of urban and cropland because the terrain of barren and forest is more complex, and the spatial heterogeneity is greater. Therefore, the model accuracy can be improved by dividing various terrain regions and then adopting different covariance models for various regions.

The accuracy decreased with the increase of NDVI and average LST because a high NDVI and high temperature usually represent the summer climate in China, when the cloud cover is large and the distribution is concentrated, which results in large LST gaps. The completeness has only a slight impact on the accuracy, possibly because the accuracy is influenced not only by the number of missing pixels, but also by their maximum diameter and distribution characteristics [69].

When constructing soft data, the accuracy of multiple linear regression does not affect the accuracy; this may be due to the ability of BME to fully consider the uncertainty of soft data. The average regression R^2 in this study was nearly 0.6, and in the subsequent application of the BME method to LST reconstruction, when the average regression R^2 in the construction process of soft data is greater than 0.6, it is unnecessary to adopt more complex regression methods to improve the regression accuracy.

4.2. Suggestions for Method Selection

We can learn the characteristics of each LST reconstruction method from Figure 11 and in combination with previous studies. The accuracy of spatial interpolation models is usually higher than that of temporal interpolation models [70]. The time interpolation models have some difficulties in capturing extreme values, and their accuracy is relatively low. Spatiotemporal gap-filling methods are often unable to fill all the missing values at one time, and one usually needs to iterate several times until all the missing values are filled. Spatial interpolation methods, especially the ones that consider auxiliary information, have a high accuracy, but usually take a relatively long time for calculations [25,26]. The study area in this study was large. To balance the calculation time and accuracy, we did not select the spatiotemporal covariance model, but rather the spatial covariance model, to consider the spatial dependence characteristics and the simple 15 day mean value to consider the time dependence characteristics. The spatiotemporal covariance model can be selected in small areas [34,71]. With the development of computing power and multi-core parallelism in the future, the computing speed will become faster.

Suggestions on how to choose an appropriate method to reconstruct LST are as follows: (1) If one hopes for a short computation time and simple computation steps, one can choose the time interpolation method; (2) if a high precision and simple calculation steps are required, the spatial interpolation method is recommended; (3) if one wants to balance the calculation speed and accuracy, we suggest using the spatiotemporal gap-filling method. All of the above three methods can consider introducing auxiliary data to improve the accuracy. In addition, note that the Kriging spatial interpolation method achieved good results in this study area and that Kriging is thus a simple method worth trying. The BME method that we used is a spatial interpolation method that considers auxiliary information, and its precision was very high in this study area.

4.3. Exploration of the Accuracy Improvement

In this part, we will explore methods to improve the accuracy and model applicability based on the assumptions of the method. One assumption of this method was that LST changed linearly in a short time. We thus subtracted the 15 day mean LST from the LST of the constructed day to consider the time correlation. Doing so can also make the data closer to a normal distribution and thus replace the step of trend removal in other BME studies. We selected 18 October, 2015 as an example (there were relatively more MODIS observed LSTs available on that day), as shown in Figure 12. The green part shows that the LST of all pixels on this day presents a non-normal distribution, while the purple part exhibits an approximately normal distribution after subtraction calculation (LST–the mean LST of 15 days) is performed. This operation achieved the desired effect. However, as can be seen from Figure 8, the time variation of LST exhibited both an overall trend and fluctuation. According to the limitation of this study's assumption, we may improve the accuracy by taking better account of the time correlation. We can do so by introducing the Annual Temperature Cycle (ATC) model, which is a general and smooth curve description of the LST annual change. Considering the time correlation with the LST of the estimated day minus the LST of the corresponding point on the ATC curve, theoretically, will be more accurate than considering the time correlation with the LST of the estimated day minus the LST of the 15 day mean value. We plan to conduct follow-up studies with regards to this.

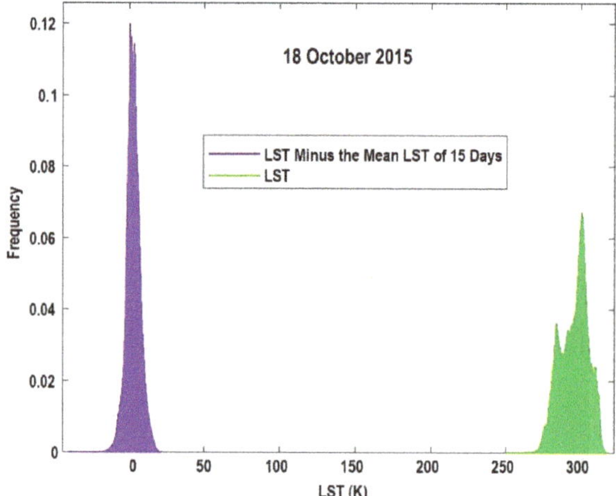

Figure 12. Data distribution of the MODIS observed LST and the difference of LST minus the 15 day mean LST on 18 October, 2015.

The second assumption of this method was that one omnidirectional covariance model of the constructed day can be used in the whole study area. Here, we want to explore the directivity of the covariance model. We constructed omnidirectional and directional covariance models for 18 October, 2015 (Table 2, Figure 13). As can be seen from the parameters of directional covariance in the study area, the overall characteristic is that the directional covariance is significantly affected by longitude and latitude, and the latitude direction changes faster than the longitude direction.

Table 2. Omnidirectional and directional covariance model parameters for 18 October, 2015.

Time (omnidirectional)	Model name	Nugget	Partial Sill	Range (km)
Daytime	spherical	0.20	0.76	12.94
Nighttime	spherical	0.20	0.79	15.59

Time (directional)	The angle between the principal axis and the horizontal axis (°)		Principal/secondary axes (km/km)	
Daytime	0		20.68/7.69	
Nighttime	3.52		18.39/10.06	

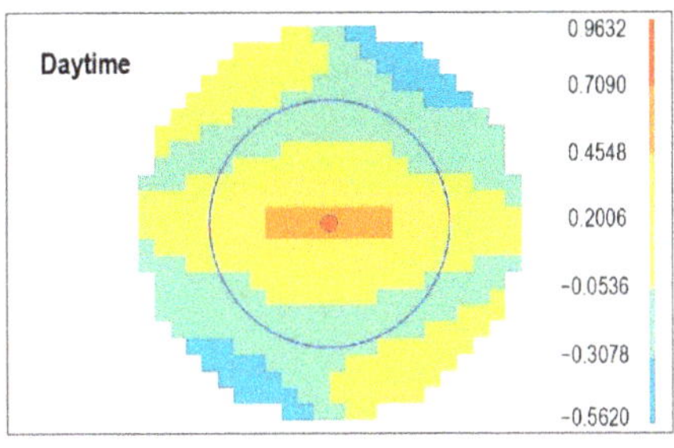
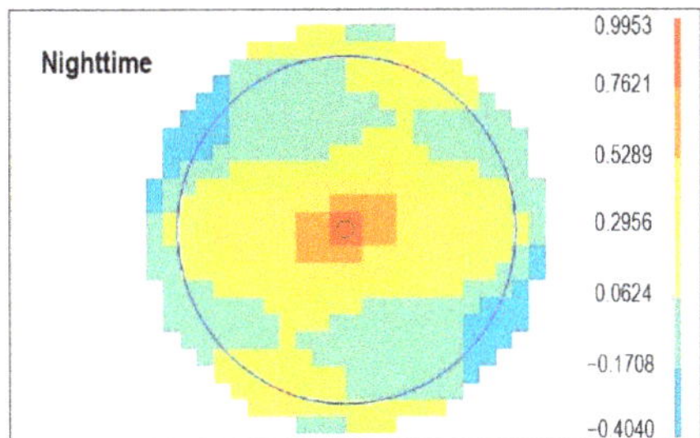

Figure 13. Directional covariance models of the daytime and nighttime on 18 October, 2015.

Omnidirectional and directional covariance models were input into the method in this study to calculate the results, and 10,000 points were randomly selected during the day and night for accuracy verification and comparison. The results show that after considering the directionality of the covariance, the accuracy in the daytime is slightly improved, and the accuracy of the night is not improved (Table 3). It is suggested that directionality can be considered during the daytime in the following research.

Table 3. Omnidirectional and directional accuracy verification results for 18 October, 2015.

Time	Model Name	MAE (K)	RMSE (K)	R^2
Daytime	omnidirectional	0.79	1.32	0.98
	directional	0.78	1.30	0.98
Nighttime	omnidirectional	0.45	0.71	0.99
	directional	0.45	0.71	0.99

In addition, large study areas will have problems with spatial covariance models that differ in different regions. First of all, however, we have not explored how the covariance model changed in different regions of China's mainland. Moreover, if we want to fill in the LST of the whole study area at once, more detailed regional division may make the model more complicated. How to apply different covariance models in different subregions of the study area is challenging and worthy of further exploration.

The third assumption of this method was that the NDVI of each day can be represented by the NDVI data of its adjacent 7 days. We used 15 day NDVI data because it had values on all pixels to ensure that soft data can be constructed on all LST missing pixels. When the LST was reconstructed on the 15th day of each month in 2015, the nearest NDVI data were selected on 9 January, 10 February, 14 March, 15 April, 17 May, 18 June, 20 July, 21 August, 22 September, 8 October, 8 November and 11 December. There were no results to prove that the closer the reconstruction date was to the obtained date of NDVI, the higher the fitting accuracy and the final LST accuracy were. We believe it was feasible to reconstruct the LST daily with a 15 day NDVI product. Lastly, we cannot accurately calculate the impact of the uncertainty of NDVI datasets on the uncertainty of the final results, which is a limitation of this method.

4.4. Recommendations for Future Studies

Due to the characteristics of the BME method, we cannot accurately determine the uncertainty of the results. At present, we can only obtain the conclusion that at a 1 km spatial resolution, the accuracy of reconstructing the daily LST of China's landmass with the BME method is acceptable.

For daytime or nighttime LST reconstruction on a certain day, one covariance model can be adopted in the whole research area on that day, which can achieve a reasonable accuracy. If one wants to use this model later, we suggest that it be directly used in the same study area and at the same spatiotemporal resolutions. If other areas or other resolutions are studied, an accuracy analysis should be conducted first to see whether it can meet the actual requirements. The soft data should also be reconstructed according to the geographic and data characteristics of the other study areas. One can try to use the four independent variables in this paper or may introduce other auxiliary data, such as soil moisture and temperature. In the future, we can try to introduce the ATC model, divide different subregions and adopt different covariance models to improve the accuracy of the BME method presented in this study.

When using the BMElib package, it is important to be aware of some parameter settings. The maximum effective distance can be set to a value similar to the range in the covariance model. In order to balance the calculation accuracy and calculation time, we recommend that the maximum number of hard data points should not exceed 50 and the maximum number of soft data points should not exceed 5.

Although clear-sky LST can be applied in some research, the real surface LST is still needed in many fields. This study calculated clear-sky LST, and if one wants to obtain cloudy-sky LST from clear-sky LST, one can refer to Zeng's method [69]. Adding microwave and ground observation data can also be considered.

5. Conclusions

The MODIS LST product has many missing values over wide areas, which hinders its practical application. In this study, we reconstructed the seamless 1 km resolution daily clear-sky LST for China's landmass based on the BME method, considering spatiotemporal correlation and taking auxiliary data as soft data. The average RMSE was 1.6 K for the daytime and 1.2 K for the nighttime, with the mean absolute error (MAE) of 1.1 K for the daytime and 0.8 K for the nighttime, and the corresponding R^2 of 0.92 for the daytime and 0.98 for the nighttime.

This method has the following advantages: (1) It simultaneously considers spatiotemporal correlation and auxiliary data and has a high accuracy in a large area. It has the ability to capture extreme values; (2) the data in this method are easy to obtain and process; (3) simple linear regression is used to construct soft data, and there is no need to adopt more complex regression methods to improve the regression accuracy, as long as the average regression R^2 is greater than 0.6; (4) even if the diameter of the missing area is large or the continuous missing time is long, this method does not need multiple step-by-step calculations to gradually fill in the missing pixels, and can estimate all the missing pixels at one time.

There are also some limitations for this method: (1) This method is not applicable when an accuracy of less than 1 K across the entire Chinese landmass is required; (2) when using the method in other study areas and spatiotemporal scales, it is necessary to first consider whether the hypothesis of LST linearity change in a short time and one omnidirectional covariance model can be applied to the entire study area are valid; (3) the method cannot quantitatively calculate the influence of the uncertainty of NDVI and DEM data on the uncertainty of the results; (4) the clear-sky LST should be converted to cloudy-sky LST if the real LST is required.

The results of this study provide a data basis for daily LST analysis and subsequent relevant studies in large areas of China. For the method, its high accuracy and great ability to capture extreme values in urban areas can help improve urban heat island research. It can also be applied to the reconstruction of missing LST values of other years, other regions and other spatial resolutions (such as Landsat), as well as the estimation of missing values of other geographical attributes.

Author Contributions: Conceptualization, Y.Z. and Y.C.; data curation, Y.Z. and Y.L.; formal analysis, Y.Z. and Y.C.; funding acquisition, Y.C. and J.L.; investigation, Y.Z.; methodology, Y.Z.; project administration, Y.C.; software, Y.Z. and Y.L.; supervision, Y.C. and J.L.; validation, Y.Z. and H.X.; visualization, Y.Z.; writing original draft, Y.Z.; writing—review and editing Y.Z., Y.C., Y.L., H.X. and J.L.

Funding: This work was supported by the National Key R&D Program on monitoring, early warning and prevention of major natural disasters under Grant 2017YFC1502406; the Natural Science Foundation of China under Grants 41571342, 41771448 and 51579135; the Beijing Natural Science Foundation under Grant 8192025; and in part by the Beijing Laboratory of Water Resources Security.

Acknowledgments: We are very grateful to Patrick Bogaert (Université Catholique de Louvain; Belgium) and Marc Serre (University of North Carolina at Chapel Hill; USA) for providing us with the MATLAB algorithm package BMElib and to Alexander Kolovos (SpaceTimeWorks, LLC; USA), Hwa-Lung Yu (National Taiwan University, Taipei; Taiwan), Steve Warmerdam and Boris Dev (San Diego State University, CA, USA) for providing us with the MATLAB visualization algorithm package SEKS-GUI. We would also like to thank the Level-1 and Atmosphere Archive & Distribution System (LAADS) Distributed Active Archive Center (DAAC) of NASA for MODIS data and the Consultative Group of International Agricultural Research-Consortium for Spatial Information (CGIAR-CSI) for SRTM DEM data. We would like to thank the anonymous reviewers for their insightful comments and substantial help in improving this article.

Conflicts of Interest: The authors declare no conflict of interest.

Appendix A

Table A1. Regression R^2 and regression coefficients of multiple linear regression for the daytime.

Date	Regression R^2	Intercept	Coefficient of NDVI	Coefficient of DEM	Coefficient of Latitude	Coefficient of Longitude
15 January, 2015	0.69	366.07	−2.10	−0.004	−1.07	−0.39
15 February, 2015	0.58	371.31	−5.23	−0.004	−1.26	−0.31
15 March, 2015	0.50	381.25	−24.29	−0.005	−1.52	−0.17
15 April, 2015	0.40	369.12	−4.05	−0.005	−0.18	−0.50
15 May, 2015	0.59	388.10	−26.26	−0.007	−0.63	−0.39
15 June, 2015	0.59	343.54	−23.48	−0.007	−0.10	−0.13
15 July, 2015	0.82	373.64	−21.95	−0.008	−0.16	−0.35
15 August, 2015	0.59	320.09	−28.35	−0.005	−0.07	0.10
15 September, 2015	0.39	317.19	−14.30	−0.003	−0.02	−0.03
15 October, 2015	0.50	349.92	−16.69	−0.004	−0.52	−0.19
15 November, 2015	0.67	370.45	−16.05	−0.004	−1.38	−0.21
15 December, 2015	0.59	358.56	−8.76	−0.003	−1.65	−0.14

Table A2. Regression R^2 and regression coefficients of multiple linear regression for the nighttime.

Date	Regression R^2	Intercept	Coefficient of NDVI	Coefficient of DEM	Coefficient of Latitude	Coefficient of Longitude
15 January, 2015	0.87	337.62	−18.39	−0.005	−0.97	−0.18
15 February, 2015	0.89	336.19	−15.67	−0.006	−0.98	−0.13
15 March, 2015	0.85	330.99	−5.02	−0.005	−1.13	−0.06
15 April, 2015	0.87	338.49	−9.03	−0.005	−0.58	−0.25
15 May, 2015	0.83	350.82	−3.16	−0.007	−0.66	−0.32
15 June, 2015	0.86	300.98	−0.42	−0.005	−0.30	0.03
15 July, 2015	0.90	326.60	−2.70	−0.006	−0.23	−0.21
15 August, 2015	0.85	308.03	−2.69	−0.005	−0.55	0.07
15 September, 2015	0.87	295.45	−2.60	−0.004	−0.34	0.07
15 October, 2015	0.87	310.69	−2.56	−0.005	−0.57	−0.03
15 November, 2015	0.85	331.74	−1.86	−0.005	−1.25	−0.06
15 December, 2015	0.79	324.41	−12.13	−0.005	−0.98	−0.09

Table A3. Names and parameters of the spatial covariance model for the daytime.

Date	Model Name	Nugget	Partial Sill	Range (km)
15 January, 2015	exponential	0.34	0.66	4.03
15 February, 2015	exponential	0.67	0.32	11.59
15 March, 2015	exponential	0.38	0.62	5.62
15 April, 2015	spherical	0.50	0.48	5.29
15 May, 2015	exponential	0.40	0.60	7.62
15 June, 2015	exponential	0.57	0.43	9.25
15 July, 2015	spherical	0.69	0.31	10.13
15 August, 2015	gaussian	0.41	0.57	9.78
15 September, 2015	spherical	0.44	0.55	14.39
15 October, 2015	exponential	0.32	0.68	13.78
15 November, 2015	spherical	0.52	0.46	13.28
15 December, 2015	exponential	0.68	0.32	18.56

Table A4. Names and parameters of the spatial covariance model for the nighttime.

Date	Model Name	Nugget	Partial Sill	Range (km)
15 January, 2015	exponential	0.34	0.66	6.34
15 February, 2015	spherical	0.35	0.65	12.08
15 March, 2015	spherical	0.57	0.43	10.17
15 April, 2015	spherical	0.56	0.43	9.57
15 May, 2015	spherical	0.39	0.61	17.30
15 June, 2015	exponential	0.45	0.55	3.56
15 July, 2015	exponential	0.53	0.47	10.54
15 August, 2015	gaussian	0.31	0.67	10.75
15 September, 2015	exponential	0.39	0.61	9.57
15 October, 2015	spherical	0.29	0.71	14.79
15 November, 2015	exponential	0.50	0.50	12.71
15 December, 2015	exponential	0.30	0.70	9.28

References

1. Norman, J.M.; Becker, F. Terminology in thermal infrared remote sensing of natural surfaces. *Agric. For. Meteorol.* **1995**, *77*, 153–166. [CrossRef]
2. Wan, Z.; Dozier, J. A generalized split-window algorithm for retrieving land-surface temperature from space. *IEEE Trans. Geosci. Remote Sens.* **1996**, *34*, 892–905.
3. Li, Z.-L.; Tang, B.-H.; Wu, H.; Ren, H.; Yan, G.; Wan, Z.; Trigo, I.F.; Sobrino, J.A. Satellite-derived land surface temperature: Current status and perspectives. *Remote Sens. Environ.* **2013**, *131*, 14–37. [CrossRef]
4. Wan, Z. New refinements and validation of the collection-6 MODIS land-surface temperature/emissivity product. *Remote Sens. Environ.* **2014**, *140*, 36–45. [CrossRef]
5. Carlson, T.N.; Gillies, R.R.; Schmugge, T.J. An interpretation of methodologies for indirect measurement of soil water content. *Agric. For. Meteorol.* **1995**, *77*, 191–205. [CrossRef]
6. Norman, J.M.; Kustas, W.P.; Humes, K.S. Source approach for estimating soil and vegetation energy fluxes in observations of directional radiometric surface temperature. *Agric. For. Meteorol.* **1995**, *77*, 263–293. [CrossRef]
7. Zhang, L.; Lemeur, R.; Goutorbe, J.P. A one-layer resistance model for estimating regional evapotranspiration using remote sensing data. *Agric. For. Meteorol.* **1995**, *77*, 241–261. [CrossRef]
8. Bodas-Salcedo, A.; Ringer, M.A.; Jones, A. Evaluation of the Surface Radiation Budget in the Atmospheric Component of the Hadley Centre Global Environmental Model (HadGEM1). *J. Clim.* **2008**, *21*, 4723–4748. [CrossRef]
9. Kustas, W.; Anderson, M. Advances in thermal infrared remote sensing for land surface modeling. *Agric. For. Meteorol.* **2009**, *149*, 2071–2081. [CrossRef]
10. Gallo, K.; Dan, T.; Yu, Y. Evaluation of the Relationship between Air and Land Surface Temperature under Clear and Cloudy-Sky Conditions. *J. Appl. Meteorol. Climatol.* **2011**, *50*, 767–775. [CrossRef]
11. Kloog, I.; Nordio, F.; Coull, B.A.; Schwartz, J. Predicting spatiotemporal mean air temperature using MODIS satellite surface temperature measurements across the Northeastern USA. *Remote Sens. Environ.* **2014**, *150*, 132–139. [CrossRef]
12. Zhang, P.; Bounoua, L.; Imhoff, M.L.; Wolfe, R.E.; Thome, K. Comparison of MODIS Land Surface Temperature and Air Temperature over the Continental USA Meteorological Stations. *Can. J. Remote Sens.* **2014**, *40*, 110–122. [CrossRef]
13. Shi, L.; Kloog, I.; Zanobetti, A.; Liu, P.; Schwartz, J.D. Impacts of temperature and its variability on mortality in New England. *Nat. Clim. Chang.* **2015**, *5*, 988. [CrossRef] [PubMed]
14. Ma, H.; Liang, S.; Xiao, Z.; Shi, H. Simultaneous inversion of multiple land surface parameters from MODIS optical–thermal observations. *ISPRS J. Photogramm. Remote Sens.* **2017**, *128*, 240–254. [CrossRef]
15. Gillespie, A.R.; Matsunaga, T.; Rokugawa, S.; Hook, S.J. Temperature and emissivity separation from advanced spaceborne thermal emission and reflection radiometer (ASTER) images. In *Infrared Spaceborne Remote Sensing IV, Proceedings of SPIE's 1996 International Symposium on Optical Science, Engineering, and Instrumentation, Denver, CO, United States, 4–9 August 1996*; SPIE: Bellingham, WA, USA, 1996; pp. 82–94.

16. Wan, Z. *Collection-5 MODIS Land Surface Temperature Products Users' Guide*; University of California: Santa Barbara, CA, USA, 2007.

17. Jiang, G.M.; Li, Z.L. Split-window algorithm for land surface temperature estimation from MSG1-SEVIRI data. *Int. J. Remote Sens.* **2008**, *29*, 6067–6074. [CrossRef]

18. MODIS Web. Available online: https://modis.gsfc.nasa.gov/ (accessed on 5 November 2019).

19. Neteler, M. Estimating Daily Land Surface Temperatures in Mountainous Environments by Reconstructed MODIS LST Data. *Remote Sens.* **2010**, *10*, 333. [CrossRef]

20. Fan, X.-M.; Liu, H.-G.; Liu, G.-H.; Li, S.-B. Reconstruction of MODIS land-surface temperature in a flat terrain and fragmented landscape. *Int. J. Remote Sens.* **2014**, *35*, 7857–7877. [CrossRef]

21. Zeng, C.; Shen, H.; Zhong, M.; Zhang, L.; Wu, P. Reconstructing MODIS LST Based on Multitemporal Classification and Robust Regression. *IEEE Geosci. Remote Sens. Lett.* **2015**, *12*, 512–516. [CrossRef]

22. Crosson, W.L.; Al-Hamdan, M.Z.; Hemmings, S.N.J.; Wade, G.M. A daily merged MODIS Aqua–Terra land surface temperature data set for the conterminous United States. *Remote Sens. Environ.* **2012**, *119*, 315–324. [CrossRef]

23. Xu, Y.; Shen, Y. Reconstruction of the land surface temperature time series using harmonic analysis. *Comput. Geosci.* **2013**, *61*, 126–132. [CrossRef]

24. Yang, J.; Wang, Y.; August, P. Estimation of land surface temperature using spatial interpolation and satellite-derived surface emissivity. *J. Environ. Inform.* **2004**, *4*, 37–44. [CrossRef]

25. Ke, L.; Ding, X.; Song, C. Reconstruction of Time-Series MODIS LST in Central Qinghai-Tibet Plateau Using Geostatistical Approach. *IEEE Geosci. Remote Sens. Lett.* **2013**, *10*, 1602–1606. [CrossRef]

26. Metz, M.; Rocchini, D.; Neteler, M. Surface Temperatures at the Continental Scale: Tracking Changes with Remote Sensing at Unprecedented Detail. *Remote Sens.* **2014**, *6*, 3822. [CrossRef]

27. Weiss, D.J.; Atkinson, P.M.; Bhatt, S.; Mappin, B.; Hay, S.I.; Gething, P.W. An effective approach for gap-filling continental scale remotely sensed time-series. *ISPRS J. Photogramm. Remote Sens.* **2014**, *98*, 106–118. [CrossRef] [PubMed]

28. Yu, W.; Ma, M.; Wang, X.; Tan, J. Estimating the land-surface temperature of pixels covered by clouds in MODIS products. *J. Appl. Remote Sens.* **2014**, *8*, 083525. [CrossRef]

29. Sun, L.; Chen, Z.; Gao, F.; Anderson, M.; Song, L.; Wang, L.; Hu, B.; Yang, Y. Reconstructing daily clear-sky land surface temperature for cloudy regions from MODIS data. *Comput. Geosci.* **2017**, *105*, 10–20. [CrossRef]

30. Li, X.; Zhou, Y.; Asrar, G.R.; Zhu, Z. Creating a seamless 1km resolution daily land surface temperature dataset for urban and surrounding areas in the conterminous United States. *Remote Sens. Environ.* **2018**, *206*, 84–97. [CrossRef]

31. Duan, S.-B.; Li, Z.-L.; Leng, P. A framework for the retrieval of all-weather land surface temperature at a high spatial resolution from polar-orbiting thermal infrared and passive microwave data. *Remote Sens. Environ.* **2017**, *195*, 107–117. [CrossRef]

32. Duan, S.-B.; Li, Z.-L.; Tang, B.-H.; Wu, H.; Tang, R. Generation of a time-consistent land surface temperature product from MODIS data. *Remote Sens. Environ.* **2014**, *140*, 339–349. [CrossRef]

33. Jin, M.; Dickinson, R.E. A generalized algorithm for retrieving cloudy sky skin temperature from satellite thermal infrared radiances. *J. Geophys. Res. Atmos.* **2000**, *105*, 27037–27047. [CrossRef]

34. Kou, X.; Jiang, L.; Bo, Y.; Yan, S.; Chai, L. Estimation of Land Surface Temperature through Blending MODIS and AMSR-E Data with the Bayesian Maximum Entropy Method. *Remote Sens.* **2016**, *8*, 105. [CrossRef]

35. Lu, L.; Venus, V.; Skidmore, A.; Wang, T.; Luo, G. Estimating land-surface temperature under clouds using MSG/SEVIRI observations. *Int. J. Appl. Earth Obs. Geoinf.* **2011**, *13*, 265–276. [CrossRef]

36. Shwetha, H.R.; Kumar, D.N. Prediction of high spatio-temporal resolution land surface temperature under cloudy conditions using microwave vegetation index and ANN. *ISPRS J. Photogramm. Remote Sens.* **2016**, *117*, 40–55. [CrossRef]

37. Zhang, X.; Pang, J.; Li, L. Estimation of Land Surface Temperature under Cloudy Skies Using Combined Diurnal Solar Radiation and Surface Temperature Evolution. *Remote Sens.* **2015**, *7*, 905–921. [CrossRef]

38. Scarino, B.; Minnis, P.; Palikonda, R.; Reichle, R.H.; Morstad, D.; Yost, C.; Shan, B.; Liu, Q. Retrieving Clear-Sky Surface Skin Temperature for Numerical Weather Prediction Applications from Geostationary Satellite Data. *Remote Sens.* **2013**, *5*, 342. [CrossRef]

39. Lai, J.; Zhan, W.; Huang, F.; Voogt, J.; Bechtel, B.; Allen, M.; Peng, S.; Hong, F.; Liu, Y.; Du, P. Identification of typical diurnal patterns for clear-sky climatology of surface urban heat islands. *Remote Sens. Environ.* **2018**, *217*, 203–220. [CrossRef]

40. Sandholt, I.; Rasmussen, K.; Andersen, J. A simple interpretation of the surface temperature/vegetation index space for assessment of surface moisture status. *Remote Sens. Environ.* **2002**, *79*, 213–224. [CrossRef]

41. Amani, M.; Salehi, B.; Mahdavi, S.; Masjedi, A.; Dehnavi, S. Temperature-Vegetation-soil Moisture Dryness Index (TVMDI). *Remote Sens. Environ.* **2017**, *197*, 1–14. [CrossRef]

42. Christakos, G.; Li, X. Bayesian Maximum Entropy Analysis and Mapping: A Farewell to Kriging Estimators? *Math. Geosci.* **1998**, *30*, 435–462.

43. Christakos, G. *Modern Spatiotemporal Geostatistics*; Oxford University Press: Oxford, UK; New York, NY, USA, 2000.

44. Christakos, G.; Serre, M.L.; Kovitz, J.L. BME representation of particulate matter distributions in the state of California on the basis of uncertain measurements. *J. Geophys. Res.* **2001**, *106*, 9717–9731. [CrossRef]

45. Kolovos, A.; Christakos, G.; Serre, M.L.; Miller, C.T. Computational Bayesian maximum entropy solution of a stochastic advection-reaction equation in the light of site-specific information. *Water Resour. Res.* **2002**, *38*, 51–54. [CrossRef]

46. Heywood, B.; Brierley, A.; Gull, S. A quantified Bayesian Maximum Entropy estimate of Antarctic krill abundance across the Scotia Sea and in small-scale management units from the CCAMLR-2000 survey. *CCAMLR Sci.* **2006**, *13*, 97–116.

47. Brus, D.; Bogaert, P.; Heuvelink, G. Bayesian Maximum Entropy prediction of soil categories using a traditional soil map as soft information. *Eur. J. Soil Sci.* **2007**, *59*, 166–177. [CrossRef]

48. Lee, S.J.; Wentz, E.A. Applying Bayesian Maximum Entropy to extrapolating local-scale water consumption in Maricopa County, Arizona. *Water Resour. Res.* **2008**, *44*. [CrossRef]

49. Lee, S.-J.; Yeatts, K.B.; Serre, M.L. A Bayesian Maximum Entropy approach to address the change of support problem in the spatial analysis of childhood asthma prevalence across North Carolina. *Spat. Spatio-Temporal Epidemiol.* **2009**, *1*, 49–60. [CrossRef]

50. Lee, S.-J.; Wentz, E.A.; Gober, P. Space–time forecasting using soft geostatistics: A case study in forecasting municipal water demand for Phoenix, Arizona. *Stoch. Environ. Res. Risk Assess.* **2010**, *24*, 283–295. [CrossRef]

51. Money, E.S.; Sackett, D.K.; Aday, D.D.; Serre, M.L. Using River Distance and Existing Hydrography Data Can Improve the Geostatistical Estimation of Fish Tissue Mercury at Unsampled Locations. *Environ. Sci. Technol.* **2011**, *45*, 7746–7753. [CrossRef]

52. Reyes, J.M.; Serre, M.L. An LUR/BME Framework to Estimate PM2.5 Explained by on Road Mobile and Stationary Sources. *Environ. Sci. Technol.* **2014**, *48*, 1736–1744. [CrossRef]

53. Lee, S.-J.; Chang, H.; Gober, P. Space and time dynamics of urban water demand in Portland, Oregon and Phoenix, Arizona. *Stoch. Environ. Res. Risk Assess.* **2015**, *29*, 1135–1147. [CrossRef]

54. Shi, Y.; Zhou, X.; Yang, X.; Shi, L.; Ma, S. Merging Satellite Ocean Color Data with Bayesian Maximum Entropy Method. *IEEE J. Sel. Top. Appl. Earth Obs. Remote Sens.* **2015**, *8*, 3294–3304. [CrossRef]

55. Sun, X.-L.; Wu, Y.-J.; Lou, Y.-L.; Wang, H.-L.; Zhang, C.; Zhao, Y.-G.; Zhang, G.-L. Updating digital soil maps with new data: A case study of soil organic matter in Jiangsu, China. *Eur. J. Soil Sci.* **2015**, *66*, 1012–1022. [CrossRef]

56. Yang, Y. Improving Environmental Prediction by Assimilating Auxiliary Information. *J. Environ. Inform.* **2015**, *26*, 91–105. [CrossRef]

57. Kolovos, A.; Smith, L.M.; Schwab-McCoy, A.; Gengler, S.; Yu, H.-L. Emerging patterns in multi-sourced data modeling uncertainty. *Spat. Stat.* **2016**, *18*, 300–317. [CrossRef]

58. Yang, Y.; Zhang, C.; Zhang, R. BME prediction of continuous geographical properties using auxiliary variables. *Stoch. Environ. Res. Risk Assess.* **2016**, *30*, 9–26. [CrossRef]

59. Yu, H.L.; Ku, S.C. A GIS tool for spatiotemporal modeling under a knowledge synthesis framework. *Stoch. Environ. Res. Risk Assess.* **2016**, *30*, 665–679. [CrossRef]

60. He, J.; Kolovos, A. Bayesian maximum entropy approach and its applications: A review. *Stoch. Environ. Res. Risk Assess.* **2017**, *32*, 859–877. [CrossRef]

61. Xiao, L.; Lang, Y.; Christakos, G. High-resolution spatiotemporal mapping of PM2.5 concentrations at Mainland China using a combined BME-GWR technique. *Atmos. Environ.* **2018**, *173*, 295–305. [CrossRef]

62. Li, A.; Bo, Y.; Zhu, Y.; Guo, P.; Bi, J.; He, Y. Blending multi-resolution satellite sea surface temperature (SST) products using Bayesian maximum entropy method. *Remote Sens. Environ.* **2013**, *135*, 52–63. [CrossRef]

63. Gao, S.; Zhu, Z.; Liu, S.; Jin, R.; Yang, G.; Tan, L. Estimating the spatial distribution of soil moisture based on Bayesian maximum entropy method with auxiliary data from remote sensing. *Int. J. Appl. Earth Obs. Geoinf.* **2014**, *32*, 54–66. [CrossRef]

64. Tang, Q.; Bo, Y.; Zhu, Y. Spatiotemporal fusion of multiple-satellite aerosol optical depth (AOD) products using Bayesian maximum entropy method. *J. Geophys. Res. Atmos.* **2016**, *121*, 4034–4048. [CrossRef]

65. Qin, D.; Ding, Y. *Climate and Environmental Change in China 1951–2012*; Springer-Verlag: Berlin/Heidelberg, Germany, 2016.

66. Coops, N.C.; Duro, D.C.; Wulder, M.A.; Han, T. Estimating afternoon MODIS land surface temperatures (LST) based on morning MODIS overpass, location and elevation information. *Int. J. Remote Sens.* **2007**, *28*, 2391–2396. [CrossRef]

67. Zhao, W.; Duan, S.-B.; Li, A.; Yin, G. A practical method for reducing terrain effect on land surface temperature using random forest regression. *Remote Sens. Environ.* **2019**, *221*, 635–649. [CrossRef]

68. Christakos, G.; Bogaert, P.; Serre, M. *Temporal GIS: Advanced Functions for Field-Based Applications*; Springer Science & Business Media: Berlin/Heidelberg, Germany, 2012.

69. Zeng, C.; Long, D.; Shen, H.; Wu, P.; Cui, Y.; Hong, Y. A two-step framework for reconstructing remotely sensed land surface temperatures contaminated by cloud. *ISPRS J. Photogramm. Remote Sens.* **2018**, *141*, 30–45. [CrossRef]

70. Pede, T.; Mountrakis, G. An empirical comparison of interpolation methods for MODIS 8-day land surface temperature composites across the conterminous Unites States. *ISPRS J. Photogramm. Remote Sens.* **2018**, *142*, 137–150. [CrossRef]

71. Christakos, G.; Yang, Y.; Wu, J.; Zhang, C.; Mei, Y.; He, J. Improved space-time mapping of PM2.5 distribution using a domain transformation method. *Ecol. Indic.* **2018**, *85*, 1273–1279. [CrossRef]

Article

Estimation of All-Weather 1 km MODIS Land Surface Temperature for Humid Summer Days

Cheolhee Yoo [1], Jungho Im [1,*], Dongjin Cho [1], Naoto Yokoya [2], Junshi Xia [2] and Benjamin Bechtel [3]

[1] School of Urban and Environmental Engineering, Ulsan National Institute of Science and Technology, Ulsan 44919, Korea; yoclhe@unist.ac.kr (C.Y.); djcho@unist.ac.kr (D.C.)
[2] Geoinformatics Unit, RIKEN Center for Advanced Intelligence Project (AIP), Mitsui Building, 15th Floor, 1-4-1 Nihonbashi, Chuo-ku, Tokyo 103-0027, Japan; naoto.yokoya@riken.jp (N.Y.); junshi.xia@riken.jp (J.X.)
[3] Department of Geography, Ruhr-University Bochum, 44807 Bochum, Germany; benjamin.bechtel@rub.de
* Correspondence: ersgis@unist.ac.kr; Tel.: +82-52-217-2824

Received: 9 March 2020; Accepted: 26 April 2020; Published: 28 April 2020

Abstract: Land surface temperature (LST) is used as a critical indicator for various environmental issues because it links land surface fluxes with the surface atmosphere. Moderate-resolution imaging spectroradiometers (MODIS) 1 km LSTs have been widely utilized but have the serious limitation of not being provided under cloudy weather conditions. In this study, we propose two schemes to estimate all-weather 1 km Aqua MODIS daytime (1:30 p.m.) and nighttime (1:30 a.m.) LSTs in South Korea for humid summer days. Scheme 1 (S1) is a two-step approach that first estimates 10 km LSTs and then conducts the spatial downscaling of LSTs from 10 km to 1 km. Scheme 2 (S2), a one-step algorithm, directly estimates the 1 km all-weather LSTs. Eight advanced microwave scanning radiometer 2 (AMSR2) brightness temperatures, three MODIS-based annual cycle parameters, and six auxiliary variables were used for the LST estimation based on random forest machine learning. To confirm the effectiveness of each scheme, we have performed different validation experiments using clear-sky MODIS LSTs. Moreover, we have validated all-weather LSTs using bias-corrected LSTs from 10 in situ stations. In clear-sky daytime, the performance of S2 was better than S1. However, in cloudy sky daytime, S1 simulated low LSTs better than S2, with an average root mean squared error (RMSE) of 2.6 °C compared to an average RMSE of 3.8 °C over 10 stations. At nighttime, S1 and S2 demonstrated no significant difference in performance both under clear and cloudy sky conditions. When the two schemes were combined, the proposed all-weather LSTs resulted in an average R^2 of 0.82 and 0.74 and with RMSE of 2.5 °C and 1.4 °C for daytime and nighttime, respectively, compared to the in situ data. This paper demonstrates the ability of the two different schemes to produce all-weather dynamic LSTs. The strategy proposed in this study can improve the applicability of LSTs in a variety of research and practical fields, particularly for areas that are very frequently covered with clouds.

Keywords: MODIS; AMSR2; annual cycle parameters; random forest; cloudy sky LST

---

1. Introduction

Land surface temperature (LST) is the radiative temperature of the land surface, which plays a crucial role in understanding various environmental problems such as heatwaves, drought, wildfire, air quality, and urban heat islands [1–7]. Since LST reflects the energy flux stability at the boundary of the surface and atmosphere, it is also used as a major parameter in modeling global physical processes, including hydrological and biogeochemical cycles [8–10]. Therefore, it is important to obtain accurate LST over large areas on both high spatial and temporal domains.

With the continued development of remote sensing technology, LST has been retrieved from satellite data for large areas with high temporal and spatial resolution. Thermal infrared (TIR) sensors are the most widely used in producing satellite-based LST. Several algorithms, such as single-channel, split-window, and temperature and emissivity separation (TES) techniques, have been developed to provide TIR-based LST [11]. One of the most well-known TIR-based LST datasets is the moderate-resolution imaging spectroradiometer (MODIS) LST onboard the Terra and Aqua satellites. MODIS offers global LST products within a 1–2 K accuracy range, with a relatively high spatial resolution of 1 km, four times a day (two daytime and two nighttime LSTs). In addition, LST products are provided by several other TIR sensors with different specifications in both low earth orbit and geostationary orbit satellites: The visible infrared imaging radiometer suite (VIIRS), spinning enhanced visible and infrared imager (SEVIRI), and advanced spaceborne thermal emission and reflection radiometer (ASTER). Unfortunately, TIR-based LST is significantly affected by weather and atmospheric conditions; particularly, the surface temperature under clouds is not available. Some studies have been conducted to fill the gaps in LST data caused by clouds [12–26].

Previous studies aimed at overcoming the lack of TIR-based LST data under cloudy areas can be divided into four groups. The first group reconstructs LSTs in cloudy areas by combining spatially, temporally, or spatiotemporally neighboring clear sky LSTs [13,18,22,25]. In particular, recent studies have looked at modeling by combining multiple algorithms, such as regression and interpolation, using spatial and temporal information of multi-temporal LSTs [16,21,26]. The second group not only uses multiple LSTs, but also auxiliary variables that are highly correlated with LST, to estimate cloudy sky LSTs, using statistical methods such as regression kriging and spline interpolation. However, the critical limitation in the methods of these first two groups is that they assume that LST under cloudy weather conditions is not different from that under clear sky conditions. In general, clouds reduce incoming shortwave radiation during daytime by blocking the Sun, and increase downward longwave radiation during nighttime. Thus, nighttime LSTs in cloudy conditions are only slightly lower than those under clear skies, while the difference in daytime LST is more significant [27]. Consequently, it is essential to model LSTs under cloudy conditions.

The third group uses physical modeling approaches like surface–energy balance (SEB) theory, which is adopted to derive cloudy LSTs from spatially neighboring clear sky LSTs. The effect of the clouds is simulated using a correction term that takes into account surface insolation, air temperature, and wind speed [15,17,23,24]. The SEB techniques, however, require complex parameterization with air temperature and wind speed as input data. Although the variation of LST is assumed to be based on insolation during the daytime, the method is not able to be applied to nighttime [17].

The fourth group uses passive microwave (PMW)-based data to overcome the issue of cloudy areas in TIR-based LST data. PMW-based data are less affected by water vapor and clouds than TIR-based LST data. Brightness temperature (BT), measured by the advanced microwave Sscanning radiometer-earth observing system (AMSR-E) and advanced microwave scanning radiometer 2 (AMSR2) sensors, are frequently used as PMW-based data to estimate LST. Although PMW-based BT has limitations of coarse spatial resolution (10–25 km), it could be used as supporting data in estimating missing values of TIR-based LSTs under cloudy conditions [14]. For example, Shwetha and Kumar resampled 25-km AMSR-E/AMSR2-based BTs directly into 1 km, using them as input variables for artificial neural networks with auxiliary data of elevation, latitude, and longitude to model the all-weather 1 km LST [19]. Meanwhile, many studies have derived the PMW-based LST using the original resolution of BT (i.e., 10–25 km), rather than resampling it to 1 km and then downscaling it to a high resolution to merge with TIR-based LST [12,20]. PMW-based methods simulate cloudy sky LSTs based on the fact that PMW can penetrate clouds. However, the previous studies have limitations in terms of spatial accuracy in merging coarse PMW-based data with high-resolution TIR-based LST.

In South Korea, summers can often be scorching, causing a variety of disasters, including heatwaves and tropical nights. During these hot summers, Northeast Asia, especially South Korea, is usually covered by clouds transported by the East Asia monsoon [28]. Therefore, the reconstruction methods using temporally or spatially neighboring clear sky LSTs could not be successfully applied in this area in summer due to a very high cloud cover rate. Moreover, daytime LST on humid summer days (i.e., July and August) in South Korea is generally high under clear sky conditions, but it drops sharply in cloudy weather, such as during the rainy season or typhoon periods. Previous studies, however, have failed to consider the variability (i.e., rapid change) of LST under cloudy conditions. In addition, many studies have used air temperature rather than LST as in situ data to validate their cloudy sky LST predictions [16,29,30]. However, it should be noted that air temperature and LST often show different patterns in regions with heterogeneous land surfaces [31,32].

This study proposes two different schemes for estimating all-weather 1 km MODIS LSTs for humid summer days over South Korea, based on machine learning, using multiple datasets made up of AMSR2 BTs, and the annual cycle parameters (ACPs) of satellite TIR-derived LSTs. The first scheme (S1) is a two-step approach that first estimates 10 km LSTs and then downscales the LSTs from 10 km to 1 km. The second scheme (S2) is a one-step algorithm that directly estimates the 1 km all-weather LSTs. The primary objective of this study is to investigate how well the two schemes that we propose simulate dynamic humid summer LSTs under clear- and cloudy sky conditions through a series of validation processes. The remainder of this paper is organized as follows. Section 2 presents the study area and the data we used. Section 3 introduces the methods in detail, including the framework of our two proposed schemes. In Section 4, the distribution of clear and cloudy sky LSTs in the summer season are analyzed using in situ station data for daytime and nighttime. Then, the two different schemes are evaluated by a series of validations, especially using in situ LSTs for both clear and cloudy conditions. Finally, Section 5 presents the conclusion of this study.

2. Study Area and Data

2.1. Study Area

The study area is the mainland of South Korea with an area of approximately 99,728 km^2 (latitude 34° N–38.5° N and longitude 126° E–129.5° E) (Figure 1). South Korea generally has a humid, continental climate affected by the Asian monsoons, with a large amount of precipitation in summer during the rainy season (usually from the end of June to the end of July). The annual mean temperature is about 10–15 °C; August, the hottest month, has a mean temperature of 23–26 °C. Humidity ranges from 60%–75% on a national scale, with summers (July and August) rising to 70%–85%. The southern coast is subject to late-summer typhoons. As seen in Figure 1, the dominant land-cover categories of the study area are forest (67.7%), agricultural land (22.2%), urban areas (4.6%), and others, including grass, water, barren, and wetlands (5.5%).

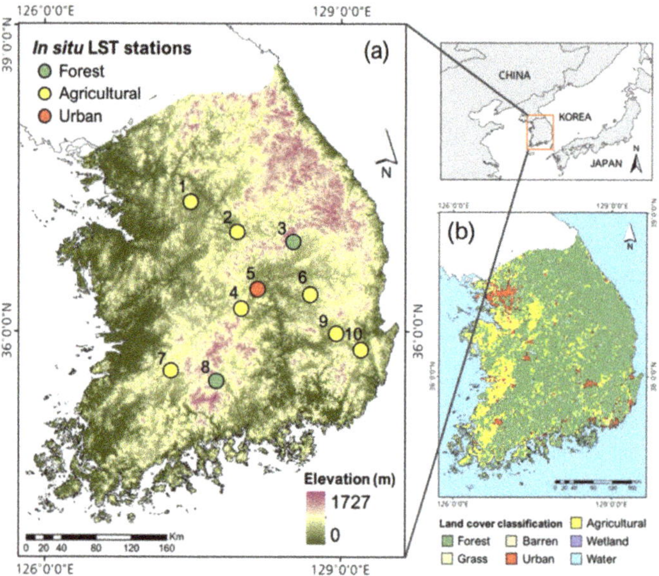

Figure 1. (**a**) Study area and in situ reference data locations with 1 km elevation derived from Shuttle Radar Topography Mission (SRTM) digital elevation model (DEM), and (**b**) The landcover provided by Ministry of Environment of South Korea (http://egis.me.go.kr).

2.2. Saetellite Data

The AMSR2 onboard the global change observation mission 1st-water (GCOM-W1) satellite, launched in May 2012, provides global PMW-based BT data. It acquires a set of daytime and nighttime microwave data twice a day: The equator crossing time is 1:30 p.m. for the ascending pass, and 1:30 a.m. for the descending pass. The AMSR-2 has seven frequencies, with both vertical and horizontal polarizations, and approximately 62×35, 62×35, 42×24, 22×14, 19×11, 12×7, and 5×3 km spatial resolution at 6.9, 7.3, 10.7, 18.7, 23.8, 36.5, and 89.0 GHz, respectively. Among these, we used the four frequencies (36.5, 23.8, 18.7, and 10.7 GHz) mostly used for the estimation of LST in the previous studies [30]. Low frequency data resampled into a 10 km resolution were downloaded from the Japan Aerospace Exploration Agency (https://gcom-w1.jaxa.jp) for 2013–2018. We used daily MODIS daytime and nighttime Aqua LST data (MYD11A1) because the equatorial-crossing times of Aqua MODIS are nearly the same as those of AMSR2 (1:30 p.m.–daytime and 1:30 a.m.–nighttime). The MYD11A1 LST data, which have 1 km spatial resolution, were retrieved using a generalized split-window algorithm [33]. The MYD11A1 products from 2013–2018 were downloaded from Earthdata Search (https://search.earthdata.nasa.gov/search). South Korea's elevation was retrieved from the shuttle radar topography mission (SRTM) digital elevation model (DEM), with 30 m spatial resolution (https://earthexplorer.usgs.gov). Global man-made impervious surface (GMIS) data with 30 m spatial resolution derived from Landsat images for the year of 2010 [34] were obtained to get the fractional impervious surface in this study.

2.3. In Situ *LST Data*

In situ LSTs (1 a.m./p.m. and 2 a.m./p.m.) from 2013 to 2018, obtained from the automated surface observing systems (ASOSs) operated by the Korea Meteorological Administration, were used as reference data. As shown in Figure 1a, a total of 10 ASOSs were selected based on the following conditions: First, the stations close to the coastline were excluded, because satellite-based LST data could be contaminated from the influence of ocean water included in the grid; and second, the stations that have high bias were also excluded from the reference data, after applying the bias-correction method described in Section 3.1.

3. Methods

3.1. LST Pre-Processing

The 1 km MODIS LSTs with quality control (QC) flags of 'cloud' were considered to be pixels under cloudy sky conditions, and all others were clear sky LSTs. To train the model and validate the estimated LST under cloudy areas, we used in situ LSTs measured at ASOSs. However, there are spatial scale differences between the MODIS LST (1 km) and the point-based ASOS LST; thus, a systematic bias could occur. To solve this problem, we fitted in situ LST and MODIS LST under clear sky conditions using polynomial regression by station [35], to bias-correct in situ LSTs for both daytime and nighttime (Equation (1)).

$$Y = aX^2 + bX + C \tag{1}$$

where X is the in situ LST and Y is the MODIS LST under clear sky conditions. a, b, and c represent the coefficients. We used the quadratic function because many ASOSs have observed that in situ LSTs rise more rapidly than MODIS LSTs at high temperatures in the summertime.

The MODIS and in situ LSTs from March to October, starting from 2013 through to 2018 at each station were used for the bias-correction. Only MODIS LSTs with good quality were used based on the QC flags, and the hourly in situ LSTs were linearly interpolated to the MODIS view time. We excluded the winter season (December through February) because the range of winter LSTs is not compatible with that of summer LSTs, which are the target variable of this study. Finally, we selected a total of 10 ASOSs for validation, with a calibration error (RMSE) < 2.5 °C for both daytime and nighttime in July–August (summer season) from the bias-correction analysis. Figure 1 shows the geographic distribution of the ten ASOSs used in this study. Appendix A describes the statistical results of before and after bias correction of the 1 km in situ LSTs at the 10 stations.

In addition, clear sky 1 km MODIS LSTs were aggregated to 10 km based on the 10 km AMSR2 grid area. At this stage, only 1 km MODIS LSTs where 95% or more clear sky LSTs exist in a 10×10 km^2 area were used for aggregation. We also fitted the in situ LSTs and the 10 km MODIS LSTs under clear sky conditions using polynomial regression in the same way as the bias correction of the 1 km LSTs at the selected 10 ASOS stations to produce the bias-corrected in situ LSTs at 10 km scale.

3.2. Pre-Processing of Input Variables

Table 1 describes the input variables used for the estimation of LST under cloudy conditions. A total of 10 AMSR2 BTs were projected onto MODIS LSTs, whilst AMSR2 pixels from near the coastline were masked to eliminate the effects of ocean water. Due to the shift of the flight overpass path, the areas of missing AMSR2 BT values were also excluded.

The 1 km MODIS annual cycle parameters (ACPs), including mean annual surface temperature (MAST), yearly amplitude surface temperature (YAST), and LST phase shift relative to the spring equinox (theta) were calculated for each of daytime and nighttime based on the following equation [36].

$$f(d) = MAST + YAST * \sin\left(\frac{d2\pi}{365} + theta\right) \tag{2}$$

where d represents the day of the cycle relative to the spring equinox. The ACPs were created for each year of the study period and averaged to construct one MAST, YAST, and theta each for daytime and nighttime.

We aggregated the 30 m SRTM elevation and 30 m GMIS impervious surface fraction to the 1 km resolution of MODIS. Latitude and longitude values were extracted from the information contained in the MODIS tiles. This study converted DOY to values ranging from −1 to 1 within a one-year period, using a sine function that considers seasonality (i.e., setting the middle of summer as 1 and the middle of winter as −1) [37].

Table 1. Description of input variables used in this study.

Type	Variables	Acronym
AMSR2 BT	36.5 GHz horizontal polarization	36H
	36.5 GHz vertical polarization	36V
	23.8 GHz horizontal polarization	23H
	23.8 GHz vertical polarization	23V
	18.7 GHz horizontal polarization	18H
	18.7 GHz vertical polarization	18V
	10.7 GHz horizontal polarization	10H
	10.7 GHz vertical polarization	10V
Annual cycle parameters	Mean annual LST (K)	MAST
	Mean annual amplitude of LST (K)	YAST
	Phase shift relative to spring equinox on the Northern hemisphere	theta
Auxiliary variables	Elevation (m)	Elev
	Impervious surface fraction (%)	Imp
	Latitude (°)	Lat
	Longitude (°)	Lon
	Converted day of year	DOY
	Year as a discrete value	Year

3.3. Random Forest (RF)

In this study, we applied machine-learning random forest (RF) to estimate the LST for all-weather conditions. RF has been widely used to conduct a variety of classification and regression tasks [38–43]. RF comprises classification and regression trees, producing a variety of independent trees to make a final decision through two randomizations: (1) Random selection of training samples and (2) random selection of predictor variables at each node of a tree [44]. This randomization is aimed at solving the overfitting problem and reducing the sensitivity of a model that comes from training sample configurations [45]. When a variable is randomly permuted, RF calculates the relative variable importance from out-of-bag (OOB) samples, measuring the mean squared error (MSE) of the OOB portion of samples in each tree. The same process is implemented whenever each input variable is perturbed, and the difference between the two MSEs of all trees is averaged. The larger the increase in the percentage of MSE (%incMSE) of a variable, the greater the contribution of the variable. It should be noted that the relative variable importance is considered as the sum of local contributions, not global importance [46–48]. The RF was implemented using the R statistical software through the "ranger" add-on package, with default model parameter settings (e.g., the number of trees set as 500). The "ranger" is a fast implementation of RF, or recursive partitioning, and is known to be suited to high-dimensional data with large datasets [49].

3.4. Schemes for Estimating All-Weather LSTs

In this study, we proposed two schemes (S1 and S2) to estimate all-weather MODIS LSTs. The S1 is based on a two-step approach (see Figure 2). The first step is estimating 10 km LSTs for daytime and nighttime. We used the following independent variables: Four different frequencies of the AMSR2 of two polarization BTs (10H/V, 18H/V, 23H/V, and 36H/V), along with five 10 km resampled auxiliary variables (Elev, Lat, Lon, DOY, and year). The aggregated 10 km MODIS LSTs under the clear sky condition were used as a target variable. Moreover, we used the 10 km bias-corrected in situ LSTs under the cloudy sky condition from 10 stations as a dependent variable together to train the model with the real LST characteristics under the cloudy skies. RF was applied to predict LSTs by training the samples from March to October between 2013 and 2018. Finally, the 10 km LSTs were produced based on the developed model using the input variables under all-weather conditions.

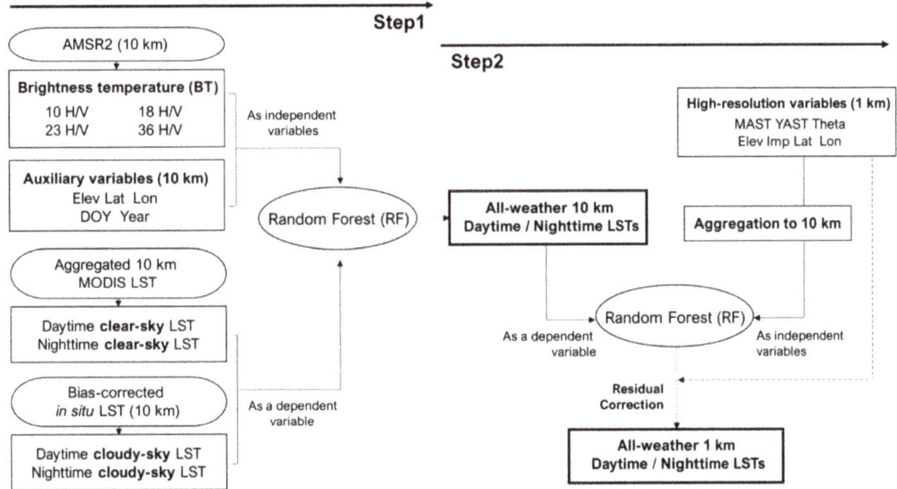

Figure 2. Flowchart of scheme 1(S1) suggested in this study. Procedures are divided into two main steps: Estimating 10 km land surface temperatures (LSTs) (**left**) and the spatial downscaling of LSTs to 1 km (**right**).

The second step involved the spatial downscaling of LSTs from 10 km to 1 km. We used RF to develop the downscaling model for each date. The Elev, Imp, Lat, Lon, and ACPs (i.e., MAST, YAST, and theta) were used as input variables, considering the surface properties and spatial distribution characteristics of 1 km LSTs. At first, the 1 km MAST, YAST, and theta with 1 km auxiliary factors—Elev, Imp, Lat, and Lon—were aggregated to 10 km for independent variables for the RF model, and the 10 km LSTs, constructed in the first step, were used as the dependent variable. After the model was trained for each date, the original high-resolution 1 km input variables were put into the RF model to finally produce the 1 km all-weather LSTs, after the residual correction proposed by Hutengs and Vohland [50].

The second scheme S2 is a one-step algorithm that directly estimates 1 km all-weather LSTs (Figure 3). To create this scheme, first we downscaled all 10 km AMSR2 BTs to 1 km using bilinear resampling. The resampled BTs together with 1 km auxiliary variables (i.e., MAST, YAST, theta, Elev, Imp, Lat, Lon, DOY, and Year) were used as input variables in RF. We used the 1 km clear sky MODIS LSTs and the 1 km bias-corrected in situ LSTs under cloudy skies from ten stations as target variables for daytime and nighttime separately. RF models were built for each year with the samples of the entire study area from March through October being used for training. We put the input variables of all-weather conditions into the developed RF model to produce 1 km LSTs for both clear and cloudy sky conditions.

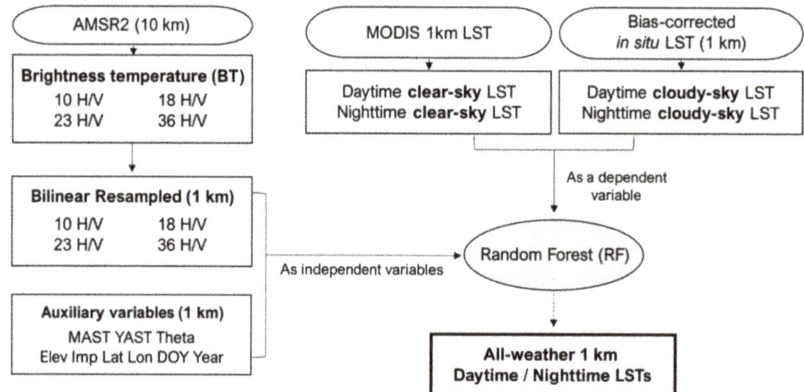

Figure 3. Flowchart of scheme 2(S2) for directly estimating 1 km all-weather LSTs in one-step.

3.5. A Series of Validations

For the first step of the first scheme, S1, two types of validations were implemented using the clear sky 10 km aggregated MODIS LSTs focusing on the summertime (i.e., July and August). In the first validation, we randomly divided July and August samples into training and validations sets (80%/20% split). In the humid summer, however, most of the area in South Korea is often covered by clouds for the whole day. So, for the second validation, we randomly divided the samples of July and August by date: 80% for training and the remaining 20% for validation. In the second step of S1, LSTs downscaled into 1 km by RF were validated using all clear sky MODIS LSTs in the summer. At this stage, we compared the performance of RF with other downscaling techniques in two ways: (1) A bilinear resampling of 10 km LSTs to 1 km; (2) a lapse-rate technique using Equation (3), which assumes that the temperature difference comes from elevation, suggested by Dual et al. [12]:

$$LST_{1\ km} = TLR \times (Elv_{1\ km} - Elv_{10\ km}) + LST_{10\ km} \tag{3}$$

where LST_{1km} is the downscaled LST at 1 km scale; LST_{10km} represents the predicted 10 km LSTs in step 1; Elv_{1km} is the surface elevation at 1 km scale; Elv_{10km} is the surface elevation at 10 km scale; and the TLR is the temperature lapse rate, which is defined as the rate of temperature decrease with elevation in the troposphere (average TLR value is 6.5 K/km; [51]). For the second scheme, S2, first and second validations similar as in S1 were performed.

Moreover, the estimated all-weather 1 km LSTs, including those under cloudy skies, were validated using bias-corrected in situ LSTs of 10 ASOSs in summer. Among the 10 ASOS stations, data at one station were used as validation and the others were used to train the model with in situ cloudy LSTs as targets. This leave one-station out cross-validation (LOOCV) was repeated for all 10 stations. The coefficient of determination (R^2; Equation (4)), RMSE (Equation (5)), normalized RMSE (nRMSE; Equation (6)), and bias (Equation (7)) were used to evaluate the performance of the models for each validation step.

$$R^2 = 1 - \frac{\sum_{i=1}^{n}(y_i - \hat{y}_i)^2}{\sum_{i=1}^{n}(y_i - \overline{y})^2} \ , \ \overline{y} = \frac{1}{n}\sum_{i=1}^{n} y_i \ , \tag{4}$$

$$RMSE\ (^{\circ}C) = \sqrt{\sum_{i=1}^{n} \frac{(\hat{y}_i - y_i)^2}{n}} \ , \tag{5}$$

$$nRMSE\ (\%) = \frac{RMSE}{(y_{max} - y_{min})} \times 100 \ , \tag{6}$$

$$\text{Bias (°C)} = \sum_{i=1}^{n} \frac{(\hat{y}_i - y_i)}{n}, \tag{7}$$

where y_i is the measured value, $\hat{y}_i$ is the predicted value, y_{max} and y_{min} represent the maximum and minimum values of observations, and n is the number of samples.

4. Results and Discussion

4.1. Analysis of Clear and Cloudy Sky in Situ LSTs in Summer

Figure 4 depicts the boxplots of bias-corrected LSTs for the 10 in situ stations under clear sky and cloudy sky conditions. It should be noted that the average LSTs in clear skies were higher than those of cloudy skies for all stations in the daytime. Although high LSTs also appear under cloudy sky conditions, LSTs are considerably lower when compared to clear sky LSTs. The boxplot (Figure 4) shows that the upper quartile of cloudy sky LSTs is close to the lower-quartile of clear sky LSTs for most in situ stations. One possible reason for this is an increasing cloud-cover on rainy days in humid areas in the summer, where precipitation reduces the surface sensible heat [12,52]. When the sky is heavily cloudy in the daytime, most of the Sun's energy does not reach the surface, preventing the heating up of the Earth. Otherwise, at nighttime, clear sky and cloudy sky LST distributions do not differ significantly for all the stations. There is no incoming radiation from the Sun at night, therefore LSTs do not rise significantly. The average nighttime LSTs—early overnight—under cloudy skies have a nearly identical range as, or only slightly lower than, the clear sky LSTs.

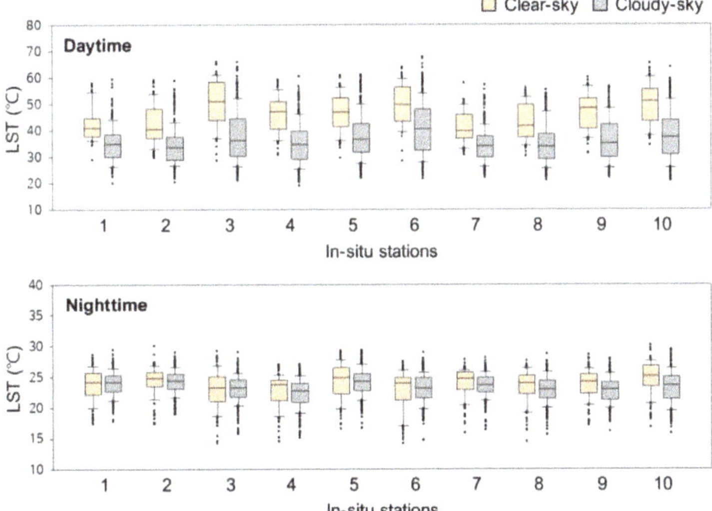

Figure 4. Boxplots of bias-corrected in situ LSTs under clear and cloudy skies for daytime and nighttime for each station. Refer to Figure 1 for station numbers.

Among the 372 days, the total number of summer days (July–August) for the six years (2013–2018), the size of good-quality clear sky samples was very small, especially for the daytime (see the sample size in Tables A1 and A2). This implies that the temporal or spatial interpolation approach to reconstructing LSTs in cloudy conditions could not be effectively applied to the summer season because of a large amount of cloud contamination.

4.2. Evaluation of Two Schemes Using Clear Sky LSTs

Figure 5 shows the results of the first (upper) and second (lower) validations for summer daytime and nighttime for Scheme 1(S1). The R^2 values for estimating 10 km daytime and nighttime LSTs were 0.82 and 0.85, respectively, in the first validation. The RMSE values have a daytime LST of 1.40 °C and a nighttime LST of 0.95 °C. Considering their similar nRMSE, the main reason for the RMSE difference between daytime and nighttime is likely the range difference in the temperature distribution. In the second validation, the R^2 (RMSE) values for estimating 10 km daytime and nighttime LSTs were 0.79 (1.55 °C) and 0.88 (1.14 °C), respectively. The accuracy of the two RF-based validations corresponds to the MODIS LST validation error of 1–2 K [53–55]. In particular, the second validation results were similar to the first validation in terms of accuracy (i.e., R^2 and nRMSE), even with a separate validation dataset by date. These results imply that the constructed 2-step RF-based model is robust for humid areas where clouds often cover most regions in summer. However, since we performed both validations using the MODIS clear sky LSTs, the effect of clouds was not considered. Thus, accuracy assessment using in situ LST data under cloudy conditions was needed.

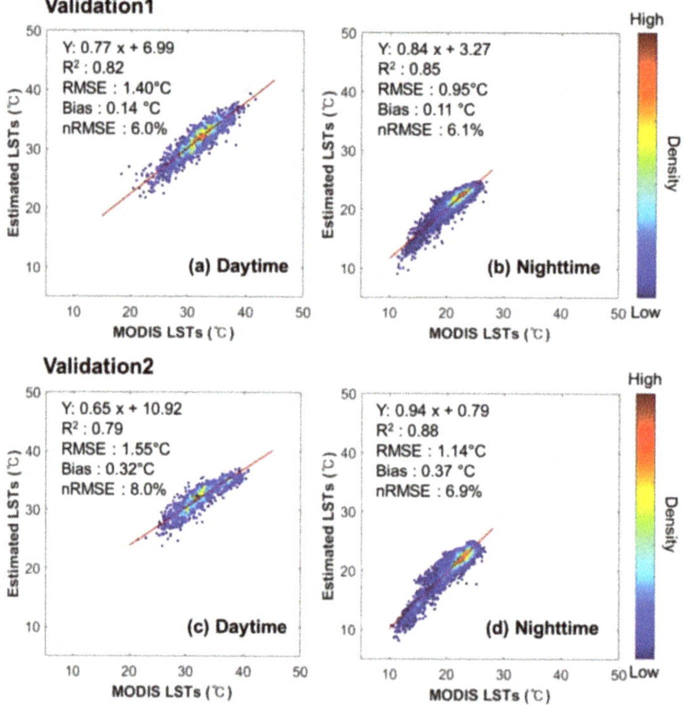

Figure 5. Density scatter plots between the estimated and 10 km moderate-resolution imaging spectroradiometer (MODIS) LSTs for scheme 1(S1) from two validation results for daytime and nighttime. The color ramp from blue to red corresponds to increasing point density. Black dashed lines show the regression line and red solid lines represent the line of identity (y = x).

Figure 6 shows the validation results of the second step for Scheme 1, using clear sky MODIS LSTs estimated with the bilinear resampling (Bilinear), the lapse rate approach (Lapse rate), and the proposed RF-based algorithm (RF) among the different land covers. RF outperformed the other two techniques with higher correlation and lower RMSE in most land covers for both daytime and nighttime. For daytime, RF showed R^2 of 0.6–0.8 and RMSE of 1.6–3.0 °C with some variation according to land cover. This accuracy is considered significant with an average range value of RMSE similar to approximately 2.0 K, which is the target accuracy in several daytime LST retrieval studies [56–58]. The reason why urban areas showed lower performance than other land covers might be that the relatively high LSTs in urban areas were not accurately simulated when downscaling 10 km LST to 1 km, considering the fact that RF underestimated the downscaled LSTs (mean bias: −1.5 °C, not shown). Moreover, it has been reported that MODIS daytime LSTs in urban areas have relatively high uncertainty [16,59]. Note that the RF outperformed the lapse rate approach, which assumes the dependency of cloudy LSTs only on the altitude for all land covers in the daytime [12]. The RF appears to simulate the surface thermal heterogeneity well in daytime, using not only Elev, but also the Imp and ACPs—MAST, YAST, and theta—as input variables.

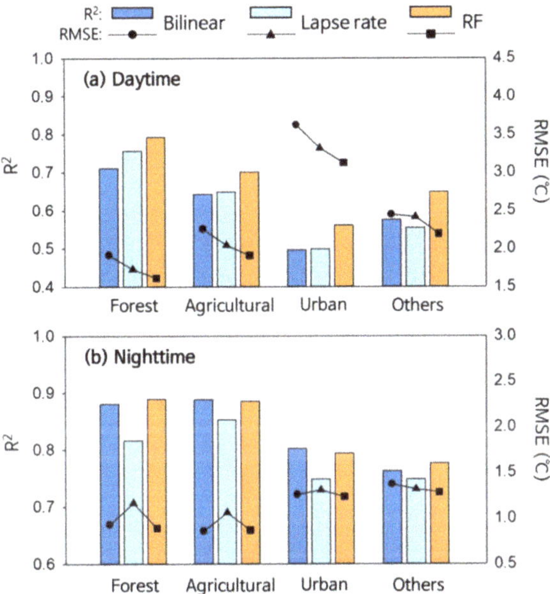

Figure 6. Model performance of three downscaling approaches (bilinear resampling (bilinear), lapse rate, and random forest (RF)) in the second step for scheme 1 by land cover for daytime (**a**) and nighttime (**b**).

For nighttime, the RF model returned highly accurate results, with R^2~0.8 to 0.9 and RMSE < 1.5 °C with some variations by land cover. The lapse rate approach showed worse performance than the other two methods for most land cover types. Duan et al. used an average lapse rate of 6.5 K/km for both day and night [12]. However, the lapse rate could be applied to the air temperature of the troposphere, not the LSTs, which implies that the LST difference by altitude does not seem to be significant on the local scale. Interestingly, the bilinear and RF models did not show much difference for nighttime, where RF showed slightly better performance than the bilinear interpolation in terms of R^2 and RMSE for most land covers. This suggests that the nighttime LST is thermal-homogenous enough for bilinear interpolation to achieve good results.

Table 2 summarizes the results of the first and second validations for a 1 km clear sky LST estimation in S2. The first validation showed excellent performance resulting in R^2 > 0.9 and RMSE < 1 °C for both daytime and nighttime. The second validation, however, dividing samples by date, yielded relatively low accuracy, which is possibly due to the daily LST variations. It is not surprising that the prediction performance of S2 by time over humid areas in the summer would be less accurate than the first validation, because clouds covered most areas for specific days, which might not have been trained well. In particular, cloudy areas in the summer daytime have an LST range different from clear sky areas (Figure 4); therefore, S2-based 1 km LSTs need further investigation using in situ data under cloudy conditions.

Table 2. The two validation results of the 1 km LST estimation from scheme 2(S2) for daytime and nighttime.

Daytime	R^2	RMSE (°C)	nRMSE (%)	Bias (°C)	Sample Size
Validation1	0.93	0.97	3.2	0.00	139,876
Validation2	0.79	1.62	5.9	0.16	138,307
Nighttime	R^2	RMSE (°C)	nRMSE (%)	Bias (°C)	Sample Size
Validation1	0.96	0.53	2.7	0.02	353,114
Validation2	0.79	1.07	6.5	0.03	298,575

4.3. Spatial Distribution Analysis of All-weather LSTs from Two Schemes

Figures 7 and 8 show the all-weather average daytime and nighttime 10 km LSTs of S1 from July through August for each year. The overall spatial distribution of the daytime and nighttime LSTs appeared similar, but the range of the daytime LSTs (22.1–34.0 °C) was considerably higher than that of the nighttime LSTs (16.0–22.6 °C). This is because the differences in heat capacity and evapotranspiration by land type result in a wide range of LSTs in the daytime, affected by incoming solar radiation [22]. Elevation (see Figure 1) yielded a negative spatial relationship with LSTs for both daytime and nighttime, which is consistent with the results of Bertoldi that showed that LST decreases with increasing elevation because of complex factors such as air temperature and vegetation amount [60]. In this study, therefore, the elevation input variable contributes significantly to the LST estimation model. In terms of land cover (see Figure 1), LSTs were relatively low in both the daytime and nighttime in the forested areas, widely distributed in the study area. Meanwhile, the cropland areas showed higher temperature values than in forests, especially for the western part of the study region, which is consistent with the results of Lee et al. [61]. The urban areas also showed relatively high LSTs compared to other landcovers, because of the highly impervious surfaces in both daytime and nighttime [32,62].

The developed 10 km LSTs of S1 also showed a temporal variation over six years (2013–2018). For example, South Korea experienced unprecedented extreme temperatures (i.e., a heatwave) in the summer of 2018 [63], where the developed LSTs show distinctly high-temperature distribution for both daytime and nighttime (Figures 7f and 8f). Furthermore, 10 km LSTs were relatively high at nighttime of summer 2013 (Figure 8a), which is consistent with the analysis of Choi and Lee that showed frequent tropical night events in South Korea in summer 2013 [64]. Overall, the LSTs estimated using S1 on the 10 km scale seem to accurately describe both spatial and temporal patterns of LSTs in summer daytime and nighttime.

Figure 9 shows the mean and variance of the 2013–2018 1 km all-weather LSTs produced by S1 and S2 for summer in both the daytime and nighttime. In the daytime, it should be noted that S2 generally simulated LSTs higher in urban and agricultural areas than S1 (Figure 9a,b). The difference in the estimated LSTs between S1 and S2 might also be due to S2 overestimating the low LSTs under cloudy conditions. The bias of the second validation of daytime was relatively high (~0.16), as shown in Table 2. At nighttime, the S1 and S2 showed similar LST distributions, except urban areas where the S1 LSTs were slightly higher than the S2 ones.

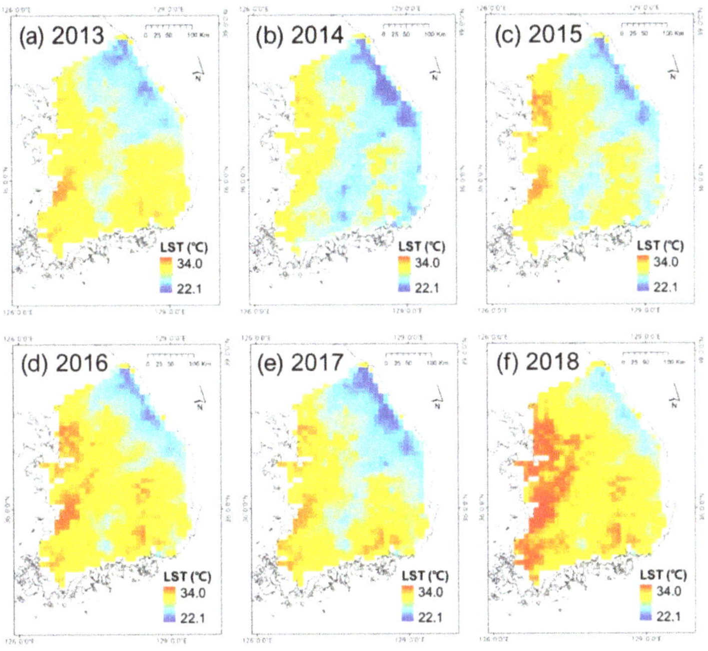

Figure 7. Spatial distribution maps of average estimated all-weather 10 km LSTs for scheme 1(S1) during July and August for each year; 2013 (**a**) to 2018 (**f**) for daytime.

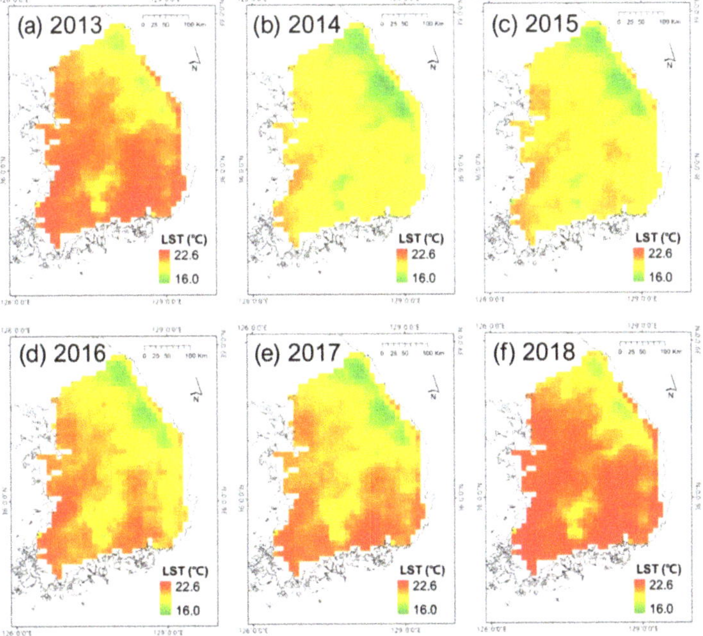

Figure 8. Spatial distribution maps of average estimated all-weather 10 km LSTs for scheme 1(S1) during July and August for each year; 2013 (**a**) to 2018 (**f**) for nighttime.

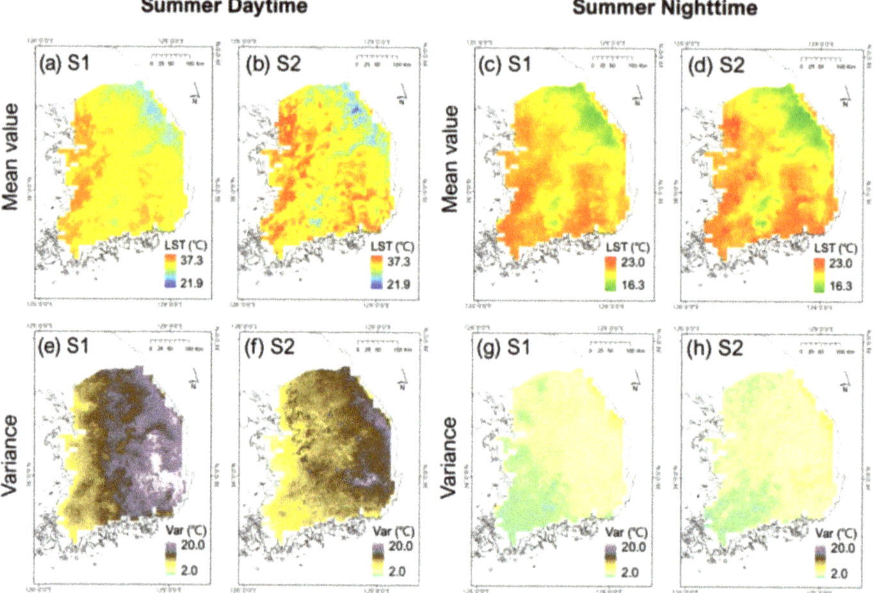

Figure 9. Spatial distribution maps of average all-weather 1 km LSTs (upper) and the variance (lower) during July and August from 2013 to 2018 produced by the two schemes (S1 and S2) for daytime (**a,b, e–f**) and nighttime (**c,d, g–h**).

Note that the variance of S1 LSTs is quite a lot higher than S2 in most areas in the daytime (Figure 9e–f). The relative variable importance (%) explains this fact as provided by RF (Figure A1). Among the several input variables, only AMSR2 BTs showed temporal changes. In S2, temporally constant variables, such as Lat, YAST, theta, and MAST, played significant roles in the RF model (Figure A1c); therefore, S2 could be relatively limited in simulating the temporal variations of LSTs. In the case of S1, AMSR2 BTs were used only to construct the 10 km LSTs first (step 1), whereas ACPs were used only for step 2. Therefore, S1 could simulate the temporal pattern of daily LSTs accurately by downscaling the proposed 10 km LSTs to 1 km for each day in step 2. At nighttime, S1 and S2 had similar LST variance for most areas (Figure 9g–h). In particular, LSTs in summer nighttime have a distinctly lower variance than those in the daytime (Figure 9e–f), which implies that the nighttime does not show significant temperature changes in summer regardless of weather conditions.

Many previous studies that predicted 1 km cloudy sky LSTs have used auxiliary variables that are indirectly related to LST, such as elevation, latitude, and longitude [12,19]. It should be noted that ACPs (i.e., MAST, YAST, and theta) have been identified as crucial variables by RF for both S1 and S2 (Figure A1). This suggests that ACPs are useful in estimating LSTs for all-weather conditions, due to their characteristics representing the spatial distribution of LSTs, without being affected by clouds.

4.4. Comparison of Two Schemes Using in Situ LSTs

The clear and cloudy sky 1 km LSTs produced from the two schemes were validated (i.e., LOOCV) using the 1 km bias-corrected in situ LSTs measured at the 10 stations (Table 3 for daytime and Table 4 for nighttime). It should be noted that S1 showed negative biases for most stations, which implies that S1 tended to underestimate 1 km LSTs under clear sky conditions. In particular, the RMSE of S1 was relatively high with a large negative bias, especially for stations 5. This might be because the station was located in built-up urban areas, where S1 LSTs were possibly underestimated, as presented in Figure 9. On the other hand, in the case of S2, the bias under clear sky conditions was close to zero and slightly positive for most stations. Therefore, S2 produced higher accuracy than S1 for high LSTs under clear sky conditions in the daytime.

Table 3. The leave one-station out cross-validation (LOOCV) results of the estimated 1 km clear and cloudy sky LSTs from scheme 1 (S1) and scheme 2 (S2) using bias-corrected in situ LSTs at 10 stations during July and August from 2013 to 2018 for daytime. Refer to Figure 1 for station numbers.

Station	Clear Skies						Cloudy Skies					
	R^2		RMSE (°C)		Bias (°C)		R^2		RMSE (°C)		Bias (°C)	
	S1	S2	S1	S2	S1	S2	S1	S2	S1	S2	S1	S2
1	0.54	0.58	2.6	1.9	−1.7	0.3	0.77	0.76	2.2	2.7	−1.2	1.9
2	0.63	0.72	2.5	1.7	−1.6	0.4	0.73	0.67	2.6	2.4	−1.5	1.1
3	0.71	0.73	3.9	2.8	−3.0	1.1	0.74	0.65	2.9	5.6	−0.1	4.6
4	0.73	0.76	1.7	1.7	−0.6	0.9	0.73	0.67	2.1	3.3	0.0	2.4
5	0.65	0.64	4.9	1.8	−4.5	0.1	0.81	0.77	3.9	3.2	−3.2	2.1
6	0.76	0.64	2.8	4.0	−0.1	2.5	0.81	0.80	2.9	5.8	1.6	5.2
7	0.63	0.65	2.2	1.5	−1.8	0.7	0.82	0.79	2.2	1.7	−1.8	0.9
8	0.56	0.64	2.2	2.3	0.1	1.2	0.81	0.77	2.1	3.5	1.2	2.9
9	0.74	0.71	3.8	2.3	−3.3	1.1	0.81	0.80	2.4	5.1	−0.2	4.4
10	0.44	0.48	1.9	2.1	−0.4	1.1	0.80	0.75	2.5	4.7	1.0	4.0

Table 4. The leave one-station out cross-validation (LOOCV) results of the estimated 1 km clear and cloudy sky LSTs from scheme 1(S1) and scheme 2(S2) using 1 km bias-corrected in situ LSTs for 10 stations during July and August from 2013 to 2018 for nighttime. Refer to Figure 1 for station numbers.

Station	Clear Skies						Cloudy Skies					
	R^2		RMSE (°C)		Bias (°C)		R^2		RMSE (°C)		Bias (°C)	
	S1	S2	S1	S2	S1	S2	S1	S2	S1	S2	S1	S2
1	0.92	0.89	0.7	0.9	0.1	0.2	0.70	0.68	1.3	1.3	−0.7	−0.8
2	0.89	0.87	1.3	1.0	−1.0	−0.5	0.68	0.64	2.2	2.0	−1.9	−1.7
3	0.87	0.82	1.0	1.2	−0.4	−0.2	0.71	0.70	1.9	1.9	−1.5	−1.5
4	0.87	0.87	0.9	0.8	0.1	0.0	0.74	0.73	1.3	1.4	−0.8	−0.9
5	0.90	0.88	1.2	1.0	−0.7	−0.3	0.63	0.69	2.1	1.8	−1.6	−1.4
6	0.87	0.85	1.0	1.0	0.3	0.1	0.60	0.60	1.5	1.6	−0.8	−0.9
7	0.83	0.81	0.9	0.9	0.5	0.2	0.69	0.69	1.0	1.1	−0.1	−0.3
8	0.76	0.71	1.0	1.1	0.2	0.1	0.70	0.71	1.1	1.2	−0.4	−0.5
9	0.84	0.84	1.0	1.0	−0.3	0.0	0.68	0.68	1.7	1.7	−1.2	−1.1
10	0.87	0.82	0.7	1.0	0.1	0.1	0.70	0.72	1.4	1.4	−0.7	−0.7

In the daytime under cloudy skies, S1 outperformed S2 for most stations. At stations 2, 3, 4, 8, and 10, S1 exhibited a significant increase in R^2 compared to S2. High temporal correlations of S1 imply the effective simulation of temporal variations of LSTs. Unlike the clear sky conditions, the RMSE of S1 was much smaller compared to S2 for most stations under cloudy skies. One possible reason for the high RMSE in S2 is that low LST values under cloudy areas could be overestimated based on the high bias of the S2 RF model under cloudy sky conditions (see also Figure 9). The RMSEs of S1 (from 2.1 to 3.9 °C in summer) for daytime under cloudy skies are comparable to or lower than those from the literature (RMSE of 4.3–8.3 °C for four stations in China [65], RMSE of 5.1–5.6 °C for two sites in Africa [17], RMSE of 1.8–2.7 °C for three stations in North China [30], and RMSE of 3.5–4.4 °C for four stations in China [12]), although those studies focused on all seasons, not only summer. It should be noted that Zhou et al. reported that the accuracy of LST estimation under the cloudy sky conditions in summer was lower than the other seasons [26].

Figure 10 represents the temporal distribution of the S1, S2, and in situ LSTs at station 9 for July–August 2017, when the LSTs dynamically changed. In the daytime, extremely high LSTs were well predicted by S2, such as for 13-July and 7-August, as opposed to S1; however, relatively low LSTs were better predicted by S1. It should be noted that there were days when the very high LSTs sharply dropped, such as between 6 July and 7 July, as well as 13 July and 15 July. S1 simulated these rapid temperature changes better than S2. Furthermore, there were many days with large amounts of precipitation, which resulted in low LSTs (i.e., 10 August and 14 August) in the humid summer. LSTs for such days were also better simulated by S1 with an improved temporal correlation.

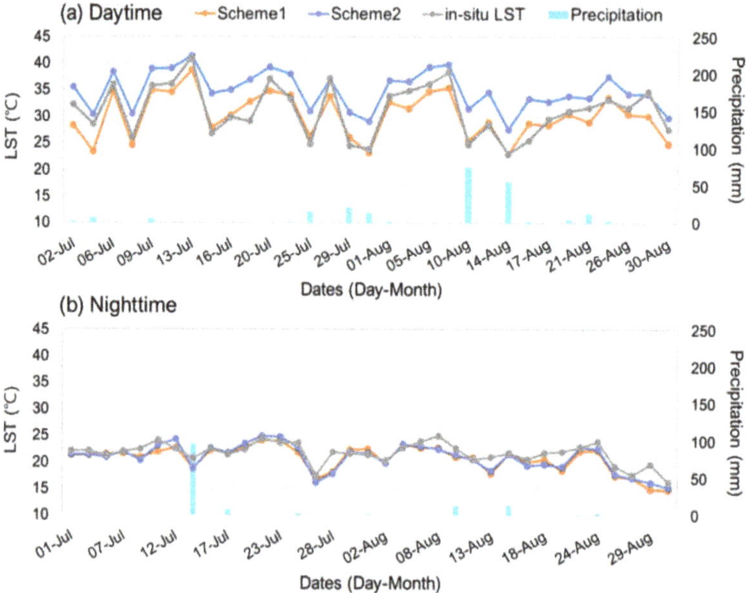

Figure 10. Time-series of the estimated LSTs for schemes 1(S1) and 2(S2), and 1 km bias-corrected in situ LSTs with daily precipitation at station 9 for (**a**) daytime and (**b**) nighttime during July and August in 2017 (except the missing days due to advanced microwave scanning radiometer (AMSR2) gaps between paths). Daily precipitation data were obtained from automated surface observing systems (ASOSs) operated by the Korea Meteorological Administration.

At nighttime, R^2 was significantly high in both S1 and S2, the RMSE was less than 1 °C, and the bias converged to zero for most stations under the clear sky conditions (Table 4). Under cloudy skies, S2 yielded higher R^2 at some stations compared to S1. Nevertheless, both S1 and S2 produced RMSE < 2 °C under the cloudy sky conditions, which suggests that the nighttime LSTs are relatively less affected by atmospheric phenomena, such as precipitation and clouds (Figure 10b).

4.5. Two Scheme Combinations

We further examined the combination of the two schemes (S1 and S2) to improve estimation performance, based on the difference of the LST distribution between clear and cloudy sky conditions, as analyzed in Section 4.1. The LSTs developed from S2 were used for days with relatively high LSTs, whereas S1 was used for days with lower temperatures. Appendix A describes the detailed combination methods of the daytime. For nighttime, the average of S1 and S2 LSTs was used since both schemes resulted in high accuracy [66].

Table 5 shows the results of the LOOCV of all-weather LSTs, as well as of S1, S2, and Scheme combined (SC) models, using bias-corrected in situ LSTs for daytime and nighttime. In the daytime SC, R^2 was relatively high and RMSE was distinctly lower at many stations when compared to S1 and S2. For nighttime, the SC exhibited significantly higher R^2 than did S1 and S2 for several stations, and the RMSE of SC was also generally close to S1 or S2, whichever was lower. The superiority of SC to S1 and S2 is consistent with previous studies' findings that the combination of different models improves performance by overcoming the limitations of each individual model [67]. Therefore, we propose all-weather LSTs by combining two different schemes with an average R^2 of 0.82 and 0.74 and with RMSE of 2.5 °C and 1.4 °C for daytime and nighttime, respectively, over the 10 in situ stations.

Table 5. The leave one-station out cross-validation (LOOCV) results of the estimated 1 km all-weather LSTs from scheme 1(S1), scheme 2(S2) and scheme combined (SC) using 1 km bias-corrected in situ LSTs at 10 stations during July and August from 2013 to 2018 for daytime. Refer to Figure 1 for station numbers.

| Station | Daytime (All-weather) | | | | | | Nighttime (All-weather) | | | | | |
| | R^2 | | | RMSE (°C) | | | R^2 | | | RMSE (°C) | | |
	S1	S2	SC	S1	S2	SC	S1	S2	SC	S1	S2	SC
1	0.78	0.77	0.82	2.2	2.5	1.9	0.76	0.73	0.75	1.1	1.2	1.2
2	0.75	0.72	0.79	2.5	2.3	2.3	0.73	0.68	0.72	2.0	1.8	1.9
3	0.80	0.73	0.82	3.2	5.0	2.6	0.73	0.71	0.73	1.7	1.7	1.7
4	0.78	0.74	0.79	2.0	3.0	2.1	0.76	0.76	0.77	1.2	1.3	1.2
5	0.83	0.79	0.81	4.1	3.0	3.3	0.70	0.73	0.73	1.9	1.7	1.7
6	0.83	0.80	0.85	2.9	5.5	3.0	0.66	0.67	0.68	1.4	1.5	1.4
7	0.83	0.81	0.82	2.2	1.7	2.1	0.74	0.74	0.75	1.0	1.0	1.0
8	0.82	0.81	0.83	2.1	3.2	2.2	0.71	0.71	0.73	1.1	1.2	1.1
9	0.84	0.83	0.82	2.8	4.5	2.5	0.73	0.73	0.74	1.5	1.5	1.5
10	0.82	0.76	0.80	2.3	4.3	2.6	0.74	0.76	0.76	1.3	1.3	1.2

5. Conclusions

In this study, we estimated all-weather 1 km MODIS LSTs for daytime and nighttime in South Korea for the humid summer days. We used eight AMSR2 BTs, three ACPs (i.e., MAST, Yast, and theta), and six auxiliary variables for the LST estimations based on RF machine learning. Both clear sky MODIS LSTs and the bias-corrected in situ LSTs under cloud sky conditions were used as a dependent variable to provide the models with the LST characteristics for clear and cloudy skies. We have proposed two schemes: A two- step approach (S1) first estimates 10 km LSTs and then involves the spatial downscaling of LSTs from 10 km to 1 km. S2 is a one-step algorithm that directly estimates the 1 km all-weather LSTs, which we have evaluated using a series of validations. In clear sky daytime, S2 slightly outperformed S1, but in cloudy sky daytime, S1 had an average R^2 of 0.78 and RMSE of 2.6°C, an improvement when compared to S2 (R^2 of 0.74 and RMSE of 3.8 °C) for the bias-corrected 10 in situ stations. At nighttime, S1 and S2 showed no significant difference in performance regardless of cloud conditions. We further examined the combination of the two schemes (S1 and S2) in order to improve estimation performance, producing promising results, with R^2 of 0.82 and 0.74 and with RMSE of 2.5 °C and 1.4 °C for daytime and nighttime, respectively, over the 10 in situ stations. This study has revealed that the two-step-based S1 was better able to simulate low LSTs in cloudy sky, humid summer daytime conditions (i.e., rainy days) than S2. Moreover, we found that ACPs appear relatively important for the estimation of LSTs in light of spatial variability. To our knowledge, this is the first study to predict all-weather 1 km MODIS LSTs that focuses on humid summer days in great detail. Nevertheless, there is still room for further validation of the constructed LSTs over built-up areas since an insufficient number of in situ stations in urban land cover were used in this study.

Although we have focused this study on South Korea, we believe that the suggested schemes could be successfully used over other regions frequently covered with clouds in humid summer seasons. Recently, new MODIS Aqua LST datasets (MYD21A1D for daytime and MYD21A1N for nighttime) were produced utilizing the ASTER temperature/emissivity separation (TES) technique suitable for hot and humid conditions, although the approach has, thus far, only been tested using a small number of measurements [68]. It is expected that the proposed technique may be used to predict MYD21 LSTs under cloudy sky conditions.

Author Contributions: Conceptualization, C.Y.; formal analysis, C.Y. and J.I.; investigation, C.Y. and D.C.; methodology, C.Y., J.I., D.C., N.Y., J.X., and B.B.; supervision, J.I.; validation, C.Y.; writing—original draft, C.Y.;

writing—review and editing, J.I., N.Y., J.X., and B.B. All authors have read and agreed to the published version of the manuscript.

Funding: This research was supported by the Korea Meteorological Administration Research and Development Program under Grant KMIPA 2017–7010, by the National Research Foundation of Korea (NRF) (Grants: NRF-2017M1A3A3A02015981; NRF-2016M3C4A7952600; NRF-2018K2A9A2A06023758), by a grant (no. 20009742) of Disaster-Safety Industry Promotion Program funded by Ministry of Interior and Safety (NOIS), Korea, and by the Ministry of Science and ICT (MSIT), Korea (IITP-2020-2018-0-01424). CY was also supported by Global PhD Fellowship Program through the National Research Foundation of Korea (NRF), funded by the Ministry of Education (NRF-2018H1A2A1062207).

Conflicts of Interest: The authors declare no conflict of interest.

Appendix A

A.1. Before-and after 1 km Bias-corrected in situ LSTs

Tables A1 and A2 show the relationship between good-quality MODIS LSTs and in situ LSTs before and after bias correction under clear skies, in daytime and nighttime, respectively. In the daytime, before bias correction, the R^2 of the time series data showed a relatively significant range of 0.46–0.76, but the RMSE and bias were quite high. The significant spatial thermal difference within a 1 km grid in the daytime, especially for the summer season, could result in the underestimation of MODIS LSTs when compared to the in situ data [69]. After the bias correction, the RMSE decreased towards 2 °C at the 10 stations in which the range of errors were similar to that of the typical MODIS LST validation results (i.e., ~2 K; [59]). The bias converged close to 0, indicating a very slightly negative signal at some stations after correction. For the nighttime, before bias correction, the temporal R^2 was relatively higher than in the daytime for most stations, and the RMSE and bias were significantly lower than the daytime. These results are consistent with previous studies' findings that the MODIS LST validation error is much lower at nighttime than in the daytime [70,71]. After the bias correction, the RMSEs were under 1.5 °C and the bias was close to zero for all stations. The reason that the nighttime has a lower bias correction error than the daytime is possibly due to the higher thermal homogeneity in one MODIS grid (i.e., 1 km resolution) at the nighttime without solar shortwave radiation.

Table A1. The relationship between clear sky MODIS LSTs and the before-and-after 1 km bias-corrected in situ LSTs during July and August from 2013 to 2018 for the 10 stations in daytime. Refer to Figure 1 for station numbers.

Station Number	Before Bias-correction			After Bias-correction			Sample Size
	R^2	RMSE (°C)	Bias (°C)	R^2	RMSE (°C)	Bias (°C)	
1	0.62	10.50	8.89	0.64	1.88	−0.06	15
2	0.75	9.01	7.29	0.74	2.00	−1.05	16
3	0.64	18.49	17.71	0.64	2.20	−0.12	26
4	0.65	14.23	13.43	0.66	1.97	−0.20	30
5	0.56	11.41	10.31	0.67	2.38	−0.09	15
6	0.61	17.45	16.92	0.60	2.12	−0.51	16
7	0.57	9.55	8.15	0.59	2.06	0.15	21
8	0.46	11.37	10.30	0.46	2.39	−1.19	27
9	0.76	11.22	10.38	0.77	1.99	−0.74	15
10	0.54	16.43	15.20	0.53	2.15	−0.43	23

Table A2. The relationship between clear sky MODIS LSTs and the before-and-after 1 km bias-corrected in situ LSTs during July and August from 2013 to 2018 for the 10 stations at nighttime. Refer to Figure 1 for station numbers.

Station Number	Before Bias-correction			After Bias-correction			Sample Size
	R^2	RMSE (°C)	Bias (°C)	R^2	RMSE (°C)	Bias (°C)	
1	0.77	3.34	3.02	0.78	1.38	0.01	58
2	0.72	2.60	2.27	0.72	1.28	0.02	34
3	0.78	2.99	2.64	0.79	1.14	−0.05	55
4	0.85	2.37	2.10	0.86	1.01	0.01	48
5	0.84	2.54	2.24	0.84	1.11	0.02	46
6	0.83	2.30	1.88	0.83	1.21	0.03	67
7	0.66	3.96	3.74	0.66	1.30	0.06	59
8	0.52	3.37	3.00	0.52	1.46	0.10	62
9	0.69	2.42	1.91	0.70	1.47	0.10	49
10	0.70	2.96	2.55	0.70	1.29	0.00	71

A.2. Variable Importance Calculated from the RF Models

Figure A1 shows the normalized relative variable importance calculated from the RF models for the two schemes. The normalization was done by sum to 100%. Daily variable importance of step 2 in S1 were averaged before normalization.

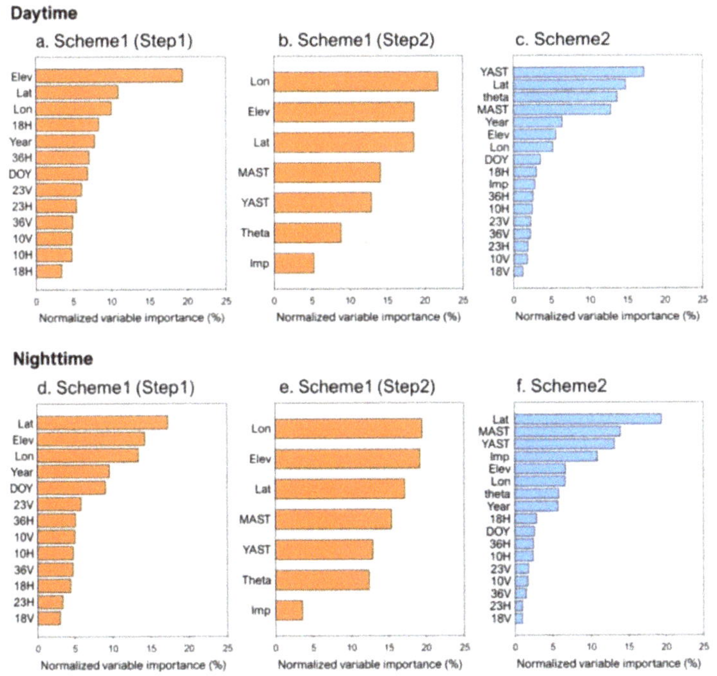

Figure A1. Normalized relative variable importance calculated from the RF models of two steps (step 1 and step 2) for scheme 1 (S1), the orange bar; and scheme 2(S2), the blue bar, for daytime (**a–c**) and nighttime (**d–f**).

A.3. Two Scheme Combination (SC) Methods of Daytime

In Section 4.1, the upper quartile of cloudy sky LST was close to the lower quartile of clear sky LST for most in situ stations in the daytime. We used the AMSR2 BT to select the dates of daytime high and low LSTs. First, all of the eight daytime AMSR2 BTs listed in Table 1 were aggregated to 1 km as a MODIS grid using bilinear resampling. Temporal correlations of AMSR2 BTs and in situ data at the 10 stations were performed over all-weather conditions in summer; we selected 10V, which showed the highest correlation, as the reference variable (Table A3). The 75th percentile (i.e., upper quartile) value of 1 km AMSR2 BT 10V was then calculated over the summer study periods (July–August, 2013–2018) for each pixel. Finally, the LSTs developed from S2 were used for days when the BT of AMSR2 10V was over the 75th percentile, and S1 was used for days when the BT of AMSR2 10V was below the 75th percentile, based on the results in Section 4.4.

Table A3. The temporal correlation (Pearson correlation coefficient; R) results between eight AMSR2 BTs and the 1 km bias-corrected in situ LSTs during July and August from 2013 to 2018 for the 10 stations in daytime. Refer to Figure 1 for station numbers.

Station	AMSR2 BTs							
	10H	**10V**	**18H**	**18V**	**23H**	**23V**	**36H**	**36V**
1	0.86	0.90	0.84	0.88	0.81	0.86	0.73	0.77
2	0.85	0.89	0.79	0.87	0.74	0.82	0.61	0.73
3	0.88	0.90	0.82	0.83	0.73	0.74	0.55	0.57
4	0.89	0.90	0.86	0.86	0.80	0.81	0.70	0.72
5	0.89	0.91	0.89	0.90	0.87	0.88	0.82	0.83
6	0.85	0.90	0.87	0.89	0.86	0.87	0.74	0.76
7	0.80	0.89	0.86	0.87	0.85	0.86	0.75	0.77
8	0.89	0.90	0.88	0.89	0.86	0.87	0.81	0.82
9	0.91	0.93	0.88	0.91	0.88	0.90	0.77	0.79
10	0.71	0.84	0.85	0.89	0.84	0.87	0.79	0.82
Average	0.85	0.90	0.85	0.88	0.82	0.85	0.73	0.76

References

1. Bayissa, Y.A.; Tadesse, T.; Svoboda, M.; Wardlow, B.; Poulsen, C.; Swigart, J.; Van Andel, S.J. Developing a satellite-based combined drought indicator to monitor agricultural drought: A case study for Ethiopia. *Gisci. Remote Sens.* **2019**, *56*, 718–748. [CrossRef]
2. Bechtel, B.; Demuzere, M.; Mills, G.; Zhan, W.; Sismanidis, P.; Small, C.; Voogt, J. SUHI analysis using Local Climate Zones—A comparison of 50 cities. *Urban Clim.* **2019**, *28*, 100451. [CrossRef]
3. Shafizadeh-Moghadam, H.; Weng, Q.; Liu, H.; Valavi, R. Modeling the spatial variation of urban land surface temperature in relation to environmental and anthropogenic factors: A case study of Tehran, Iran. *Gisci. Remote Sens.* **2020**. [CrossRef]
4. Song, Y.; Wu, C. Examining human heat stress with remote sensing technology. *Gisci. Remote Sens.* **2018**, *55*, 19–37. [CrossRef]
5. Weng, Q.; Firozjaei, M.K.; Sedighi, A.; Kiavarz, M.; Alavipanah, S.K. Statistical analysis of surface urban heat island intensity variations: A case study of Babol city, Iran. *Gisci. Remote Sens.* **2019**, *56*, 576–604. [CrossRef]
6. Ziaul, S.; Pal, S. Analyzing control of respiratory particulate matter on Land Surface Temperature in local climatic zones of English Bazar Municipality and Surroundings. *Urban Clim.* **2018**, *24*, 34–50. [CrossRef]
7. Park, S.; Kang, D.; Yoo, C.; Im, J.; Lee, M.-I. Recent ENSO influence on East African drought during rainy seasons through the synergistic use of satellite and reanalysis data. *ISPRS J. Photogramm. Remote Sens.* **2020**, *162*, 17–26. [CrossRef]
8. Neinavaz, E.; Skidmore, A.K.; Darvishzadeh, R. Effects of prediction accuracy of the proportion of vegetation cover on land surface emissivity and temperature using the NDVI threshold method. *Int. J. Appl. Earth Obs. Geoinf.* **2020**, *85*, 101984. [CrossRef]
9. Srivastava, P.K.; Han, D.; Ramirez, M.R.; Islam, T. Machine learning techniques for downscaling SMOS satellite soil moisture using MODIS land surface temperature for hydrological application. *Water Resour. Manag.* **2013**, *27*, 3127–3144. [CrossRef]

10. Wang, L.; Koike, T.; Yang, K.; Yeh, P.J.-F. Assessment of a distributed biosphere hydrological model against streamflow and MODIS land surface temperature in the upper Tone River Basin. *J. Hydrol.* **2009**, *377*, 21–34. [CrossRef]

11. Li, Z.-L.; Tang, B.-H.; Wu, H.; Ren, H.; Yan, G.; Wan, Z.; Trigo, I.F.; Sobrino, J.A. Satellite-derived land surface temperature: Current status and perspectives. *Remote Sens. Environ.* **2013**, *131*, 14–37. [CrossRef]

12. Duan, S.-B.; Li, Z.-L.; Leng, P. A framework for the retrieval of all-weather land surface temperature at a high spatial resolution from polar-orbiting thermal infrared and passive microwave data. *Remote Sens. Environ.* **2017**, *195*, 107–117. [CrossRef]

13. Fu, P.; Weng, Q. Consistent land surface temperature data generation from irregularly spaced Landsat imagery. *Remote Sens. Environ.* **2016**, *184*, 175–187. [CrossRef]

14. Huang, C.; Duan, S.-B.; Jiang, X.-G.; Han, X.-J.; Leng, P.; Gao, M.-F.; Li, Z.-L. A physically based algorithm for retrieving land surface temperature under cloudy conditions from AMSR2 passive microwave measurements. *Int. J. Remote Sens.* **2019**, *40*, 1828–1843. [CrossRef]

15. Jin, M. Interpolation of surface radiative temperature measured from polar orbiting satellites to a diurnal cycle: 2. Cloudy-pixel treatment. *J. Geophys. Res. Atmos.* **2000**, *105*, 4061–4076. [CrossRef]

16. Li, X.; Zhou, Y.; Asrar, G.R.; Zhu, Z. Creating a seamless 1 km resolution daily land surface temperature dataset for urban and surrounding areas in the conterminous United States. *Remote Sens. Environ.* **2018**, *206*, 84–97. [CrossRef]

17. Lu, L.; Venus, V.; Skidmore, A.; Wang, T.; Luo, G. Estimating land-surface temperature under clouds using MSG/SEVIRI observations. *Int. J. Appl. Earth Obs. Geoinf.* **2011**, *13*, 265–276. [CrossRef]

18. Shuai, T.; Zhang, X.; Wang, S.; Zhang, L.; Shang, K.; Chen, X.; Wang, J. A spectral angle distance-weighting reconstruction method for filled pixels of the MODIS land surface temperature product. *IEEE Geosci. Remote Sens. Lett.* **2014**, *11*, 1514–1518. [CrossRef]

19. Shwetha, H.; Kumar, D.N. Prediction of high spatio-temporal resolution land surface temperature under cloudy conditions using microwave vegetation index and ANN. *ISPRS J. Photogramm. Remote Sens.* **2016**, *117*, 40–55. [CrossRef]

20. Sun, D.; Li, Y.; Zhan, X.; Houser, P.; Yang, C.; Chiu, L.; Yang, R. Land Surface Temperature Derivation under All Sky Conditions through Integrating AMSR-E/AMSR-2 and MODIS/GOES Observations. *Remote Sens.* **2019**, *11*, 1704. [CrossRef]

21. Sun, L.; Chen, Z.; Gao, F.; Anderson, M.; Song, L.; Wang, L.; Hu, B.; Yang, Y. Reconstructing daily clear-sky land surface temperature for cloudy regions from MODIS data. *Comput. Geosci.* **2017**, *105*, 10–20. [CrossRef]

22. Xu, Y.; Shen, Y.; Wu, Z. Spatial and temporal variations of land surface temperature over the Tibetan Plateau based on harmonic analysis. *Mt. Res. Dev.* **2013**, *33*, 85–94. [CrossRef]

23. Yu, W.; Ma, M.; Wang, X.; Tan, J. Estimating the land-surface temperature of pixels covered by clouds in MODIS products. *J. Appl. Remote Sens.* **2014**, *8*, 083525. [CrossRef]

24. Zeng, C.; Long, D.; Shen, H.; Wu, P.; Cui, Y.; Hong, Y. A two-step framework for reconstructing remotely sensed land surface temperatures contaminated by cloud. *ISPRS J. Photogramm. Remote Sens.* **2018**, *141*, 30–45. [CrossRef]

25. Zeng, C.; Shen, H.; Zhong, M.; Zhang, L.; Wu, P. Reconstructing MODIS LST based on multitemporal classification and robust regression. *IEEE Geosci. Remote Sens. Lett.* **2014**, *12*, 512–516. [CrossRef]

26. Zhou, W.; Peng, B.; Shi, J. Reconstructing spatial–temporal continuous MODIS land surface temperature using the DINEOF method. *J. Appl. Remote Sens.* **2017**, *11*, 046016. [CrossRef]

27. Dai, A.; Trenberth, K.E.; Karl, T.R. Effects of clouds, soil moisture, precipitation, and water vapor on diurnal temperature range. *J. Clim.* **1999**, *12*, 2451–2473. [CrossRef]

28. Lim, Y.-K.; Kim, K.-Y.; Lee, H.-S. Temporal and spatial evolution of the Asian summer monsoon in the seasonal cycle of synoptic fields. *J. Clim.* **2002**, *15*, 3630–3644. [CrossRef]

29. Fu, P.; Xie, Y.; Weng, Q.; Myint, S.; Meacham-Hensold, K.; Bernacchi, C. A physical model-based method for retrieving urban land surface temperatures under cloudy conditions. *Remote Sens. Environ.* **2019**, *230*, 111191. [CrossRef]

30. Zhang, X.; Zhou, J.; Göttsche, F.-M.; Zhan, W.; Liu, S.; Cao, R. A method based on temporal component decomposition for estimating 1-km all-weather land surface temperature by merging satellite thermal infrared and passive microwave observations. *IEEE Trans. Geosci. Remote Sens.* **2019**, *57*, 4670–4691. [CrossRef]

31. Rhee, J.; Im, J. Estimating high spatial resolution air temperature for regions with limited in situ data using MODIS products. *Remote Sens.* **2014**, *6*, 7360–7378. [CrossRef]

32. Yoo, C.; Im, J.; Park, S.; Quackenbush, L.J. Estimation of daily maximum and minimum air temperatures in urban landscapes using MODIS time series satellite data. *ISPRS J. Photogramm. Remote Sens.* **2018**, *137*, 149–162. [CrossRef]

33. Wan, Z.; Dozier, J. A generalized split-window algorithm for retrieving land-surface temperature from space. *IEEE Trans. Geosci. Remote Sens.* **1996**, *34*, 892–905.

34. Brown de Colstoun, E.C.; Huang, C.; Wang, P.; Tilton, J.C.; Tan, B.; Phillips, J.; Niemczura, S.; Ling, P.-Y.; Wolfe, R.E. *Global Man-made Impervious Surface (GMIS) Dataset From Landsat*; NASA Socioeconomic Data and Applications Center (SEDAC): Palisades, NY, USA, 2017.

35. Christensen, J.H.; Boberg, F.; Christensen, O.B.; Lucas-Picher, P. On the need for bias correction of regional climate change projections of temperature and precipitation. *Geophys. Res. Lett.* **2008**, 35. [CrossRef]

36. Bechtel, B. Robustness of annual cycle parameters to characterize the urban thermal landscapes. *IEEE Geosci. Remote Sens. Lett.* **2012**, *9*, 876–880. [CrossRef]

37. Stolwijk, A.; Straatman, H.; Zielhuis, G. Studying seasonality by using sine and cosine functions in regression analysis. *J. Epidemiol. Community Health* **1999**, *53*, 235–238. [CrossRef] [PubMed]

38. Park, S.; Im, J.; Park, S.; Yoo, C.; Han, H.; Rhee, J. Classification and mapping of paddy rice by combining Landsat and SAR time series data. *Remote Sens.* **2018**, *10*, 447. [CrossRef]

39. Park, S.; Park, H.; Im, J.; Yoo, C.; Rhee, J.; Lee, B.; Kwon, C. Delineation of high resolution climate regions over the Korean Peninsula using machine learning approaches. *PLoS ONE* **2019**, *14*, e0223362. [CrossRef]

40. Yoo, C.; Han, D.; Im, J.; Bechtel, B. Comparison between convolutional neural networks and random forest for local climate zone classification in mega urban areas using Landsat images. *ISPRS J. Photogramm. Remote Sens.* **2019**, *157*, 155–170. [CrossRef]

41. Cho, D.; Yoo, C.; Im, J.; Cha, D.H. Comparative assessment of various machine learning-based bias correction methods for numerical weather prediction model forecasts of extreme air temperatures in urban areas. *Earth Space Sci.* **2020**, *7*, e2019EA000740. [CrossRef]

42. McLaren, K.; McIntyre, K.; Prospere, K. Using the random forest algorithm to integrate hydroacoustic data with satellite images to improve the mapping of shallow nearshore benthic features in a marine protected area in Jamaica. *Gisci. Remote Sens.* **2019**, *56*, 1065–1092. [CrossRef]

43. Mutowo, G.; Mutanga, O.; Masocha, M. Including shaded leaves in a sample affects the accuracy of remotely estimating foliar nitrogen. *Gisci. Remote Sens.* **2019**, *56*, 1114–1127. [CrossRef]

44. Breiman, L. Random forests. *Mach. Learn.* **2001**, *45*, 5–32. [CrossRef]

45. Liaw, A.; Wiener, M. Classification and regression by randomForest. *R News* **2002**, *2*, 18–22.

46. Strobl, C.; Boulesteix, A.-L.; Zeileis, A.; Hothorn, T. Bias in random forest variable importance measures: Illustrations, sources and a solution. *BMC Bioinform.* **2007**, *8*, 25. [CrossRef]

47. Wei, P.; Lu, Z.; Song, J. Variable importance analysis: A comprehensive review. *Reliab. Eng. Syst. Saf.* **2015**, *142*, 399–432. [CrossRef]

48. Yoo, S.; Im, J.; Wagner, J.E. Variable selection for hedonic model using machine learning approaches: A case study in Onondaga County, NY. *Landsc. Urban Plan.* **2012**, *107*, 293–306. [CrossRef]

49. Wright, M.N.; Ziegler, A. ranger: A fast implementation of random forests for high dimensional data in C++ and R. *arXiv* **2015**, arXiv:1508.04409. [CrossRef]

50. Hutengs, C.; Vohland, M. Downscaling land surface temperatures at regional scales with random forest regression. *Remote Sens. Environ.* **2016**, *178*, 127–141. [CrossRef]

51. Minder, J.R.; Mote, P.W.; Lundquist, J.D. Surface temperature lapse rates over complex terrain: Lessons from the Cascade Mountains. *J. Geophys. Res. Atmos.* **2010**, 115. [CrossRef]

52. Lo, M.H.; Famiglietti, J.S. Precipitation response to land subsurface hydrologic processes in atmospheric general circulation model simulations. *J. Geophys. Res. Atmos.* **2011**, 116. [CrossRef]

53. Wan, Z.; Li, Z.-L. MODIS land surface temperature and emissivity. In *Land Remote Sensing and Global Environmental Change*; Springer: New York, NY, USA, 2010; pp. 563–577.

54. Wang, K.; Liang, S. Evaluation of ASTER and MODIS land surface temperature and emissivity products using long-term surface longwave radiation observations at SURFRAD sites. *Remote Sens. Environ.* **2009**, *113*, 1556–1565. [CrossRef]

55. Wang, K.; Wan, Z.; Wang, P.; Sparrow, M.; Liu, J.; Haginoya, S. Evaluation and improvement of the MODIS land surface temperature/emissivity products using ground-based measurements at a semi-desert site on the western Tibetan Plateau. *Int. J. Remote Sens.* **2007**, *28*, 2549–2565. [CrossRef]

56. Fan, X.; Tang, B.-H.; Wu, H.; Yan, G.; Li, Z.-L. Daytime land surface temperature extraction from MODIS thermal infrared data under cirrus clouds. *Sensors* **2015**, *15*, 9942–9961. [CrossRef]

57. Guillevic, P.; Göttsche, F.; Nickeson, J.; Hulley, G.; Ghent, D.; Yu, Y.; Trigo, I.; Hook, S.; Sobrino, J.; Remedios, J. Land surface temperature product validation best practice protocol. Version 1.0. In *Best Practice for Satellite-Derived Land Product Validation*; Land Product Validation Subgroup (WGCV/CEOS): Washington, DC, USA, 2017; p. 31.

58. Zheng, Y.; Ren, H.; Guo, J.; Ghent, D.; Tansey, K.; Hu, X.; Nie, J.; Chen, S. Land Surface Temperature Retrieval from Sentinel-3A Sea and Land Surface Temperature Radiometer, Using a Split-Window Algorithm. *Remote Sens.* **2019**, *11*, 650. [CrossRef]

59. Stroppiana, D.; Antoninetti, M.; Brivio, P.A. Seasonality of MODIS LST over Southern Italy and correlation with land cover, topography and solar radiation. *Eur. J. Remote Sens.* **2014**, *47*, 133–152. [CrossRef]

60. Bertoldi, G.; Notarnicola, C.; Leitinger, G.; Endrizzi, S.; Zebisch, M.; Della Chiesa, S.; Tappeiner, U. Topographical and ecohydrological controls on land surface temperature in an alpine catchment. *Ecohydrology* **2010**, *3*, 189–204. [CrossRef]

61. Lee, X.; Goulden, M.L.; Hollinger, D.Y.; Barr, A.; Black, T.A.; Bohrer, G.; Bracho, R.; Drake, B.; Goldstein, A.; Gu, L. Observed increase in local cooling effect of deforestation at higher latitudes. *Nature* **2011**, *479*, 384–387. [CrossRef]

62. Meng, Q.; Zhang, L.; Sun, Z.; Meng, F.; Wang, L.; Sun, Y. Characterizing spatial and temporal trends of surface urban heat island effect in an urban main built-up area: A 12-year case study in Beijing, China. *Remote Sens. Environ.* **2018**, *204*, 826–837. [CrossRef]

63. Im, E.-S.; Thanh, N.-X.; Kim, Y.-H.; Ahn, J.-B. 2018 summer extreme temperatures in South Korea and their intensification under 3° C global warming. *Environ. Res. Lett.* **2019**, *14*, 094020. [CrossRef]

64. Choi, N.; Lee, M.-I. Spatial variability and long-term trend in the occurrence frequency of heatwave and tropical night in Korea. *Asia Pac. J. Atmos. Sci.* **2019**, *55*, 101–114. [CrossRef]

65. Xu, S.; Cheng, J.; Zhang, Q. Reconstructing All-Weather Land Surface Temperature Using the Bayesian Maximum Entropy Method Over the Tibetan Plateau and Heihe River Basin. *IEEE J. Sel. Top. Appl. Earth Obs. Remote Sens.* **2019**, *12*, 3307–3316. [CrossRef]

66. Delle Monache, L.; Stull, R. An ensemble air-quality forecast over western Europe during an ozone episode. *Atmos. Environ.* **2003**, *37*, 3469–3474. [CrossRef]

67. Healey, S.P.; Cohen, W.B.; Yang, Z.; Brewer, C.K.; Brooks, E.B.; Gorelick, N.; Hernandez, A.J.; Huang, C.; Hughes, M.J.; Kennedy, R.E. Mapping forest change using stacked generalization: An ensemble approach. *Remote Sens. Environ.* **2018**, *204*, 717–728. [CrossRef]

68. Hulley, G.; Malakar, N.; Freepartner, R. *Moderate Resolution Imaging Spectroradiometer (MODIS) Land Surface Temperature and Emissivity Product (MxD21) Algorithm Theoretical Basis Document Collection-6*; JPL Publication: Pasadena, CA, USA, 2016; pp. 12–17.

69. Simó, G.; García-Santos, V.; Jiménez, M.A.; Martínez-Villagrasa, D.; Picos, R.; Caselles, V.; Cuxart, J. Landsat and local land surface temperatures in a heterogeneous terrain compared to modis values. *Remote Sens.* **2016**, *8*, 849. [CrossRef]

70. Wan, Z.; Zhang, Y.; Zhang, Q.; Li, Z.-l. Validation of the land-surface temperature products retrieved from Terra Moderate Resolution Imaging Spectroradiometer data. *Remote Sens. Environ.* **2002**, *83*, 163–180. [CrossRef]

71. Wang, W.; Liang, S.; Meyers, T. Validating MODIS land surface temperature products using long-term nighttime ground measurements. *Remote Sens. Environ.* **2008**, *112*, 623–635. [CrossRef]

 *remote sensing*

Article

Monitoring 10-m LST from the Combination MODIS/Sentinel-2, Validation in a High Contrast Semi-Arid Agroecosystem

Juan M. Sánchez [1,*], Joan M. Galve [1], José González-Piqueras [1], Ramón López-Urrea [2], Raquel Niclòs [3] and Alfonso Calera [1]

[1] Regional Development Institute, University of Castilla-La Mancha, Campus Universitario s/n, 02071 Albacete, Spain; joanmiquel.galve@uclm.es (J.M.G.); jose.gonzalez@uclm.es (J.G.-P.); alfonso.calera@uclm.es (A.C.)

[2] Instituto Técnico Agronómico Provincial de Albacete and FUNDESCAM, Parque Empresarial Campollano, 2ª Avda. N°61, 02007 Albacete, Spain; rlu.itap@dipualba.es

[3] Earth Physics and Thermodynamics Department, University of Valencia, C/ Dr. Moliner 50, 46100 Burjassot, Spain; raquel.niclos@uv.es

* Correspondence: juanmanuel.sanchez@uclm.es

Received: 25 March 2020; Accepted: 1 May 2020; Published: 4 May 2020

Abstract: Downscaling techniques offer a solution to the lack of high-resolution satellite Thermal InfraRed (TIR) data and can bridge the gap until operational TIR missions accomplishing spatio-temporal requirements are available. These techniques are generally based on the Visible Near InfraRed (VNIR)-TIR variable relations at a coarse spatial resolution, and the assumption that the relationship between spectral bands is independent of the spatial resolution. In this work, we adopted a previous downscaling method and introduced some adjustments to the original formulation to improve the model performance. Maps of Land Surface Temperature (LST) with 10-m spatial resolution were obtained as output from the combination of MODIS/Sentinel-2 images. An experiment was conducted in an agricultural area located in the Barrax test site, Spain (39°03′35″ N, 2°06′ W), for the summer of 2018. Ground measurements of LST transects collocated with the MODIS overpasses were used for a robust local validation of the downscaling approach. Data from 6 different dates were available, covering a variety of croplands and surface conditions, with LST values ranging 300–325 K. Differences within ±4.0 K were observed between measured and modeled temperatures, with an average estimation error of ±2.2 K and a systematic deviation of 0.2 K for the full ground dataset. A further cross-validation of the disaggregated 10-m LST products was conducted using an additional set of Landsat-7/ETM+ images. A similar uncertainty of ±2.0 K was obtained as an average. These results are encouraging for the adaptation of this methodology to the tandem Sentinel-3/Sentinel-2, and are promising since the 10-m pixel size, together with the 3–5 days revisit frequency of Sentinel-2 satellites can fulfill the LST input requirements of the surface energy balance methods for a variety of hydrological, climatological or agricultural applications. However, certain limitations to capture the variability of extreme LST, or in recently sprinkler irrigated fields, claim the necessity to explore the implementation of soil moisture or vegetation indices sensitive to soil water content as inputs in the downscaling approach. The ground LST dataset introduced in this paper will be of great value for further refinements and assessments.

Keywords: Downscaling; thermal infrared; land surface temperature; disaggregation; Copernicus

1. Introduction

Time series of fine spatial and temporal resolution Thermal Infrared Images (TIR) are essential in a variety of agricultural applications, water resources management or irrigation scheduling, based on surface energy balance modeling [1–4]. However, spatio-temporal resolution of the operational TIR satellite sensors results are insufficient for some applications and services, including agriculture. The importance of high-resolution TIR images is being claimed [5–9]. The limitation in the TIR domain remains, since the revisit time for high spatial resolution TIR sensors is typically poor, while the spatial resolution for those with a high revisit frequency is too coarse. In practice, the spatial resolution requirements of satellite-derived surface temperature for agricultural applications are <50 m to face certain physical limitations related to the sensor's point spread function in TIR observations [2,10,11]. As for the temporal resolution, daily TIR observations are desired, although this requirement could be relaxed to 3 days as a minimum threshold [7,11].

The Copernicus conceptual mission LSTM [12] could complement other planned high-resolution TIR missions (e.g., the JPL-NASA Landsat 9-10 or the Indian-French TRISHNA mission [13]) and fulfill the spatio-temporal requirements stated above. In the meantime, downscaling methods are contributing to filling this gap by downscaling the TIR coarse resolution to finer resolutions [3,14–17]. Several techniques have been proposed in the literature to enhance the spatial resolution of the TIR domain over vegetated areas by linking TIR and reflectance information in the Visible Near Infrared (VNIR) [18–21]. These techniques are generally based on the assumption that there exists a relation between the vegetation cover and the LST. According to these approaches, a relation between the TIR and VNIR bands is first obtained at coarse spatial resolution, and then applied at the finer resolution of the VNIR bands, assuming that this relation is scale invariant.

The Normalized Difference Vegetation Index (NDVI) or the Fractional Vegetation Cover (FVC) are the most commonly used inputs in sharpening techniques, although some studies have recently explored the possibility of implementing other combinations of reflectance values that can better characterize the surface response [17,22,23]. There are also some efforts attempting to integrate soil moisture delineated vegetation indices [24], and even radar-derived soil moisture [25] in the formulations of the LST downscaling.

For years, the Moderate Resolution Imaging Spectroradiometer (MODIS) or the Advanced Along-Track Scanning Radiometer (AATSR) were combined with Landsat or Advanced Spaceborne Thermal Emission and Reflection Radiometer (ASTER) imagery to downscale LST from 1000 m × 1000 m to ~1 ha (10,000 m^2) scale. Higher resolution VNIR sensors, such as Formosat or the Satellite pour l'Observation de la Terre (SPOT), have been also used to improve the disaggregated LST pixel size [10,26].

The synergistic use of Copernicus Sentinel-2 (S2) and Sentinel-3 (S3) imagery could offer the desired solution of high spatial and temporal resolution [8,26,27]. Although no TIR information is provided, the Sentinel-2A and -2B tandem offers a 3–5-day repeat cycle, and a 10–20 m spatial resolution in the VNIR bands. Revisit time for S3 reduces to 1–2 days, with a spatial resolution of 1000 m for their thermal channels. The relationship between TIR and VNIR bands could be extracted from S3 and then applied to S2 VNIR data. Sobrino et al. [27] explored the conceptual combination of the MultiSpectral Instrument (MSI), on board the Sentinel-2, and the Ocean and Land Color Instrument/Sea and Land Surface Temperature Radiometer (OLCI/SLSTR), on board the Sentinel-3, to show an improvement in LST products derived from AATSR at that time, before the Sentinel-3 data were available. High spatial resolution data from S2 was used to improve the characterization of the sub-pixel heterogeneity through a better parameterization of surface emissivity, although no downscaling was applied by these authors. A machine learning algorithm was proposed by Guzinski and Nieto [8] to sharpen low-resolution TIR observations from S3 using high-resolution VNIR S2 imagery. Huryna et al. [28] applied the methodology introduced by Agam et al. [19] to the combination of S3–S2 imagery. However, the methodology was tested using Terra/MODIS or Landsat observations in both works, due to the lack of high-resolution TIR data to use for cross-validation.

Despite the extraordinary growth of downscaling studies in the past decade, the assessment of the thermal sharpening techniques has been traditionally conducted by cross-validation with derived LST products at original Landsat or ASTER TIR spatial resolutions, 60–100 m and 90 m, respectively [1,8,17,21,28]. Comprehensive ground validations of disaggregated LST are quite scarce, due to the lack of robust datasets covering high contrast heterogeneous areas.

This paper continues the work initiated by Bisquert et al. [1]. These authors tested the application of different downscaling techniques in an experimental site in Barrax (Albacete, Spain) from the combination MODIS-Landsat to provide LST at fine spatial and temporal resolutions, to fulfill the requirements in the estimation of surface energy fluxes and evapotranspiration in the agricultural areas of semi-arid regions, where small land holdings dominate. Bisquert et al. [1] analyzed both classical methods based on the VNIR-LST regression, as well as more advanced approaches based on Neural Networks (NN) or Data Mining (DM). Linear, quadratic and exponential relationships, proposed in the literature, were tested and results were compared to those obtained by applying NN and regression trees in a DM approach using reflectance values from all the spectral bands available. These authors observed that NN and DM, as well as the nonlinear regression tested, have the risk of overfitting, being very sensitive to noise in the samples. They concluded that the simpler NDVI-LST linear regression led to the better results in this case. Bisquert et al. [1] explored the technique results for the different land covers in the Barrax area, and found the largest uncertainties for irrigated croplands, especially in summer when cover heterogeneity and irrigation effects are stressed. As a follow-up, Bisquert et al. [26] extracted disaggregated LST maps at a 10-m spatial resolution for the first time, using high-resolution SPOT-5 images in the framework of the Spot-5 Take 5 project. Results shown in [26] were encouraging for the further application of the model to operational S2 images.

In this context, the objective of this paper is to revise and adapt the downscaling technique to the combination MODIS-S2 to derive operational LST maps with a spatial resolution of 10 m. Some adjustments to the original formulation of the approach were introduced to reduce the model uncertainty by adding an additional image-based parametrization of the residual as a function of the VNIR response. Ground LST data from an experimental campaign carried out in the summer of 2018 were used for the model evaluation. The variety of croplands and the contrast in the surface conditions during the experiment in the selected area allowed a comprehensive analysis of the performance of the downscaling technique, not achieved before. Strengths and limitations of the models were discussed, and also some guidelines for the optimal use of this technique with Sentinel-3 and Sentinel-2 imagery are given.

This paper is structured as follows. Section 2 describes the study site, the field measurements and the satellite imagery used, as well as the downscaling methodology. Results of the ground validation and distributed assessment are shown in Section 3. Interpretation of the results and comparison with previous studies comprise Section 4. Finally, Section 5 summarizes the main conclusions of this work.

2. Materials and Methods

2.1. Study Site and Measurements

This work was conducted in the semi-arid area of Barrax, southeast Spain (39°03′ N, 2°06′ W). This is a very flat area with an average altitude of 700 m a.s.l, close to Albacete (Figure 1), traditionally used by ESA (European Space Agency) as a test site in different international campaigns [29–31]. Irrigated and rainfed crops combine in this agroecosystem, with field size ranging from small terrains below 1 ha to large pivots over 50 ha (Figure 1). This large variety makes Barrax a perfect target to assess the performance of a downscaling technique and explore its strengths and weaknesses.

Ground measurements of LST (LST_g) were registered in "Las Tiesas" experimental farm during the summer of 2018, concurrent with EOS-Terra/MODIS overpasses, and covering a total of 10 different crops in 13 independent fields (Figure 1). The temperatures were measured using four hand-held infrared radiometers (IRTs) Apogee MI-210. These radiometers have a broad thermal band (8–14 μm), with a 22° field of view and an accuracy of ±0.2 K, according to the manufacturer (Apogee Instruments, Inc.). In fact, the similar Apogee SI-121 radiometers (same radiometer, but with a field of view of 18° and without datalogger) were calibrated against a National Institute of Standards and Technology (NIST) blackbody, during a comparison of TIR radiometers carried out in Miami by the Committee on Earth Observation Satellites (CEOS), and the accuracy was established at 0.2 K [32]. Special care was taken with the ground measurements in the sparse crops (vineyard and almond orchards), by averaging soil and canopy component temperatures to obtain representative values of the target LST. The radiometers were manually carried back and forth along transects on the fields pointing at nadir view, at a height of 1.5–2 m above the ground surface. Temperatures were registered at a rate of 5–10 measurements/min, in transect distances of 30–50 m/min, and then covering several hectares with each IRT. The 10-min averages centered at the satellite overpass time were considered. Radiometric temperatures were corrected from atmospheric and emissivity effects [33]. Downwelling sky radiance was measured with each radiometer and emissivity data were obtained through the Temperature-Emissivity Separation (TES) procedure [34,35] from in situ thermal radiance measurements using a multispectral radiometer CIMEL Electronique CE 312-2 [36].

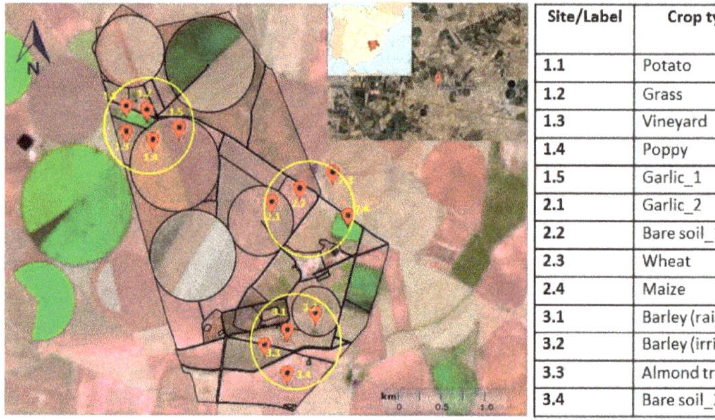

Site/Label	Crop type	Field Size (ha)
1.1	Potato	1.1
1.2	Grass	2.2
1.3	Vineyard	4.5
1.4	Poppy	25.4
1.5	Garlic_1	25.4
2.1	Garlic_2	27.4
2.2	Bare soil_1	7.7
2.3	Wheat	34.5
2.4	Maize	2.9
3.1	Barley (rainfed)	11.3
3.2	Barley (irrigated)	14.3
3.3	Almond trees	11.1
3.4	Bare soil_2	30.3

Figure 1. Overview of "Las Tiesas" experimental farm. Measurement sites are located over a S2 false color composition corresponding to date 25 July 2018. Labels for the different study fields are explained in the adjacent Table, together with indication of crop type and field size.

2.2. Satellite Images

Terra/MODIS images with near nadir observations of the study site (field of view <25°) were selected to minimize the bowtie effect [37]. Six different dates were used for this work (Table 1). MODIS VNIR and TIR data were extracted from the MOD09GQ, and MOD11_L2 products, respectively, downloaded from the NASA Earthdata Search tool. MOD09GQ offers surface Red and NIR reflectivity values at a 250 m spatial resolution. MOD11_L2 product provides LST, atmospherically corrected with a split-window algorithm, at a 1000 m spatial resolution [38].

Sentinel-2A and Sentinel-2B images concurrent or within ±1-day timing difference with MODIS were used (see dates in Table 1). S2 Level-2A products were downloaded from the Copernicus Open Access Hub, and they contain 10-m surface reflectance values. Bands 4 and 8 were used to compose the Normalized Difference Vegetation Index (NDVI). Figure 2 shows an example of the spatial distribution of the NDVI over the study site. Note the wide range in NDVI values available in the area during the experiment. Plot in Figure 3 shows the NDVI values for the different crop fields in the selected dates.

Table 1. List of satellite images used in this study, with indication of overpass time. Number of crop fields where LST_g data were measured per date is included (N). Meteorological conditions in the area at the overpass time are also listed: air temperature (T_a), relative humidity (H_r) and wind speed (u). All dates correspond to the year 2018.

Terra/MODIS			Sentinel-2	Landsat-7/ETM+	LST_g Data	Ta	Hr	u
Date	*Time*	*Viewing Angle (°)*	*(A/B)*	*(Path/Row)-Time*	*(N)*	*(°C)*	*(%)*	*(ms⁻¹)*
22 June	11:14	19	A (22 June)	no image	9	28.2	33.1	2.4
5 July	11:17	10	B (5 July)	no image	9	23.9	37.1	3.9
9 July	11:02	0	A (10 July)	(199/33)-10:32	7	30.1	33.0	1.5
16 July	11:08	12	B (15 July)	(200/33)-10:38	8	24.3	37.3	6.4
23 July	11:14	24	A (23 July)	no image	9	29.7	34.1	1.5
25 July	11:02	2	B (25 July)	(199/33)-10:32	9	28.6	37.6	2.0

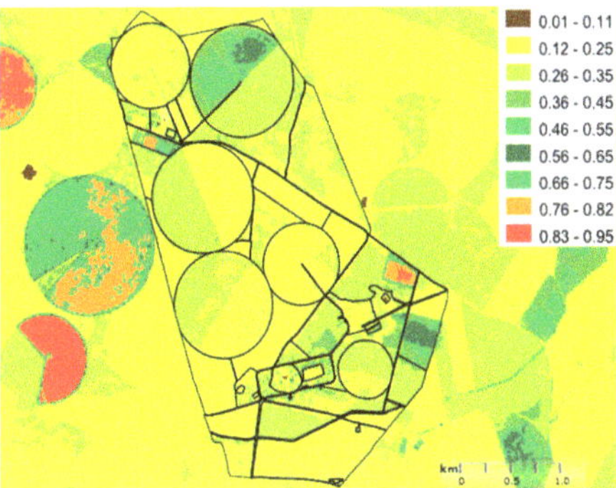

Figure 2. Sentinel-2 Normalized Difference Vegetation Index (NDVI) image of "Las Tiesas" farm, corresponding to date 25 July 2018.

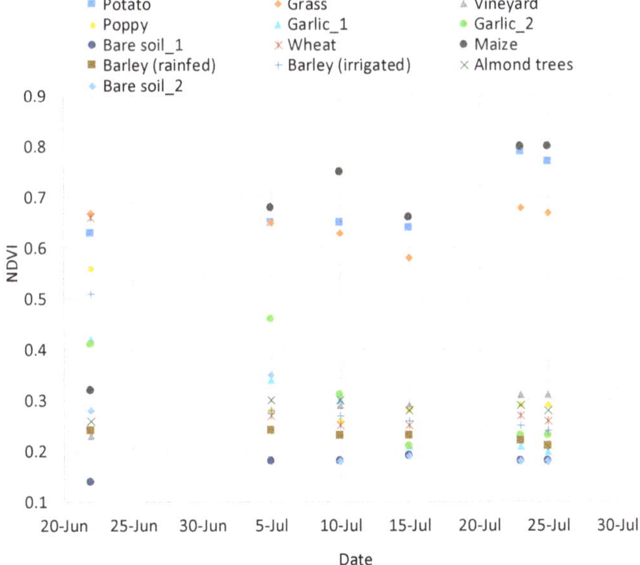

Figure 3. Distribution of the Sentinel-2 NDVI values for the different sites labeled in Figure 1, and the images/dates listed in Table 1.

Landsat-7/ETM+ overpasses were also available for 3 of the selected dates (9, 16 and 25 July). Although ETM+ TIR band has a resolution of 60 m, a cubic convolution resampling to 30 m is applied for user distribution. Thus, these images, with a 30-m spatial resolution, were used as a reference for an extended validation of the disaggregated LST. For the VNIR bands the Landsat Surface Reflectance (CDR) product was used, whereas the original TIR data in band 6 were corrected from atmospheric and emissivity effects following the method proposed by Galve et al. [39].

2.3. Downscaling Approach

Bisquert et al. [1] tested different downscaling methods with pairs of Landsat/MODIS images in this Barrax area. A modification of the sharpening method presented by Agam et al. [19] showed the best results. This method is based on the linear relationship established between NDVI and LST at the MODIS 1000 m resolution ($NDVI_{MOD}$ and LST_{MOD}, respectively). The approach outlined by Bisquert et al. [1] has been revised and adapted to the combination MODIS-S2 and used in this work as a basis to derive 10-m LST maps.

The flowchart in Figure 4 shows the main steps and calculations of this downscaling algorithm that can be summarized as follows:

1. The aggregation of the VNIR bands was carried out by averaging the reflectance values in the red and NIR bands of the 10-m S2 pixels, and 250-m MODIS pixels within an equivalent 1000 m MODIS pixel;

2. NDVI values were calculated from both S2 ($NDVI_{S2}$) and MODIS ($NDVI_{MOD}$) VNIR data at 1000 m resolution;

3. Differences between S2 and MODIS VNIR data due to spectral resolution, atmospheric correction, viewing angle or pixel footprint were corrected through a normalization extracted from the 1000 m NDVI, then applied to 10-m S2 NDVI ($NDVI_N$);

4. The 1000 m coarse spatial resolution required a previous selection of "pure" pixels for the NDVI-LST adjustment. This selection was based on a confidence value calculated from the comparison between $NDVI_{MOD}$ and aggregated $NDVI_N$. This confidence value was computed as the ratio between the standard deviation from the 4×4 pixels belonging to each 1000 m pixel, and its mean value, as suggested by [18]. Pixels with confidence values within the lowest quartile were selected in this step;

5. A linear regression was established between $NDVI_{MOD}$ and LST_{MOD} at 1000 m, using data from those "pure" pixels, and then applied to the $NDVI_N$ values to obtain a prime estimate of 10-m LST (LST_{prime});

6. The Bisquert et al. [1] algorithm included a residual (R_{LST}) correction to account for the local conditions, and to correct the possible deviations produced by the NDVI-LST equation. This residue was calculated as the difference between the original and predicted LST at a coarse resolution, and this residue value was then added equally to all high-resolution pixels composing a coarse pixel. Since this residual correction leads to some boxy effect, Bisquert et al. [1] used a Gaussian filter to smooth. This final step was revised, and a modification is introduced in this work by adding a smoothing based on a linearization between the residual R_{LST} and the $NDVI_{MOD}$ itself from 1000 m data. This linear relationship between the residue and the NDVI was then applied to 10-m $NDVI_N$ (Figure 4);

7. Finally, 10-m LST values were obtained by adding this residual R_{LST} to original 10-m LST_{prime} data from step 5. This new protocol to derive the residue values was expected to reduce the LST deviation, particularly in small size fields surrounded by a different cover, and then contribute to an overall improvement in the model performance.

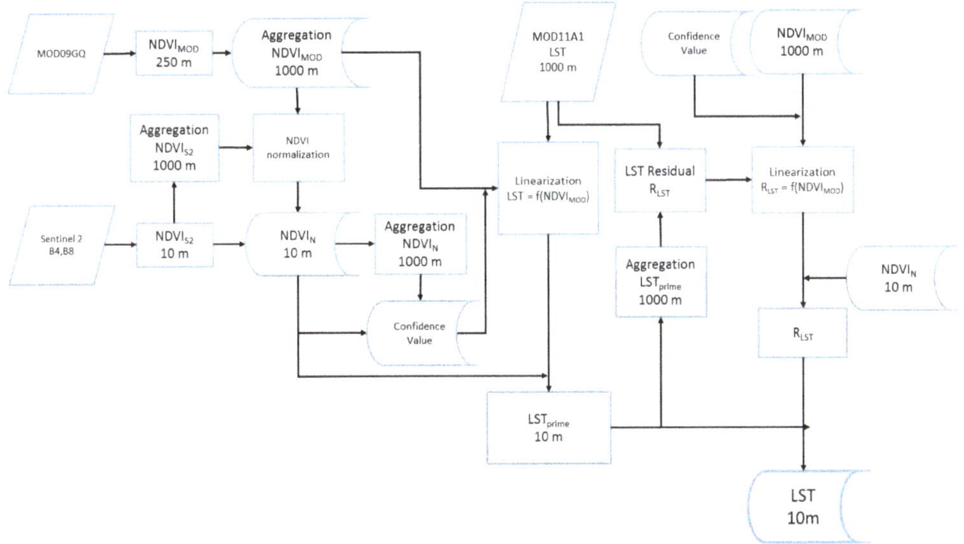

Figure 4. Flowchart of the downscaling methodology, including the different processing steps, inputs and outputs. Variable descriptions are included in the text.

This downscaling procedure was applied to pairs of MODIS-S2 images. When no Sentinel-2 image was available concurrent with the MODIS overpass, close in time images (±1 day) were used, under the assumption of minimum changes in NDVI. Note the normalization procedure applied in step 3 reduced possible differences at this point.

The assessment of the MODIS-S2 downscaling method was carried at both local and distributed scales, by comparison with ground measurements and Landsat-7/ETM+ LST products, respectively. In this last case, the comparison was established at the 30-m spatial resolution provided by U.S. Geological Survey (USGS). Following Gao et al. [20], the aggregation of 10-m S2 LST was done through the Stefan-Boltzmann law, with the assumption of similar emissivity values for adjacent pixels. Some differences may arise due to the 20–30 min delay in acquisition time between MODIS and ETM+ sensors. A normalization procedure was applied to minimize these discrepancies in LST values [1]. A linear regression between the aggregated Landsat and MODIS images at 1000 m was obtained, and then applied at 30-m spatial resolution, for each pair of MODIS-ETM+ images.

The model performance was quantified in terms of classical statistical metrics, such as the determination coefficient (r^2), the root mean square difference (RMSD), the systematic difference parameter (Bias), the mean absolute deviation (MAD), or the mean absolute deviation in percentage (MADP) [40]. Following Schneider et al. [41], other statistics considered more robust and less influenced by outliers were also calculated, such as the median bias (Me), robust standard deviation (RSD) and robust RMSD (R-RMSD). The skewness and kurtosis were also included, which quantitatively describe the distribution of the differences between the estimated and observed values.

3. Results

3.1. Ground Validation

The field scale assessment was performed using the ground data as a reference. IRT radiometric temperatures were corrected from atmospheric and emissivity effects, and average values for each 10-min transect/field were calculated. A total of 51 LST_g data were available for this study (Table 1). Measured LST_g values were in the range 297–327 K. The lowest values were observed for fully vegetated crops (grass, potato, or maize), whereas the largest values corresponded to bare soil, soil dominated crops (vineyard and almond orchard) and senescence cereals. Standard deviation of LST_g data per crop field were <±1.5 K in 90% of the dataset, with a maximum value of ±1.9 K, showing the thermal homogeneity of the fields and the thermal stability during the 10-min interval.

The methodology described above was applied to the six pairs of MODIS-S2 images listed in Table 1. Mean values of 5 × 5 high-resolution pixel arrays, centered in the location of the ground transects, were calculated and plotted against LST_g (see Figure 5). Disaggregated LST values ranged from 302 to 322 K, pointing to a certain limitation of the disaggregation technique to reproduce extreme low and high temperatures. Values of the standard deviation for the 5 x 5 pixel averages were <± 2.0 K.

All parcels in this study were provided with sprinkler irrigation system, except the vineyard and almond orchard, where drip irrigation was supplied. Irrigation was scheduled and frequently applied during the study period. For a few hours after an irrigation event, a cooling effect occurs consequence of the wetted surface. This effect is stressed when sprinklers are used. This was the case of 20% of our dataset, with 12 ground transects collected just a few hours/minutes after irrigation events. These points are plotted with non-filled circles in Figure 5. Note the evident overestimation of the disaggregated LST compared to LST_g values, with differences >10 K in some cases. These results reinforce Bisquert et al. [26] findings pointing a shortcoming of the method over wet soil areas.

Certain limitations were also observed for the highest LST values. By stablishing a threshold of 325 K, only five data were excluded corresponding to fallow and tilled barley or poppy.

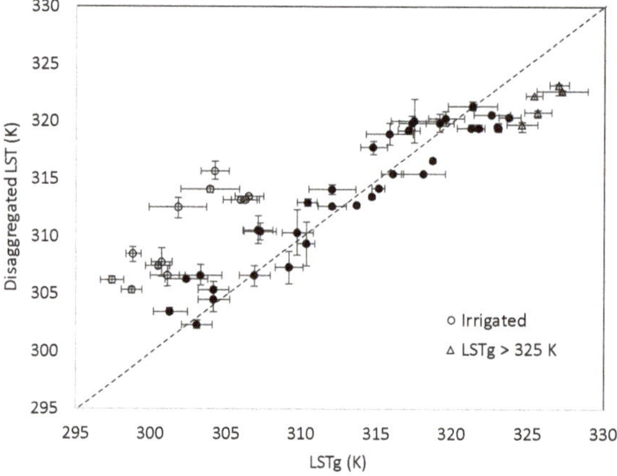

Figure 5. Linear regression between disaggregated LST_{10m} and ground-measured values (LST_g). Crop fields under recently irrigated conditions are labeled with a non-filled circle. Outlier high LST values are labeled with triangles. Dashed line represents the 1:1 agreement.

Focusing on LST data lower than 325 K, and excluding points corresponding to recently irrigated conditions, a good agreement (r^2 = 0.90) is observed between disaggregated LST_{10m} and LST_g values (Figure 5). Differences range between ±4.0 K, with a systematic deviation of 0.2 K and a RMSD value of ±2.2 K (see Table 2). The kurtosis values (~−1) indicate a behavior close to the normal distribution, while the negligible skewness observed indicates a LST-difference distribution closely centered at 0.

The plot in Figure 6 superposes results obtained running the Bisquert et al. [1] algorithm. Good agreement is also observed by this original formulation, although some scatter is added, with an increase in the RMSD value up to ±2.7 K in this case.

MODIS LST values are superposed to plot in Figure 6 too, showing a large scatter (RMSD=±8.0 K) and discrepancies >10 K. This deviation is stressed for low temperatures registered in small size vegetated parcels that are surrounded by bare soil or other croplands with higher LST. This effect was already observed by Bisquert et al. [26].

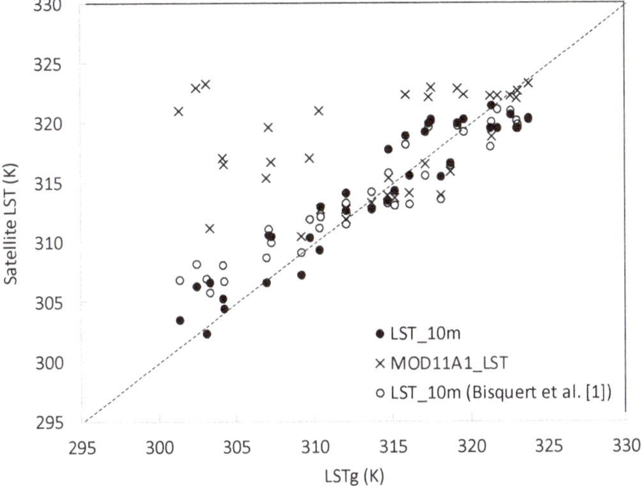

Figure 6. Disaggregated LST_{10m} (dots) and Moderate Resolution Imaging Spectroradiometer (MODIS) LST (crosses) versus ground LST measurements (LST_g). Non-filled dots represent results using the original Bisquert et al. [1] formulation of the downscaling approach. Error bars are not included in this figure for cleanliness. Dashed line represents the 1:1 agreement.

Table 2. Quantitative analysis of the differences between disaggregated LST or MODIS LST, and ground-measured LST data. The statistics include: mean bias (Bias); standard deviation (SD); mean absolute deviation (MAD); Mean Absolute Deviation in Percentage (MADP), obtained as the MAD divided by the mean observed value; root mean square difference (RMSD); coefficient of determination (r^2); median bias (Me); robust standard deviation (RSD); robust RMSD (R–RMSD); skewness (S); and kurtosis (K).

N = 34	Min (K)	Max (K)	Bias (K)	SD (K)	MAD (K)	MADP (%)	RMSD (K)	r^2	Me (K)	RSD (K)	R-RMSD (K)	S	K
LST_{10m}	−3.6	3.9	0.2	2.2	1.9	0.6	2.2	0.90	0.2	2.8	2.8	−0.04	−1.2
LST_{10m} (Bisquert et al. [1])	−4.5	5.7	0.4	2.7	2.3	0.7	2.7	0.90	0.5	3.3	3.4	0.06	−0.8
LST_MOD	−4.2	20.5	4.4	6.8	5.4	1.7	8.0	0.10	1.2	7.7	7.8	1.1	0.3

3.2. Distributed Assessment

Beyond the ground validation at a field scale, the model performance was assessed at a larger distributed scale by using the three concurrent Landsat-7/ETM+ images as a reference. The Single-Band Atmospheric Correction (SBAC) tool, recently introduced by Galve et al. [39], was used in this work for the correction of the TIR data.

Prior to the *downscaling* assessment, the feasibility of the Landsat-derived LST data needs to be tested. Ground data were also used to evaluate the Landsat-7/ETM+ LST estimates. The plot in Figure 7 shows the comparison between estimated and ground-measured LST values for a total of 21 data available for the three Landsat dates/images. Differences ranged within ± 3.5 K, except four cases corresponding to 1.1 and 1.2 sites. Note that these are small size parcels (<2 ha), for which the spatial resolution of Landsat 7/ETM+ is not fine enough, resulting in an overestimation of the LST values in these vegetated targets (potato and grass). Excluding these data from the analysis, a good matching with the 1:1 line was observed, with a coefficient of determination of $r^2 = 0.96$. An average error of RMSD = ±1.8 K was obtained. These results are in agreement with those reported by Galve et al. [39], using data from 2015–2016 in this same agricultural area. A RMSD value of ±1.6 K was obtained by these authors using ground LST data measured in six of the crop fields within "Las Tiesas" experimental farm also used in the present work.

Note significant differences in terms of LST between MODIS and high-resolution sensors (Landsat or ASTER) up to 2–3 °C have been reported, induced by difference in the retrieval algorithm, atmospheric correction, sensor performance, acquisition time, view geometry, or spectral response function [10,42–45]. Weng et al. [44] pointed out that comparison of thermal data from different sensors requires some pre-processing procedure. In this work, the differences in the spectral characteristics and overpass time (20–30 min difference) between Landsat and MODIS were minimized by applying a normalization process to the Landsat bands [1]. The disaggregated LST_{10m} were aggregated to the equivalent 30 m Landsat pixels (LST_{30m}) by a 3×3 pixel averaging, based on the Stefan-Boltzmann law, following Gao et al. [20].

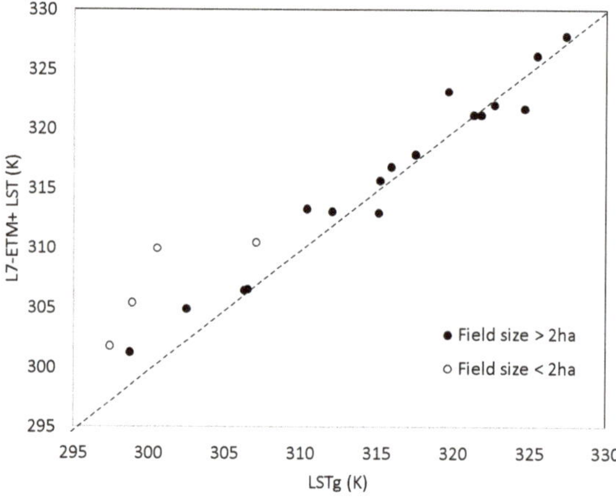

Figure 7. Linear adjustment between L7-ETM+ estimates and ground-measured LST values. Data collected in Sites 1.1 and 1.2 (field extension <2 ha) are labeled with non-filled dots. Dashed line represents the 1:1 agreement.

Figure 8 shows the comparison between the original 30-m LST derived from L7-ETM+, and LST disaggregation products at 10-m spatial resolution, for a subset of 10×10 km^2 centered in the "Las Tiesas" experimental farm. Visual inspection points the significant improvement in the capacity to discriminate the different field borders. Although the real potential of the downscaling approach is revealed when focusing on parcels <5 ha, where 3×3 thermal pixels of L7-ETM+ can be hardly fit in, being these areas the main responsible of the scatter ($r^2 = 0.82$) observed in the regression plot in Figure 9.

To quantify the performance of the downscaling approach at a full scene perspective, pixel-to-pixel differences were calculated at the 30 m spatial resolution for the selected subset of 10×10 km^2 (Figure 9). Statistical metrics of the differences are listed in Table 3. Considering more than 270,000 pixel/data, an average RMSD of ±2.0 K was obtained, with a minor overestimation of 0.3 K.

Table 3. Quantitative analysis of the differences plotted in Figure 9. Statistical metrics as defined in Table 2.

	N	Bias (K)	SD (K)	MAD (K)	MADP (%)	RMSD (K)	r^2	Me (K)	RSD (K)	R-RMSD (K)
9 July	67826	0.7	2.5	1.9	0.6	2.6	0.55	0.5	2.8	2.8
16 July	116015	−0.4	1.4	1.2	0.4	1.4	0.63	−0.6	1.7	1.8
25 July	90804	0.9	1.8	1.5	0.5	2.0	0.75	0.8	2.2	2.3
Average	274643	0.3	1.9	1.4	0.5	2.0	0.82	0.010	2.1	2.1

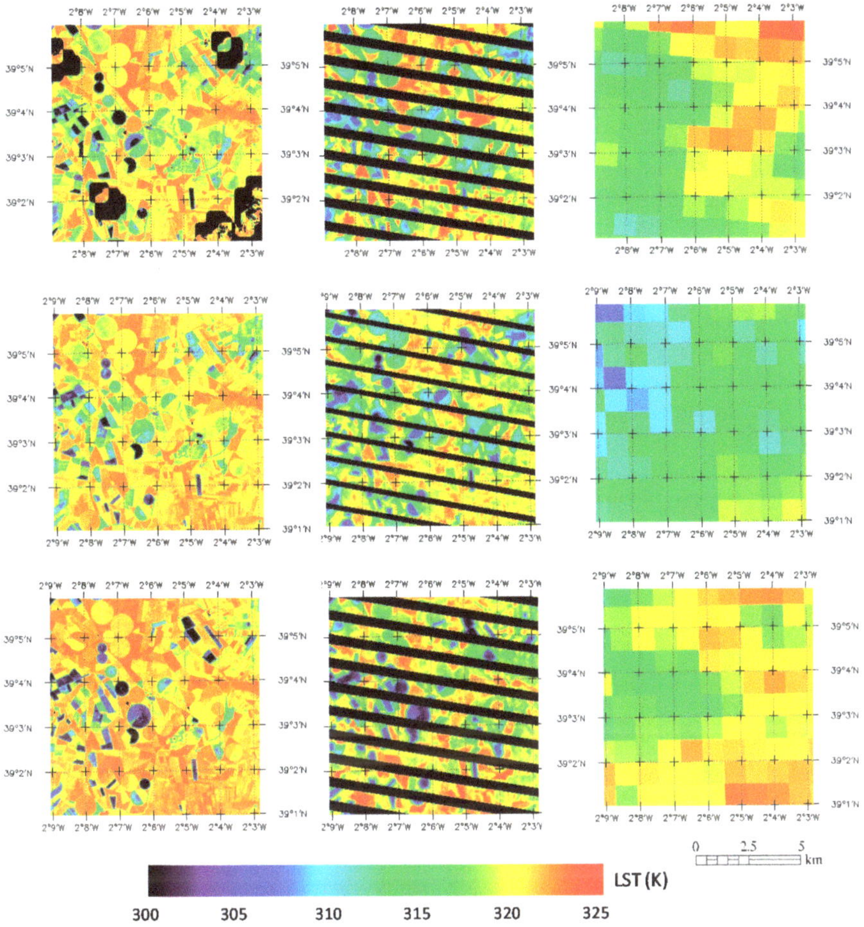

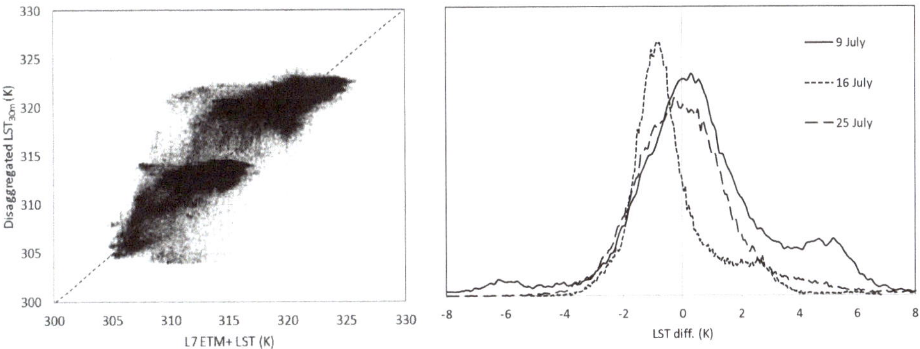

Figure 8. Disaggregated LST$_{10m}$ (left column), LST derived from original 30 m L7-ETM+ Thermal InfraRed (TIR) band (center), and MOD11A1_LST product (right column). Examples corresponding to dates 9 July 2018 (up), 16 July 2018 (middle), and 25 July 2018 (bottom).

Figure 9. Disaggregated LST$_{30m}$ versus L7-ETM+ LST for the full dataset (left). Dashed line represents the 1:1 agreement. Histograms of pixel-to-pixel differences between disaggregated LST$_{30m}$ products and LST estimates from L7-ETM+ images for three concurrent MODIS and Landsat-7 overpasses (right).

4. Discussions

Agromomy management decisions based on TIR data need confidence in LST estimates. An absolute uncertainty <1.5 K is traditionally reported as a requirement [7,46]. The translation of this uncertainty to ET accuracy depends on the model but ranges between 10% and 20% [47,48]. Based on this threshold, the results obtained in this work (average RMSD = ±2.2 K) are encouraging. Additionally, the 10-m pixel size and the revisit frequency of the MODIS data, much better than the 3–5 days revisit frequency of S2A and S2B satellites, can fulfill the LST input requirements of the surface energy balance methods for a variety of hydrological, climatological or agricultural applications. At this point, no significant differences in the model performance were observed in connection with the collocation delay between the MODIS overpass and the Sentinel-2 image used, i.e., ±1-day mismatch seems not to have had an effect on the disaggregation.

With the new treatment of the residuals (R_{LST}) introduced in this work, as part of the downscaling scheme, an improvement around 20% was obtained in the performance of the original formulation of the model [1] that simply included a Gaussian filter, as suggested by Anderson et al. [2].

Ground validation results are in agreement with those obtained by Bisquert et al. [26] using MODIS-Spot 5 pairs in this case. These authors reported a bias of 0.2 K and a RMSD of ±2.4 K based on the comparison between disaggregated 10-m LST and ground-measured LST values for 10 different dates and five different fields. Regarding the distributed assessment, our results are similar to the RMSD value of ±2.6 K reported by Bisquert et al. [26] at a scene scale. In a first work using MODIS-Landsat combination [1], these authors reported an average RMSD = ±2.0 K for disaggregated 60-m LST in this case.

In this context, Agam et al. [19] tested a sharpening model (TsHARP) over extensive corn/soybean fields in central Iowa, USA. RMSD values between ±0.7 and ±1.4 °C were obtained by sharpening down simulated MODIS thermal maps at 1000 m to 250 m and between ±1.8 and ±2.4 °C by sharpening simulated thermal Landsat maps from 60 and 120 m to a VNIR 30 m resolution. Also using TsHARP, Duan and Li [49] disaggregated MODIS LST from 1000 m to 90 m, with an uncertainty of ±2.7 °C. Jeganathan et al. [14] tested TsHARP from MODIS over a heterogeneous agricultural landscape in India, and found uncertainties ranging ±2–3 K, using ASTER thermal data as a reference. Eswar et al. [23] used a thermal sharpening technique with five different indices to downscale MODIS LST from 960 m to 120 m, and compared this with the Landsat 7 LST data at different sites in India. These authors found that NDVI/FVC showed better result for wet areas, whereas the Normalized Difference Water Index (NDWI) was found better for dry areas. Yang et al. [17] used the multiple linear regression models to downscale the aggregated Landsat TIRS (360 m) image to 90 m, using a relation of LST with multiple scale factors in an area of mixed land covers (water, vegetation, bare soil, impervious surface), and then compared with the pure Landsat LST. The result found was satisfactory with coefficient of determination of 0.87 and RMSD of ±1.13 K. Merlin et al. [10] used a time series of higher resolution Formosat-2 images to test a new disaggregation procedure of kilometric thermal data over an irrigated cropping area in northwestern Mexico during an agricultural season. RMSD values about ±3 °C were obtained by these authors.

Many of these previous studies already pointed larger uncertainties in disaggregated LST over irrigated lands [1,10,26]. This is a weekness that remains in the present work since, although affected, VNIR reflectivity data does not fully capture the cooling effect produced in a wetted surface. Therefore, the downscaling technique still fails at reproducing LST values for spots with an undergoing irrigation or recently irrigated targets. In an attempt to face these limitations, some works incorporate additional reflectance information in the regression algorithms. Gosh and Joshi [22] tested several regression algorithms using EO1-Hyperion hyperspectral data over different land use land cover scenes. These authors used three pairs of coincident Hyperion and Landsat 7-ETM+ images as a reference for the assessment. Liu et al. [24] compared the performances of a thermal disaggregation technique, based on three different indices: temperature vegetation dryness index (TVDI), NDVI and fractional vegetation cover (FVC), over a humid agriculture region. These authors found the smallest

RMSD using TVDI, with an improvement of 0.2 K in comparison to the results obtained using NDVI or FVC. A similar reduction of the uncertainty in 0.2 K was obtained by Amazirh et al. [25], thanks to the inclusion of Sentinel-1 radar data, linked to surface soil moisture, in a new formulation to improve the LST disaggregation methodology. These authors used Sentinel-1 imagery to derive 100-m resolution LST, and the results were compared with Landsat LST, used as a reference over two heterogeneous sites (irrigated and rainfed). However, average RMSD values for the six dates of study resulted over ±3.0 K, with even worse accuracy during summer. So, further efforts are still required to improve this soil moisture integration. Further works should also explore the inclusion of additional Sentinel-2 bands in the shortwave infrared (SWIR) in the sharpening scheme, since they might account for vegetation and soil water content [50].

Another finding is this work is the difficulty of the downscaling approaches to reproduce excepcionally high LST when these conditions are constrained to small parcels in the image, and there is a lack of coarse original MODIS pixels showing this homogeneous thermal conditions. The modeled relationship LST-VNIR reflectivities may not fully cover these conditions, leading to an underestimation of LST.

Focusing on the combination of S3-S2 images, very few quantitative studies have been conducted. In a first attempt, Huryna et al. [28] applied TsHARP sharpening. These authors reported LST differences of ±1.3–1.5 K, when compared with sharpened to 60-m S3 temperature with reference Landsat 8 temperature at 60 m, with a positive bias of 0.3–0.6 K, depending on the study site. However, the lack of local measurements prevented these authors from conducting a ground assessment. Further research should merge Sentinel-2 and Sentinel-3 imagery and conduct robust assessment of the downscaling results. Collections of ground LST measurements under a variety of surface and environmental conditions are then required, and the dataset gathered in the framework of this work is potentially attractive for this aim.

5. Conclusions

This work adds to the previous literature dealing with thermal infrared downscaling. The 10-m LST maps generated from the combination MODIS-S2 can contribute to fill the gap until high spatial-temporal resolution TIR images are available. The linear relation NDVI-LST was adopted as a basis for the downscaling approach. Results obtained encourage the parametrization of the residual as a function of the NDVI as a key step in the algorithm (an improvement of 0.5 K was achieved). The variety of surface conditions and the wide range of NDVI and LST values in the semi-arid area of Barrax allowed a robust assessment of the downscaling approach. An average estimation error of ±2.2 K in LST_{10m} resulted from the ground validation. This evaluation was reinforced by the pixel-to-pixel comparison of rescaled LST_{30m} with Landsat-7/ETM+ LST estimates, showing a similar RMSD of ±2.0 K for the distributed assessment.

Findings in this study highlight the limitations of the methodology to capture the variability of extreme LST, and the problems in recently sprinkler irrigated fields. Results indicate the need for caution, since disaggregated LST under these conditions may result artificially higher than expected.

Despite the weaknesses, this work gives promising insights for the adaptation of this methodology to the tandem S3-S2 in coming works. Further research could also benefit from the ground LST dataset introduced in this paper for a comprehensive performance assessment.

Finally, note that the benefits of this research may extend to other applications, such as monitoring volcanic activity and wildfire, estimating evapotranspiration or assessing drought severity.

Author Contributions: Conceptualization, J.M.S.; Methodology, J.M.S. and J.M.G.; Resources, J.G.-P., R.L.-U. and A.C.; Software, J.M.G.; Validation, J.M.S., J.M.G. and R.N.; Writing—original draft, J.M.S.; Writing—review & editing, J.G.-P., R.L.-U., R.N. and A.C. All authors have read and agreed to the published version of the manuscript.

Funding: This research was funded by the Spanish Economy and Competitiveness Ministry, together with FEDER funds (project CGL2016-80609-R), the Education and Science Council (JCCM, Spain) (project

SBPLY/17/180501/000357), the European Commission (SUPROMED project, N° 1813), and Juan de la Cierva Research Grant of Galve (FJCI-2015-24876).

Acknowledgments: We thank the ITAP-FUNDESCAM and the Thermal Remote Sensing Group of University of Valencia for the logistic support in the experimental campaigns (MINECO-FEDER, EU projects AGL2017-83738-C3-3-R and CGL2015-64268-R).

Conflicts of Interest: The authors declare no conflict of interest.

References

1. Bisquert, M.; Sánchez, J.M.; Caselles, V. Evaluation of Disaggregation Methods for Downscaling Modis Land Surface Temperature to Landsat Spatial Resolution in Barrax Test Site. *IEEE J. Sel. Top. Appl. Earth Obs. Remote Sens.* **2016**, *9*, 1430–1438. [CrossRef]

2. Anderson, M.C.; Kustas, W.P.; Alfieri, J.G.; Gao, F.; Hain, C.; Prueger, J.H.; Evett, S.; Colaizzi, P.; Howell, T.; Chávez, J.L. Mapping Daily Evapotranspiration at Landsat Spatial Scales During the BEAREX'08 Field Campaign. *Adv. Water Resour.* **2012**, *50*, 162–177. [CrossRef]

3. Ha, W.; Gowda, P.; Howell, T. A Review of Downscaling Methods for Remote Sensing-Based Irrigation Management: Part I. *Irrig. Sci.* **2013**, *31*, 831–850. [CrossRef]

4. Guzinski, R.; Anderson, M.C.; Kustas, W.P.; Nieto, H.; Sandholt, I. Using a Thermal Based Two Source Energy Balance Model with Time-Differencing to Estimate Surface Energy Fluxes with Day-Night MODIS Observations. *Hydrol. Earth Syst. Sci.* **2013**, *17*, 2809–2825. [CrossRef]

5. Fisher, J.B. The future of evapotranspiration: Global Requirements for Ecosystem Functioning, Carbon and Climate Feedbacks, Agricultural Management, and Water Resources. *Water Resour. Res.* **2017**, *53*, 2618–2626. [CrossRef]

6. Hulley, G.; Hook, S.; Fisher, J.; Lee, C. Ecostress, A NASA Earth-Ventures Instrument for Studying Links Between the Water Cycle and Plant Health over the Diurnal Cycle. In Proceedings of the 2017 IEEE International Geoscience and Remote Sensing Symposium (IGARSS), Fort Worth, TX, USA, 23–28 July 2017; pp. 5494–5496. [CrossRef]

7. Sobrino, J.A.; Jiménez-Muñoz, J.C.; Soria, G.; Ruescas, A.B.; Danne, O.; Brockmann, C.; Ghent, D.; Remedios, J.; North, P.; Merchant, C.; et al. Synergistic Use of MERIS and AATSR as a Proxyfor Estimating Land Surface Temperature from Sentinel-3 Data. *Remote Sens. Environ.* **2016**, *179*, 149–161. [CrossRef]

8. Guzinski, R.; Nieto, H. Evaluating the Feasibility of Using Sentinel-2 and Sentinel-3 Satellites for High-Resolution Evapotranspiration Estimations. *Remote Sens. Environ.* **2019**, *221*, 157–172. [CrossRef]

9. He, B.-J.; Zhao, Z.-Q.; Shen, L.-D.; Wang, H.-B.; Li, L.-G. An Approach to Examining Performances of Cool/Hot Sources in Mitigating/Enhancing Land Surface Temperature under Different Temperature Backgrounds Based on Landsat 8 Image. *Sustain. Cities Soc.* **2019**, *44*, 416–427. [CrossRef]

10. Merlin, O.; Duchemin, B.; Hagolle, O.; Jacob, F.; Coudert, B.; Chehbouni, G.; Dedieu, G.; Garatuza, J.; Kerr, Y. Disaggregation of MODIS Surface Temperature over an Agricultural Area Using a Time Series of Formosat-2 Images. *Remote Sens. Environ.* **2010**, *114*, 2500–2512. [CrossRef]

11. Lagouarde, J.-P. The MISTIGRI Thermal Infrared Project: Scientific Objectives and Mission Specifications. *Int. J. Remote Sens.* **2013**, *34*, 3437–3466. [CrossRef]

12. Koetz, B.; Berger, M.; Blommaert, J.; Del Bello, U.; Drusch, M.; Duca, R.; Gascon, F.; Ghent, D.; Hoogeveen, J.; Hook, S.; et al. *Copernicus High Spatio-Temporal Resolution Land Surface Temperature Mission: Mission Requirements Document*; ESA: Noordwijk, The Netherlands, 2019.

13. Lagouarde, J.-P. The Indian-French Trishna Mission: Earth Observation in the Thermal Infrared with High Spatio-Temporal Resolution. In Proceedings of the IGARSS 2018—2018 IEEE International Geoscience and Remote Sensing Symposium, Valencia, Spain, 22–27 July 2018; pp. 4078–4081. [CrossRef]

14. Jeganathan, C.; Hamm, N.A.S.; Mukherjee, S.; Atkinson, P.M.; Raju, P.L.N.; Dadhwal, V.K. Evaluating a Thermal Image Sharpening Model over a Mixed Agricultural Landscape in India. *Int. J. Appl. Earth Obs. Geoinform.* **2011**, *13*, 178–191. [CrossRef]

15. Zhan, W.; Chen, Y.; Zhou, J.; Wang, J.; Liu, W.; Voogt, J.; Zhu, Z.; Quan, J.; Li, J. Disaggregation of Remotely Sensed Land Surface Temperature: Literature Survey, Taxonomy, Issues, and Caveats. *Remote Sens. Environ.* **2013**, *131*, 119–139. [CrossRef]

16. Cammalleri, C.; Anderson, M.C.; Gao, F.; Hain, C.R.; Kustas, W.P. A Data Fusion Approach for Mapping Daily Evapotranspiration at Field Scale. *Water Resour. Res.* **2013**, *49*, 4672–4686. [CrossRef]

17. Yang, Y.; Li, X.; Pan, X.; Zhang, Y.; Cao, C. Downscaling Land Surface Temperature in Complex Regions by Using Multiple Scale Factors with Adaptive Thresholds. *Sensors* **2017**, *17*, 744. [CrossRef]

18. Kustas, W.P.; Norman, J.M.; Anderson, M.C.; French, A.N. Estimating Subpixel Surface Temperatures and Energy Fluxes from the Vegetation Index–Radiometric Temperature Relationship. *Remote Sens. Environ.* **2003**, *85*, 429–440. [CrossRef]

19. Agam, N.; Kustas, W.P.; Anderson, M.C.; Li, F.; Neale, C.M.U. A Vegetation Index Based Technique for Spatial Sharpening of Thermal Imagery. *Remote Sens. Environ.* **2007**, *107*, 545–558. [CrossRef]

20. Gao, F.; Kustas, W.; Anderson, M. A Data Mining Approach for Sharpening Thermal Satellite Imagery over Land. *Remote Sens.* **2012**, *4*, 3287–3319. [CrossRef]

21. Bindhu, V.M.; Narasimhan, B.; Sudheer, K.P. Development and Verification of a Non-Linear Disaggregation Method (NL-Distrad) to Downscale MODIS Land Surface Temperature to the Spatial Scale of Landsat Thermal Data to Estimate Evapotranspiration. *Remote Sens. Environ.* **2013**, *135*, 118–129. [CrossRef]

22. Ghosh, A.; Joshi, P.K. Hyperspectral Imagery for Disaggregation of Land Surface Temperature with Selected Regression Algorithms over Different Land Use Land Cover Scenes. *ISPRS J. Photogramm. Remote Sens.* **2014**, *96*, 76–93. [CrossRef]

23. Eswar, R.; Sekhar, M.; Bhattacharya, B.K. Disaggregation of LST over India: Comparative Analysis of Different Vegetation Indices. *Int. J. Remote Sens.* **2016**, *37*, 1035–1054. [CrossRef]

24. Liu, K.; Wand, S.; Li, X.; Li, Y.; Zhang, B.; Zhai, R. The Assessment of Different Vegetation Indices for Spatial Disaggregating of Thermal Imagery over the Humid Agricultural Region. *Int. J. Remote Sens.* **2019**, *41*, 1907–1926. [CrossRef]

25. Amazirh, A.; Merlin, O.; Er-Raki, S. Including Sentinel-1 Radar Data to Improve the Disaggregation of MODIS Land Surface Temperature Data. *ISPRS J. Photogramm. Remote Sens.* **2019**, *150*, 11–26. [CrossRef]

26. Bisquert, M.; Sánchez, J.M.; López-Urrea, R.; Caselles, V. Estimating High Resolution Evapotranspiration from Disaggregated Thermal Images. *Remote Sens. Environ.* **2016**, *187*, 423–433. [CrossRef]

27. Sobrino, J.; Jiménez-Muñoz, J.; Brockmann, C.; Ruescas, A.; Danne, O.; North, P.; Heckel, A.; Davies, W.; Berger, M.; Merchant, C.; et al. Land Surface Temperature Retrieval from Sentinel 2 and 3 Missions. In Proceedings of the Sentinel-3 OLCI/SLSTR and MERIS/(A)ATSR Workshop, Frascati, Italy, 15–19 October 2012; Ouwehand, L., Ed.; ESA Communications: Oakville, ON, Canada, 2013. ISBN 9789290922759 (SP-711).

28. Huryna, H.; Cohen, Y.; Karnieli, A.; Panov, N.; Kustas, W.P.; Agam, N. Evaluation of Tsharp Utility for Thermal Sharpening of Sentinel-3 Satellite Images Using Sentinel-2 Visual Imagery. *Remote Sens.* **2019**, *11*, 2304. [CrossRef]

29. Moreno, J.F. "The SPECTRA Barrax campaign (SPARC): Overview and First Results from CHRIS Data". In Proceedings of the Second CHRIS/Proba Workshop, ESA SP-578, Frascati, Italy, 28–30 April 2004.

30. Sobrino, J.A. Thermal Remote Sensing in the Framework of the SEN2FLEX Project: Field Measurements, Airborne Data and Applications. *Int. J. Remote Sens.* **2008**, *29*, 4961–4991. [CrossRef]

31. Latorre, C. Seasonal Monitoring of FAPAR Over the Barrax Cropland Site in Spain, In Support of the Validation of PROBA-V Products at 333 m. In Proceedings of the Recent Advances in Quantitative Remote Sensing, Torrent, Spain, 22–26 September 2014; pp. 431–435.

32. Niclòs, R.; Pérez-Planells, L.l.; Coll, C.; Valiente, J.A.; Valor, E. Evaluation of the S-NPP VIIRS Land Surface Temperature Product Using Ground Data Acquired by an Autonomous System at a Rice Paddy. *ISPRS J. Photogramm. Remote Sens.* **2018**, *135*, 1–12. [CrossRef]

33. Sánchez, J.M.; López-Urrea, R.; Rubio, E.; González-Piqueras, J.; Caselles, V. Assessing Crop Coefficients of Sunflower and Canola Using Two-Source Energy Balance and Thermal Radiometry. *Agric. Water Manag.* **2014**, *137*, 23–29. [CrossRef]

34. Gillespie, A.; Rokugawa, S.; Matsunaga, T.; Cothern, J.S.; Hook, S.; Kahle, A.B. A Temperature and Emissivity Separation from Advanced Spaceborne Thermal Emission and Reflection Radiometer (ASTER) Images. *IEEE Trans. Geosci. Remote Sens.* **1998**, *36*, 1113–1126. [CrossRef]

35. Sánchez, J.M.; French, A.N.; Mira, M.; Hunsaker, D.J.; Thorp, K.R.; Valor, E.; Caselles, V. Thermal Infrared Emissivity Dependence on Soil Moisture in Field Conditions. *IEEE Trans. Geosci. Remote Sens.* **2011**, *49*, 4652–4659. [CrossRef]

36. Legrand, M.; Pietras, C.; Brogniez, G.; Haeffelin, M.; Abuhassan, N.K.; Sicard, M.A. high-Accuracy Multiwavelength Radiometer for in Situ Measurements in the Thermal Infrared. Part I: Characterization of the Instrument. *J. Atmos. Ocean Technol.* **2000**, *17*, 1203–1214. [CrossRef]
37. Wolfe, R.E.; Nishihama, M.; Fleig, A.J.; Kuyper, J.A.; Roy, D.P.; Storey, J.C.; Patt, F.S. Achieving Sub-Pixel Geolocation Accuracy in Support of MODIS Land Science. *Remote Sens. Environ.* **2002**, *83*, 31–49. [CrossRef]
38. Wan, Z. New Refinements and Validation of the MODIS Land Surface Temperature/Emissivity Products. *Remote Sens. Environ.* **2008**, *112*, 59–74. [CrossRef]
39. Galve, J.M.; Sánchez, J.M.; Coll, C.; Villodre, J. A New Single-Band Pixel-By-Pixel Atmospheric Correction Method to Improve the Accuracy in Remote Sensing Estimates of LST. Application to Landsat 7-ETM+. *Remote Sens.* **2018**, *10*, 826. [CrossRef]
40. Willmott, C.J. Some Comments on the Evaluation of Model Performance. *Bull. Am. Meteorol. Soc.* **1982**, *63*, 1309–1313. [CrossRef]
41. Schneider, P.; Ghent, D.; Corlett, G.; Prata, F.; Remedios, J. *Land Surface Temperature Validation Protocol*; ESA: Noordwijk, The Netherlands, 2012.
42. Essa, W.; Verbeiren, B.; van der Kwast, J.; Batelaan, O. Improved DisTrad for Downscaling Thermal MODIS Imagery over Urban Areas. *Remote Sens.* **2017**, *9*, 1243. [CrossRef]
43. Liu, Y.; Hiyama, T.; Yamaguchi, Y. Scaling of Land Surface Temperature Using Satellite Data: A Case Examination on ASTER and MODIS Products over a Heterogeneous Terrain Area. *Remote Sens. Environ.* **2006**, *105*, 115–128. [CrossRef]
44. Weng, Q.; Fu, P.; Gao, F. Generating Daily Land Surface Temperature at Landsat Resolution by Fusing Landsat and MODIS Data. *Remote Sens. Environ.* **2014**, *145*, 55–67. [CrossRef]
45. Brown, M.E.; Pinzon, J.E.; Didan, K.; Morisette, J.T.; Tucker, C.J. Evaluation of the Consistency of Long-Term NDVI Time Series Derived from AVHRR, SPOT-Vegetation, Seawifs, MODIS, and Landsat ETM+ Sensors. *IEEE Trans. Geosci. Remote Sens.* **2006**, *44*, 1787–1793. [CrossRef]
46. Ghent, D.J.; Corlett, G.K.; Göttsche, F.M.; Remedios, J.J. Global Land Surface Temperature from the Along Track Scanning Radiometers. *J. Geophys. Res. Atmos.* **2017**, *122*, 12167–12193. [CrossRef]
47. Allen, R.; Tasumi, M.; Trezza, R. Satellite-Based Energy Balance for Mapping Evapotranspiration with Internalized Calibration (METRIC)\97 Model. *J. Irrig. Drain. Eng.* **2007**, *133*, 380–394. [CrossRef]
48. Sánchez, J.M.; Kustas, W.P.; Caselles, V.; Anderson, M. Modelling Surface Energy Fluxes over Maize Using a Two-Source Patch Model and Radiometric Soil and Canopy Temperature Observations. *Remote Sens. Environ.* **2008**, *112*, 1130–1143. [CrossRef]
49. Duan, S.-B.; Li, Z.L. Spatial Downscaling of MODIS Land Surface Temperatures Using Geographically Weighted Regression: Case Study in Northern China. *IEEE Trans. Geosci. Remote Sens.* **2016**, *54*, 6458–6469. [CrossRef]
50. Sadeghi, M.; Babaeian, E.; Tuller, M.; Jones, S.B. The Optical Trapezoid Model: A Novel Approach to Remote Rensing of Soil Moisture Applied to Sentinel-2 and Landsat-8 Observations. *Remote Sens. Environ.* **2017**, *198*, 52–68. [CrossRef]

Article

Modelling High-Resolution Actual Evapotranspiration through Sentinel-2 and Sentinel-3 Data Fusion

Radoslaw Guzinski [1],*,† , Hector Nieto [2],† , Inge Sandholt [3] and Georgios Karamitilios [3]

[1] DHI GRAS A/S, Agern Alle 5, 2970 Hørsholm, Denmark
[2] COMPLUTIG, Colegios 2, 28801 Alcalá de Henares, Spain; hector.nieto@complutig.com
[3] SANDHOLT, Sankt Nikolaj Vej 8, 1953 Frederiksberg C, Denmark; inge@sandholt.eu (I.S.); georgios.karamitilios@sandholt.eu (G.K.)
* Correspondence: rmgu@dhigroup.com
† These authors contributed equally to this work.

Received: 17 April 2020; Accepted: 28 April 2020; Published: 1 May 2020

Abstract: The Sentinel-2 and Sentinel-3 satellite constellation contains most of the spatial, temporal and spectral characteristics required for accurate, field-scale actual evapotranspiration (ET) estimation. The one remaining major challenge is the spatial scale mismatch between the thermal-infrared observations acquired by the Sentinel-3 satellites at around 1 km resolution and the multispectral shortwave observations acquired by the Sentinel-2 satellite at around 20 m resolution. In this study we evaluate a number of approaches for bridging this gap by improving the spatial resolution of the thermal images. The resulting data is then used as input into three ET models, working under different assumptions: TSEB, METRIC and ESVEP. Latent, sensible and ground heat fluxes as well as net radiation produced by the models at 20 m resolution are validated against observations coming from 11 flux towers located in various land covers and climatological conditions. The results show that using the sharpened high-resolution thermal data as input for the TSEB model is a sound approach with relative root mean square error of instantaneous latent heat flux of around 30% in agricultural areas. The proposed methodology is a promising solution to the lack of thermal data with high spatio-temporal resolution required for field-scale ET modelling and can fill this data gap until next generation of thermal satellites are launched.

Keywords: evapotranspiration; data fusion; field-scale; machine-learning; physical model; Sentinel-2; Sentinel-3

1. Introduction

The fluxes of water (e.g., evapotranspiration—ET) and energy (e.g., of latent and sensible heat) at the surface of the Earth are critical to quantify for many applications in the fields of climatology, meteorology, hydrology and agronomy. Easy access to reliable estimations of ET is considered a key requirement within natural resource management, and if ET can be estimated accurately enough it holds a vast potential to assist in the current attempts of meeting the UN Sustainable Development Goals (SDG), e.g., SDG2—zero hunger, or SDG6—clean water and sanitation (https://sustainabledevelopment.un.org, last accessed 10 December 2018).

Water and energy fluxes show large spatio-temporal variability since they are highly dependent not only on the meteorological conditions, but also on different characteristics and properties of the land surface, such as soil moisture/water availability, land cover type and amount of vegetation biomass and its health. Remote sensing data can provide spatially-distributed information about relevant land surface states and properties used to model the relevant fluxes and hence this technology addresses a key limitation of conventional point scale observations when estimating fluxes at watershed and regional scales. In particular, thermal remote sensing has been widely used for assessing land surface turbulent fluxes [1]. While there are a variety of existing remote sensing ET methods and data options available [2,3], none is fully satisfying the user needs for reliable, operational and easy accessible estimates and tools able to derive ET at agricultural-parcel scale. The limitations have so far primarily been centred on the lack of suitable satellite-based input data sources.

With the recent launch of Sentinel-2 and Sentinel-3 satellites, the data foundation for producing operational ET maps has been set since as a constellation they contain most of the required spatial, temporal and spectral characteristics [4]. Sentinel-3 Sea and Land Surface Temperature Radiometer (SLSTR) instrument acquires daily thermal infrared (TIR) information of the surface at ca. 1 km scale [5]. However, the reliable estimation of ET in agricultural and heterogeneous landscapes requires that the model's spatial resolution matches the dominant landscape feature scale, usually tens or hundreds of meters. Sentinel 2, with a spatial resolution ranging from 10 to 60 m and 5 day revisit time with Sentinel 2A & B combined [6], can resolve part of these scaling issues, although it lacks a TIR instrument at high spatial resolution such as in the Landsat missions. Therefore sharpening [7–9] and/or disaggregation methods [10] are required to bridge the spatial gap between the currently available Sentinel constellation's thermal-infrared (referred to as "thermal" in the reminder of the paper) and optical-shortwave (referred to as "shortwave" in the reminder of the paper) observational capabilities in order to optimally exploit the synergies of both types of sensors for field-scale ET estimations. The aim of this study is to develop an optimal combination of thermal sharpening and ET modelling methods for the derivation field-scale ET with combined Sentinel-2 and Sentinel-3 observations.

Several data fusion methods have been proposed to merge low resolution thermal infrared imagery with high resolution shortwave imagery in order to obtain estimates of surface temperature (T_{rad}) and/or ET at high spatial resolution. In this study we focus on different, but possibly complementary, approaches: empirical and semi-empirical methods that exploit relationships between shortwave bands and thermal or ET data (hereinafter called image sharpening methods); and physically-based ET downscaling methods (hereinafter called ET disaggregation).

Thermal image sharpening uses information from the thermal and shortwave images themselves to calibrate empirical or semi-empirical models. Those models relate coarse resolution T_{rad} (or ET) with coarse resolution (or fine resolution aggregated to coarse resolution) shortwave bands, and then apply the calibrated model to the fine scale shortwave image, producing either a sharpened T_{rad}, or directly an ET product.

One of the first attempts to sharpen T_{rad} was TsHARP [11], who tested different regression models between T_{rad} and NDVI. Since then, TsHARP has been utilised as reference method for developing and testing other sharpening methods [8,12,13]. The Data Mining Sharpening (DMS) approach [8] used local and global regression trees between reflective bands and T_{rad} of homogeneous samples at coarse scale (based on coefficient of variation threshold). Residual analysis was performed to ensure energy conservation (based on emitted radiances) between original resolution and sharpened images. To avoid overfitting of regression trees such as in DMS the use of random forests was proposed instead [14]. Following with the machine learning algorithms, Yang et al. [15] used an Artificial Neural Network with Genetic Algorithm and Self-Organizing Feature Mapping trained with different land surface parameters for each land cover class (vegetation, bare soil, urban and water). A different approach used an unmixing method to derive brightness temperature and emissivity at fine scale [16]. The unmixed brightness temperature and emissivity were then the inputs to a generalized split-window algorithm to retrieve fine resolution T_{rad}.

The use of a contextual algorithm can also be applied in sharpening, such as is the case of DISPATCH-LST (DISaggregation based on Physical And Theoretical scale CHange) by Merlin et al. [7] who used shortwave information on fractional vegetation cover and fractional photosynthetically active vegetation cover in contextual scatterplots of fractional green vegetation cover versus T_{rad} and albedo versus T_{rad} to define minimum and maximum soil and canopy endmember temperatures.

Finally, two or more different methods can be used together and combined through weighted averaging, such as in Chen et al. [17], who combined TsHARP and a Thin Plate Spline interpolation by weighting their corresponding residuals. Besides of the fact that all methods described above can be used as well to sharpen ET, other studies have already suggested methods to directly downscale coarse scale ET using shortwave data [18–21]. In any case, shortwave images provide limited information related to some surface energy balance processes, such as turbulent transport, soil moisture, and meteorological forcing. Therefore ancillary variables could be included in T_{rad} or ET sharpening such as land cover maps (to account for different aerodynamic roughness), local meteorology, or surface geometry [22].

A previous study [4] found that using a "disaggregation" approach [10,23] significantly enhanced the accuracy of turbulent fluxes derived with sharpened T_{rad}. That approach ensures spatial consistency between fluxes derived at fine and coarse spatial scales by first estimating them at the coarse scale at which the thermal observations were acquired. In the following step, the low-resolution air temperature is varied to adjust the flux estimates for all high-resolution pixels falling within one low-resolution pixel. This is repeated until a consistency between the two scales is obtained. This approach assumes that since the coarse scale estimates are derived with T_{rad} at original spatial resolution they are of higher accuracy. The disaggregation was shown to improve ET model skill when compared with outputs produced at either coarse or fine resolution alone [4,23,24].

The sharpened T_{rad} can be used as input to land-surface energy flux models. The latent heat flux λE (or energy used for ET) can be estimated as the residual of the surface energy budget, using estimates of the net radiation (R_n), soil heat flux (G) and sensible heat flux (H). The thermal-based ET models were originally formulated for computing H, which is governed by the bulk resistance equation for heat transfer [25], and is driven by the gradient between an ensemble surface temperature, called the "aerodynamic surface temperature" (T_0), and the surface layer air temperature. Besides of the estimation of that surface-to-air temperature gradient, the estimation of H requires the modelling of an aerodynamic resistance term, which can be viewed as a simplification of the complex turbulent transport of heat, momentum and water vapour, by using a similarity with Ohm's law for electric transport. These resistances therefore represent how efficiently a scalar (heat, momentum or water vapour) is transported from one point to another following a gradient (i.e., vertical differences of temperature and/or vapour pressure). Several formulations and/or parametrizations have been proposed to describe these turbulent transport processes but generally they include variables related to surface aerodynamic roughness, wind speed as well as wind attenuation through the canopy, and atmospheric stability [26].

The challenge in resistance energy balance models is that T_0 cannot be directly estimated by remote sensing [27,28]. Hence, remote sensing ET models differ from each other on how the existing difference between the radiometric temperature (T_{rad}) observed by satellite sensors and T_0 is considered. Single-source or bulk transfer schemes for modelling H treat soil and canopy as a single flux source and often employ an additional resistance term (R_{AH}, usually dependent on the Stanton number kB^{-1}) because heat transport is less efficient than momentum transport from land surface (see e.g., Garratt and Hicks [29] or Verhoef et al. [30]). Appropriately calibrated, one-source energy balance (OSEB) models have shown satisfactory estimates of surface energy fluxes in heterogeneous landscapes [31–34]. However, due to the difficulty in robustly and parsimoniously parametrizing R_{AH} for OSEB schemes at different landscapes, climates, and observational configurations [35], the two-source energy balance (TSEB) modelling approach was developed [36]. TSEB models partition the surface energy fluxes and the radiometric temperature between nominal soil and canopy sources, and include a more physical representation of processes related to T_{rad} and T_0 without requiring any additional input information beyond that needed by single-source models using more sophisticated kB^{-1} parametrizing. However, because direct measurements of canopy (T_C) and soil (T_S) temperatures rarely are available, in most applications these component temperatures are derived from a measurement of the bulk surface radiometric temperature T_{rad}. Partitioning of T_{rad} between T_C and T_S requires some assumptions related to the evaporative efficiency of soil or canopy [36–38].

Finally, like all remote sensing retrievals, satellite radiometric temperature is prone to uncertainty due to sensor noise, surface emissivity and atmospheric effects. To overcome this issue in ET estimation, several methods have been proposed based on either contextual models [39–41], by constraining the ET range between hot (no ET) and cold (potential ET) pixels [31,32], or using time-differenced morning temperature rise [42,43]. Regarding the contextual methods, all of them require homogeneous forcing and coupling between land surface/atmosphere which is a disadvantage when applied at large scales. In addition, those models assume that the coldest pixel in the image means potential transpiration, and the hottest pixel means zero transpiration which is not always the case (e.g., in humid and sub-humid areas).

In this study we will evaluate three different ET models driven by Sentinel-2 and Sentinel-3 imagery: METRIC [32] is a one source energy balance model that is less sensitive to heat transfer coefficient parametrizing than other OSEB model such as SEBS [33]; TSEB-PT [36] as a widely used two source energy balance model; and ESVEP [44] as a hybrid contextual-two source energy balance model.

2. Materials and Methods

In this section we first describe the evaluated ET models (Section 2.1), before presenting the validation sites (Section 2.2) and the input data sources (Section 2.3).

2.1. Description of ET Models

The energy balance can be expressed as (1)

$$R_n \approx G + H + \lambda E \tag{1}$$

where net radiation R_n is a key element in the energy budget of the land surface as it determines the available energy that the land utilises for water evapotranspiration (latent heat flux, λE) and for heating up the overlying air layer (sensible heat flux, H). Equation (1) assumes that other energy terms (heat advection, heat storage in the canopy layer, and energy for photosynthesis) are negligible. Since ET is the combined process of soil evaporation and canopy transpiration, R_n can be also be partitioned into soil ($R_{n,S}$) and canopy net radiation ($R_{n,C}$), with both sensible and latent heat flux also partitioned between soil (i.e., evaporation process) and canopy (transpiration).

Using remote sensing data to derive R_n has proven to be a sound alternative to ground-based measurements of both shortwave and longwave net radiation. Different approaches have been proposed to estimate surface albedo, ranging from empirical relationships between ground measured albedo and the different reflective bands in satellite [45] to more physically based methods relying on modeling the surface anisotropic effects [46,47]. Indeed, one of the major challenges when estimating albedo with satellite remote sensing data is that such sensors typically measure the outgoing radiance at a given direction while the estimation of albedo needs to account for the outgoing radiance in all the directions of the hemisphere [48,49]. Methods based on the modelling of those bidirectional effects have proven to be effective to overcome this challenge. In this study, we use a method for retrieving soil and canopy shortwave net radiation, proposed by Kustas and Norman [50], based on the different spectral behaviour of soil and vegetation for the photosynthetically active radiation (PAR) and near infrared (NIR) regions of the spectrum. Such approach is based on the radiative transfer model (RTM) described in Campbell and Norman [51] to obtain estimates of soil and canopy albedo as well as canopy transmittance in the PAR and NIR. This approach requires as inputs Leaf Area Index and leaf inclination distribution [52], the different bihemispherical reflectances and transmittances of soil and a single leaf, and the proportion of diffuse irradiance. However, this approach assumes homogeneous canopies and it requires some corrections when dealing with clumped canopies [53], which were also used in this study.

On the other hand, longwave net radiation is primarily driven by the thermal radiation emitted by the surface, which depends on surface emissivity and skin temperature following the Stefan-Boltzman law. Besides, Kirchoff's law can be applied to derive the atmospheric longwave radiation that is absorbed by the surface. When running OSEB models (i.e., METRIC), only those two components are taken into account. However, when using TSEB models (i.e., TSEB-PT and ESVEP), surface anisotropy can also be accounted for when estimating the net longwave radiation, considering that leaves and soil have different temperatures and hence emit different amounts of thermal radiation [50].

Soil heat flux G is usually assumed to be a ratio of the soil net radiation. Choudhury et al. [54], Bastiaanssen et al. [31] suggested that G is ca. 35% of net radiation of the soil around midday hours and this is the approach used in this study by TSEB models. Since net radiation of the soil cannot be computed when using OSEB models, a specific equation suggested by Bastiaanssen et al. [31] is used instead.

In all three evaluated models, λE is estimated as residual of Equation (1). The main difference between the models is in the way in which H is calculated, as briefly described in the sections below.

2.1.1. Mapping Evapotranspiration at High Resolution with Internalized Calibration, METRIC

Sensible heat flux in METRIC [32] is derived in a contextual manner by finding hot and cold pixels (Equation (2)).

$$H = \rho C_p \frac{\delta T}{R_{AH}} \tag{2a}$$

$$\delta T = c + m\, T_{rad} \tag{2b}$$

where δT is the estimated gradient between aerodynamic and air temperature, estimated as a linear equation function of T_{rad} with c and m parameters are linearly solved from expressing Equation (2b) from two cold and hot endpoints:

$$m = \frac{\delta T_{hot} - \delta T_{cold}}{T_{hot} - T_{cold}} \tag{3a}$$

$$c = \delta T_{hot} - m\, T_{hot} \tag{3b}$$

METRIC scales λE between these two hot (T_{hot}) and cold (T_{cold}) endmembers based on a linear relationship between actual ET and reference ET using the standardised ASCE Penman-Monteith equation for an ideal alfalfa field [55]. Therefore, METRIC, as opposed to SEBAL [31], does not assume zero sensible heat flux at the cold pixel, which can have a positive impact on model accuracy at well watered areas under large vapour pressure deficit conditions. According to Allen et al. [32], cold pixels yield a 5% larger ET than the reference ET ($\lambda E_{cold} = 1.05\lambda E_{ref}$), but earlier in the season and off-season, cold pixel ET is instead a function of fractional cover or NDVI: $\lambda E_{cold}/\lambda E_{ref} = f(NDVI)$. On the other hand, METRIC overcomes the issue of estimating kB^{-1} by computing R_{AH} using the profile at two different heights above z_{0H}. Finally the authors stated the need for either computing an "excess resistance" in aerodynamically rough and dry surfaces when using the δT calibration performed over agricultural areas, or calibrating different δT slopes at different land covers/environmental conditions [32].

For Equation (2) to hold true, δT and H require constant wind speed at the application domain, so the model uses wind speed at blending height to overcome this issue. It also requires constant irradiance and air temperature, i.e., δT changes are only either due to root-zone soil moisture or aerodynamic roughness. Furthermore, the model requires heterogeneity in hydrologic and vegetation conditions and therefore we applied METRIC over two different vegetation domains, short vegetation (crops, grass and shrubs) and tall vegetation (broadleaved and conifer forests as well as wooded savannas). Finally, METRIC is sensitive to the definition of hot and cold pixels. Several different methodologies to find those endmember values were proposed, which can be especially challenging in heterogeneous areas where pixels become mixed at coarse spatial resolution. In our case we adopted the Exhaustive Search Algorithm solution proposed by Bhattarai et al. [56].

2.1.2. Priestley-Taylor Two-Source Energy Balance Model, TSEB-PT

Two-source energy balance models treat the land surface as two layers, soil and canopy, contributing to the energy and water fluxes (Equation (4))

$$R_{n,C} = H_C + \lambda E_C \tag{4a}$$

$$R_{n,S} = H_S + \lambda E_S + G \tag{4b}$$

where soil (canopy) sensible heat flux is computed from the gradient between the soil (canopy) temperature (T_S and T_C respectively) and the air temperature at the sink-source height (equivalent to T_0). In the TSEB-PT model [28,36,53], an electrical circuit analogy is used in which H from soil and canopy are estimated based on three aerodynamic resistances to heat transport arranged in a series network. Since T_C and T_S are unknown *a priori*, they are estimated in an iterative process in which it is first assumed that green canopy (expressed as the fraction of LAI that is green, f_g) transpires a potential rate based on Priestley–Taylor formulation [36]:

$$\lambda E_C = \alpha_{PT} f_g \frac{\Delta}{\Delta + \gamma} R_{n,C}, \quad \alpha_{PT} = 1.26 \tag{5}$$

where α_{PT} is the Priestley and Taylor [57] coefficient, Δ is the slope of the vapour pressure to air temperature curve and γ is the psychrometric constant. Then the canopy transpiration is sequentially reduced (i.e., $\alpha_{PT} < 1.26$) until realistic fluxes are obtained ($\lambda E_C \geq 0$ and $\lambda E_S \geq 0$).

TSEB-PT probably is the model that requires most accurate retrievals of physical inputs (LAI and T_{rad}), and studies already reported larger uncertainty in senescent vegetation (i.e., $f_g < 1$) and very dense (high LAI) or tall vegetation [43,58]. It is more complex than METRIC and therefore has a large number of parameters and modelling options. Finally, the Priestley–Taylor formulation was shown to produce larger uncertainty in high advection conditions, cases in which initializing λE_C with a Penman-Monteith formulation showed better results [37]. Combining TSEB-PT model with the disaggregation approach (described in Section 1) results in a disTSEB model [23].

2.1.3. End-Member-Based Soil and Vegetation Energy Partitioning, ESVEP

ESVEP is based on a trapezoid $T_{rad} - f_{cover}$ framework, in which it considers fluxes acting in a "parallel" soil and canopy system [44]. As in TSEB-PT, ESVEP partitions T_{rad} as a linear weight of emitted radiance. Other similar models to ESVEP are HTEM [59] and TTEM [60], but ESVEP solves the trapezoid in a pixel-per-pixel basis overcoming the need for homogeneous weather forcing and roughness (Equation (6a)).

$$T_{S,max} = \frac{r_a (R_{n,S} - G)}{\rho_a C_p} + T_A \tag{6a}$$

$$T_{C,max} = \frac{r_a R_{n,C}}{\rho_a C_p} \frac{\gamma \left(1 + r_{b,dry}/r_a\right)}{\Delta + \gamma \left(1 + r_{b,dry}/r_a\right)} - \frac{vpd}{\Delta + \gamma \left(1 + r_{b,dry}/r_a\right)} + T_A \tag{6b}$$

$$T_{S,min} = \frac{r_a (R_{n,S} - G)}{\rho_a C_p} \frac{\gamma}{\Delta + \gamma} - \frac{vpd}{\Delta + \gamma} + T_A \tag{6c}$$

$$T_{C,min} = \frac{r_a R_{n,C}}{\rho_a C_p} \frac{\gamma (1 + r_{b,wet}/r_a)}{\Delta + \gamma (1 + r_{b,wet}/r_a)} - \frac{vpd}{\Delta + \gamma (1 + r_{b,wet}/r_a)} + T_A \tag{6d}$$

where r_a is the aerodynamic resistance, $r_{b,dry}$ and $r_{b,wet}$ are resistances of dry and wet canopy respectively, ρ_a is the density or air, C_p is specific heat capacity at constant pressure, γ is psychrometric constant and vpd is vapour pressure deficit of the air.

2.2. Validation Sites

Year 2017 measurements from eleven eddy covariance (EC) sites were used in this study, covering a wide range of land cover types and climates. Sites are summarised in Table 1 and data used in validation included the four components of net radiation R_n (shortwave/longwave downwelling/upwelling), soil heat flux G, sensible heat flux H, and latent heat flux λE. In addition, friction velocity, Monin-Obukhov length, and wind direction data from the EC system was used to estimate the satellite pixel footprint contribution [61,62] to the turbulent fluxes at the satellite overpass. Validation sites comprise 5 agricultural sites, both irrigated and rainfed, including row crops (e.g., Sierra Loma vineyard) and an olive grove (Taous). In addition, two sites over grassland, one humid meadow (Grillenburg) and a semi-arid steppe (Kendall Grassland), one in conifer (Hyltemossa) and one in broadleaved forests (Harvard Forest) are also included in the validation list. Finally, complex heterogeneous landscapes are represented by two wooded savannas (Dahra and Majadas de Tiétar). From all these sites, 3 are in Mediterranean climate, and two more in semi-arid climates, whereas the rest of the sites are located in temperate climates.

Table 1. Description of eddy covariance sites used for validation. Sites are listed in alphabetic order. Z shows the EC measurement height in meters, while the contact person for the EC tower is credited in PI column.

Site (Abrevation)	Land Cover	Climate	Location	Z (m)	PI/Reference
Choptank (CH)	Cropland, irrigated (rotation of corn and soybean)	Temperate	United States 39.058743 N 75.851282 W	5	William P. Kustas (ARS-USDA)
Dahra (DA)	Savanna	Semi-arid	Senegal 15.40278 N 15.43222 W	9	Torbern Tagesson (Univ. Copenhagen) [63]
Grillenburg (GR)	Grassland, meadow	Temperate	Germany 50.950013 N 13.512583 E	3	Christian Bernhofer (T.U. Dresden)
Harvard Forest (HF)	Broadleaved forest	Temperate	United States 42.536874 N 72.172578 W	30	J. William Munger (Harvard Univ.)
Hyltemossa (HTM)	Conifer forest (spruce)	Temperate	Sweden 56.097584 N 13.419056 E	27	Michal Heliasz (Lund Univ.)
Kendall Grassland (KG)	Grassland, steppe	Semi-arid	United States 31.73652 N 109.94185 W	6.4	Russell Scott (ARS-USDA) [64,65]
Klingenberg (KL)	Cropland (spring barley and catch crops)	Temperate	Germany 50.8931 N 13.5224 E	3.5	Christian Bernhofer (T.U. Dresden)
Majadas de Tieétar (MT)	Savanna	Mediterranean	Spain 39.940332 N 5.774647 W	15.5	Arnaud Carrara (CEAM)
Selhausen (SE)	Cropland (sugar beets and winter barley)	Temperate	Germany 50.870623 N 6.449653 E	2.3	Marius Schmidt (Jülich)
Sierra Loma (SL) (previously known as Borden)	Cropland, irrigated (vineyard)	Mediterranean	United States 38.289355 N 121.11779 W	5	William P. Kustas (ARS-USDA) & Joseph Alfieri (ARS-USDA) [66,67]
Taous (TA)	Cropland, rainfed (olive)	Mediterranean	Tunisia 34.93111 N 10.60153 E	9.5	Gilles Boulet (CESBIO) & Dalenda Boujnah (Institut de l'Olivier)

Error metrics included mean bias error ($\sum(Obs. - Pred.)/N$), root mean squared error ($RMSE = \sqrt{\sum(Obs. - Pred.)^2/N}$), relative RMSE ($RMSE/\overline{Obs}$), and Pearson correlation coefficient between observed and predicted. Due to the lack of energy closure in the eddy covariance data, unless otherwise stated all metrics were computed after adding the energy balance residual (residual = $R_{n,EC} - G_{EC} - \lambda E_{EC} - H_{EC}$) to the latent heat flux, taking the assumption that errors in measurements of λE are larger than errors in the measurements of H [68].

2.3. Input Data Sources

The input data required to run the evapotranspiration models came from three main and two ancillary sources. The main sources were: Sentinel-2 shortwave observations, Sentinel-3 thermal observations and European Center for Medium-range Weather Forecasts (ECMWF) ERA-5 meteorological reanalysis data. The ancillary sources were: a digital elevation model (DEM) from the Shuttle Radar Topography Mission (SRTM) satellite, and land cover map created as part of the ESA Climate Change Initiative (CCI).

2.3.1. Satellite Data

The main satellite data inputs come from the Sentinel-2 (both A and B) and Sentinel-3 (A only) satellites. Sentinel-2 and Sentinel-3 were chosen as the primary satellite data sources for this study for a number of reasons. Firstly, as mentioned previously, when treated as a constellation those satellites are able to satisfy the need for temporally dense observations at high spatial resolution and with good radiometric accuracies. Secondly, they are part of the Copernicus programme, meaning that there is a guaranteed long-term continuity of data into the future. Thirdly, the data from those satellites is free and open and will remain so, again due to being part of the Copernicus programme.

High-resolution shortwave observations needed to characterise the state of vegetation in the evapotranspiration model as well as to sharpen TIR data were obtained by the MultiSpectral Instrument (MSI) on board of the Sentinel-2A & 2B satellites. MSI acquires reflectance information in 13 spectral bands (with central wavelength ranging from 444 nm to 2202 nm) with a spatial resolution of 10 m, 20 m, or 60 m (depending the spectral band) and global geometric revisit of at least 5 days when both satellites are considered [6]. The MSI sensor has 3 spectral bands in the leaf-pigment sensitive red-edge part of the electromagnetic spectrum and two bands in water-content sensitive shortwave-infrared part of the spectrum, in addition to the more traditional visible and near-infrared bands, which makes it well suited for vegetation mapping and monitoring [69]. For each of the validation sites, all Sentinel-2 images for year 2017 were visually scanned and the ones which were cloud, fog and shadow free in the closest vicinity of the flux towers locations were selected for processing.

L1C top of the atmosphere images were converted to bottom-of-atmosphere (BOA) reflectances (L2A) using the Sen2Cor atmospheric correction processor [70] v2.5.5. BOA reflectance values were then used as input to the Biophysical Processor [71] available in the SNAP software v6.0.1 (step.esa.int—last accessed 28 November 2018) in order to obtain effective values of green Leaf Area Index (LAI), Fractional Vegetation Cover (FVC), Fraction of Absorbed Photosynthetically Active Radiation (FAPAR), Canopy Chlorophyll Content (CCC) and Canopy Water Content (CWC). The outputs of the SNAP Biophysical Processor have been validated in a number of studies, with good performance in herbaceous crops [72,73] but overestimation of LAI in bare-soil cases and underestimation of LAI in forests [74]. Those inaccuracies could have an impact on the results of this study in semi-arid and forested areas.

The fraction of vegetation which is green and transpiring (f_g) was estimated based on Fisher et al. [75] (Equation (7)):

$$f_g = FAPAR/FIPAR \tag{7}$$

where FIPAR is the fraction of photosynthetically active radiation intercepted by green and brown vegetation. FAPAR was obtained from the biophysical processor as described above, while FIPAR was derived iteratively from Equation (8) of Campbell and Norman [51]:

$$FIPAR = 1 - \exp\frac{-0.5PAI}{\cos\theta} \tag{8}$$

where θ is the solar zenith angle at the time of the S2 overpass, and PAI is the plant area index with initial PAI equal to LAI and in subsequent iterations

$$PAI = LAI/f_g \tag{9}$$

until f_g converges. Two assumptions made in Equation (8) are that all intercepted PAR comes from the solar beams, and that both FAPAR and FIPAR are computed from a canopy with a spherical leaf inclination distribution. Indeed, from the the average leaf angle histogram, from which the training database was built in Weiss and Baret [71], most training cases in the Biophysical processor correspond to a spherical distribution (mode at 60° leaf angle). Equation (9) was subsequently used within the land surface models to convert LAI, which was assumed to represent green LAI [71], into PAI. Afterwards, PAI, leaf bi-hemispherical reflectance and transmittance, together with constant values for soil reflectance in the visible (VIS = [400–700] nm, $\rho_{soil,VIS}$ = 0.15) and near infrared (NIR = [700–2500] nm, $\rho_{soil,NIR}$ = 0.25) were used to quantify the shortwave net radiation of the soil and canopy. Leaf chlorophyll concentration (i.e., $C_{a+b} = CCC/LAI$) was used to derive the leaf bihemispherical reflectance and transmittance in the visible spectrum after a curve fitting of 45,000 ProspectD [76] simulations. Likewise, equivalent water thickness (i.e., $C_w = CWC/LAI$) was used to retrieve leaf bihemispherical reflectance and transmittance in the NIR spectral region.

The thermal data needed to drive the evapotranspiration model was obtained from the Sea and Land Surface Temperature Radiometer (SLSTR) on board of the Sentinel-3A satellite [5]. SLSTR contains 3 thermal infrared (TIR) channels (with two dynamic range settings—for fire monitoring and for sea/land surface temperature monitoring) with 1 km spatial resolution and less than two days temporal resolution with one satellite (less than one day with both A and B satellites) at the equator. For each selected S2 scene, all the S3 scenes falling on the day of S2 overpass or within ten days before and after, were selected for processing. In the current study we used the L2A Land Surface Temperature (LST) product as downloaded from the Copernicus Open Access Hub (https://scihub.copernicus.eu/—last accessed 10 September 2019). The accuracy of this product is reported to be below 1 K when comparing against in situ radiometer measurements and independent operational reference products [77].

Finally, the parameters in the ET models that could not be directly retrieved from shortwave observations (e.g., vegetation height or leaf inclination angle) were set based on a land cover map and a look-up table (see Table 2). The CCI landcover map from 2015 [78], which was produced with a global coverage and 300 m spatial resolution, was used as the initial input layer before being reclassified into the smaller number of classes as shown in Table 2. Out of the parameters set according to the look-up table, the vegetation height (h_C) has the largest influence on the modelled fluxes as it effects aerodynamic roughness [79,80]. Therefore in herbaceous classes where it can change throughout the growing season (grasslands and croplands) it was scaled with PAI using a power law, with maximum value $h_{C,max}$ indicated in Table 2 reached at a prescribed maximum PAI PAI_{max} (5 in croplands and 4 in grasslands) and a minimum value set to 10% of the maximum value.

Table 2. Land cover based Look-Up-Table for ancillary parameters used in ET models. CCI-LC is the land cover code for the ESA's CCI land cover legend (http://maps.elie.ucl.ac.be/CCI/viewer/download/CCI-LC_Maps_Legend.pdf—last accessed 13 April 2020); $h_{C,max}$ is the maximum canopy height occurring when PAI reaches PAI_{max}; f_C is fraction of the ground occupied by a clumped canopy ($f_C = 1$ for a homogeneous canopy); w_C is canopy shape parameter, representing the canopy width to canopy height ratio; l_w is the average leaf size; χ Campbell [52] leaf angle distribution parameter.

CCI-LC	$h_{C,max}$ (m)	PAI_{max} (–)	f_C (–)	w_C/h_C (–)	l_w (m)	χ
0	0	0	0	0	0	0
10	1.2	5	1	1	0.02	0.5
11	1	5	1	1	0.02	0.5
12	2	5	0.5	2	0.1	1
20	1.2	5	1	1	0.02	0.5
30	1.2	5	0.5	1	0.05	0.5
40	1.2	5	0.5	1	0.1	0.5
50	10	5	1	1	0.15	1
60	10	5	1	1	0.15	1
61	10	5	1	1	0.15	1
62	10	5	0.4	1	0.15	1
70	20	5	1	2	0.05	1
71	20	5	1	2	0.05	1
72	20	5	0.4	2	0.05	1
80	20	5	1	2	0.05	1
81	20	5	1	2	0.05	1
82	20	5	0.4	2	0.05	1
90	15	5	1	1.5	0.1	1
100	8	5	0.75	1.5	0.15	0.8
110	8	5	0.25	1	0.02	0.5
120	1.5	4	1	1	0.05	1
121	1.5	4	1	1	0.05	1
122	1.5	4	1	1	0.05	1
130	0.5	4	1	1	0.02	0.5
140	0.05	1	1	1	0.001	1
150	2	2	0.15	1	0.05	1
151	10	5	0.15	1	0.1	1
152	1.5	4	0.15	1	0.05	1
153	0.5	4	0.15	1	0.02	0.5
160	10	5	1	1	0.1	1
170	10	5	1	1	0.1	1
180	1	5	1	1	0.02	0.5
190	20	0	0	0	0	0
200	0	0	0	0	0	0
201	0	0	0	0	0	0
202	0	0	0	0	0	0
210	0	0	0	0	0	0
220	0	0	0	0	0	0

2.3.2. Meteorological Data Source

The meteorological data used in this study consists of air temperature at 2 m, dew point temperature at 2 m, wind speed at 100 m, surface pressure, total column water vapour (TCWV), aerosol optical thickness (AOT) at 550 nm and surface geopotential. Those inputs are obtained from the ECMWF ERA5 reanalysis ensemble means dataset [81]. The only exception was AOT which come from the Copernicus Atmosphere Monitoring Service (CAMS) forecast dataset [82], since it is not included in ERA5. Inputs at the time of the satellite overpass are computed by linear interpolation between the previous and posterior reanalysis timestep. Due to the low spatial resolution of the air temperature and wind speed fields (tens of kilometers) they are assumed to represent the surface conditions derived from conditions above the blending height (100 m above the surface) rather then the actual surface conditions. Therefore, air temperature at 100 m is calculated using the 2 m estimate, ECMWF surface geopotential, SRTM DEM and lapse rate for moist air. Those 100 m estimates are then used as inputs into the land surface flux models. AOT together with TCWV, surface pressure, SRTM DEM elevation and solar zenith angle at the time of Sentinel-3 satellite overpass were used to estimate the instantaneous shortwave irradiance on a horizontal surface at the satellite overpass [83,84].

2.4. Thermal Data Sharpening Approach

The thermal data sharpening approach used in this study is based on ensemble of modified decision trees. The basic scheme of the method (Figure 1) is based on Gao et al. [8] and has been previously applied by Guzinski and Nieto [4] to sharpen thermal data to be used as input to evapotranspiration models. Each S3 scene is matched with an S2 scene acquired at most ten days before or after the S3 acquisition and the regression model used for sharpening is derived specifically for each scene pair.

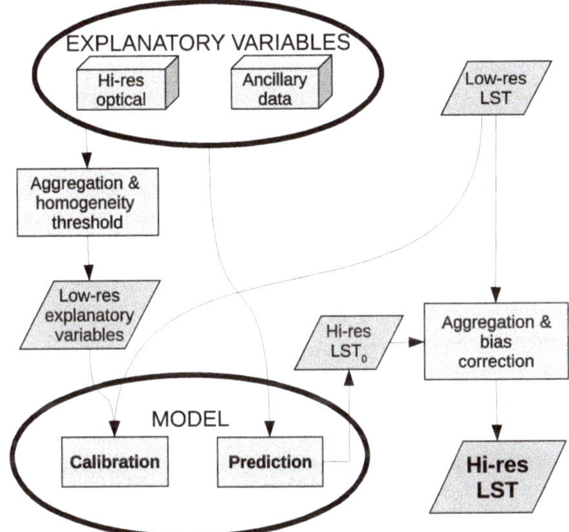

Figure 1. General thermal sharpening workflow. Explanatory variables include both shortwave bands as well any other ancillary explanatory variable, such as elevation, land cover type or exposure. Model could be any regression model, such as multivariate linear regression or machine learning techniques.

Briefly, the atmospherically corrected Sentinel-2 shortwave data (all the 10 m and 20 m spectral bands) with a spatial resolution of 20 m is resampled to match the pixel sampling of the SLSTR sensor (around 1 km spatial resolution). Concurrently, the SRTM DEM is used to derive slope and aspect maps which, together with S3 overpass time, are used to estimate the solar irradiance incident angle of a flat tilted surface. The DEM and the solar angle maps are also resampled to the SLSTR resolution. A multivariate regression model is then trained with the three resampled datasets used as predictors and the T_{rad} used as the dependent variable. The selection of training samples is performed automatically by estimating the coefficient of variation (CV) of all the high-resolution pixels falling within one low-resolution pixel and selecting 80% of pixels with lowest CV. The regression model is based on bagging ensemble [85] of decision trees. The decision trees are additionally modified such that all samples within a regression tree leaf node are fitted with a multivariate linear model, as proposed by [8].

The regression models are trained on the whole S2 tile (100 km by 100 km) as well as on subsets of 30 by 30 S3 pixels in a moving window fashion. Once they are trained they are also applied on the whole scene and on each window. The bias between the predicted high-resolution T_{rad} pixels aggregated to the low-resolution and the original low-resolution T_{rad} is calculated and the outputs of the whole-scene and moving-window regressions are combined based on a weight inversely proportional to the bias [8]. Finally, the T_{rad} predicted by the regression model is corrected by comparing the emitted longwave radiance of the sharpened fine T_{rad} versus the original coarse T_{rad}. A bias-corrected T_{rad} is therefore re-calculated by adding an offset all fine scale pixels falling within coarse scale pixel in order to remove any residual bias. This is done to ensure the conservation of energy between the two thermal images with different spatial resolutions [8]. The output of the sharpening is a 20 m representation of the LST.

This image sharpening approach relies on the direct or indirect relationship that different regions of the shortwave spectrum have with the radiometric temperature and/or the ET process. For instance the temperature of denser canopies, with higher contrast between visible and near-infrared bands, is lower than the temperature of bare soils [86,87]. In addition, surfaces with higher water content (i.e., larger absorption in the short-wave infrared) have a larger evaporative capability and hence lower temperature [88]. Likewise, higher chlorophyll concentrations (i.e., larger absorption in the red and red-edge regions) might lead to higher light and water use efficiency and hence lower temperatures. The information contained in the DEM (i.e., the altitude and solar illumination conditions) also has a direct relationship with radiometric temperature, with sunlit areas having higher temperatures than shaded ones and lower altitude sufraces having higher temperatures than higher altitude surfaces.

3. Results

The overall performance of the tested models using sharpened temperatures from Decision Trees regressor (hereinafter $T_{rad,DT}$) is shown in Table 3. Scatter plots of modelled versus measured fluxes for all the validation sites are in the Supplement. We removed all the cases in which the S3 image was contaminated by clouds in the vicinity of the flux towers or in which the SLSTR view zenith angle was larger than 45 degrees. In addition, we filtered all cases where estimated $R_n \leq 50$ W m^{-2}, assuming that noisy outputs will be produced under low available energy, as well as those yielding unrealistic fluxes during daytime (≤ -500 W m^{-2} and ≥ 1000 W m^{-2}). After filtering the data, more than 400 cases were available overall for the following analyses. However, it is worth noting that ESVEP yielded significantly fewer valid retrievals. This issue might be due to the fact that ESVEP's end-member estimation equations were designed and parametrised for herbaceous crops [44] while in this study they were applied to varied land-covers. All models returned a similar performance regarding the estimation of R_n, with mean bias between -10 and -24 W m^{-2}, RMSE ranging between 49 and 59 W m^{-2} and r above 0.91. This similar behaviour is explained by the fact that all models share the same approach and same inputs in modelling net shortwave radiation, which is the component with larger magnitude of R_n. Likewise, G showed similar behaviour as well, but in this case G_{METRIC} is computed differently as it is a function of surface R_n [31,32] as opposed to TSEB and ESVEP where, as two-source models, G is computed from $R_{n,S}$ [36,44].

The main differences in model performance are therefore in the estimation of turbulent fluxes (i.e., sensible and latent heat fluxes), and TSEB (TSEB-PT and disTSEB) usually produced most accurate estimates in terms of RMSE (≈ 80 W m^{-2}, 45% relative error, in H; and ≈ 90 W m^{-2}, 45% relative error, in λE) and higher correlation between observed and predicted values (≈ 0.67 for H and ≈ 0.76 for λE). disTSEB performs slightly better than TSEB-PT but the difference is not significant. For METRIC and ESVEP, the RMSE values are in all cases higher than 120 W m^{-2} (going as high as 220 W m^{-2} in case of H modelled with ESVEP) and with lower correlation (≤ 0.47).

The choice of closing the energy balance gap in field measurements by assigning it to λE has influence on the above results. Therefore, in Table 4 we also present the accuracy statistics of the turbulent fluxes when Bowen ratio is preserved during the energy gap closure procedure. The overall ranking of the models is preserved with the TSEB models still obtaining the lowest RMSE and highest correlation coefficients. However, the differences between the models (particularly in case of RMSE) are not as large as in Table 3. In particular the RMSE of the TSEB models increases significantly while there is a decrease in r, while the influence of closure method on the other two models is much weaker with the RMSE of ESVEP even decreasing slightly. In subsequent analysis we always assign the residual energy to λE.

Table 3. Error metrics for METRIC, TSEB-PT, disTSEB (TSEB-PT with flux disaggregation) and ESVEP modelled fluxes using Decision Tree sharpened temperatures and closing the energy balance gap in field measurements by assigning residual energy to latent heat flux. N, number of valid cases; $\overline{\text{Obs.}}$; mean of observed values (W m^{-2}); bias, mean difference between predicted and observed (W m^{-2}); MAE, Mean Absolute Error (W m^{-2}), RMSE, Root Mean Square Error (W m^{-2}); rRMSE, Relative RMSE (–); r, Pearson correlation coefficient (–).

Variable	Model	N	$\overline{\text{Obs.}}$	Bias	MAE	RMSE	rRMSE	r
H	METRIC	450	177	49	103	156	0.885	0.238
	TSEB-PT	467	178	−47	65	81	0.454	0.670
	disTSEB	452	178	−38	62	76	0.429	0.671
	ESVEP	386	166	81	140	220	1.323	0.380
λE	METRIC	417	201	−29	100	128	0.637	0.472
	TSEB-PT	459	194	22	72	89	0.457	0.756
	disTSEB	442	196	24	72	88	0.451	0.769
	ESVEP	326	221	−51	111	140	0.635	0.420
Rn	METRIC	505	446	−10	39	51	0.113	0.920
	TSEB-PT	505	446	−14	44	56	0.125	0.908
	disTSEB	480	449	−10	40	49	0.110	0.927
	ESVEP	496	446	−24	46	59	0.132	0.907
G	METRIC	498	76	−1	40	50	0.668	0.497
	TSEB-PT	498	76	14	44	54	0.718	0.452
	disTSEB	473	77	3	41	50	0.657	0.505
	ESVEP	491	76	22	47	60	0.790	0.410

Table 4. Error metrics for METRIC, TSEB-PT, disTSEB (TSEB-PT with flux disaggregation) and ESVEP modelled fluxes using Decision Tree sharpened temperatures and closing the energy balance gap in field measurements by preserving the Bowen ratio. N, number of valid cases; $\overline{\text{Obs.}}$; mean of observed values (W m^{-2}); bias, mean difference between predicted and observed (W m^{-2}); MAE, Mean Absolute Error (W m^{-2}), RMSE, Root Mean Square Error (W m^{-2}); rRMSE, Relative RMSE (–); r, Pearson correlation coefficient (–).

Variable	Model	N	$\overline{\text{Obs.}}$	Bias	MAE	RMSE	rRMSE	r
H	METRIC	450	231	−6	115	164	0.709	0.151
	TSEB-PT	467	235	−103	118	145	0.618	0.417
	disTSEB	452	233	−93	111	138	0.592	0.422
	ESVEP	386	225	22	157	219	0.972	0.254
λE	METRIC	417	141	31	92	125	0.885	0.477
	TSEB-PT	459	134	82	94	124	0.919	0.698
	disTSEB	442	137	83	95	125	0.916	0.699
	ESVEP	326	158	12	102	134	0.847	0.388

In order to evaluate the model sensitivity and uncertainty to different vegetation types, we have split the results of Table 3 into four main vegetation types, depending on differences in aerodynamic roughness, horizontal homogeneity and/or seasonal dynamics/senescence (i.e., croplands, grasslands, savannas and forests, Table 5). Similar to the overall results, the TSEB models output most accurate turbulent fluxes across all four vegetation types. They obtain the best results for H in grassland (RMSE $\approx$ 70 W m^{-2}, r $\approx$ 0.8) and for λE in cropland (RMSE $\approx$ 80 W m^{-2}, r $\approx$ 0.75). In grassland and cropland TSEB-PT and disTSEB produce very similar fluxes while in savanna disTSEB improves the accuracy of modelled H and λE by up to 10 W m^{-2}. METRIC has its best overall performance in savanna (RMSE of 132 W m^{-2} and r of 0.43 for H; RMSE of 99 W m^{-2} and r of 0.61 for λE) followed by cropland while ESVEP produces inaccurate H in all vegetation types (rRMSE > 1) and its best overall λE in grassland. It should also be noted that RMSE of R_n is for all models double in savanna ($\approx$65 W m^{-2}) than in the other land cover types. This is due to vegetation being most sparse at those sites meaning that uncertainties in estimation of albedo and emissivity of soil have the biggest influence on shortwave and longwave net radiation respectively. Finally, very few valid cases are available to evaluate the forest sites and hence the results are not very conclusive, with the TSEB models again outperforming the METRIC and ESVEP.

Table 5. Error dependence on land cover for METRIC, TSEB-PT, disTSEB (TSEB-PT with flux disaggregation) and ESVEP modelled fluxes using Decision Trees sharpened temperatures. N, number of valid cases; $\overline{\text{Obs.}}$; mean of observed values (W m^{-2}); bias, mean difference between predicted and observed (W m^{-2}); MAE, Mean Absolute Error (W m^{-2}), RMSE, Root Mean Square Error (W m^{-2}); rRMSE, Relative RMSE (–); r, Pearson correlation coefficient (–).

Variable	Land Cover	Model	N	$\overline{\text{Obs.}}$	Bias	MAE	RMSE	rRMSE	r
	cropland	METRIC	187	158	61	92	115	0.726	0.285
		TSEB-PT	189	158	−57	71	86	0.546	0.503
		disTSEB	177	157	−48	68	83	0.532	0.440
		ESVEP	166	147	78	125	190	1.292	0.309
	grassland	METRIC	103	195	18	132	204	1.050	0.164
		TSEB-PT	110	197	−24	58	73	0.369	0.788
		disTSEB	108	196	−26	58	71	0.365	0.792
H		ESVEP	92	185	50	120	198	1.070	0.437
	savanna	METRIC	148	189	37	83	132	0.700	0.425
		TSEB-PT	151	187	−50	62	78	0.416	0.671
		disTSEB	150	187	−35	55	68	0.364	0.701
		ESVEP	114	176	107	177	275	1.564	0.364
	forest	METRIC	12	160	260	278	370	2.306	−0.030
		TSEB-PT	17	202	−45	74	98	0.485	0.661
		disTSEB	17	202	−35	76	94	0.468	0.657
		ESVEP	14	193	110	156	184	0.953	0.692
	cropland	METRIC	179	256	−80	105	135	0.527	0.550
		TSEB-PT	183	254	11	67	82	0.322	0.748
		disTSEB	169	261	13	69	83	0.320	0.738
		ESVEP	145	269	−83	111	142	0.527	0.525
	grassland	METRIC	91	136	88	122	147	1.079	0.491
		TSEB-PT	108	128	55	79	92	0.718	0.786
		disTSEB	106	127	64	82	96	0.756	0.794
λE		ESVEP	79	134	31	93	110	0.825	0.446
	savanna	METRIC	140	165	−38	78	99	0.602	0.610
		TSEB-PT	151	160	8	71	88	0.549	0.642
		disTSEB	150	161	5	63	79	0.490	0.723
		ESVEP	89	215	−67	117	151	0.702	0.131
	forest	METRIC	7	337	−64	121	173	0.512	0.763
		TSEB-PT	17	282	55	109	138	0.488	0.909
		disTSEB	17	282	51	118	143	0.509	0.899
		ESVEP	13	256	−93	177	198	0.774	0.770

Table 5. *Cont.*

Variable	Land Cover	Model	N	$\overline{\text{Obs.}}$	Bias	MAE	RMSE	rRMSE	r
Rn	cropland	METRIC	222	441	−4	27	38	0.086	0.955
		TSEB-PT	222	441	−10	32	42	0.096	0.950
		disTSEB	200	450	−6	29	37	0.081	0.962
		ESVEP	218	444	−22	37	49	0.111	0.942
	grassland	METRIC	113	475	−15	36	45	0.094	0.929
		TSEB-PT	113	475	−9	34	43	0.090	0.936
		disTSEB	111	472	−7	33	42	0.089	0.937
		ESVEP	110	474	−12	34	42	0.089	0.939
	savanna	METRIC	153	425	−15	62	69	0.163	0.815
		TSEB-PT	153	425	−23	72	79	0.187	0.763
		disTSEB	152	424	−17	60	68	0.159	0.830
		ESVEP	153	425	−36	69	80	0.189	0.790
	forest	METRIC	17	511	−5	20	27	0.053	0.994
		TSEB-PT	17	511	−1	18	24	0.047	0.995
		disTSEB	17	511	−1	18	24	0.047	0.995
		ESVEP	15	486	−6	17	24	0.049	0.996
G	cropland	METRIC	222	43	28	41	50	1.166	0.303
		TSEB-PT	222	43	46	56	64	1.509	0.253
		disTSEB	200	43	34	48	56	1.297	0.228
		ESVEP	218	42	55	62	74	1.754	0.255
	grassland	METRIC	113	144	−66	68	75	0.518	0.799
		TSEB-PT	113	144	−43	49	60	0.419	0.677
		disTSEB	111	143	−46	50	61	0.429	0.710
		ESVEP	110	147	−42	49	59	0.403	0.647
	savanna	METRIC	153	78	2	19	23	0.293	0.792
		TSEB-PT	153	78	12	25	30	0.381	0.622
		disTSEB	152	78	−3	27	32	0.418	0.472
		ESVEP	153	78	23	28	34	0.435	0.724
	forest	METRIC	10	−2	25	25	32	21.275	0.959
		TSEB-PT	10	−2	9	9	11	6.989	0.905
		disTSEB	10	−2	8	8	9	6.083	0.770
		ESVEP	10	−2	13	13	14	9.383	0.981

The agriculture class was further split into herbaceous and woody types, with results shown in Table 6. The former sub-class represents crops such as corn, soybean or wheat while the latter represents olive groves and vineyards. TSEB models produce the most consistent results for both types of crops, although somewhat surprisingly the RMSE of λE in woody crops (76–79 W m^{-2}) is significantly lower than in herbaceous crops (91–93 W m^{-2}), while opposite is the case for RSME of H (69–71 W m^{-2} in herbaceous crops and 91–94 W m^{-2}). rRMSE of λE in both agricultural sub-classes was 0.32 which is of the same magnitude as energy closure gap at the validation sites (e.g., the mean value at CH was 0.34 at the times at which fluxes were modelled). METRIC is very clearly performing better in woody crops, while ESVEP obtains better results for H in herbaceous crops and better results for λE in woody crops. It is also worth noting that R_n and G showed larger relative errors in woody crops than in herbaceous crops, since woody canopies are more complex and therefore more difficult to capture by the models and/or parametrizations used [89,90].

Table 6. Crop type dependent errors for METRIC, TSEB and ESVEP modelled fluxes using Decision Tree sharpened temperatures. N, number of valid cases; $\overline{Obs.}$; mean of observed values (W m^{-2}); bias, mean difference between predicted and observed (W m^{-2}); MAE, Mean Absolute Error (W m^{-2}), RMSE, Root Mean Square Error (W m^{-2}); rRMSE, Relative RMSE (–); r, Pearson correlation coefficient (–).

Variable	Land Cover	Model	N	$\overline{Obs.}$	Bias	MAE	RMSE	rRMSE	r
H	herbaceous	METRIC	66	135	115	120	144	1.068	0.452
		TSEB-PT	67	134	−39	55	71	0.528	0.509
		disTSEB	67	134	−31	55	69	0.517	0.470
		ESVEP	62	133	46	78	107	0.805	0.440
	woody	METRIC	121	171	31	77	96	0.558	0.320
		TSEB-PT	122	172	−67	80	94	0.547	0.515
		disTSEB	110	171	−58	75	91	0.533	0.424
		ESVEP	104	155	98	152	225	1.452	0.258
λE	herbaceous	METRIC	58	289	−151	157	189	0.656	0.215
		TSEB-PT	59	288	−42	78	93	0.324	0.662
		disTSEB	59	288	−33	77	91	0.316	0.676
		ESVEP	55	285	−118	126	149	0.523	0.605
	woody	METRIC	121	241	−46	80	99	0.411	0.738
		TSEB-PT	124	238	36	61	76	0.318	0.840
		disTSEB	110	247	38	65	79	0.321	0.823
		ESVEP	90	259	−62	101	137	0.529	0.522
Rn	herbaceous	METRIC	68	461	−23	30	38	0.082	0.976
		TSEB-PT	68	461	−36	40	47	0.102	0.975
		disTSEB	68	461	−29	35	42	0.091	0.976
		ESVEP	67	460	−36	39	47	0.102	0.975
	woody	METRIC	154	433	5	26	38	0.087	0.952
		TSEB-PT	154	433	1	29	40	0.092	0.952
		disTSEB	132	444	6	26	33	0.075	0.968
		ESVEP	151	437	−16	36	50	0.115	0.929
G	herbaceous	METRIC	68	48	7	42	51	1.062	0.447
		TSEB-PT	68	48	34	55	64	1.351	0.296
		disTSEB	68	48	22	49	60	1.261	0.234
		ESVEP	67	46	34	51	62	1.328	0.395
	woody	METRIC	154	40	38	41	49	1.221	0.507
		TSEB-PT	154	40	51	57	64	1.591	0.472
		disTSEB	132	40	41	48	53	1.314	0.490
		ESVEP	151	41	64	67	79	1.955	0.458

Finally, Table 7 lists the model performance depending on whether sites are under Mediterranean and semi-arid climate (i.e., water limited sites), or sites under temperate climate (i.e., energy limited sites). First of all it is worth noting that due to cloud coverage conditions, more valid cases are obtained over semi-arid conditions than in temperate areas. TSEB models showed similar range of errors in both climatic conditions, with RMSE in λE at around 85 W and 99 W m^{-2} for semi-arid and temperate conditions, and correspondingly around 80 and 70 W m^{-2} for H. ESVEP and METRIC yielded more varying results between climates, with METRIC producing more accurate estimates of both H and λE in semi-arid conditions and ESVEP showing better performance for H in temperate climates and better performance for λE in semi-arid climates.

Table 7. Climate dependence of errors for METRIC, TSEB and ESVEP modelled fluxes using Decision Trees sharpened temperatures. N, number of valid cases; $\overline{\text{Obs.}}$; mean of observed values (W m^{-2}); bias, mean difference between predicted and observed (W m^{-2}); MAE, Mean Absolute Error (W m^{-2}), RMSE, Root Mean Square Error (W m^{-2}); rRMSE, Relative RMSE (–); r, Pearson correlation coefficient (–).

Variable	Climate	Model	N	$\overline{\text{Obs.}}$	Bias	MAE	RMSE	rRMSE	r
H	semi arid	METRIC	354	190	28	95	149	0.780	0.270
		TSEB-PT	365	191	−51	69	84	0.438	0.676
		disTSEB	350	191	−41	64	78	0.411	0.670
		ESVEP	296	177	87	155	242	1.370	0.337
	temperate	METRIC	96	126	126	133	182	1.441	0.356
		TSEB-PT	102	134	−32	53	71	0.530	0.635
		disTSEB	102	134	−26	52	69	0.513	0.637
		ESVEP	90	132	62	92	122	0.927	0.586
λE	semi arid	METRIC	335	178	-6	90	114	0.641	0.534
		TSEB-PT	366	170	32	71	86	0.504	0.763
		disTSEB	349	171	33	70	85	0.498	0.776
		ESVEP	245	202	−30	103	134	0.665	0.391
	temperate	METRIC	82	295	−123	140	174	0.589	0.419
		TSEB-PT	93	289	−19	79	99	0.343	0.739
		disTSEB	93	289	−10	79	98	0.341	0.750
		ESVEP	81	278	−115	135	157	0.564	0.646
Rn	semi arid	METRIC	401	442	−9	42	53	0.121	0.893
		TSEB-PT	401	442	−11	47	59	0.133	0.877
		disTSEB	376	446	−8	42	52	0.116	0.902
		ESVEP	398	444	−23	48	62	0.140	0.876
	temperate	METRIC	104	462	−13	29	38	0.082	0.976
		TSEB-PT	104	462	−23	35	43	0.093	0.975
		disTSEB	104	462	−17	31	40	0.086	0.976
		ESVEP	98	453	−26	35	43	0.094	0.977
G	semi arid	METRIC	401	84	−3	41	52	0.616	0.431
		TSEB-PT	401	84	11	44	54	0.643	0.398
		disTSEB	376	86	0	41	50	0.582	0.467
		ESVEP	398	84	21	49	61	0.727	0.316
	temperate	METRIC	97	40	7	36	45	1.104	0.480
		TSEB-PT	97	40	26	43	55	1.362	0.399
		disTSEB	97	40	16	38	51	1.272	0.358
		ESVEP	93	39	29	42	53	1.355	0.477

4. Discussion

4.1. ET Model Intercomparison

Overall results listed in Table 3 show that TSEB models produced more robust estimates of both sensible and latent heat fluxes, with lower errors around 80 to 90 W m^{-2} and larger correlation coefficient, while at the same time returning more valid cases than the other two models, METRIC and ESVEP. Those errors are within the expected and reported errors in literature, e.g., Kalma et al. [2] showed errors in λE ranging between 24 and 105 W m^{-2} for a wide range of models, Chirouze et al. [58] reported errors for TSEB > 100 W m^{-2} in a semi-arid area of Mexico, and 50 W m^{-2} errors are reported in Tang et al. [35]. Choi et al. [91] found TSEB-PT and METRIC produced similar errors of 54 W m^{-2} in a watershed in Iowa, US. However it is worth noting that most of the reported errors in these studies [35,58,91,92] used actual surface temperature at high spatial resolution (e.g., Landsat or ASTER), whereas in this study we used low resolution temperature sharpened to high spatial resolution, which provides an additional input uncertainty to the models. For that reason, Section 4.2 is dedicated to this issue in depth.

TSEB-PT was developed trying to solve some of the issues in sparse vegetation and semi-arid conditions previously raised by less complex models [36], and therefore it adapts better to a wider range of climatic and vegetation conditions [1] as it was shown in Tables 5 and 7. METRIC, on the other hand, was primarily designed for standard crops and requires concomitant presence of stressed and well watered-full vegetation conditions within the scene itself. This more often happens in semi-arid climate where irrigated crops and rainfed crops and natural vegetation are present. Those cases in which, either due to the increased presence of clouds (i.e., fewer available pixels in the scene) or in regions where these hot and cold pixels cannot be simultaneously found, METRIC would produce more uncertain retrieval, as already pointed by Choi et al. [91] and Tang et al. [35] (in humid or sub-humid areas), or even would not produce any valid data. Similarly, ESVEP was designed and tested in an agricultural area located in a subhumid and monsoon climate [44] and therefore certain assumptions and parameterizations taken in that model might not transfer well to other vegetation or climatic conditions.

Despite of TSEB being the model with the largest required amount of input data, this study proposed several new approaches to retrieve some of those inputs operationally, with special focus on exploiting the spectral capabilities of Sentinel-2, in particular the bands in the red-edge region that is sensitive to leaf pigments. A simple empirical approach relating leaf bihemispherical reflectance and transmittance with the leaf biochemical properties resulted in accurate estimates of net radiation. More importantly, due to the larger uncertainty of TSEB models over senescent vegetation, we derived a method to obtain both total LAI and its green fraction based on Fisher et al. [75] FAPAR/FIPAR relationship. Nevertheless, more research is needed to systematically derive other vegetation properties such as canopy height/aerodynamic roughness or vegetation clumping.

Finally, it is worth pointing out that even *in situ* EC measurements are prone to uncertainty as is confirmed for instance by the usual energy imbalance in those systems. Particularly we found a larger disagreement between observed and predicted net radiation in Dahra (see Figures in the Supplement). We hypothesise that this could be due to two possible reasons. Firstly, our modelled irradiance, with depends on TCWV and aerosol optical thickness, could be more noisy at Dahra than the other sites, due to unaccounted dust aerosols in that site placed over the Sahel. The second issue might be the actual R_n measurements, as in this site only a NR-lite (Kipp & Zonen, Netherlands) is available to measure global R_n that might be less accurate than the radiometers at the other sites, which are measuring the four components of radiation. In addition, Harvard Forest site lacks *in situ* G measurements, which effects the energy balance closure correction. This issue together with the fact that very few cases are available in forests (Table 5), leads us to avoid strong conclusions regarding the performance of the models in forested areas.

4.2. Sharpening and Disaggregation

As was previously mentioned, thermal sharpening relates empirically or semi-empirically coarse resolution surface temperature with fine resolution multispectral and other ancillary data. This technique could be a sound alternative to the lack high resolution thermal imagery for operational activities. However, previous studies in thermal sharpening have reported some uncertainties when compared to actual T_{rad} temperatures, with errors ranging up to 3.5 K [7–9,11,12,14]. Therefore, for some applications requiring ET estimates at higher accuracy (i.e., precision agriculture), sharpening might not be considered as a suitable substitute of T_{rad} but complementary to it, such as in the fusion approach by Knipper et al. [93].

In order to reduce flux retrieval errors with sharpened T_{rad} inputs, we also tested a flux disaggregation method [10,23]. Our results listed in Tables 3–7 show that disTSEB model, i.e., coarse S3 TSEB-PT fluxes disaggregated with fluxes derived with TSEB-PT and fine resolution sharpened T_{rad}, yielded only modest improvement (5 W m^{-2} reduction in RMSE in case of H and only 1 W m^{-2} in case of λE) to the TSEB-PT model, i.e., running TSEB directly on the sharpened T_{rad} imagery. The one exception was at the savanna sites (see Table 5) where using disaggregation reduced the errors in H by around 12% and errors in λE by around 10%. However, previous studies have shown the robustness of this approach to overcome limitations of the likely less reliable fine resolution T_{rad} images [4,93,94]. Furthermore, coarse input data must be produced beforehand for thermal sharpening and hence it is readily available for running the models at coarse resolution, which indeed is computationally inexpensive given the much lower number of pixels within a scene. Therefore, flux disaggregation would still be recommended when running TSEB-PT with sharpened temperatures.

In addition, the sharpening of a coarse resolution T_{rad} image using fine resolution images acquired on different days, with a maximum of 10 days offset, might lead to additional uncertainties. This is caused by the fact that some changes in either land cover properties, (e.g., vegetation growth, harvests, fires) or moisture conditions (e.g., rainfall or irrigation) might happen between the Sentinel 2 and 3 acquisitions. Figure 2 shows that at a general level (all validation sites taken together) this does not appear to be a significant issue as the error does not increase as the day offset between thermal and shortwave acquisitions gets larger. Particularly relevant in this analysis is H since it is the energy component that is directly related to T_{rad}, and hence more prone to errors in sharpening. However, more studies should be conducted to look at the effect of the day offset in particular situations, e.g., in crops during senescence or with localized irrigation patterns. It might be also worth to investigate using high-resolution radar data (e.g., from Sentinel-1), which is sensitive to soil moisture, in the thermal data sharpening approach [95]. Furthermore, the Landsat family of satellites could also be utilised during the sharpening since they acquire thermal data at around 100 m spatial resolution although at 8 days (two satellites) to 16 days (single satellite) temporal resolution. Using observations from those satellites would both increase the temporal density of the high-resolution data but also capture physical processes and properties which are not reflected in the shortwave data, such as near soil surface soil moisture and soil evaporative efficiency.

Finally, some studies have reported larger errors than in this study, but they were using coarser resolution imagery [43]. This is probably due to the scale mismatch between the coarse pixel estimate and the footprint of the EC towers' measurements. We evaluated this by comparing fluxes modelled at Sentinel-2 (i.e., with sharpened T_{rad}) and Sentinel-3 spatial resolutions against measurements form towers. This was done for all the sites put together and also for the validation sites split into two categories: those in which the tower is located in a landscape feature too small to have significant effect on the original resolution Sentinel-3 T_{rad} (category "small" containing CH, KL, GR and SE sites), and those where the opposite is true (category "large" containing SL, DA, HF, HTM, MT TA and KG sites). The results (Table 8) indicate that using sharpened T_{rad} is most important when modelling H in the "small" category. However, the correlation of high-resolution fluxes against tower measurements is in almost all the cases higher than that of low-resolution fluxes and rRMSE is lower or the same in case of turbulent fluxes. Therefore, even though sharpened T_{rad} might be more prone to errors than actual high-resolution T_{rad}, it is still a good option for downscaling fluxes for model validation [4], addressing therefore the vegetation cover variability within coarse resolution pixels. Nevertheless, there is still an open question on how feasible thermal sharpening is for early detection of water stress at small scales, compared to using high resolution thermal imagery. This issue is especially relevant for precision irrigation tasks and therefore future studies should address this topic.

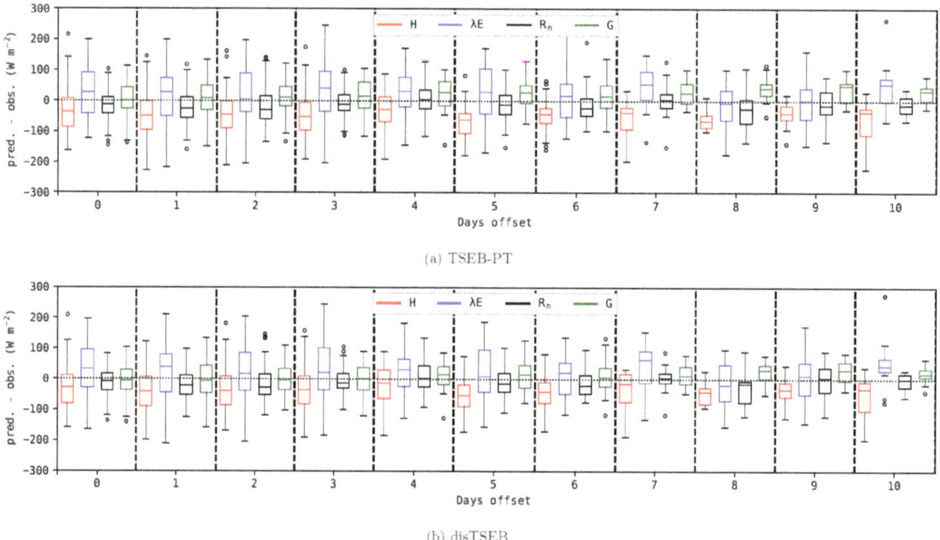

(a) TSEB-PT

(b) disTSEB

Figure 2. Error (modelled–measured) distribution of fluxes modelled with TSEB-PT and disTSEB models using sharpened T_{rad} depending on offset days between a Sentinel-3 T_{rad} image and the fine-scale Sentinel-2 multispectral image. Error computed for all sites together.

Table 8. Landscape feature size dependence of errors for low and high resolution TSEB-PT modelled fluxes using Decision Trees sharpened temperatures. N, number of valid cases; $\overline{Obs.}$; mean of observed values (W m^{-2}); bias, mean difference between predicted and observed (W m^{-2}); MAE, Mean Absolute Error (W m^{-2}), RMSE, Root Mean Square Error (W m^{-2}); rRMSE, Relative RMSE (–); r, Pearson correlation coefficient (–).

Feature Size	Variable	Resolution	N	$\overline{Obs.}$	Bias	MAE	RMSE	rRMSE	r
all	H	Sentinel-2	467	178	−47	65	81	0.454	0.670
		Sentinel-3	456	176	−38	73	89	0.509	0.548
	λE	Sentinel-2	459	194	22	72	89	0.457	0.756
		Sentinel-3	446	196	13	71	92	0.467	0.726
	Rn	Sentinel-2	505	446	−14	44	56	0.125	0.908
		Sentinel-3	481	446	−12	42	56	0.126	0.902
	G	Sentinel-2	498	76	14	44	54	0.718	0.452
		Sentinel-3	474	74	13	43	51	0.690	0.491
small	H	Sentinel-2	85	121	−29	48	65	0.534	0.514
		Sentinel-3	88	120	4	65	88	0.736	0.186
	λE	Sentinel-2	76	290	−35	73	88	0.303	0.706
		Sentinel-3	79	290	−26	70	88	0.303	0.694
	Rn	Sentinel-2	87	452	−27	38	46	0.101	0.964
		Sentinel-3	90	451	−11	30	37	0.083	0.970
	G	Sentinel-2	87	45	28	47	58	1.283	0.301
		Sentinel-3	90	46	8	45	57	1.250	0.016

Table 8. *Cont.*

Feature Size	Variable	Resolution	N	Obs.	Bias	MAE	RMSE	rRMSE	r
large	H	Sentinel-2	382	191	−51	69	84	0.441	0.673
		Sentinel-3	368	189	−49	75	90	0.475	0.620
	λE	Sentinel-2	383	175	33	72	89	0.507	0.770
		Sentinel-3	367	176	21	72	93	0.525	0.719
	Rn	Sentinel-2	418	445	−11	45	58	0.130	0.897
		Sentinel-3	391	445	−13	45	60	0.135	0.884
	G	Sentinel-2	411	82	11	43	53	0.652	0.444
		Sentinel-3	384	81	15	42	50	0.614	0.515

4.3. Effects of Ancillary Inputs

Ancillary data is required to characterise the canopy structure, since it affects both the radiation transmission through the canopy [53], and hence albedo and radiation partitioning, as well as the surface aerodynamic properties [79]. In this study we have used a static land cover map at global scale to assign some standard values to each land cover type (Table 2). However, the large difference in spatial resolution between the S2 data and CCI map can lead to visible artefacts in the output fluxes when modelled at 20 m resolution, especially on the edges of two classes with different vegetation properties (e.g., croplands and forests). However, those spatial artefacts seem not to have any influence on the validation results. Nevertheless, some discrepancies were found between the land cover type flagged by the map and the actual type at the validation sites. In Majadas de Tiétar, CCI-LC flagged the site as cropland (CCI-LC = 11), thus $h_{c,MAX} = 0.5$ m, $f_c = 1$ and $l_w = 0.02$ m, but actually this site is a savanna with 8 m tress at 20% coverage (CCI-LC = 30). In addition, the prescribed values that were assigned in Table 2 are very general, as they are trying to fit a global-based land cover legend. Therefore they can significantly deviate from the site's actual values. Indeed, all croplands were assumed to be not clumped ($f_c = 1$) although row crops, like the vineyard in Sierra Loma, or orchards like the olive grove in Taous have very different canopy structure compared to a standard crop. Therefore, a significant improvement could be expected if a more area-specific surface characteristics parametrization was used, either using some ancillary remote sensing like SAR imagery or LiDAR or a regional/local oriented land cover classification.

To conclude, atmospheric forcing from numerical weather prediction models might add some uncertainty to the ET model compared to using local meteorological data, specially for precision agriculture where access to local agrometeorological stations is possible. In this study we relied on ERA5 reanalysis data and despite large discrepancy between spatial resolution of ERA5 (tens of kilometres) and point scale measurements from the towers there is a strong agreement for the most important meteorological parameters (Figure 3). Instantaneous shortwave irradiance, which was computed at the Sentinel 3 overpass time using ECMWF AOT and TCWV dataset, showed no systematic bias but a RMSE of 29 W m^{-2}. This in turns directly effects on the accuracy of R_n (Tables 3–7), and indirectly that of λE since it is estimated as a residual of the energy balance. Therefore, errors in λE could be significantly reduced if more accurate inputs of irradiance were used, especially over temperate areas (i.e., radiation limited) which are more sensitive to uncertainty in available energy. On the other hand, the errors in both air temperature (RMSE = 1.8 K) and windspeed (RMSE = 1.3 m s^{-1}) affect mainly on the retrievals of sensible heat flux. This issue become more relevant in the estimation of λE over semi-arid (i.e., water limited) areas. For near-real-time applications it is necessary to use forecast or analysis data, instead of the ensemble mean reanalysis data, and those issues could become more evident.

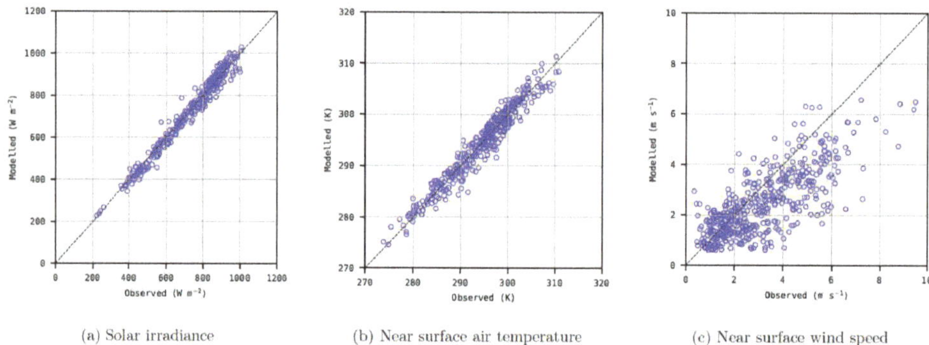

(a) Solar irradiance   (b) Near surface air temperature   (c) Near surface wind speed

Figure 3. Scatterplot between the input ERA5 ensemble meant reanalysis data and in situ measurements for the main atmospheric forcings. Data from all validation sites is shown on the same plot.

5. Conclusions

The multispectral shortwave images acquired by the Sentinel-2 satellites at 10–20 m spatial resolution are highly suitable for characterizing vegetation in order to derive inputs required for evapotranspiration models. At the same time, thermal data acquired daily by Sentinel-3 satellites is also suitable as input to the ET models. However, its low spatial resolution (1 km) needs to be increased if the models are to be run at the scale of predominant landscape features, which is usually on the order of tens of meters. This study evaluated three thermal-based remote sensing ET models (METRIC, ESVEP and TSEB-PT) and a thermal sharpening method (ensembles of modified Decision Trees) in order to derive methodology for operational estimates of water and energy fluxes using Sentinel data and applicable for the whole globe. Further evolution of the thermal sharpening methodology by using other data sources with high spatial resolution and variable temporal resolutions, e.g., Sentinel-1 radar [95] or Landsat thermal observations [96], is planned.

TSEB-PT produced overall the most accurate estimates in terms of sensible heat and latent heat (i.e., evapotranspiration) fluxes, being robust in different land covers and climates. Additional disaggregation step further improved TSEB-PT output accuracy in savanna ecosystems. Without any site-specific tuning and relying only on global datasets the methodology achieved RMSE of 80–90 W m^{-2} for modelled instantaneous H and λE across eleven validation sites located in different land cover classes and climatic conditions. In an agricultural setting the modelled fluxes were more accurate with rRMSE of λE of around 0.3 which is of the same magnitude as uncertainty of the measured turbulent fluxes from the validation dataset. Until a new generation of thermal satellites are launched [97], the proposed methodology will be useful solution for overcoming the lack of thermal data with high spatio-temporal resolution required for operational ET modelling at field scale.

Supplementary Materials: The following are available at http://www.mdpi.com/2072-4292/12/9/1433/s1, Figure S1: Model scatterplot of predicted vs. EC observed using sharpened Trad with Decision Trees for Sierra Loma, Figure S2: Model scatterplot of predicted vs. EC observed using sharpened Trad with Decision Trees for Choptank, Figure S3: Model scatterplot of predicted vs. EC observed using sharpened Trad with Decision Trees for Dahra, Figure S4: Model scatterplot of predicted vs. EC observed using sharpened Trad with Decision Trees for Grillenburg, Figure S5: Model scatterplot of predicted vs. EC observed using sharpened Trad with Decision Trees for Harvard, Figure S6: Model scatterplot of predicted vs. EC observed using sharpened Trad with Decision Trees for Hyltemossa, Figure S7: Model scatterplot of predicted vs. EC observed using sharpened Trad with Decision Trees for Klingenberg, Figure S8: Model scatterplot of predicted vs. EC observed using sharpened Trad with Decision Trees for Majadas, Figure S9: Model scatterplot of predicted vs. EC observed using sharpened Trad with Decision Trees for Selhausen, Figure S10: Model scatterplot of predicted vs. EC observed using sharpened Trad with Decision Trees for Taous, Figure S11: Model scatterplot of predicted vs. EC observed using sharpened Trad with Decision Trees for Kendall Grassland.

Author Contributions: Conceptualization, R.G. and H.N.; Data curation, R.G., H.N., I.S. and G.K.; Formal analysis, R.G., H.N., I.S. and G.K.; Investigation, R.G. and H.N.; Methodology, R.G. and H.N.; Software, R.G. and

H.N.; Writing—original draft, R.G. and H.N.; Writing—review & editing, I.S. All authors have read and agreed to the published version of the manuscript.

Funding: This research was funded by the European Space Agency contract number 4000121772/17/I-NB.

Acknowledgments: We would like to thank all the PIs (listed in Table 1) of the flux towers used in this study for making their data available for model validation. We gratefully thank the TERestrial ENvironmental Observations (TERENO) for providing data at Selhausen site. Technical support from Pascal Fanise (CESBIO) and Zoubeir Majoub (Institut de l'Olivier) for the Taous tower dataset is gratefully acknowledged. Data collection at the Kendall Grassland site (AmeriFlux site: US-Wkg) was supported by USDA-ARS and by the U.S. Department of Energy. This study was conducted as part of the Sen-ET project (http://esa-sen4et.org/).

Conflicts of Interest: The authors declare no conflict of interest.

References

1. Kustas, W.P.; Anderson, M.C. Advances in thermal infrared remote sensing for land surface modeling. *Agric. For. Meteorol.* **2009**, *149*, 2071–2081. [CrossRef]
2. Kalma, J.; McVicar, T.; McCabe, M. Estimating Land Surface Evaporation: A Review of Methods Using Remotely Sensed Surface Temperature Data. *Surv. Geophys.* **2008**, *29*, 421–469. [CrossRef]
3. Karimi, P.; Bastiaanssen, W.G.M. Spatial evapotranspiration, rainfall and land use data in water accounting—Part 1: Review of the accuracy of the remote sensing data. *Hydrol. Earth Syst. Sci.* **2015**, *19*, 507–532. [CrossRef]
4. Guzinski, R.; Nieto, H. Evaluating the feasibility of using Sentinel-2 and Sentinel-3 satellites for high-resolution evapotranspiration estimations. *Remote Sens. Environ.* **2019**, *221*, 157–172. [CrossRef]
5. Donlon, C.; Berruti, B.; Buongiorno, A.; Ferreira, M.H.; Féménias, P.; Frerick, J.; Goryl, P.; Klein, U.; Laur, H.; Mavrocordatos, C.; et al. The Global Monitoring for Environment and Security (GMES) Sentinel-3 mission. *Remote Sens. Environ.* **2012**, *120*, 37–57. [CrossRef]
6. Drusch, M.; Del Bello, U.; Carlier, S.; Colin, O.; Fernandez, V.; Gascon, F.; Hoersch, B.; Isola, C.; Laberinti, P.; Martimort, P.; et al. Sentinel-2: ESA's Optical High-Resolution Mission for GMES Operational Services. *Remote Sens. Environ.* **2012**, *120*, 25–36. [CrossRef]
7. Merlin, O.; Duchemin, B.; Hagolle, O.; Jacob, F.; Coudert, B.; Chehbouni, G.; Dedieu, G.; Garatuza, J.; Kerr, Y. Disaggregation of MODIS surface temperature over an agricultural area using a time series of Formosat-2 images. *Remote Sens. Environ.* **2010**, *114*, 2500–2512. [CrossRef]
8. Gao, F.; Kustas, W.P.; Anderson, M.C. A Data Mining Approach for Sharpening Thermal Satellite Imagery over Land. *Remote Sens.* **2012**, *4*, 3287–3319. [CrossRef]
9. Bindhu, V.; Narasimhan, B.; Sudheer, K. Development and verification of a non-linear disaggregation method (NL-DisTrad) to downscale MODIS land surface temperature to the spatial scale of Landsat thermal data to estimate evapotranspiration. *Remote Sens. Environ.* **2013**, *135*, 118–129. [CrossRef]
10. Norman, J.; Anderson, M.; Kustas, W.; French, A.; Mecikalski, J.; Torn, R.; Diak, G.; Schmugge, T.; Tanner, B. Remote sensing of surface energy fluxes at 10^{-1} m pixel resolutions. *Water Resour. Res.* **2003**, *39*. [CrossRef]
11. Agam, N.; Kustas, W.P.; Anderson, M.C.; Li, F.; Neale, C.M. A vegetation index based technique for spatial sharpening of thermal imagery. *Remote Sens. Environ.* **2007**, *107*, 545 – 558. [CrossRef]
12. Zhou, J.; Liu, S.; Li, M.; Zhan, W.; Xu, Z.; Xu, T. Quantification of the Scale Effect in Downscaling Remotely Sensed Land Surface Temperature. *Remote Sens.* **2016**, *8*, 975. [CrossRef]
13. Bisquert, M.; Sánchez, J.M.; Caselles, V. Evaluation of Disaggregation Methods for Downscaling MODIS Land Surface Temperature to Landsat Spatial Resolution in Barrax Test Site. *IEEE J. Sel. Top. Appl. Earth Obs. Remote Sens.* **2016**, *9*, 1430–1438. [CrossRef]
14. Hutengs, C.; Vohland, M. Downscaling land surface temperatures at regional scales with random forest regression. *Remote Sens. Environ.* **2016**, *178*, 127–141. [CrossRef]
15. Yang, G.; Pu, R.; Huang, W.; Wang, J.; Zhao, C. A Novel Method to Estimate Subpixel Temperature by Fusing Solar-Reflective and Thermal-Infrared Remote-Sensing Data With an Artificial Neural Network. *IEEE Trans. Geosci. Remote Sens.* **2010**, *48*, 2170–2178. [CrossRef]
16. Mitraka, Z.; Chrysoulakis, N.; Doxani, G.; Del Frate, F.; Berger, M. Urban Surface Temperature Time Series Estimation at the Local Scale by Spatial-Spectral Unmixing of Satellite Observations. *Remote Sens.* **2015**, *7*, 4139–4156. [CrossRef]

17. Chen, X.; Li, W.; Chen, J.; Rao, Y.; Yamaguchi, Y. A Combination of TsHARP and Thin Plate Spline Interpolation for Spatial Sharpening of Thermal Imagery. *Remote Sens.* **2014**, *6*, 2845–2863. [CrossRef]
18. Peng, Z.Q.; Xin, X.; Jiao, J.J.; Zhou, T.; Liu, Q. Remote sensing algorithm for surface evapotranspiration considering landscape and statistical effects on mixed pixels. *Hydrol. Earth Syst. Sci.* **2016**, *20*, 4409–4438. [CrossRef]
19. Ke, Y.; Im, J.; Park, S.; Gong, H. Downscaling of MODIS One Kilometer Evapotranspiration Using Landsat-8 Data and Machine Learning Approaches. *Remote Sens.* **2016**, *8*, 215. [CrossRef]
20. Bisquert, M.; Sánchez, J.M.; López-Urrea, R.; Caselles, V. Estimating high resolution evapotranspiration from disaggregated thermal images. *Remote Sens. Environ.* **2016**, *187*, 423–433. [CrossRef]
21. Tan, S.; Wu, B.; Yan, N.; Zhu, W. An NDVI-Based Statistical ET Downscaling Method. *Water* **2017**, *9*, 995. [CrossRef]
22. Zhan, W.; Chen, Y.; Zhou, J.; Wang, J.; Liu, W.; Voogt, J.; Zhu, X.; Quan, J.; Li, J. Disaggregation of remotely sensed land surface temperature: Literature survey, taxonomy, issues, and caveats. *Remote Sens. Environ.* **2013**, *131*, 119–139. [CrossRef]
23. Guzinski, R.; Nieto, H.; Jensen, R.; Mendiguren, G. Remotely sensed land-surface energy fluxes at sub-field scale in heterogeneous agricultural landscape and coniferous plantation. *Biogeosciences* **2014**, *11*, 5021–5046. [CrossRef]
24. Semmens, K.A.; Anderson, M.C.; Kustas, W.P.; Gao, F.; Alfieri, J.G.; McKee, L.; Prueger, J.H.; Hain, C.R.; Cammalleri, C.; Yang, Y.; et al. Monitoring daily evapotranspiration over two California vineyards using Landsat 8 in a multi-sensor data fusion approach. *Remote Sens. Environ.* **2016**, *185*, 155–170. [CrossRef]
25. Brutsaert, W. *Hydrology. An Introduction*; Cambridge University Press: Cambridge, UK, 2005.
26. Brutsaert, W. Aspects of bulk atmospheric boundary layer similarity under free-convective conditions. *Rev. Geophys.* **1999**, *37*, 439–451. [CrossRef]
27. Lhomme, J.P.; Monteny, B.; Amadou, M. Estimating sensible heat flux from radiometric temperature over sparse millet. *Agric. For. Meteorol.* **1994**, *68*, 77–91. [CrossRef]
28. Kustas, W.P.; Nieto, H.; Morillas, L.; Anderson, M.C.; Alfieri, J.G.; Hipps, L.E.; Villagarcía, L.; Domingo, F.; García, M. Revisiting the paper "Using radiometric surface temperature for surface energy flux estimation in Mediterranean drylands from a two-source perspective". *Remote Sens. Environ.* **2016**, *184*, 645–653. [CrossRef]
29. Garratt, J.; Hicks, B. Momentum, heat and water vapour transfer to and from natural and artificial surfaces. *Q. J. R. Meteorol. Soc.* **1973**, *99*, 680–687. [CrossRef]
30. Verhoef, A.; De Bruin, H.; Van Den Hurk, B. Some practical notes on the parameter kB^{-1} for sparse vegetation. *J. Appl. Meteorol.* **1997**, *36*, 560–572. [CrossRef]
31. Bastiaanssen, W.; Menenti, M.; Feddes, R.; Holtslag, A. A remote sensing surface energy balance algorithm for land (SEBAL). 1. Formulation. *J. Hydrol.* **1998**, *212*, 198–212. [CrossRef]
32. Allen, R.; Tasumi, M.; Trezza, R. Satellite-Based Energy Balance for Mapping Evapotranspiration with Internalized Calibration (METRIC)\97Model. *J. Irrig. Drain. Eng.* **2007**, *133*, 380–394. [CrossRef]
33. Su, Z. The Surface Energy Balance System (SEBS) for estimation of turbulent heat fluxes. *Hydrol. Earth Syst. Sci. Discuss.* **2002**, *6*, 85–100. [CrossRef]
34. Kustas, W.P.; Choudhury, B.J.; Kunkel, K.E.; Gay, L.W. Estimate of the aerodynamic roughness parameters over an incomplete canopy cover of cotton. *Agric. For. Meteorol.* **1989**, *46*, 91–105. [CrossRef]
35. Tang, R.; Li, Z.L.; Jia, Y.; Li, C.; Sun, X.; Kustas, W.P.; Anderson, M.C. An intercomparison of three remote-sensing-based energy balance models using Large Aperture Scintillometer measurements over a wheat-corn production region. *Remote Sens. Environ.* **2011**, *115*, 3187–3202. [CrossRef]
36. Norman, J.M.; Kustas, W.P.; Humes, K.S. Source approach for estimating soil and vegetation energy fluxes in observations of directional radiometric surface temperature. *Agric. For. Meteorol.* **1995**, *77*, 263–293. [CrossRef]
37. Colaizzi, P.D.; Kustas, W.P.; Anderson, M.C.; Agam, N.; Tolk, J.A.; Evett, S.R.; Howell, T.A.; Gowda, P.H.; O'Shaughnessy, S.A. Two-source energy balance model estimates of evapotranspiration using component and composite surface temperatures. *Adv. Water Resour.* **2012**, *50*, 134–151. [CrossRef]

38. Boulet, G.; Mougenot, B.; L'homme, J.P.; Fanise, P.; Lili-Chabaane, Z.; Olioso, A.; Bahir, M.; Rivalland, V.; Jarlan, L.; Merlin, O.; et al. The SPARSE model for the prediction of water stress and evapotranspiration components from thermal infra-red data and its evaluation over irrigated and rainfed wheat. *Hydrol. Earth Syst. Sci.* **2015**, *19*, 4653–4672. [CrossRef]

39. Roerink, G.; Su, Z.; Menenti, M. S-SEBI: A simple remote sensing algorithm to estimate the surface energy balance. *Phys. Chem. Earth Part B Hydrol. Ocean. Atmos.* **2000**, *25*, 147–157. [CrossRef]

40. Sandholt, I.; Rasmussen, K.; Andersen, J. A simple interpretation of the surface temperature/vegetation index space for assessment of surface moisture status. *Remote Sens. Environ.* **2002**, *79*, 213–224. [CrossRef]

41. Merlin, O.; Chirouze, J.; Olioso, A.; Jarlan, L.; Chehbouni, G.; Boulet, G. An image-based four-source surface energy balance model to estimate crop evapotranspiration from solar reflectance/thermal emission data (SEB-4S). *Agric. For. Meteorol.* **2014**, *184*, 188–203. [CrossRef]

42. Stisen, S.; Sandholt, I.; Nørgaard, A.; Fensholt, R.; Jensen, K. Combining the triangle method with thermal inertia to estimate regional evapotranspiration—Applied to MSG-SEVIRI data in the Senegal River basin. *Remote Sens. Environ.* **2008**, *112*, 1242–1255. [CrossRef]

43. Guzinski, R.; Anderson, M.; Kustas, W.; Nieto, H.; Sandholt, I. Using a thermal-based two source energy balance model with time-differencing to estimate surface energy fluxes with day-night MODIS observations. *Hydrol. Earth Syst. Sci.* **2013**, *17*, 2809–2825. [CrossRef]

44. Tang, R.; Li, Z.L. An End-Member-Based Two-Source Approach for Estimating Land Surface Evapotranspiration From Remote Sensing Data. *IEEE Trans. Geosci. Remote. Sens.* **2017**, *55*, 5818–5832. [CrossRef]

45. Russell, M.J.; Nunez, M.; Chladil, M.A.; Valiente, J.A.; Lopez-Baeza, E. Conversion of nadir, narrowband reflectance in red and near-infrared channels to hemispherical surface albedo. *Remote Sens. Environ.* **1997**, *61*, 16–23. [CrossRef]

46. Franch, B.; Vermote, E.; Sobrino, J.; Julien, Y. Retrieval of Surface Albedo on a Daily Basis: Application to MODIS Data. *IEEE Trans. Geosci. Remote. Sens.* **2014**, *52*, 7549–7558. [CrossRef]

47. Lucht, W.; Schaaf, C.; Strahler, A. An algorithm for the retrieval of albedo from space using semiempirical BRDF models. *IEEE Trans. Geosci. Remote. Sens.* **2000**, *38*, 977–998. [CrossRef]

48. Jacob, F.; Olioso, A.; Weiss, M.; Baret, F.; Hautecoeur, O. Mapping short-wave albedo of agricultural surfaces using airborne PolDER data. *Remote Sens. Environ.* **2002**, *80*, 36–46. [CrossRef]

49. Jacob, F.; Olioso, A. Derivation of diurnal courses of albedo and reflected solar irradiance from airborne POLDER data acquired near solar noon. *J. Geophys. Res. Atmos.* **2005**, *110*. [CrossRef]

50. Kustas, W.P.; Norman, J.M. Evaluation of soil and vegetation heat flux predictions using a simple two-source model with radiometric temperatures for partial canopy cover. *Agric. For. Meteorol.* **1999**, *94*, 13–29. [CrossRef]

51. Campbell, G.; Norman, J. *An Introduction to Environmental Biophysics*, 2nd ed.; Springer: New York, NY, USA, 1998.

52. Campbell, G.S. Extinction coefficients for radiation in plant canopies calculated using an ellipsoidal inclination angle distribution. *Agric. For. Meteorol.* **1986**, *36*, 317–321. [CrossRef]

53. Kustas, W.; Norman, J. A two-source energy balance approach using directional radiometric temperature observations for sparse canopy covered surfaces. *Agron. J.* **2000**, *92*, 847–854. [CrossRef]

54. Choudhury, B.; Idso, S.; Reginato, R. Analysis of an empirical model for soil heat flux under a growing wheat crop for estimating evaporation by an infrared-temperature based energy balance equation. *Agric. For. Meteorol.* **1987**, *39*, 283–297. [CrossRef]

55. Walter, I.A.; Allen, R.G.; Elliott, R.; Jensen, M.E.; Itenfisu, D.; Mecham, B.; Howell, T.A.; Snyder, R.; Brown, P.; Echings, S.; et al. ASCE's Standardized Reference Evapotranspiration Equation. In Proceedings of the Watershed Management and Operations Management 2000, Collins, CO, USA, 20–24 June 2000.

56. Bhattarai, N.; Quackenbush, L.J.; Im, J.; Shaw, S.B. A new optimized algorithm for automating endmember pixel selection in the SEBAL and METRIC models. *Remote Sens. Environ.* **2017**, *196*, 178–192. [CrossRef]

57. Priestley, C.H.B.; Taylor, R.J. On the Assessment of Surface Heat Flux and Evaporation Using Large-Scale Parameters. *Mon. Weather Rev.* **1972**, *100*, 81–92. [CrossRef]

58. Chirouze, J.; Boulet, G.; Jarlan, L.; Fieuzal, R.; Rodriguez, J.C.; Ezzahar, J.; Er-Raki, S.; Bigeard, G.; Merlin, O.; Garatuza-Payan, J.; et al. Intercomparison of four remote-sensing-based energy balance methods to retrieve surface evapotranspiration and water stress of irrigated fields in semi-arid climate. *Hydrol. Earth Syst. Sci.* **2014**, *18*, 1165–1188. [CrossRef]

59. Yang, Y.; Shang, S. A hybrid dual-source scheme and trapezoid framework-based evapotranspiration model (HTEM) using satellite images: Algorithm and model test: HTEM. *J. Geophys. Res. Atmos.* **2013**, *118*, 2284–2300. [CrossRef]

60. Long, D.; Singh, V.P. A Two-source Trapezoid Model for Evapotranspiration (TTME) from satellite imagery. *Remote Sens. Environ.* **2012**, *121*, 370–388. [CrossRef]

61. Detto, M.; Montaldo, N.; Albertson, J.D.; Mancini, M.; Katul, G. Soil moisture and vegetation controls on evapotranspiration in a heterogeneous Mediterranean ecosystem on Sardinia, Italy. *Water Resour. Res.* **2006**, *42*. [CrossRef]

62. Hsieh, C.I.; Katul, G.; wen Chi, T. An approximate analytical model for footprint estimation of scalar fluxes in thermally stratified atmospheric flows. *Adv. Water Resour.* **2000**, *23*, 765–772. [CrossRef]

63. Tagesson, T.; Fensholt, R.; Guiro, I.; Rasmussen, M.O.; Huber, S.; Mbow, C.; Garcia, M.; Horion, S.; Sandholt, I.; Holm-Rasmussen, B.; et al. Ecosystem properties of semiarid savanna grassland in West Africa and its relationship with environmental variability. *Glob. Chang. Biol.* **2015**, *21*, 250–264. [CrossRef]

64. Scott, R.L.; Hamerlynck, E.P.; Jenerette, G.D.; Moran, M.S.; Barron-Gafford, G.A. Carbon dioxide exchange in a semidesert grassland through drought-induced vegetation change. *J. Geophys. Res. Biogeosci.* **2010**, *115*. [CrossRef]

65. Scott, R. *AmeriFlux US-Wkg Walnut Gulch Kendall Grassland*; United States Department of Agriculture: Washington, DC, USA, 2016; doi:10.17190/AMF/1246112. [CrossRef]

66. Kustas, W.P.; Anderson, M.C.; Alfieri, J.G.; Knipper, K.; Torres-Rua, A.; Parry, C.K.; Nieto, H.; Agam, N.; White, W.A.; Gao, F.; et al. The Grape Remote Sensing Atmospheric Profile and Evapotranspiration Experiment. *Bull. Am. Meteorol. Soc.* **2018**, *99*, 1791–1812. [CrossRef]

67. Alfieri, J.G.; Kustas, W.P.; Prueger, J.H.; McKee, L.G.; Hipps, L.E.; Gao, F. A multi-year intercomparison of micrometeorological observations at adjacent vineyards in California's Central Valley during GRAPEX. *Irrig. Sci.* **2019**, *37*, 345–357. [CrossRef]

68. Foken, T.; Aubinet, M.; Finnigan, J.J.; Leclerc, M.Y.; Mauder, M.; Paw U, K.T. Results of A Panel Discussion about The Energy Balance Closure Correction For Trace Gases. *Bull. Am. Meteorol. Soc.* **2011**, *92*, ES13–ES18. [CrossRef]

69. Immitzer, M.; Vuolo, F.; Atzberger, C. First Experience with Sentinel-2 Data for Crop and Tree Species Classifications in Central Europe. *Remote Sens.* **2016**, *8*, 166. [CrossRef]

70. Louis, J.; Debaecker, V.; Pflug, B.; Main-Knorn, M.; Bieniarz, J.; Mueller-Wilm, U.; Cadau, E.; Gascon, F. Sentinel-2 Sen2Cor: L2A Processor for Users. In *Proceedings Living Planet Symposium 2016*; Ouwehand, L., Ed.; Spacebooks Online: Prague, Czech Republic, 2016; Volume SP-740, pp. 1–8.

71. Weiss, M.; Baret, F. *S2ToolBox Level 2 Products: LAI, FAPAR, FCOVER*; Algorithm Theoretical Basis Document; INRA: Paris, France, 2016.

72. Pan, H.; Chen, Z.; Ren, J.; Li, H.; Wu, S. Modeling Winter Wheat Leaf Area Index and Canopy Water Content With Three Different Approaches Using Sentinel-2 Multispectral Instrument Data. *IEEE J. Sel. Top. Appl. Earth Obs. Remote Sens.* **2019**, *12*, 482–492. [CrossRef]

73. Xie, Q.; Dash, J.; Huete, A.; Jiang, A.; Yin, G.; Ding, Y.; Peng, D.; Hall, C.C.; Brown, L.; Shi, Y.; et al. Retrieval of crop biophysical parameters from Sentinel-2 remote sensing imagery. *Int. J. Appl. Earth Obs. Geoinf.* **2019**, *80*, 187–195. [CrossRef]

74. Vinué, D.; Camacho, F.; Oliver-Villanueva, J.V.; Coll, E. Validation of Sentinel-2 L2B LAI and FAPAR products from SNAP over forests and crops in a Mediterranean environment. In Proceedings of the 2nd Sentinel-2 Validation Team Meeting, Frascati, Italy, 29–31 January 2018.

75. Fisher, J.B.; Tu, K.P.; Baldocchi, D.D. Global estimates of the land-atmosphere water flux based on monthly AVHRR and ISLSCP-II data, validated at 16 FLUXNET sites. *Remote Sens. Environ.* **2008**, *112*, 901–919. [CrossRef]

76. Féret, J.B.; Gitelson, A.; Noble, S.; Jacquemoud, S. PROSPECT-D: Towards modeling leaf optical properties through a complete lifecycle. *Remote Sens. Environ.* **2017**, *193*, 204–215. [CrossRef]

77. The Sentinel-3 Mission Performance Center. S3 SLSTR Cyclic Performance Report, 2020. Avaiable online: https://sentinel.esa.int/documents/247904/4069145/Sentinel-3-MPC-RAL-SLSTR-Cyclic-Report-055-036.pdf (accessed on 13 April 2020).

78. Bontemps, S.; Defourny, P.; Radoux, J.; Van Bogaert, E.; Lamarche, C.; Achard, F.; Mayaux, P.; Boettcher, M.; Brockmann, C.; Kirches, G.; et al. Consistent global land cover maps for climate modelling communities: Current achievements of the ESA's land cover CCI. In Proceedings of the ESA Living Planet Symposium, Edinburgh, UK, 9–13 September 2013; pp. 9–13.

79. Raupach, M.R. Simplified expressions for vegetation roughness length and zero-plane displacement as functions of canopy height and area index. *Bound.-Layer Meteorol.* **1994**, *71*, 211–216. [CrossRef]

80. Schaudt, K.; Dickinson, R. An approach to deriving roughness length and zero-plane displacement height from satellite data, prototyped with BOREAS data. *Agric. For. Meteorol.* **2000**, *104*, 143–155. [CrossRef]

81. Hersbach, H. The ERA5 Atmospheric Reanalysis. In Proceedings of the AGU Fall Meeting Abstracts, San Francisco, CA, USA, 12–16 December 2016.

82. Morcrette, J.J.; Boucher, O.; Jones, L.; Salmond, D.; Bechtold, P.; Beljaars, A.; Benedetti, A.; Bonet, A.; Kaiser, J.W.; Razinger, M.; et al. Aerosol analysis and forecast in the European Centre for Medium-Range Weather Forecasts Integrated Forecast System: Forward modeling. *J. Geophys. Res. Atmos.* **2009**, *114*. [CrossRef]

83. Ineichen, P.; Perez, R. A new airmass independent formulation for the Linke turbidity coefficient. *Sol. Energy* **2002**, *73*, 151–157. [CrossRef]

84. Ineichen, P. Conversion function between the Linke turbidity and the atmospheric water vapor and aerosol content. *Sol. Energy* **2008**, *82*, 1095–1097. [CrossRef]

85. Breiman, L. Bagging predictors. *Mach. Learn.* **1996**, *24*, 123–140. [CrossRef]

86. Goward, S.N.; Waring, R.H.; Dye, D.G.; Yang, J.L. Ecological Remote-Sensing at OTTER: Satellite Macroscale Observations. *Ecol. Appl.* **1994**, *4*, 322–343. [CrossRef]

87. Prince, S.D.; Goetz, S.J.; Dubayah, R.O.; Czajkowski, K.P.; Thawley, M. Inference of surface and air temperature, atmospheric precipitable water and vapor pressure deficit using Advanced Very High-Resolution Radiometer satellite observations: Comparison with field observations. *J. Hydrol.* **1998**, *213*, 230–249. [CrossRef]

88. Ceccato, P.; Flasse, S.; Tarantola, S.; Jacquemoud, S.; Grégoire, J.M. Detecting vegetation leaf water content using reflectance in the optical domain. *Remote Sens. Environ.* **2001**, *77*, 22–33. [CrossRef]

89. Agam, N.; Kustas, W.P.; Alfieri, J.G.; Gao, F.; McKee, L.M.; Prueger, J.H.; Hipps, L.E. Micro-scale spatial variability in soil heat flux (SHF) in a wine-grape vineyard. *Irrig. Sci.* **2019**, *37*, 253–268. [CrossRef]

90. Parry, C.K.; Nieto, H.; Guillevic, P.; Agam, N.; Kustas, W.P.; Alfieri, J.; McKee, L.; McElrone, A.J. An intercomparison of radiation partitioning models in vineyard canopies. *Irrig. Sci.* **2019**, *37*, 239–252. [CrossRef]

91. Choi, M.; Kustas, W.P.; Anderson, M.C.; Allen, R.G.; Li, F.; Kjaersgaard, J.H. An intercomparison of three remote sensing-based surface energy balance algorithms over a corn and soybean production region (Iowa, U.S.) during SMACEX. *Agric. For. Meteorol.* **2009**, *149*, 2082–2097. [CrossRef]

92. González-Dugo, M.; Neale, C.; Mateos, L.; Kustas, W.; Prueger, J.; Anderson, M.; Li, F. A comparison of operational remote sensing-based models for estimating crop evapotranspiration. *Agric. For. Meteorol.* **2009**, *149*, 1843–1853. [CrossRef]

93. Knipper, K.R.; Kustas, W.P.; Anderson, M.C.; Alfieri, J.G.; Prueger, J.H.; Hain, C.R.; Gao, F.; Yang, Y.; McKee, L.G.; Nieto, H.; et al. Evapotranspiration estimates derived using thermal-based satellite remote sensing and data fusion for irrigation management in California vineyards. *Irrig. Sci.* **2018**, 1–19. [CrossRef]

94. Anderson, M.C.; Norman, J.M.; Kustas, W.P.; Li, F.; Prueger, J.H.; Mecikalski, J.R. Effects of Vegetation Clumping on Two-Source Model Estimates of Surface Energy Fluxes from an Agricultural Landscape during SMACEX. *J. Hydrometeorol.* **2005**, *6*, 892–909. [CrossRef]

95. Amazirh, A.; Merlin, O.; Er-Raki, S. Including Sentinel-1 radar data to improve the disaggregation of MODIS land surface temperature data. *ISPRS J. Photogramm. Remote Sens.* **2019**, *150*, 11–26. [CrossRef]

96. Olivera-Guerra, L.; Mattar, C.; Merlin, O.; Durán-Alarcón, C.; Santamaría-Artigas, A.; Fuster, R. An operational method for the disaggregation of land surface temperature to estimate actual evapotranspiration in the arid region of Chile. *ISPRS J. Photogramm. Remote Sens.* **2017**, *128*, 170–181. [CrossRef]

97. Koetz, B.; Bastiaanssen, W.; Berger, M.; Defourney, P.; Bello, U.D.; Drusch, M.; Drinkwater, M.; Duca, R.; Fernandez, V.; Ghent, D.; et al. High Spatio-Temporal Resolution Land Surface Temperature Mission—A Copernicus Candidate Mission in Support of Agricultural Monitoring. In Proceedings of the IGARSS 2018—2018 IEEE International Geoscience and Remote Sensing Symposium, Valencia, Spain, 23–27 July 2018; pp. 8160–8162.

Article

An Improved Approach for Downscaling Coarse-Resolution Thermal Data by Minimizing the Spatial Averaging Biases in Random Forest

Sammy M. Njuki *, Chris M. Mannaerts and Zhongbo Su

Department of Water Resources, Faculty of Geo-Information Science and Earth Observation (ITC), University of Twente, Hengelosestraat 99, 7514 AE Enschede, The Netherlands; c.m.m.mannaerts@utwente.nl (C.M.M.); z.su@utwente.nl (Z.S.)
* Correspondence: s.m.njuki@utwente.nl

Received: 19 August 2020; Accepted: 22 October 2020; Published: 25 October 2020

Abstract: Land surface temperature (LST) plays a fundamental role in various geophysical processes at varying spatial and temporal scales. Satellite-based observations of LST provide a viable option for monitoring the spatial-temporal evolution of these processes. Downscaling is a widely adopted approach for solving the spatial-temporal trade-off associated with satellite-based observations of LST. However, despite the advances made in the field of LST downscaling, issues related to spatial averaging in the downscaling methodologies greatly hamper the utility of coarse-resolution thermal data for downscaling applications in complex environments. In this study, an improved LST downscaling approach based on random forest (RF) regression is presented. The proposed approach addresses issues related to spatial averaging biases associated with the downscaling model developed at the coarse resolution. The approach was applied to downscale the coarse-resolution Satellite Application Facility on Land Surface Analysis (LSA-SAF) LST product derived from the Spinning Enhanced Visible and Infrared Imager (SEVIRI) sensor aboard the Meteosat Second Generation (MSG) weather satellite. The LSA-SAF product was downscaled to a spatial resolution of ~30 m, based on predictor variables derived from Sentinel 2, and the Advanced Land Observing Satellite (ALOS) digital elevation model (DEM). Quantitatively and qualitatively, better downscaling results were obtained using the proposed approach in comparison to the conventional approach of downscaling LST using RF widely adopted in LST downscaling studies. The enhanced performance indicates that the proposed approach has the ability to reduce the spatial averaging biases inherent in the LST downscaling methodology and thus is more suitable for downscaling applications in complex environments.

Keywords: LST; downscaling; LSA-SAF; Sentinel 2; random forest; DEM; spatial averaging biases

1. Introduction

Land surface temperature (LST) plays a critical role in surface energy balance and partitioning and hence drives water and biogeochemical cycles [1–3]. The dynamics induced by this cycling, through land–atmosphere interactions, play an essential role in the evolution of weather and climate [4]. Consequently, LST has been widely used as a critical parameter in the estimation of an array of geophysical variables, such as evapotranspiration [5–7], soil moisture [8,9], vegetation water stress [10,11], and urban heat fluxes [12,13], to support various applications, ranging from but not limited to agriculture, climate change, urban climate, forest fire monitoring, energy and water management.

Cognizant to the fundamental role of LST in the understanding of various geophysical processes, and the need to capture their spatial-temporal evolution, space-borne missions often include thermal infra-red (TIR) sensors dedicated to LST mapping. However, the applicability of LST products derived from these sensors for operational purposes is hampered by the inherent spatial-temporal trade-off in TIR imagery. Downscaling, also referred to as disaggregation, has come to the fore as a cheaper alternative solution to the spatial-temporal trade-off problem associated with TIR imagery [14,15]. Downscaling can be defined as the synergistic utilization of the complementary nature of the high-resolution visible near-infrared (VNIR) imagery and the coarse resolution TIR imagery, sometimes supported by ancillary information to discern the spatial distribution of thermal elements at the resolution of the VNIR imagery. Although various downscaling methods have been proposed, statistical downscaling methods are often preferred over other downscaling methods owing to their simplicity [14,15]. Statistical downscaling techniques based on the well-established relationship between LST and vegetation indices (VI) [16], such as the disaggregation procedure for radiometric surface temperature (DisTrad) [17], and the algorithm for sharpening thermal imagery (TsHARP) [18] have shown excellent performance in relatively homogeneous vegetation canopies. However, the assumptions made in the LST-VI feature space-based approach are rarely satisfied in fragmented landscapes undermining their performance [14,15,19].

Consequently, statistical downscaling methods that utilize multiple predictor variables such as multiple linear regression and machine-learning algorithms are often preferred in heterogeneous landscapes over the LST-VI feature space-based approaches [15,20]. Machine learning-based methods, in particular, have consistently shown superior performance in comparison to other downscaling methods in complex environments. This is due to their ability to learn the complex, often non-linear statistical relationships that exist between the predictor variables and LST in such complex landscapes [15,21]. In a study by Li et al. [21], the ability of random forest (RF), artificial neural networks (ANN), support vector machines (SVM) and TsHARP to downscale the Moderate Resolution Imaging Spectroradiometer (MODIS) LST in two complex environments around the city of Beijing in China, were compared. In both regions, all the machine learning-based algorithms produced higher downscaling accuracies in comparison to the TsHARP method. Also, their study obtained better downscaling results with RF and ANN compared to SVM. In another study, Hutengs and Vohland, [22] applied RF to downscale MODIS LST product from a spatial resolution of ~1 km to 250 m of the MODIS VNIR reflectance in the Jordan River valley, a region characterized by complex terrain. The results of their work showed an improved performance of up to 19% for the RF-based downscaling in comparison to the TsHARP method. In yet another study, Pan et al. [23] utilized remote-sensing indices derived from Landsat 8 Optical Land Imager (OLI) to downscale MODIS LST using RF in an urban area located in an oasis–desert transition in China. They reported better accuracies in the downscaled LST based on RF as compared to LST derived from Landsat 8. Besides providing superior robust downscaling performance, machine learning-based downscaling methods have contributed immensely in our understanding of the influence of different predictors on LST under varying environments, such as in arid regions [23,24], complex terrain [22,25], and urban areas [26], among others.

Since its first application in the downscaling of LST by Hutengs and Vohland, [22], RF has become popular in the downscaling of TIR data. Being a non-parametric ensemble learning method [27], the model is less prone to overfitting and has the ability to handle high dimensional multicollinear data [28,29], which explains its superior performance and popularity in downscaling applications. However, although the ensemble decision tree with bagging approach adopted in constructing the RF model reduces the risk of overfitting, the model constructed is the average of the randomly constructed decision trees. Thus, RF is unable to predict data beyond the range of data presented during the model training. Since the methodology adopted in statistical downscaling methods involves a transfer of the model built at the coarse resolution to predict LST at the fine resolution, RF regression-based downscaling approaches will introduce biases related to spatial averaging in the downscaled LST. This issue has not been addressed in downscaling studies since it is generally assumed that the model error correction procedure in the downscaling methodology is sufficient to correct for biases in the model. However, as noted in Bindhu, Narasimhan and Sudheer, [30] and Essa et al. [31], the assumption made in the model error correction procedure is rarely satisfied in complex, heterogeneous landscapes. The procedure is based on the assumption that the model error computed at the coarse resolution pixel remains constant for the fine-resolution LST pixels that constitute the coarse resolution pixel. In such environments, the validity of this assumption will tend to decrease as the differences in the spatial resolution between the original LST and the downscaled LST increase. On the other hand, the spatial averaging biases inherent to RF will increase with increasing differences in the spatial resolution. Thus, a downscaling approach that takes into account the effect of spatial averaging in RF regression-based downscaling methodology is needed. Such an approach could harness the potential of TIR data derived from the high temporal but low spatial resolution images from geostationary satellites such as MSG, the Geostationary Operational Environmental Satellite (GOES), and Himawari-8 to provide high spatial-temporal TIR data, alleviating the spatial-temporal trade-off problem associated with TIR imagery.

In this study, an improved universally applicable RF regression-based downscaling approach, which minimizes the spatial averaging biases related to the model trained at the coarse resolution, is presented. The approach is applied to enhance the spatial resolution of the Satellite Application Facility on Land Surface Analysis (LSA-SAF) LST product from a spatial resolution of ~3 km at the sub-satellite view to ~100 m based on Sentinel 2 and Advanced Land Observing Satellite (ALOS) digital elevation model (DEM)-derived predictor variables. The downscaled LST maps are then validated against LST maps derived from Landsat 8. The approach is tested in a region in Kenya, comprising the Kenyan Rift valley system, the Kenyan highlands and the arid and semi-arid lands (ASAL) in the northern part of the country.

2. Materials and Methods

2.1. Methodology

2.1.1. Proposed Downscaling Approach

The conventional approach of downscaling LST in RF is represented by the steps described by Equations (1)–(4). This approach has been widely adopted in various LST downscaling studies (e.g., [21–24]). The studies differ mainly in terms of the chosen predictor variables and the target coarse-resolution LST product.

The proposed approach follows the steps of the conventional approach from Equations (1)–(4). On the other hand, Equations (5)–(9) describe the extension of the conventional approach of downscaling LST in RF to derive the proposed improved approach for downscaling LST using RF regression.

In the first step, an RF regression linking model is trained using the coarse resolution LSA-SAF LST and predictor variables derived from Sentinel 2 and ALOS DEM. The linking model is defined using Equation (1) and describes the relationship between the predictor variables and the target LST at the coarse-scale.

$$LST_{CR} = F(Var\_CR_i \ldots, Var\_CR_n) \tag{1}$$

where, LST_{CR} is the coarse resolution LSA-SAF LST, F is the linking model at the coarse resolution, Var_CR is the predictor variable aggregated to the resolution of the LSA-SAF LST, while i and n denotes the ith and nth selected predictor variable, respectively.

The linking model developed in Equation (1) is scene dependent, but the relationship it describes is assumed to be valid across multiple spatial scales within the scene, thus for each scene, a new model should be trained. Based on this assumption, the linking model is applied to the predictor variables at the fine spatial resolution to predict LST at the fine spatial resolution using Equation (2).

$$LST_{FHR} = F(Var\_HR_i \ldots, Var\_HR_n) \tag{2}$$

where LST_{FHR} is the high-resolution LST modeled using the coarse resolution model F, and Var_HR is the predictor variable at the fine spatial resolution.

However, since the selected predictor variables cannot fully account for all the variations in LST, the linking model has an inherent error; thus, the derived high-resolution LST should be corrected for this error. The model error correction procedure involves a spatial aggregation of the modeled high-resolution LST to the coarse resolution and subsequently computing the model error (ΔLST_{CR}) as the difference between the original coarse resolution LSA-SAF LST and the aggregated LST from the model, as shown in Equation (3).

$$\Delta LST_{CR} = LST_{CR} - \overline{LST_{FHR}} \tag{3}$$

where $\overline{LST_{FHR}}$ is the predicted LST from Equation (2) aggregated to the coarse resolution. Prior to aggregation to the coarse scale, the high-resolution LST was first converted to radiance using the Stefan–Boltzmann law. Subsequently, the high-resolution radiance was aggregated to the coarse resolution and then converted to the aggregated LST at the coarse resolution by inverting the Stefan–Boltzmann equation. The emissivity used in the Stefan–Boltzmann equation was derived from fractional vegetation cover using the method of Jimenez-Munoz et al. [32]. The fractional vegetation cover was derived from the Normalized Difference Vegetation Index (NDVI) using the method of Carlson and Ripley [33]. The bare soil and full vegetation cover emissivity values adopted in the vegetation fraction based emissivity method are 0.97 Sobrino et al. [34], and 0.99 Sobrino et al. [35], respectively. This step was included in the aggregation procedure to avoid introducing biases in the aggregated LST owing to the non-linear relationship between radiance and LST.

In the model error correction step, it is assumed that the contribution to the model error by the fine resolution pixels that constitute the coarse resolution pixel is constant. Based on this assumption, the model error is resampled to the original resolution of the predictor variables and added to the high-resolution LST obtained using Equation (2) to derive the final downscaled LST as shown in Equation (4).

$$LST_{HR} = LST_{FHR} + \Delta LST_{HR} \tag{4}$$

where LST_{HR} is the final downscaled LST based on the conventional approach, and ΔLST_{HR} is the disaggregated model error.

However, in spatially heterogeneous environments and in particular, where the difference in spatial scales between the fine resolution predictor variables and the coarse resolution LST is large, this assumption is rarely satisfied.

The other shortcoming in the downscaling approach, especially when regression tree-based machine-learning algorithms such as RF are applied, is the assumption that the model developed in Equation (1) is valid across multiple spatial scales. The training data used to derive the model at the coarse-scale is an average representation of the spatial variability within the scene. Thus, as the difference in spatial scales between the fine resolution predictor variables and the coarse resolution LST increases, more spatial details which were not present at the coarse-scale will emerge. Since regression tree-based machine-learning algorithms are poor at predicting data outside the range of data present during training, the model trained at the coarse resolution will rarely capture such spatial details.

To counter this problem, the high-resolution LST obtained in Equation (4) is used to construct a new RF regression model and predict new data range-enhanced LST at the fine resolution using Equation (5).

$$LST_{fHR} = f(Var\_HR_i \ldots, Var\_HR_n) \tag{5}$$

where LST_{fHR} is the data range enhanced high-resolution LST, and f is the RF regression model trained at the fine resolution.

Unlike the LST predicted using Equation (2), the LST from Equation (5) is less affected by issues related to averaged data ranges in the training data. However, the LST from Equation (5) still needs to be corrected for the model error, since the selected predictor variables cannot entirely account for the variations in LST. Also, it should be adjusted for the averaging effect related to the constant redistribution of the model error in the fine resolution LST obtained using Equation (4). Thus, the final corrected LST is represented by Equation (6).

$$LST_{NHR} = LST_{fHR} + \Delta LST_{HR} + \Delta LST_{AVG} \tag{6}$$

where LST_{NHR} is the final downscaled LST using the proposed approach and ΔLST_{AVG} is the averaging effect in the model error correction procedure.

The difference between the LST from Equation (4) and LST from Equation (5) is assumed to be primarily related to the averaging effect in the model error correction procedure. Thus, the averaging effect can be approximated using Equation (7).

$$\Delta LST_{AVG} = LST_{fHR} - LST_{HR} \tag{7}$$

Substituting Equation (4) into Equation (7) and rearranging the terms yields:

$$\Delta LST_{HR} + \Delta LST_{AVG} = LST_{fHR} - LST_{FHR} \tag{8}$$

Equation (8) indicates that the corrections for the model error and the averaging errors can be approximated as the difference between the LST predicted using Equation (5), and LST predicted using Equation (2). Thus, Equation (6) can be rewritten as follows:

$$LST_{NHR} = 2 \times LST_{fHR} - LST_{FHR} \tag{9}$$

Equation (9) represents the final proposed approach for downscaling LST in complex environments using RF regression.

Besides enhancing the spatial details in the LST predicted at the fine resolution, the proposed approach indirectly incorporates the coarse-resolution model error correction as shown in Equation (8). Thus, the impact of the coarse-resolution model error correction on the downscaled LST should be less pronounced since the approach does not assume a constant redistribution of the coarse-resolution model error.

Due to the indirect incorporation of the model error correction as opposed to the direct addition of the model error in the conventional method (Equation (4)), the proposed approach will also predict LST even for cloud-contaminated pixels during the overpass time of the thermal sensor. However, the LST predicted for the cloud contaminated pixels will be highly uncertain since the RF regression models used to derive LST_{fHR} and LST_{FHR} do not contain any information on LST under cloudy conditions. Thus, we strongly recommend masking out the cloud-contaminated pixels using the cloud mask of the thermal image.

2.1.2. Random Forest Model Description

Random forests [27] are an ensemble decision tree-based supervised machine learning algorithm. The ensemble consists of an average over several randomized and de-correlated decision trees adapted for either classification or regression. For regression purposes, non-linear multivariate regression trees are constructed, with a set of decision rules being used to determine the splitting when building the trees. A bootstrap sample which contains about two-thirds of the observations is selected during model training to grow each tree, and by using a randomly selected subset of predictors, the chance of model overfitting is reduced. Splits are achieved by minimizing a cost function between the target and the predictors resulting in a regression tree.

A third of the unseen observations during model training are used to compute the popular out-of-bag (OOB) error estimate in RF [27]. A prediction is constructed for each of the unseen data instances using only those regression trees in which the respective data instance was not used in training. The average error over all OOB predictions is then used to derive the overall OOB score, providing a convenient way of assessing the performance of the model. On the other hand, improvements recorded on the splitting decision at each split node and for each tree are used to derive variable importance rankings, enabling the assessment of the contribution of each predictor to the final model. Other advantages include the ability of RF to account for correlations among features [29,36], to handle continuous and categorical data simultaneously, and the relatively small number of model parameters needed [37].

In this study, the RF regression framework for the downscaling of LST was implemented in Python using the scikit-learn (sklearn) package [38].

2.1.3. Selection of Predictor Variables

Multiple predictors are often needed to adequately describe variations in LST in fragmented landscapes owing to the complex interactions between LST and various factors related to climate, land use and land cover (LULC), and topography [22]. However, besides the computation demand that comes with increasing the number of predictors in machine-learning algorithms, it has been shown that an increase in the number of predictors increases the risk of model overfitting [39]. Additionally, despite the ability of RF to handle high-dimensional and correlated variables, it has been shown that high-dimensional variables can lead to inflated OOB accuracy scores and model instability [40].

Given these shortcomings, an objective feature selection approach based on the recursive feature elimination with cross-validation (RFECV) implemented through the sklearn package was chosen for selecting predictor variables. RFECV is a combination of the iterative feature elimination criterion proposed by Guyon et al. [39] and a cross-validation (CV) data splitting strategy. The iterative feature elimination involves the removal of the least ranked feature at each step based on weights computed based on a minimization objective function. In this study, an RFECV with the mean squared error (MSE) as the objective function and a CV of 3 (default in sklearn) was adopted. The predictors selected through the RFECV criterion were then used to train the RF regression linking model.

2.1.4. Validation of Downscaled Land Surface Temperature (LST) Maps

Owing to the scarcity of in situ LST measurements, checking for product consistency against results obtained from other models or LST products is a generally accepted indirect validation approach in LST-related studies [1,41]. As such, per pixel comparison between LST maps derived through downscaling and LST maps derived from Landsat 8 was performed to check for consistency between the downscaled LST against the one derived from Landsat 8. The performance of the downscaling approaches was evaluated in terms of the root mean squared error (RMSE), the coefficient of determination (R^2) and bias, described by Equations (10)–(12), respectively.

$$RMSE = \sqrt{\frac{1}{n}\sum_{i=1}^{n}(LST_m - LST_r)^2} \tag{10}$$

where LST_m is the modeled LST, LST_r is the reference LST and n is the total number of pixels.

$$R^2 = 1 - \frac{\sum(LST_r - LST_m)^2}{\sum\left(LST_r - \overline{LST_r}\right)^2} \tag{11}$$

where $\overline{LST_r}$ is the mean reference LST.

$$Bias = \frac{\sum_{i=1}^{n}(LST_m - LST_r)}{n} \tag{12}$$

2.2. Study Area and Data

2.2.1. Study Area

The study area is located in Kenya and lies roughly between 35°E and 38°E, and 3°S and 3°N as shown in Figure 1. The study area coverage was influenced mainly by the availability of coinciding orbital paths for both Sentinel 2 and Landsat 8. The red dotted line in Figure 1 indicates the separation between the two orbital paths of Sentinel 2, 092 to the right, and 135 to the left, which makes up a mosaic of the study area. On the other hand, the orbital path for Landsat 8 in Figure 1 corresponds to orbital path number 168, and is a mosaic of tile numbers 059, 060 and 061, and forms the validation area for the downscaling. Although the orbital path of Landsat 8 falls exclusively on orbital path 092 of Sentinel 2, the study area was extended to cover path 135 to increase the thermal contrast within the study area. Since the time lag between the overpass dates for path 092 and 135 is three days at maximum, it was assumed that no significant discrepancies would arise on the reflectance mosaic obtained from the two orbits. The total study area consists of a mosaic of approximately 24 full tiles of Sentinel 2.

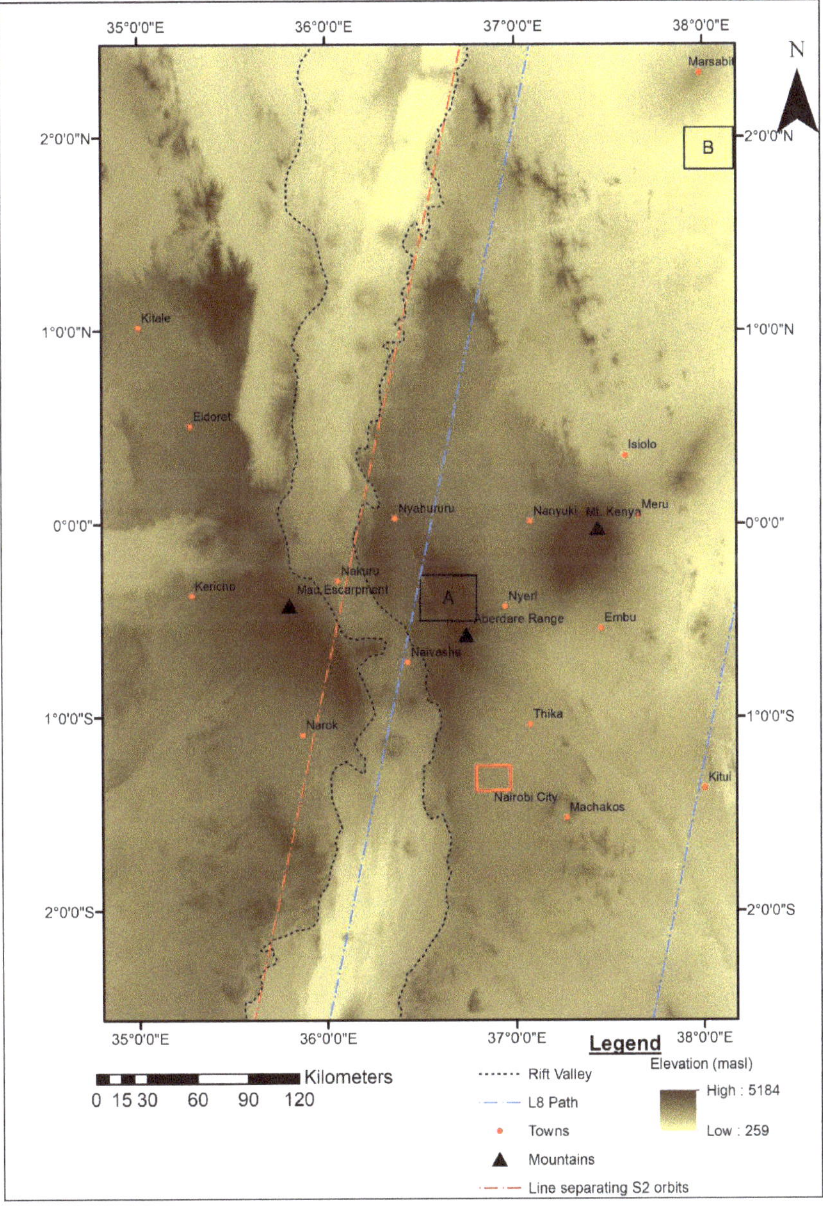

Figure 1. Map showing the study area and the coverage of data used.

The study area is characterized by pronounced topographical differences with elevation ranging from as low as 259 m above sea level (masl) in the northern parts to as high as roughly 5184 masl on top of Mt. Kenya. The study area is straddled from north to south by the eastern branch (Kenyan Rift valley system) of the East African Rift System (EARS). The Kenyan Rift Valley system cuts in between the Mau Escarpment and the Aberdares Mountain Ranges at the central part of the study area, creating areas with sharp elevation drop between the mountains and the floor of the valley. Variations in climatic conditions within the study area are mainly topography-induced. Mean daily air temperature range from less than 10 °C in the mountains to above 30 °C in the low lying ASAL regions. The study area is also characterized by highly fragmented LULC due to the highly mixed land use related to small scale farming activities.

2.2.2. Data Acquisition and Processing

Table 1 shows the datasets used in this study, their characteristics, and their respective sources.

Table 1. Datasets used in this study, their source and their characteristics.

Dataset	Temporal Resolution	Spatial Resolution	Processing Level	Source
Sentinel 2	5 days	20 m	Level 1C (TOA)	https://scihub.copernicus.eu/dhus/#/home
LSA-SAF LST	15 min	~3 km	Operational product	https://landsaf.ipma.pt/en/products/land-surface-temperature/
ALOS DEM	Once	30 m	Void filled	https://www.eorc.jaxa.jp/ALOS/en/aw3d30/data/index.htm
Landsat 8 OLI (band 4 and 5)	16 days	30 m	C1 Level-2 (BOA)	https://earthexplorer.usgs.gov/
Landsat 8 TIR (band 10 and 11)	16 days	100 m	C1 Level-1 (TOA)	https://earthexplorer.usgs.gov/
Total column water vapor	Hourly	30 km	Model output	https://cds.climate.copernicus.eu/cdsapp#!/dataset/reanalysis-era5-single-levels?tab=form

The LSA-SAF LST product [42] was used as the coarse target resolution LST for building the model. The LST product from LSA-SAF is an operational LST product disseminated through the Satellite Application Facility on Land Surface Analysis (Land-SAF) of the European Organization for the Exploitation of Meteorological Satellites (EUMETSAT) in near real-time at time intervals of 15 min and at a spatial resolution of approximately 3 km at sub-satellite. It is derived from the split window channels of the SEVIRI sensor onboard the MSG suite of weather satellites based on the generalized split-window algorithm proposed by Dozier, [43] adopted for the SEVIRI sensor [42,44]. The product has consistently been shown to meet the set target RMSE of 2 K [45,46].

The ALOS World 3D-30m (AW3D30) DEM [47], was used to derive topography-related predictors, namely Elevation, Slope, and Aspect. ALOS DEM is considered the most accurate of the current freely available DEM's averaging an RMSE of 1.78 m for the vertical height [48]. The ALOS DEM-derived predictor variables were then aggregated to the sampling resolution (100 m) of Landsat 8 TIR bands, which was adopted as the base resolution for the downscaling model in order to match the spatial resolution of the validation LST maps derived from Landsat 8.

Sentinel 2 Level 1C products represent the top of atmosphere (TOA) reflectance and were downloaded and converted to surface or bottom of atmosphere (BOA) reflectance by correcting for atmospheric effects using the Sen2Cor algorithm [49]. A total of 11 bands were obtained after atmospheric correction since the coastal aerosol, and the cirrus bands do not contain surface information, thus not included as part of the Level 2A product produced by the Sentinel Application Platform (SNAP) software. The bands obtained after atmospheric correction include Blue, Green, Red, Near Infrared (NIR), Red Edge 1 (RE1), Red Edge 2 (RE2), Red Edge 3 (RE3), Narrow Near Infrared (NNIR), Shortwave Infrared 1 (SWIR1), Shortwave Infrared 2 (SWIR2), and the Water vapor (wvp band). The cloud and cloud shadow masks produced during the atmospheric correction procedure were applied to the remaining surface reflectance bands to obtain cloud-free pixels. Cloud-free pixels from two consecutive overpasses were then used to create composite images, with reduced effects of cloud contamination. The dates for the Sentinel 2 images used to generate the composite images, the LSASAF LST acquisition dates, and the corresponding Landsat 8 overpasses are shown in Table 2.

Table 2. Acquisition dates for images used in the study.

LSA-SAF	Landsat 8 (Path 168)	Sentinel 2 (Orbit 092)	Sentinel 2 (Orbit 135)
13 January 2018—07: 45	13 January 2018—07: 42	[14 January 2018 and 19 January 2018]	[12 January 2018 and 17 January 2018]
29 January 2018—07: 45	29 January 2018—07: 42	[24 January 2018 and 29 January 2018]	[27 January 2018 and 1 February 2018]
1 February 2019—07: 45	29 January 2018—07:42	[29 January 2019 and 3 February 2019]	[1 February 2019 and 6 February 2019]
21 March 2019—07: 45	29 January 2018—07: 42	[20 March 2019 and 25 March 2019]	[18 March 2019 and 23 March 2019]

The composite surface reflectance images were then resampled and co-registered to match the sampling spatial resolution (100 m) of Landsat 8 TIR bands. The resampled surface reflectance bands were then used directly as predictor variables. Additionally, the surface reflectance bands were used to derive the additional remote sensing predictor variables shown in Table 3.

Landsat 8 imagery and Total Column Water Vapour (TCW) product from the European Centre for Medium-Range Weather Forecasts (ECMWF) reanalysis (ERA-5) were used to derive LST maps for validating the downscaled LST maps. Landsat 8 LST maps were derived based on the generalized split-window algorithm proposed by Li and Jiang, [41] for Landsat 8 data. Before the derivation of the LST maps from Landsat 8, all the datasets were resampled to the sampling resolution (100 m) of Landsat 8's TIR bands. It should be noted that, although stray light contamination had plagued the TIR bands of Landsat 8 [50], an operational stray light correction procedure was implemented in 2017 to correct for these artifacts [51]. Gerace and Montanaro, [51] report that the fidelity of the corrected bands is within the range of that of MODIS TIR channels. Li and Jiang, [41] reported an error of less than 1 K using their generalized split-window algorithm in comparison to the MOD11_L2 V6 MODIS LST product, indicating the ability of the algorithm to retrieve LST from the TIR bands of Landsat 8 accurately.

Table 3. Sentinel 2-derived remote-sensing indicators.

Remote Sensing Indicator	Targeted Surface/Characteristics	Formulation	Author
Normalized Difference Vegetation Index (NDVI)	Vegetation	$\dfrac{(NIR - Red)}{(NIR + Red)}$	[52]
Enhanced Vegetation Index (EVI)	Vegetation	$2.5 \times \dfrac{(NIR - Red)}{(NIR + 6 \times Red - 7.5 \times Blue + 1)}$	[53]
Soil-Adjusted Vegetation Index (SAVI)	Vegetation	$\dfrac{(NIR - Red)}{(NIR + Red + 0.5)} \times (1 + 0.5)$	[54]
Fraction Vegetation Cover (FVC)	Vegetation cover/density	$\left(\dfrac{(NDVI - NDVI_s)}{(NDVI_v - NDVI_s)}\right)^2$ Where $NDVI_v = 0.86$ and $NDVI_s = 0.20$	[33,55]
Bare Soil Index (BSI)	Bare soil surfaces	$\dfrac{(SWIR1 + Red) - (NIR + Blue)}{(SWIR1 + Red) + (NIR + Blue)}$	[56]

Table 3. *Cont.*

Remote Sensing Indicator	Targeted Surface/Characteristics	Formulation	Author
Normalized Difference Built-up Index (NDBI)	Built-up areas	$\dfrac{\text{SWIR1} - \text{NIR}}{\text{SWIR1} + \text{NIR}}$	[57]
Normalized Difference Water Index (NDWI)	Water bodies	$\dfrac{(\text{Green} - \text{NIR})}{(\text{Green} + \text{NIR})}$	[58]
Normalized Multi-band Drought Index (NMDI)	Soil and Vegetation moisture stress	$\dfrac{\text{NIR} - (\text{SWIR1} - \text{SWIR2})}{\text{NIR} + (\text{SWIR1} - \text{SWIR2})}$	[59]
Normalized Difference Water (Moisture) Index (NDMI)	Vegetation Water Content	$\dfrac{(\text{NIR} - \text{SWIR1})}{(\text{NIR} + \text{SWIR1})}$	[60]

3. Results

3.1. Selection of Predictor Variables

The results of the predictor variable selection process are shown in Figure 2 and Table 4. Figure 2 shows the influence of increasing the number of predictor variables on the model's performance. The model performance increases drastically when predictor variables are increased from 1 to about 12 variables, as indicated by the decrease in the RMSE with increasing variables. Beyond 12 variables, increasing the number of variables yields minimal effect on the model performance.

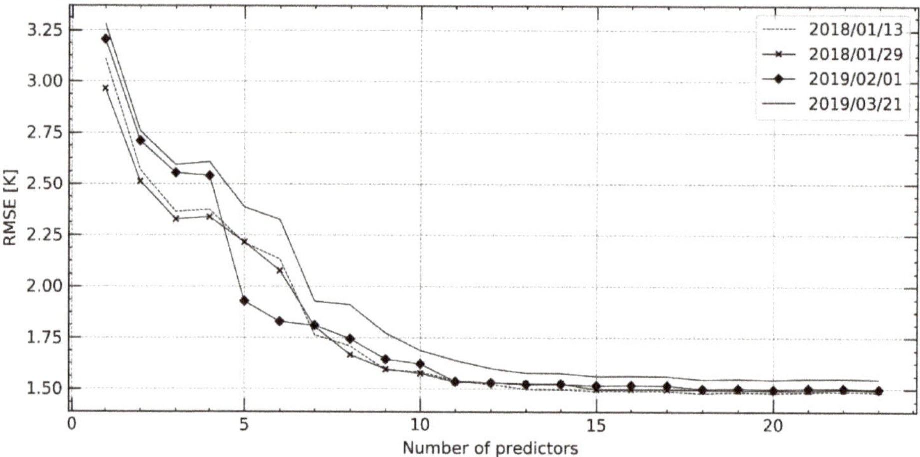

Figure 2. Variation of the root mean squared error (RMSE) obtained through recursive feature elimination with cross-validation (RFECV) with an increasing number of predictors.

Table 4 shows the total number of variables selected, the RMSE values obtained during the variable selection process using RFECV, and the OOB score values for models trained using all the variables and the RFECV selected variables, respectively. In three of the four days, RFECV variable selection resulted in 20 variables except for 13 January 2018, when 18 variables were selected. Overall, the performance of the models based on the two sets of predictor variables is more or less the same.

Table 4. Error metrics obtained with all variables and RFECV selected variables.

Date	All Variables		Selected Variables		Total Selected Variables
	OOB Score [-]	RMSE [K]	OOB Score [-]	RMSE [K]	
13 January 2018	0.93	1.49	0.93	1.48	18
29 January 2018	0.95	1.47	0.96	1.46	20
1 February 2019	0.95	1.48	0.95	1.48	20
21 March 2019	0.93	1.54	0.94	1.53	20
Average	0.94	1.50	0.95	1.49	20

Figure 3 shows the individual variables selected through RFECV for each day and their respective percentage contribution to the final model. Although the RFECV variable selection did not show much variation in the number of selected variables across the days considered, the individual variables selected vary across the days, as observed in Figure 3. Topography-related variables and indices are constantly selected, and only reflectance variables change during the selection process across the four days. In terms of the contribution of the variables to the final model, vegetation indices, i.e., NDVI, Enhanced Vegetation Index (EVI), Soil-Adjusted Vegetation Index (SAVI), and the Fraction Vegetation Cover (FVC), have the highest contribution to the final model. The least contribution is observed in reflectance variables. Elevation follows the vegetation-related indices in the contribution to the final model. From Figure 3, it is also observed that the overall trend in terms of the percentage contribution by the topography-related variables is mostly unchanged.

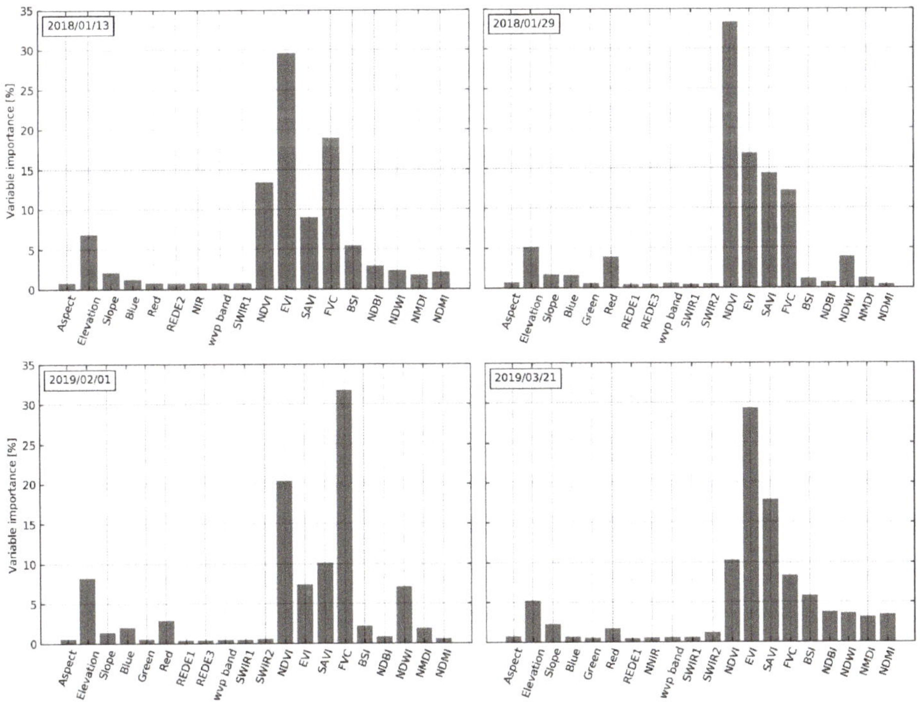

Figure 3. Histograms showing the contribution of each selected predictor variable to the final model for each day corresponding to the four Landsat 8 overpass dates used in the study.

3.2. The Prediction Ability of the Selected Model

Figure 4 shows the scatterplot between the original LSA-SAF LST and the predicted LST obtained using the RF regression model built on the selected variables at the coarse-scale, with the data split into

70% to 30% training and validation sets, respectively. The spread between the points is quite linear and follows the one-to-one line.

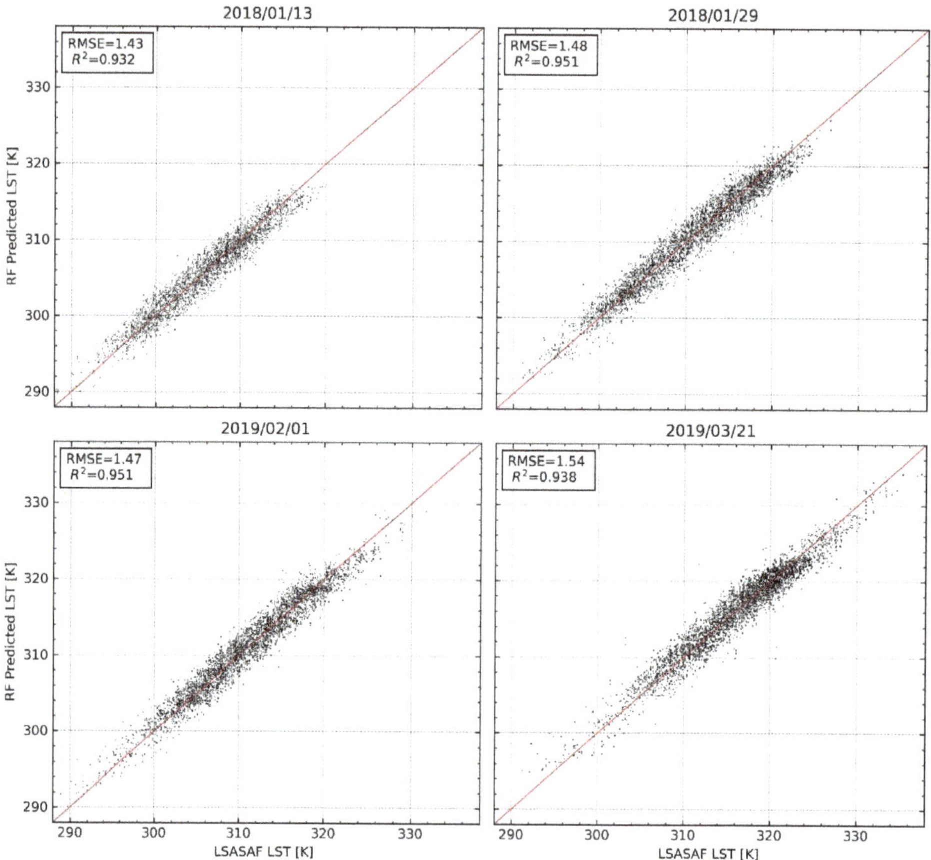

Figure 4. Relationship between LSA-SAF LST and random forest (RF) model-predicted LST at 3km for each day corresponding to the four Landsat 8 overpass dates used in the study.

To investigate the ability of the developed model to predict LST at high resolutions, LST predicted by the model was compared to LST derived from Landsat 8. The results of the comparison are shown in Figure 5. The results indicate a linear relationship between the predicted LST and Landsat 8 LST. However, compared to results at the coarse scale (Figure 4), a decrease in the model performance is observed in Figure 5. There is a reduction in the model performance both in terms of the R2 and the RMSE. Notably, a significant departure from the linear trend is observed in both the upper and lower limits of the scatterplots. At the lower limits, Landsat 8 LST decreases without a corresponding decrease in the predicted LST. In contrast, at the upper limits, Landsat 8 LST increases without a corresponding increase in the predicted LST. Although the model performance at the high resolution is considerably lower, the trend in the model performance in terms of both the R2 and the RMSE observed in Figure 5 is consistent with the trend in the model performance seen in Figure 4.

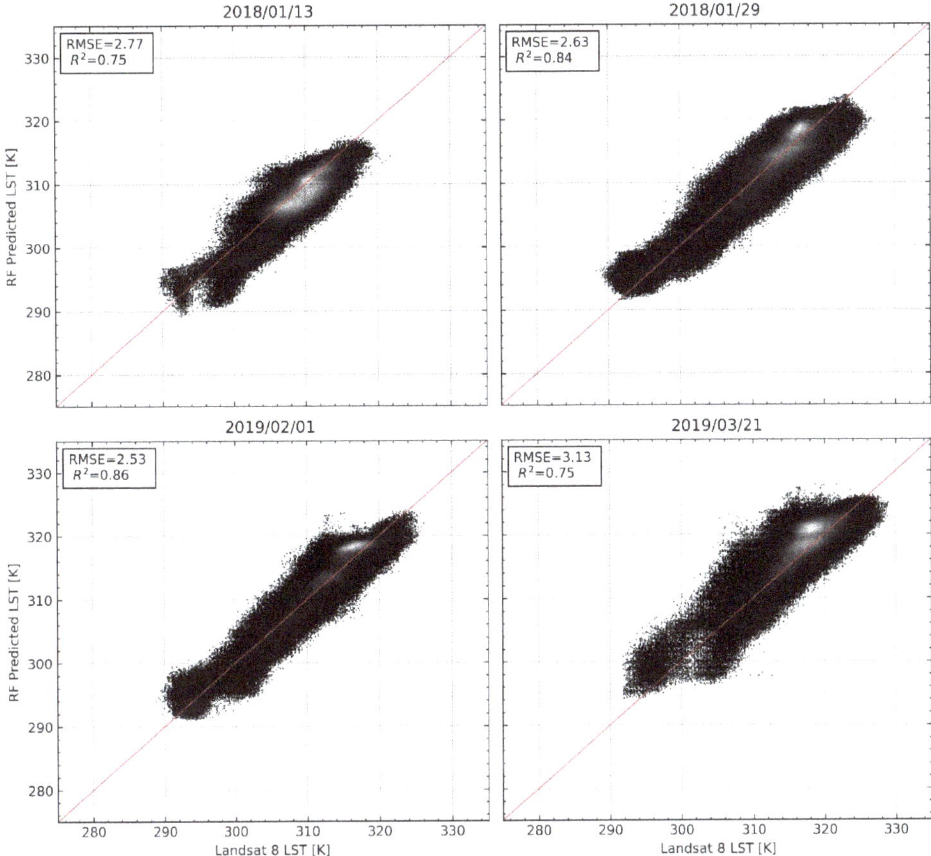

Figure 5. Relationship between Landsat 8-derived LST and RF-predicted LST at 100 m for each day corresponding to the four Landsat 8 overpass dates used in the study.

3.3. Downscaling Results

Figure 6 shows LST maps derived from Landsat 8 and the downscaled LST maps obtained using the conventional RF approach and the proposed RF approach. The downscaled LST maps from both approaches show consistency with Landsat 8-derived LST maps both in terms of the spatial details and variations in the LST across the four days. One notable difference is the considerably lower amount of pixels with missing data (white patches) in the maps derived using the proposed approach in comparison to the maps obtained using the conventional approach as well as from Landsat 8. The pixels with missing data are mainly related to cloud cover, and to a lesser extent, water bodies.

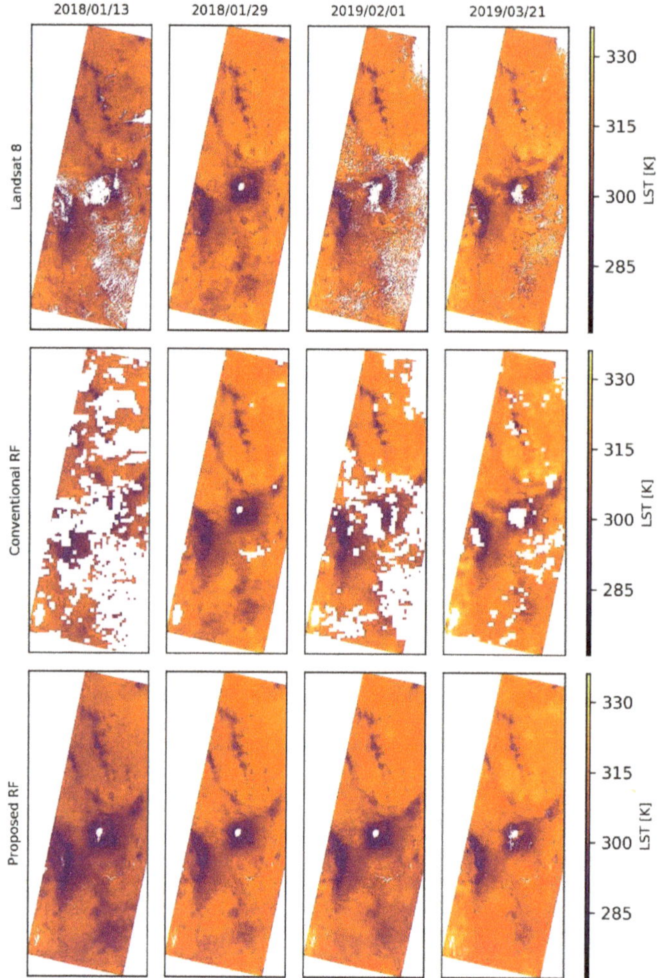

Figure 6. LST maps derived from Landsat 8 (**top**) and the downscaled LST maps (**middle** and **bottom**) corresponding to orbit 168 of Landsat 8 for the four Landsat 8 overpass days used in the study.

Figure 7 shows LST maps for a section of the Aberdares Mountain ranges, one of the areas in the study area that is characterized by complex terrain. The downscaled LST maps in Figure 7 show more spatial details than the coarse-scale LSA-SAF LST maps. The spatial details in the downscaled maps are also consistent with the ones observed in the Landsat 8 LST maps. However, the maps obtained using the proposed approach show enhanced details, which are more consistent with the ones observed in the Landsat 8 LST maps, especially in the transition zones between the mountains and the surrounding areas.

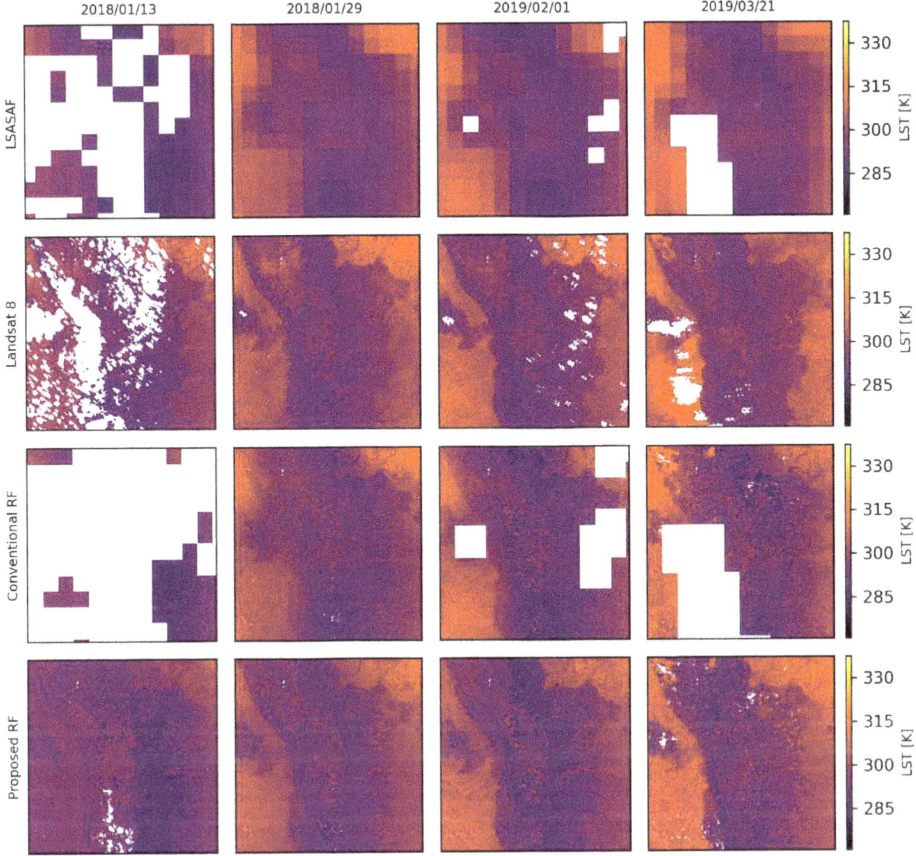

Figure 7. A zoom-in to the area labeled A in Figure 1 showing variations in LST in the complex terrain of the Aberdares Mountain ranges.

Figure 8 shows an ASAL area in the semi-arid northern part of Kenya. Similar to the observations in Figure 7, enhanced spatial details are observed in the downscaled images in comparison to the LSA-SAF LST maps. In addition, more spatial details and better spatial consistency with Landsat 8 LST maps are observed in the maps derived using the proposed approach than in the conventional approach. However, it is also observed that the downscaled maps from both approaches show slightly higher LST than that observed in the Landsat 8 LST maps.

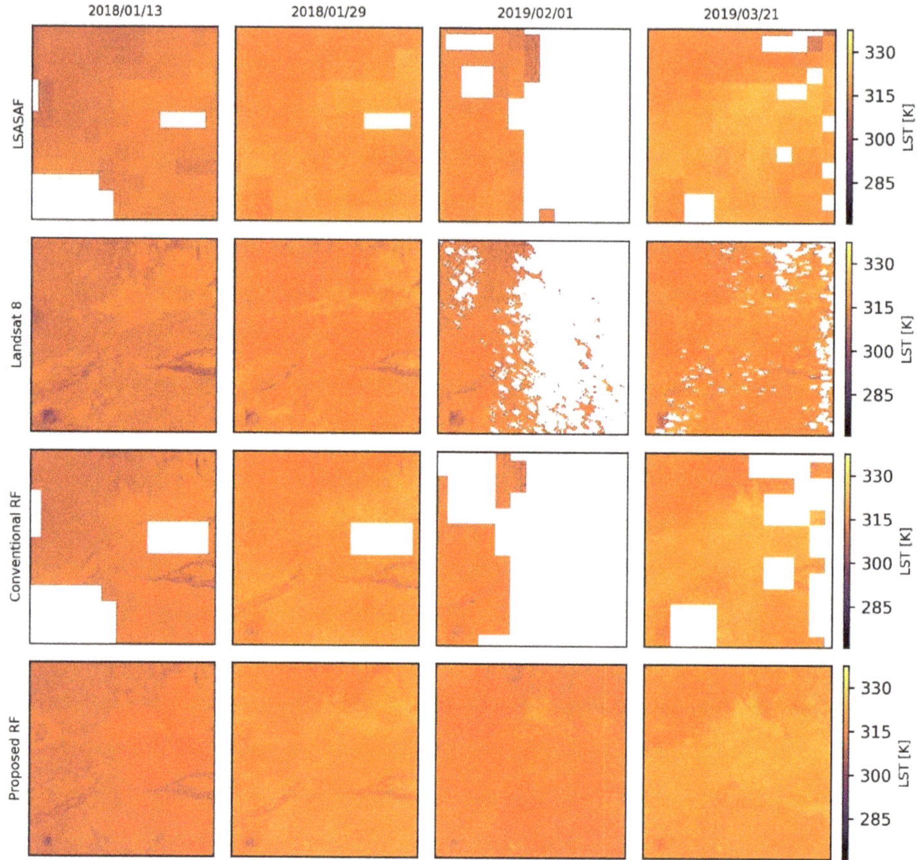

Figure 8. A zoom-in to the area labeled B in Figure 1 showing variations in LST in an arid and semi-arid land (ASAL) area.

Figure 9 shows the linear regression results between the Landsat 8 derived LST and the downscaled LST from the two approaches. The regression results confirm the spatial consistency observed between Landsat 8 LST maps and downscaled LST maps in Figure 6. A strong linear relationship is observed between the Landsat 8 LST and the downscaled LST using both downscaling approaches. However, the proposed approach shows a slightly better performance in terms of the R^2 in comparison to the conventional approach on 29 January 2018 and 1 February 2019 while a slight deterioration in the R^2 is observed on 13 January 2018 and 21 March 2019. In terms of the RMSE, the proposed approach performs marginally better than the conventional approach across all the days. In terms of the bias, the proposed approach shows a better performance than the conventional approach across all the days apart from 13 January 2018. With regards to the spread of points relative to the one-to-one line, the proposed approach shows less scatter, especially at the lower temperatures. However, the proposed approach shows more scatter in the upper-temperature ranges in some of the days, e.g., on 21 March 2019. Based on the scatterplots, it is also evident that, in general, both downscaling approaches tend to overestimate the LST observed by Landsat 8 at the higher LST values.

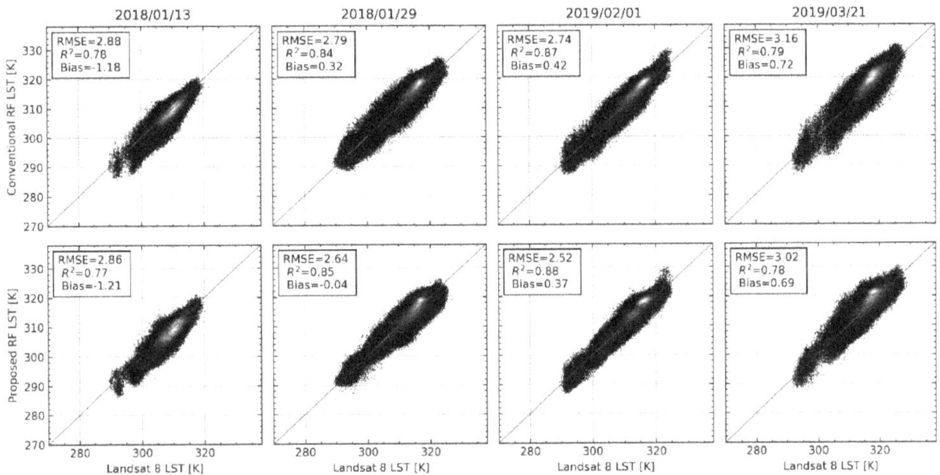

Figure 9. Relationship between Landsat 8 LST and the downscaled LST based on the two approaches for each day corresponding to the four Landsat 8 overpass dates used in the study.

Table 5 shows the statistical results for three categories of vegetation cover classified based on NDVI ranges.

Table 5. Statistical results across three vegetation cover categories based on NDVI.

Vegetation Coverage	Downscaling Approach	Statistical Metric	Date				
			13 January 2018	29 January 2018	1 February 2019	21 March 2019	Average
Sparsely Vegetated [0 < NDVI < 0.2]	Conventional	R^2	0.58	0.59	0.66	0.59	0.61
		RMSE	2.50	2.55	2.46	2.96	2.62
		Bias	−0.73	0.76	1.12	1.13	0.57
	Proposed	R^2	0.52	0.52	0.61	0.53	0.55
		RMSE	2.53	2.62	2.48	2.94	2.64
		Bias	−0.78	0.29	0.96	1.00	0.37
Partially Vegetated [0.2 < NDVI < 0.5]	Conventional	R^2	0.44	0.52	0.58	0.47	0.50
		RMSE	3.76	3.19	2.99	3.94	3.47
		Bias	−2.43	−0.81	−0.64	−1.49	−1.34
	Proposed	R^2	0.46	0.65	0.65	0.53	0.57
		RMSE	3.46	2.04	1.56	2.87	2.48
		Bias	−2.41	−0.57	−0.41	−1.17	−1.14
Fully Vegetated [0.5 < NDVI < 1]	Conventional	R^2	0.33	0.53	0.53	0.31	0.43
		RMSE	4.44	3.29	3.57	4.49	3.95
		Bias	−2.99	−0.54	−1.29	−1.70	−1.63
	Proposed	R^2	0.51	0.77	0.74	0.56	0.65
		RMSE	3.51	2.26	1.63	3.12	2.63
		Bias	−2.67	−0.14	−0.82	−0.75	−1.10

Based on the statistical results presented in Table 5, the least performance based on the conventional approach is observed in the fully vegetated areas. In contrast, the best performance based on the conventional approach is observed in the sparsely vegetated areas. The proposed approach shows a significant improvement in the performance of the downscaled LST in partial and full vegetation canopies based on both the R^2 and the RMSE in comparison to the conventional approach. In terms of the R^2, an average increase of 0.07 and 0.22 is observed in the partially and fully vegetated areas, respectively. In terms of the RMSE, an average reduction of 0.99 and 1.32 K is obtained in the partially and fully vegetated areas, respectively. However, in the sparsely vegetated areas, the proposed approach shows a slight deterioration. An average reduction of 0.06 in the R^2 and an increase of 0.02 K in the RMSE is observed in the sparsely vegetated areas. The proposed approach shows a lower bias across the three classes of vegetation coverage in comparison to the conventional approach. Based on the bias, it is also observed that both approaches overestimate the Landsat 8 LST in the sparsely vegetated areas and underestimate Landsat 8 LST in both the partially and fully vegetated areas.

4. Discussion

The main goal of this study was to reduce the spatial averaging bias inherent in LST downscaling methodology using RF regression in complex environments. As noted in Kustas and Anderson, [2], in complex environments, the partitioning of available energy at the land surface is influenced by many other factors besides soil moisture and vegetation cover. Thus, as highlighted in Hutengs and Vohland, [22], Yang et al. [24] and Zhao et al. [25], additional predictor variables, especially those that have an influence on albedo, surface emissivity, and solar insolation are vital for downscaling LST in complex environments. The results shown in Figure 2 are a confirmation of the added benefits of using multiple predictors variables in the downscaling of LST in complex environments. However, based on the results in Figure 2 and Table 4, it is evident that there is a limit to the number of predictor variables that are beneficial to the model performance. As observed in Figure 2, it appears that the inflection point in terms of the contribution of the predictor variables to the model performance is about 12 variables. Thus, most of the predictor variables beyond the inflection point are likely redundant. The redundancy may likely be due to correlated predictor variables or simply because some predictor variables have little correlation with LST. Although RF is not highly affected by feature correlation, it is beneficial to carry out variable selection since, as pointed out in Guyon et al. [39] and Millard and Richardson, [40], besides reducing the computation demand, the choice of predictors in machine learning reduces the dimensionality problem and thus lowers the risk of model overfitting. Based on the results presented in Figure 3, it appears that reflectance bands constitute most of the redundant predictors. They are the most affected during the process of variable selection in RFECV, and in most cases, have the least contribution to the final model, indicating that they are weaker LST predictors. Similar findings have been reported in Yang et al. [24] and are attributed to the fact that, unlike remote-sensing indices, raw reflectance bands contain less information on the characteristics of the surface. The high contribution by vegetation indices to the linking model observed in Figure 3 is an assertion to the well-established LST-VI feature space concept reported in Sandholt et al. [16]. On the other hand, the stable contribution by topography-related variables shows the significant influence of topography on LST in complex terrain and agrees with findings in Hutengs and Vohland, [22] and Zhao et al. [25].

Although the risk of overfitting in RF is low even in the presence of correlated variables [27,29,36], Strobl et al. [29] note that correlation has huge influence on the variable importance measures generated using RF. The considerable variations in each vegetation index's contributions despite their relatively constant combined contribution to the final model observed in Figure 3 across the four days support the argument by Strobl et al. [29]. Through random selection of predictor variables, the effect of collinearity on the prediction ability of RF is reduced since the contribution of the other correlated variables is significantly reduced because the impurity they can remove is already removed by the first randomly selected predictor variable. Besides, as noted in Strobl et al. [29], the random selection of predictor variables allows for the inclusion of marginalized predictors in the ensemble, thus improving the stability of the model. However, since the random permutation importance approach employed in the computation of variable importance in RF assesses the change in the model performance based on the presence or absence of the selected variable, collinear variables will tend to compensate for each other during the process of random permutation, leading to biases in the variable importance derived from RF. Thus, as observed in Figure 3 any of the correlated variables can be selected to remove the impurity when splitting the nodes making it difficult to determine the individual contribution of each predictor variable to the final model. Consequently, as argued in Matsuki et al. [36] the contribution of the correlated vegetation related variables in Figure 3 can only be interpreted as group and not independently.

Results presented in Figure 4 indicate that the linking models built using Equation (1) are capable of predicting LST in each of the respective model training scenes. A high linear relationship exists between the modeled LST against the 30% unseen LST data set aside during model training, as confirmed by the high R^2 (0.94 on average), indicating the selected predictors can largely explain the thermal variability in the respective scenes. However, upon transfer to the fine spatial resolution, the model performance deteriorates. This deterioration is primarily due to the inability of the selected predictor variables to account for other variations in LST at fine resolution [17]. However, the trend observed on the extreme edges of the scatterplots in Figure 5 cannot be attributed solely to the inability of the predictor variables to capture variations in LST at high resolution. This trend is attributed to the inability of the RF model to predict beyond the data range present in the training data. This trend indicates that the model developed at the coarse-scale largely represents the average conditions within the scene, thus cannot adequately capture the variations in LST under extreme conditions.

Based on visual comparison of the downscaled LST maps presented in Figure 6, not much difference can be discerned between the conventional approach and the proposed approach, apart from less cloud contamination in the maps derived from the proposed approach. However, the zoomed-in areas shown in Figures 7 and 8 indicate that besides lesser cloud contamination, more spatial details consistent with Landsat 8 LST maps are observed in the LST maps derived using the proposed approach. This improvement is attributed to the step implemented to enhance the data range in the RF regression model and the model error correction procedure adopted, which avoids a constant redistribution of the model error calculated at the coarse-scale. However, this step results in the proposed approach reproducing LST even under cloud-contaminated pixels, resulting in less cloud-contaminated LST maps. Although this might be appealing to the user, as suggested in the methodology, it is strongly recommended to mask out the cloud-contaminated pixels. As pointed out in Martins et al. [44], clouds alter the redistribution of the incoming shortwave and longwave radiation at the surface resulting in an entirely different energy balance at the surface in comparison to the energy balance under clear sky conditions. The relatively poor performance in terms of the spread of scatter points observed on 13 January 2018 in Figure 9 also suggests that cloud cover may influence the downscaling results since cloud masking may not detect all the cloud-contaminated pixels. Besides, as pointed out in Kustas et al. [17] and Hutengs and Vohland, [22], downscaling of LST is influenced considerably by thermal contrast within the scene, which may be affected by excessive cloud cover. The reduced thermal contrast may lead to an imbalance in the distribution of LST values in the target LST when training the model. As noted in Millard and Richardson [40], imbalances in the target variable will increase the risk of model overfitting, undermining the model's prediction ability. However, the effect may also be related to the effect of cloud cover on Landsat 8 LST, since the coarse resolution, TCW product used cannot capture the high spatial variability in atmospheric water vapor under cloudy conditions.

The statistical results shown in Table 5 and the scatterplots in Figure 9 indicate that the proposed approach is superior to the conventional approach, especially in vegetated areas. Vegetated areas mainly correspond to the lower temperatures where much of the improvement in the scatter in Figure 9 is observed. The weaker performance in vegetated areas compared to the sparsely vegetated areas based on the conventional approach, appears to contradict results from other studies, e.g., in Kustas et al. [17] and Agam et al. [18]. However, the weaker performance is mainly attributed to topography since most of the fully vegetated areas correspond to forested areas in the mountains. In contrast, the sparsely vegetated areas are mostly located in low lying areas where the effect of topography is less pronounced. As pointed out in Hutengs and Vohland, [22], the complex terrain in mountainous areas has profound implications on LST downscaling. The significant improvement observed in the fully vegetated areas using the proposed approach can be attributed to an improved ability of the proposed approach to capture the spatial details in the complex mountain environment. The enhanced spatial details are due to reduced spatial averaging errors related to the coarse model and the model error correction in the proposed approach.

The biases observed in the downscaled LST in both approaches may be partly related to systematic differences in the Landsat 8 LST and the coarse-resolution LSA SAF LST. However, the tendency by the downscaled LST to overestimate the Landsat 8 LST in the arid sparsely vegetated areas may also be partly attributed to the inability of the selected predictor variables to capture the thermal variability related to other factors such as soil moisture. In addition, it should also be noted that Landsat 8-derived LST has been shown to underestimate observed LST in arid environments [23]. The underestimation is partly related to the use of vegetation cover to obtain emissivity in the calculation of LST from Landsat 8 [41]. The constant bare surface emissivity adopted in the vegetation cover-based emissivity method cannot capture the enormous variations in emissivities depicted by bare surfaces. Thus, it is difficult to draw conclusions from the performance of the downscaling results obtained in the arid sparsely vegetated areas in the absence of reliable LST estimates.

5. Conclusions

In this study, an improved approach for downscaling coarse-resolution TIR data in complex environments based on RF regression is presented. The approach is applied to downscale the coarse-resolution LSA-SAF LST (~3 km) to ~100 m using predictor variables derived from Sentinel 2 and ALOS DEM. Visually, the LST maps obtained based on the proposed downscaling approach show more spatial details, which are more consistent with the Landsat 8-derived maps as opposed to the conventional RF-based downscaling approach.

Quantitatively, comparison with LST derived from Landsat 8 indicates that the proposed approach, in general, outperforms the conventional approach. The proposed approach shows an overall decrease in the bias across all the three vegetation cover categories. Significant improvements are also observed in the forested mountainous areas, where a 0.22 increase in the R^2 and a decrease of 1.32 K in the RMSE is obtained. In the partially vegetated areas, the R^2 improved by 0.07, while a reduction in the RMSE by 0.99 K is achieved. On the other hand, a slight decrease in performance, a 0.06 decrease in the R^2, and an increase of 0.02 K in RMSE is observed in the sparsely vegetated areas. Concrete conclusion on the performance of the downscaling results in the sparsely vegetated areas cannot, however, be made owing to uncertainties inherent in Landsat 8 LST in arid regions.

The improved performance observed based on the proposed approach indicates that the approach is suitable for correcting for spatial averaging errors related to the model trained at the coarse resolution in regions characterized by complex terrain.

Author Contributions: Conceptualization, S.M.N.; data curation, S.M.N.; formal analysis, S.M.N.; investigation, S.M.N.; methodology, S.M.N. and C.M.M.; project administration, C.M.M. and Z.S.; resources, C.M.M. and Z.S.; software, S.M.N.; supervision, C.M.M. and Z.S.; validation, S.M.N.; visualization, S.M.N.; writing—original draft, S.M.N.; writing—review and editing, S.M.N., C.M.M. and Z.S.; All authors have read and agreed to the published version of the manuscript.

Funding: This research received no external funding.

Conflicts of Interest: The authors declare no conflict of interest.

References

1. Li, Z.-L.; Tang, B.-H.; Wu, H.; Ren, H.; Yan, G.; Wan, Z.; Trigo, I.F.; Sobrino, J.A. Satellite-derived land surface temperature: Current status and perspectives. *Remote Sens. Environ.* **2013**, *131*, 14–37. [CrossRef]
2. Kustas, W.; Anderson, M. Advances in thermal infrared remote sensing for land surface modeling. *Agric. For. Meteorol.* **2009**, *149*, 2071–2081. [CrossRef]
3. Luyssaert, S.; Jammet, M.; Stoy, P.C.; Estel, S.; Pongratz, J.; Ceschia, E.; Churkina, G.; Don, A.; Erb, K.; Ferlicoq, M.; et al. Land management and land-cover change have impacts of similar magnitude on surface temperature. *Nat. Clim. Chang.* **2014**, *4*, 389–393. [CrossRef]
4. Sismanidis, P.; Bechtel, B.; Keramitsoglou, I.; Kiranoudis, C.T. Mapping the Spatiotemporal Dynamics of Europe's Land Surface Temperatures. *IEEE Geosci. Remote Sens. Lett.* **2018**, *15*, 202–206. [CrossRef]

5.   Senay, G.B.; Friedrichs, M.; Singh, R.K.; Velpuri, N.M. Evaluating Landsat 8 evapotranspiration for water use mapping in the Colorado River Basin. *Remote Sens. Environ.* **2016**, *185*, 171–185. [CrossRef]

6.   Timmermans, W.J.; Kustas, W.P.; Andreu, A. Utility of an automated thermal-based approach for monitoring evapotranspiration. *Acta Geophys.* **2015**, *63*, 1571–1608. [CrossRef]

7.   Allen, R.; Irmak, A.; Trezza, R.; Hendrickx, J.M.H.; Bastiaanssen, W.; Kjaersgaard, J. Satellite-based ET estimation in agriculture using SEBAL and METRIC. *Hydrol. Process.* **2011**, *25*, 4011–4027. [CrossRef]

8.   Taktikou, E.; Bourazanis, G.; Papaioannou, G.; Kerkides, P. Prediction of Soil Moisture from Remote Sensing Data. *Procedia Eng.* **2016**, *162*, 309–316. [CrossRef]

9.   Mishra, V.; Ellenburg, W.L.; Griffin, R.E.; Mecikalski, J.R.; Cruise, J.F.; Hain, C.R.; Anderson, M.C. An initial assessment of a SMAP soil moisture disaggregation scheme using TIR surface evaporation data over the continental United States. *Int. J. Appl. Earth Obs. Geoinf.* **2018**, *68*, 92–104. [CrossRef]

10. Holzman, M.E.; Carmona, F.; Rivas, R.; Niclòs, R. Early assessment of crop yield from remotely sensed water stress and solar radiation data. *ISPRS J. Photogramm. Remote Sens.* **2018**, *145*, 297–308. [CrossRef]

11. Hong, S.; Lakshmi, V.; Small, E.E.; Hong, S.; Lakshmi, V.; Small, E.E. Relationship between Vegetation Biophysical Properties and Surface Temperature Using Multisensor Satellite Data. *J. Clim.* **2007**, *20*, 5593–5606. [CrossRef]

12. Tran, H.; Uchihama, D.; Ochi, S.; Yasuoka, Y. Assessment with satellite data of the urban heat island effects in Asian mega cities. *Int. J. Appl. Earth Obs. Geoinf.* **2006**, *8*, 34–48. [CrossRef]

13. Dominguez, A.; Kleissl, J.; Luvall, J.C.; Rickman, D.L. High-resolution urban thermal sharpener (HUTS). *Remote Sens. Environ.* **2011**, *115*, 1772–1780. [CrossRef]

14. Ha, W.; Gowda, P.H.; Howell, T.A. A review of downscaling methods for remote sensing-based irrigation management: Part I. *Irrig. Sci.* **2013**, *31*, 831–850. [CrossRef]

15. Zhan, W.; Chen, Y.; Zhou, J.; Wang, J.; Liu, W.; Voogt, J.; Zhu, X.; Quan, J.; Li, J. Disaggregation of remotely sensed land surface temperature: Literature survey, taxonomy, issues, and caveats. *Remote Sens. Environ.* **2013**, *131*, 119–139. [CrossRef]

16. Sandholt, I.; Rasmussen, K.; Andersen, J. A simple interpretation of the surface temperature/vegetation index space for assessment of surface moisture status. *Remote Sens. Environ.* **2002**, *79*, 213–224. [CrossRef]

17. Kustas, W.P.; Norman, J.M.; Anderson, M.C.; French, A.N. Estimating subpixel surface temperatures and energy fluxes from the vegetation index–radiometric temperature relationship. *Remote Sens. Environ.* **2003**, *85*, 429–440. [CrossRef]

18. Agam, N.; Kustas, W.P.; Anderson, M.C.; Li, F.; Neale, C.M.U. A vegetation index based technique for spatial sharpening of thermal imagery. *Remote Sens. Environ.* **2007**, *107*, 545–558. [CrossRef]

19. Govil, H.; Guha, S.; Dey, A.; Gill, N. Seasonal evaluation of downscaled land surface temperature: A case study in a humid tropical city. *Heliyon* **2019**, *5*, e01923. [CrossRef] [PubMed]

20. Wu, J.; Zhong, B.; Tian, S.; Yang, A.; Wu, J. Downscaling of Urban Land Surface Temperature Based on Multi-Factor Geographically Weighted Regression. *IEEE J. Sel. Top. Appl. Earth Obs. Remote Sens.* **2019**, *12*, 2897–2911. [CrossRef]

21. Li, W.; Ni, L.; Li, Z.L.; Duan, S.B.; Wu, H. Evaluation of machine learning algorithms in spatial downscaling of modis land surface temperature. *IEEE J. Sel. Top. Appl. Earth Obs. Remote Sens.* **2019**, *12*, 2299–2307. [CrossRef]

22. Hutengs, C.; Vohland, M. Downscaling land surface temperatures at regional scales with random forest regression. *Remote Sens. Environ.* **2016**, *178*, 127–141. [CrossRef]

23. Pan, X.; Zhu, X.; Yang, Y.; Cao, C.; Zhang, X.; Shan, L. Applicability of Downscaling Land Surface Temperature by Using Normalized Difference Sand Index. *Sci. Rep.* **2018**, *8*, 9530. [CrossRef] [PubMed]

24. Yang, Y.; Cao, C.; Pan, X.; Li, X.; Zhu, X.; Yang, Y.; Cao, C.; Pan, X.; Li, X.; Zhu, X. Downscaling Land Surface Temperature in an Arid Area by Using Multiple Remote Sensing Indices with Random Forest Regression. *Remote Sens.* **2017**, *9*, 789. [CrossRef]

25. Zhao, W.; Duan, S.-B.; Li, A.; Yin, G. A practical method for reducing terrain effect on land surface temperature using random forest regression. *Remote Sens. Environ.* **2019**, *221*, 635–649. [CrossRef]

26. Xu, J.; Zhang, F.; Jiang, H.; Hu, H.; Zhong, K.; Jing, W.; Yang, J.; Jia, B. Downscaling Aster Land Surface Temperature over Urban Areas with Machine Learning-Based Area-To-Point Regression Kriging. *Remote Sens.* **2020**, *12*, 1082. [CrossRef]

27. Breiman, L. Random Forests. *Mach. Learn.* **2001**, *45*, 5–32. [CrossRef]

28. Tyralis, H.; Papacharalampous, G.; Langousis, A.; Tyralis, H.; Papacharalampous, G.; Langousis, A. A Brief Review of Random Forests for Water Scientists and Practitioners and Their Recent History in Water Resources. *Water* **2019**, *11*, 910. [CrossRef]

29. Strobl, C.; Boulesteix, A.L.; Kneib, T.; Augustin, T.; Zeileis, A. Conditional variable importance for random forests. *BMC Bioinform.* **2008**, *9*, 307. [CrossRef]

30. Bindhu, V.M.; Narasimhan, B.; Sudheer, K.P. Development and verification of a non-linear disaggregation method (NL-DisTrad) to downscale MODIS land surface temperature to the spatial scale of Landsat thermal data to estimate evapotranspiration. *Remote Sens. Environ.* **2013**, *135*, 118–129. [CrossRef]

31. Essa, W.; Verbeiren, B.; van der Kwast, J.; Batelaan, O. Improved DisTrad for Downscaling Thermal MODIS Imagery over Urban Areas. *Remote Sens.* **2017**, *9*, 1243. [CrossRef]

32. Jimenez-Munoz, J.C.; Cristobal, J.; Sobrino, J.A.; Sòria, G.; Ninyerola, M.; Pons, X. Revision of the single-channel algorithm for land surface temperature retrieval from landsat thermal-infrared data. *IEEE Trans. Geosci. Remote Sens.* **2009**, *47*, 339–349. [CrossRef]

33. Carlson, T.N.; Ripley, D.A. On the relation between NDVI, fractional vegetation cover, and leaf area index. *Remote Sens. Environ.* **1997**, *62*, 241–252. [CrossRef]

34. Sobrino, J.A.; Mattar, C.; Pardo, P.; Jiménez-Muñoz, J.C.; Hook, S.J.; Baldridge, A.; Ibañez, R. Soil emissivity and reflectance spectra measurements. *Appl. Opt.* **2009**, *48*, 3664–3670. [CrossRef]

35. Sobrino, J.A.; Jiménez-Muñoz, J.C.; Sòria, G.; Romaguera, M.; Guanter, L.; Moreno, J.; Plaza, A.; Martínez, P. Land surface emissivity retrieval from different VNIR and TIR sensors. *IEEE Trans. Geosci. Remote Sens.* **2008**, *46*, 316–327. [CrossRef]

36. Matsuki, K.; Kuperman, V.; Van Dyke, J.A. The Random Forests statistical technique: An examination of its value for the study of reading. *Sci. Stud. Read.* **2016**, *20*, 20–33. [CrossRef] [PubMed]

37. Chen, X.; Ishwaran, H. Random forests for genomic data analysis. *Genomics* **2012**, *99*, 323–329. [CrossRef] [PubMed]

38. Pedregosa, F.; Varoquaux, G.; Gramfort, A.; Michel, V.; Thirion, B.; Grisel, O.; Blondel, M.; Prettenhofer, P.; Weiss, R.; Dubourg, V.; et al. Scikit-learn: Machine Learning in Python. *J. Mach. Learn. Res.* **2011**, *12*, 2825–2830.

39. Guyon, I.; Weston, J.; Barnhill, S.; Vapnik, V. Gene Selection for Cancer Classification using Support Vector Machines. *Mach. Learn.* **2002**, *46*, 389–422. [CrossRef]

40. Millard, K.; Richardson, M. On the Importance of Training Data Sample Selection in Random Forest Image Classification: A Case Study in Peatland Ecosystem Mapping. *Remote Sens.* **2015**, *7*, 8489–8515. [CrossRef]

41. Li, S.; Jiang, G. Land Surface Temperature Retrieval from Landsat-8 Data with the Generalized Split-Window Algorithm. *IEEE Access* **2018**, *6*, 18149–18162. [CrossRef]

42. Trigo, I.; Monteiro, I.; Olesen, F.; Kabsch, E. An assessment of remotely sensed land surface temperature. *J. Geophys. Res.* **2008**, *113*, D17108. [CrossRef]

43. Dozier, J. A generalized split-window algorithm for retrieving land-surface temperature from space. *IEEE Trans. Geosci. Remote Sens.* **1996**, *34*, 892–905. [CrossRef]

44. Martins, J.P.A.; Trigo, I.F.; Ghilain, N.; Jimenez, C.; Göttsche, F.-M.; Ermida, S.L.; Olesen, F.-S.; Gellens-Meulenberghs, F.; Arboleda, A. An All-Weather Land Surface Temperature Product Based on MSG/SEVIRI Observations. *Remote Sens.* **2019**, *11*, 3044. [CrossRef]

45. Göttsche, F.-M.; Olesen, F.-S.; Trigo, F.I.; Bork-Unkelbach, A.; Martin, A.M. Long Term Validation of Land Surface Temperature Retrieved from MSG/SEVIRI with Continuous in-Situ Measurements in Africa. *Remote Sens.* **2016**, *8*, 410. [CrossRef]

46. Niclòs, R.; Galve, J.M.; Valiente, J.A.; Estrela, M.J.; Coll, C. Accuracy assessment of land surface temperature retrievals from MSG2-SEVIRI data. *Remote Sens. Environ.* **2011**, *115*, 2126–2140. [CrossRef]

47. Tadono, T.; Nagai, H.; Ishida, H.; Oda, F.; Naito, S.; Minakawa, K.; Iwamoto, H. Generation of the 30 m-mesh global digital surface model by alos prism. *ISPRS Int. Arch. Photogramm. Remote Sens. Spat. Inf. Sci.* **2016**, *41*, 157–162. [CrossRef]

48. Caglar, B.; Becek, K.; Mekik, C.; Ozendi, M. On the vertical accuracy of the ALOS world 3D-30m digital elevation model. *Remote Sens. Lett.* **2018**, *9*, 607–615. [CrossRef]

49. Louis, J.; Debaecker, V.; Pflug, B.; Main-Knorn, M.; Bieniarz, J.; Mueller-Wilm, U.; Cadau, E.; Gascon, F. Sentinel-2 Sen2Cor: L2A processor for users. In Proceedings of the ESA Living Planet Symposium, Prague, Czech Republic, 9–13 May 2016.

50. Montanaro, M.; Gerace, A.; Lunsford, A.; Reuter, D. Stray Light Artifacts in Imagery from the Landsat 8 Thermal Infrared Sensor. *Remote Sens.* **2014**, *6*, 10435–10456. [CrossRef]
51. Gerace, A.; Montanaro, M. Derivation and validation of the stray light correction algorithm for the thermal infrared sensor onboard Landsat 8. *Remote Sens. Environ.* **2017**, *191*, 246–257. [CrossRef]
52. Rouse, J.W.J.; Haas, R.H.; Schell, J.A.; Deering, D.W. Monitoring Vegetation Systems in the Great Plains with Erts. *NASA Spec. Publ.* **1974**, *351*, 309–317.
53. Huete, A.; Didan, K.; Miura, T.; Rodriguez, E.P.; Gao, X.; Ferreira, L.G. Overview of the radiometric and biophysical performance of the MODIS vegetation indices. *Remote Sens. Environ.* **2002**, *83*, 195–213. [CrossRef]
54. Huete, A.R. A soil-adjusted vegetation index (SAVI). *Remote Sens. Environ.* **1988**, *25*, 295–309. [CrossRef]
55. Meng, X.; Cheng, J.; Zhao, S.; Liu, S.; Yao, Y. Estimating Land Surface Temperature from Landsat-8 Data using the NOAA JPSS Enterprise Algorithm. *Remote Sens.* **2019**, *11*, 155. [CrossRef]
56. Rikimaru, A.; Roy, P.S.; Miyatake, S. Tropical forest cover density mapping. *Trop. Ecol.* **2002**, *43*, 9–39.
57. Zha, Y.; Gao, J.; Ni, S. Use of normalized difference built-up index in automatically mapping urban areas from TM imagery. *Int. J. Remote Sens.* **2003**, *24*, 583–594. [CrossRef]
58. McFeeters, S.K. The use of the Normalized Difference Water Index (NDWI) in the delineation of open water features. *Int. J. Remote Sens.* **1996**, *17*, 1425–1432. [CrossRef]
59. Wang, L.; Qu, J.J. NMDI: A normalized multi-band drought index for monitoring soil and vegetation moisture with satellite remote sensing. *Geophys. Res. Lett.* **2007**, *34*, L20405. [CrossRef]
60. Gao, B.C. NDWI—A normalized difference water index for remote sensing of vegetation liquid water from space. *Remote Sens. Environ.* **1996**, *58*, 257–266. [CrossRef]

Publisher's Note: MDPI stays neutral with regard to jurisdictional claims in published maps and institutional affiliations.

Article

Evaluating the Variability of Urban Land Surface Temperatures Using Drone Observations

Joseph Naughton and Walter McDonald *

Department of Civil, Construction & Environmental Engineering, Marquette University, P.O. Box 1881, Milwaukee, WI 53211, USA
* Correspondence: walter.mcdonald@marquette.edu; Tel.: +1-414-288-2117

Received: 30 May 2019; Accepted: 16 July 2019; Published: 20 July 2019

Abstract: Urbanization and climate change are driving increases in urban land surface temperatures that pose a threat to human and environmental health. To address this challenge, we must be able to observe land surface temperatures within spatially complex urban environments. However, many existing remote sensing studies are based upon satellite or aerial imagery that capture temperature at coarse resolutions that fail to capture the spatial complexities of urban land surfaces that can change at a sub-meter resolution. This study seeks to fill this gap by evaluating the spatial variability of land surface temperatures through drone thermal imagery captured at high-resolutions (13 cm). In this study, flights were conducted using a quadcopter drone and thermal camera at two case study locations in Milwaukee, Wisconsin and El Paso, Texas. Results indicate that land use types exhibit significant variability in their surface temperatures (3.9–15.8 °C) and that this variability is influenced by surface material properties, traffic, weather and urban geometry. Air temperature and solar radiation were statistically significant predictors of land surface temperature (R^2 0.37–0.84) but the predictive power of the models was lower for land use types that were heavily impacted by pedestrian or vehicular traffic. The findings from this study ultimately elucidate factors that contribute to land surface temperature variability in the urban environment, which can be applied to develop better temperature mitigation practices to protect human and environmental health.

Keywords: land surface temperature; drones; unmanned aerial vehicles; thermal remote sensing

1. Introduction

Urban areas across the world are subject to thermal stresses caused by the surface urban heat island (SUHI) effect where urban land surfaces experience higher temperatures than their surrounding rural areas. This is in large part due to the replacement of undeveloped vegetated land with anthropogenic materials that absorb more solar radiation and have different heat capacity and surface radiative properties [1]. This results in higher surface temperatures that pose a significant threat to human health [2], as well as higher storm runoff temperatures that can harm aquatic life [3–5]. These stresses are only expected to grow with increases in global temperatures and urban populations; therefore, it is critical that we understand the fundamental processes that drive land surface temperature (LST) to develop solutions that can protect human and environmental health.

To that end, thermal remote sensing is an important tool for evaluating urban land surface temperatures. This includes satellite sensors such as ASTER, MODIS and Landsat that can capture land surface temperatures at 30 m–1 km resolutions [6]. Data from these satellites have been used to extensively study urban land surface temperatures and their effects [7–14]. However, while satellite remote sensing is valuable for evaluating LST across a city scale, the spatial resolution precludes its applications to smaller spatial scales that better reflect the spatial complexity of the urban environment. To acquire higher resolution thermal data, studies have used aerial reconnaissance or downscaling techniques [15,16]; however, these are still at resolutions (4–10 m) that cannot capture changes that occur on a sub meter resolution. Furthermore, satellite remote sensing is temporally constrained to intervals between 1–14 days. Aerial flights do not have the same temporal constraints; however, doing so at on-demand temporal resolutions would not be economically practical. Therefore, these methods are inadequate for evaluating changes in urban LST that occur throughout the day or capturing the spatial heterogeneity of urban LST at small scales.

This challenge is important to overcome as urban land surfaces are spatially complex and significant variations in land cover can occur on a sub meter spatial resolution [17]. While existing research has demonstrated that the spatial configuration of land use classifications at a city scale are important (i.e. industrial, residential, forest) [18,19], less is known about the importance of the spatial configuration and variations in LST at smaller scales (i.e., sidewalks, grass medians, flowerbeds, etc.). In addition, the urban environment is dynamic and land surface temperatures can be significantly influenced by other factors besides land cover material properties [20]. Land surface temperature may therefore vary significantly across small spatial scales; however, the factors that control this variation are not well defined. Doing so requires direct measurements of surface temperatures across wide spatial and temporal scales, yet little research to date has evaluated the spatial variability in temperature among urban land use types in sub-meter resolutions. This may be due to measurement limitations, as satellite data is too coarse and in-situ temperature probes are too expensive to densely distribute across an urban landscape. Therefore, new and innovative approaches to measuring land surface temperatures at small spatial and temporal scales are needed to assess thermal variability across land use types in the urban environment.

Unmanned Aerial Vehicles (UAVs) or drones, are a technology that can meet this challenge. Recent advances in UAVs and radiometric thermal cameras have made it possible to capture land surface temperatures on-demand and at sub-meter spatial resolutions that accurately reflect the spatial complexity and detail of land surface temperatures in the urban environment [21]. UAVs also have advantages in that they can be flown on demand to capture LST at temporal resolutions unmatched by satellite or aerial imagery. While the limited battery life of around 30 minutes for quad-copter UAVs constrains the area that can be captured in a single flight, their spatial and temporal resolutions offer significant advantages for evaluating the variability of LST in the urban environment at fine spatial and temporal scales.

We therefore present a study to evaluate the variability of temperatures across urban land surfaces using a UAV. In this study, we apply a UAV and radiometric thermal camera to capture land surface temperatures at high-resolutions (13 cm) in two case study locations: Milwaukee, Wisconsin and El Paso, Texas. Using data collected throughout a calendar year, we evaluate the variability in land surface temperatures, develop models to predict mean land surface temperature based upon weather parameters and evaluate the diurnal trends in urban land surface temperature. To do so, we (1) quantify land surface temperature variability across different surface types, (2) evaluate variance in temperature across different surface types based upon meteorological and/or other derived parameters (e.g., albedo, normalized difference vegetation index, apparent thermal inertia, etc.), (3) predict land surface temperature based upon meteorological parameters and (4) assess diurnal variability in land surface temperature magnitude and uncertainty. Ultimately, this study helps to elucidate factors that contribute to land surface temperature variability in the urban environment at small spatial scales, which can then be applied to develop better temperature mitigation strategies.

2. Materials and Methods

2.1. Case Study Locations

Two case study locations were chosen for this project: (1) a portion of Marquette University's campus in Milwaukee, WI and (2) a portion University of Texas El Paso's (UTEP) campus in El Paso, Texas (Figure 1). The Marquette and UTEP case study areas were roughly 21,300 m^2 and 27,300 m^2, respectively and included a balance of both natural landscape and impervious gray surfaces. Surface types within each case study location were manually delineated using ESRI's ArcMap software. The nine surfaces types identified at Marquette and UTEP and their respective surface areas are listed in Table 1. The specific locations on each campus were chosen for their variety of surface types, similarities in land use between the two locations and suitability for drone takeoff/landing and flying. In addition, these locations provide a contrast in geography, climate and weather that are helpful in testing the generalizability of our findings. For example, Milwaukee's climate is classified by Koppen and Geiger as Dfa (Humid Continental Hot Summers With Year Around Precipitation) and receives 870 mm of precipitation annually, while El Paso is classified as BWk (Cold Desert Climate) and receives 221 mm of precipitation annually [22,23].

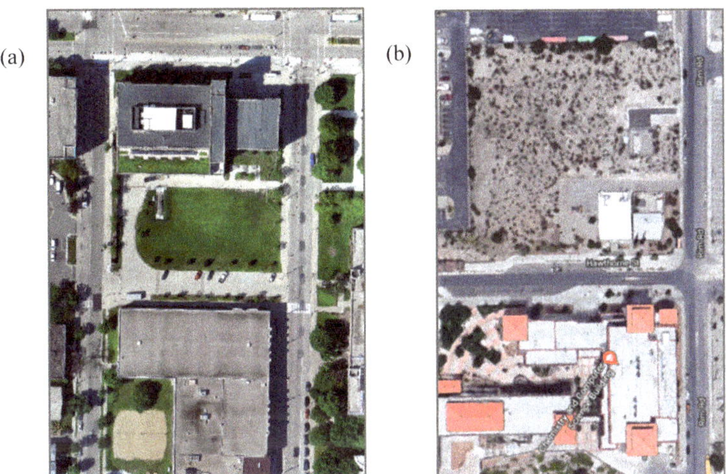

Figure 1. Visual imagery of the case study locations: Marquette University (**a**) and University of Texas El Paso (UTEP) (**b**). Visual imagery of Marquette was captured from a drone on 11 August 2018. Visual imagery of UTEP was pulled from Google Maps on 13 March 2019.

Table 1. Surface types and surface areas within each case study location.

MARQUETTE		UTEP	
Surface Type	**Surface Area (m^2)**	**Surface Type**	**Surface Area (m^2)**
Grass	2738	Rooftop (rammed earth)	503
Sidewalk	904	Desert Shrub	173
Rooftop (composite)	336	Rooftop (composite)	2047
Road (asphalt)	3299	Parking Lot (asphalt)	1350
Parking Lot (concrete)	908	Sidewalk (concrete)	9808
Rooftop (rubber)	6,057	Road (asphalt)	4253
Canopy Cover	4758	Parking Lot (concrete)	4081
Shrub/mulch	2272	Grass	1270
Solar	65	Canopy Cover	3782

2.2. Equipment

Remote sensing data was collected using a DJI Matrice 100 (M100) quadcopter UAV. The M100 was deployed at our case study locations with three types of camera payloads—visual, multispectral and infrared. These cameras include the DJI Zenmuse X3 visual (12 MP), Zenmuse X3 multispectral (Blue-Green-NIR 680–800 nm at 12 MP) and DJI Zenmuse XTR radiometric thermal (13 mm, 30 Hz and spectral bandwidth of 7–13 μm). Additionally, ground temperatures were validated using a Nubee NUB8380 Digital Infrared Thermometer.

2.3. Data Collection Methods

Two datasets were collected during the 2018 calendar year: (1) surface temperature measured at 12:00 PM across the entire year and (2) surface temperature measured on a diurnal cycle. To evaluate surface temperature across the entire year, fourteen flights in Milwaukee and one in El Paso were recorded between 26 February and 13 September 2018 (Table 2). To evaluate the diurnal cycle of temperature, three flights in Milwaukee and one in El Paso measured temperature throughout the day at 9:00 AM, 12:00 PM, 3:00 PM and 5:00 PM (Table 3). Weather data was collected at Marquette from a station on top of Engineering Hall and weather data at UTEP was collected from a weather station 10.5 km away at El Paso International Airport. Each station recorded air temperature, relative humidity, wind speed, wind direction, relative humidity, solar radiation and atmospheric pressure. Drone imagery was captured autonomously using a third-party photogrammetry software called Pix4Dcapture. Using this software, autonomous flight paths were programmed to the drone prior to each mission. Programmed flight path information included drone speed, altitude and image overlap. Drone speed was set at 54 km/h for visual and multispectral flights but set at a lower threshold of 30.6 km/h for thermal flights due to the difference in image capture speed between the two camera technologies. The flight altitude for each mission was set to the United States Federal Aviation Administration (FAA) maximum allowable limit of 120 m, which resulted in thermal imagery at a 13 cm pixel size. Finally, the image overlap was set to 85%, which provided reliable overlap for stitching an orthomosaic during data processing.

Table 2. Flight log and summary of meteorological variables recorded for Marquette and UTEP during fifteen noon flights.

Flight Number	Flight Date	Flight Time	Air Temp (°C)	Relative Humidity (%)	Wind Speed (m/s)	Wind Dir (Degrees)	Solar Rad (kW/m^2)	Pressure (kPa)
MU 1	26 February2018	12:00 PM	−1.7	54	4.0	225	0.00	102.2
MU 2	12 April 2018	12:00 PM	12.4	65.8	6.8	285	0.41	100.2
MU 3	8 May 2018	12:00 PM	26.4	22.1	4.2	218	0.81	101.7
UTEP 1	20 May 2018	12:00 PM	27.8	26	5.8	120	0.96	101.7
MU 4	13 June 2018	12:00 PM	25.3	33.3	3.4	321	0.89	101.2
MU 5	29 June 2018	12:00 PM	31.5	54.7	5.4	193	0.80	101.0
MU 6	11 July 2018	12:00 PM	25.9	44.1	2.0	91	0.78	101.9
MU 7	12 July 2018	12:00 PM	27.4	43.2	5.9	204	0.60	101.8
MU 8	17 July 2018	12:00 PM	25	38.9	3.0	38	0.74	101.6
MU 9	18 July 2018	12:00 PM	22.5	56.3	3.1	101	0.83	101.8
MU 10	25 July 2018	12:00 PM	28.6	31.9	2.5	271	0.83	101.4
MU 11	10 August 2018	12:00 PM	25.8	58.8	2.3	84	0.77	101.3
MU 12	31 August 2018	12:00 PM	25.8	49.9	4.0	158	0.09	101.6
MU 13	12 September 2018	12:00 PM	26.1	55.3	3.1	168	0.56	101.9
MU 14	13 September 2018	12:00 PM	22.8	64.9	4.2	127	0.68	102.0

Table 3. Flight log and summary of meteorological variables recorded for Marquette and UTEP during four diurnal flights.

Flight Number	Flight Date	Flight Time	Air Temp (°C)	Relative Humidity (%)	Wind Speed (m/s)	Wind Dir (Degrees)	Solar Rad (kW/m^2)	Pressure (kPa)
MU1	13 June 2018	9:00 AM	22.5	42.8	4.4	320	0.73	101.1
MU1	13 June 2018	12:00 PM	25.3	33.3	3.4	321	0.89	101.2
MU1	13 June 2018	3:00 PM	27.5	20.1	3.0	328	0.80	101.2
MU1	13 June 2018	5:00 PM	27.9	19.8	2.0	285	0.51	101.2
MU2	17 July 2018	9:00 AM	23.8	37.7	3.1	8	0.52	101.6
MU2	17 July 2018	12:00 PM	25	38.9	3.0	38	0.74	101.6
MU2	17 July 2018	3:00 PM	25	41.2	3.0	38	0.76	101.7
MU2	17 July 2018	5:00 PM	22.8	57.9	3.4	34	0.50	101.7
MU3	10 August 2018	9:00 AM	27.4	46.1	2.4	33	0.70	101.3
MU3	10 August 2018	12:00 PM	25.7	58.8	2.3	84	0.77	101.3
MU3	10 August 2018	3:00 PM	27.4	46.1	2.4	33	0.70	101.3
MU3	10 August 2018	5:00 PM	27.4	33.4	2.4	37	0.43	101.2
UTEP1	20 May 2018	9:00 AM	25	32	5.8	90	0.66	101.8
UTEP1	20 May 2018	12:00 PM	27.8	26	5.8	120	0.96	101.7
UTEP1	20 May 2018	3:00 PM	31.1	17	4.0	120	0.83	101.4
UTEP1	20 May 2018	5:00 PM	31.1	21	4.9	90	0.50	101.3

2.4. Thermal Data Processing

After data collection in the field, a series of post-processing steps were performed using Pix4D and ESRI's ArcMap to stitch the drone thermal imagery into orthomosaics, correct temperature values for emissivity and extract surface temperature data for analysis. First, Pix4D was used to stitch the captured thermal images into orthomosaics, export the orthomosaics as a 32-bit TIFF and georeference them for application within ArcMap.

Once in ArcMap, an emissivity correction was applied to each thermal orthomosaic. Emissivity is a measure of how well a material can emit energy as thermal radiation and different materials have different values of emissivity depending on their surface properties [24]. Land use classifications that were previously delineated for each case study area were used to apply emissivity values to the target surfaces. The emissivity values for each land use classification used in this study are listed in Table 4 and are based upon a review of emissivity studies. These emissivity values were then applied in the following emissivity correction equation derived from Stefan-Boltzmann Law:

$$T_{target} = \sqrt[4]{\frac{T_{sensor}^4 - (1 - \varepsilon) * T_{background}^4}{\varepsilon}} \tag{1}$$

where T_{target} is the actual temperature of the target surface [K], T_{sensor} is the temperature measured by the infrared camera [K], $T_{background}$ is the recorded air temperature [K] and ε is the emissivity value of the target surface [25]. This equation was used to correct each surface type for their respective emissivity before performing spatial data analysis.

Table 4. Emissivity values for each surface type.

Land Use Type	Emissivity Value	Reference
Grass	0.979	[26]
Shrub/mulch	0.928	[27]
Road (asphalt)	0.95	[28,29]
Parking Lot (concrete)	0.91	[29–31]
Sidewalk (concrete)	0.91	[24,29–31]
Rooftop (tar and stone)	0.973	[24]
Rooftop (black rubber)	0.859	[24]
Solar Panel	0.85	[32]
Canopy Cover	0.977	[33]

Once the thermal data were corrected for emissivity, spatial data analysis was performed in ArcMap. First, a land use feature map was created that categorized the surface types in each case study location. Then inconsistencies within these areas, such as a parked car within a parking lot, human traffic on a sidewalk or construction materials on the street, were clipped and removed for each flight. Once these inconsistencies were removed, zonal statistics was applied to compute summary statistics of each surface type such as mean and standard deviation of the land surface temperature. A complete flow-chart of the process from flight programming to developing summary statistics is shown in Figure 2. In total this process took about 3 h to complete for each flight.

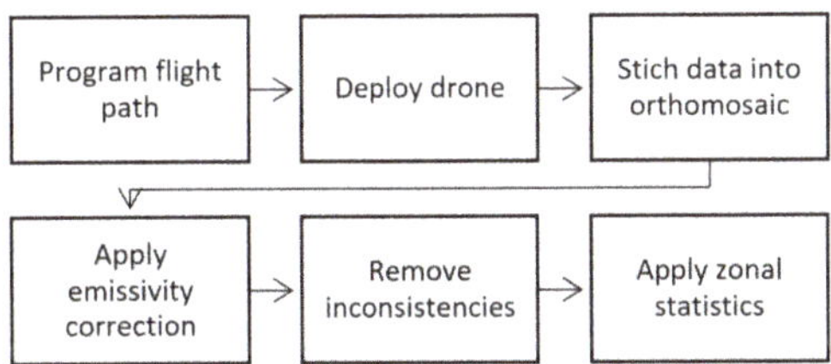

Figure 2. Flow chart of data collection and processing.

2.5. Surface Parameters

In addition to surface temperature, three other material properties were derived from visual and multispectral imagery, converted into spatial distribution rasters and averaged for each surface type. These include albedo (S), normalized difference vegetation index ($NDVI$) and apparent thermal inertia (ATI). Albedo, a measure of solar reflectance of a material, was derived from blue, green, red and near-IR image bands as shown in the following equation:

$$S = c_b b_k + c_g g_k + c_r r_k + c_i i_k \qquad (2)$$

where $c_b = 0.17$, $c_g = -0.13$, $c_r = 0.33$ and $c_i = 0.54$ are derived constants and b_k, g_k, r_k and i_k are the band reflectance's for—blue, b_k (420–492 nm); green, g_k (533–587 nm); red, r_k (604–664 nm); and near-IR, i_k (833–920 nm) [34].

Visual and multispectral imagery were also used to derive $NDVI$, which is a measure of the degree of live vegetation and is commonly used to evaluate soil moisture dynamics, erosion potential and plant and crop health. As shown in Equation (3), $NDVI$ is a function of near-IR and red band reflectance and is estimated on a scale of -1 to $+1$, with higher values indicating higher vegetative cover and greater plant health [35].

$$NDVI = \frac{(NIR - Red)}{(NIR + Red)} \qquad (3)$$

Finally, ATI was derived for each surface type from albedo (S), solar correction (SCR) and the diurnal temperature amplitude (DTA) (Equation (4)). ATI is an estimation of thermal inertia from remotely sensed observations and can be estimated from diurnal changes in temperature. Specifically, ATI is derived from solar correction (SCR), albedo (S) and the diurnal temperature amplitude (DTA), where DTA is the difference between the maximum and minimum surface temperature recorded at the time the remote images were captured and SCR is the solar correction factor (Equation (5)), which is dependent on geographic location, the local latitude (θ) and the solar declination (φ) [36].

$$ATI = \frac{SCR(1 - S)}{DTA} \qquad (4)$$

$$SCR = \sin\theta \, \sin\varphi(1 - (\tan\theta \, \tan\varphi)^2) + \cos\theta \, \cos\varphi \, \arccos(-\tan\theta \, \tan\varphi) \qquad (5)$$

2.6. Model Development

Drone observations were applied to develop empirical models of land surface temperature. These include (1) a regression model to predict spatially averaged surface temperatures at 12:00 PM based upon meteorological variables and (2) a model to assess diurnal variability and predict surface temperatures throughout a given day.

Multi-variable regression models were developed to predict spatially averaged surface temperature of the fourteen Milwaukee and single El Paso 12:00 PM flights using MATLAB and the statistical software package JMP 13 [37]. Response screening was performed for each of the respective datasets to identify the strength of relationship between surface temperature (response) and meteorological parameters (predictors). Between the two case study locations, six surface types that were common to both locations were used as response variables: grass, canopy cover, concrete parking lot, concrete sidewalk, composite rooftop and road surface. Meteorological predictor variables included air temperature, relative humidity, preceding 24 h rainfall, wind speed, wind direction, barometric pressure and solar radiation. After response screening, stepwise linear regression was then performed to predict land surface temperature based upon meteorological parameters as represented in following equation:

$$y = \beta_0 + \beta_1 x_1 + \beta_2 x_2 + \cdots + \beta_k x_k \qquad (6)$$

where y is the response variable, $\beta_0, \beta_1, \ldots, \beta_k$ are the regression coefficients and $x_1, x_2, \ldots, x_k$ are the predictor variables for k predictors [38]. These models were developed using data from both the Milwaukee and El Paso flights; therefore, to evaluate the influence and leverage of the El Paso dataset we computed Cook's D influence and hat matrix leverage statistics [38].

Finally, we explored the variation in surface temperatures as they change throughout the day (9 AM, 12 PM, 3 PM and 5 PM) and evaluated if this variation could be explained by any meteorological parameters. In addition to exploring diurnal changes in variability, we applied the data to develop a model to predict land surface temperatures throughout the day for the six land use types common to each location. To do so, we applied the drone data collected on the four diurnal flight missions to estimate land surface temperatures based upon the solar radiation and the difference between the air and land surface temperatures, which have been found to be statistically significant predictors for diurnal estimates of pavement temperatures [39].

First, we computed a parameter (g) based upon the drone-derived mean land surface temperature and measured air temperature and solar radiation:

$$g = \left(\overline{T_s} - T_a\right) * S \tag{7}$$

where $\overline{T_s}$ is the mean surface temperature of the land use, T_a is the measured air temperature and S is the measured solar radiation (kW). Next, g at a given hour i was estimated using a Gaussian peak model given by the following:

$$g_i = a * e^{-0.5*\left(\frac{i-b}{c}\right)^2} \tag{8}$$

where g_i, is the parameter g at hour i, a is the peak value, b is the critical point and c is the growth rate [40]. Using this model, the mean land surface temperature can be predicted based upon air temperature and solar radiation for any time of day using the following:

$$T_{s,i} = T_{a,i} + (g_i / S_i) \tag{9}$$

where $T_{s,i}$ is the estimated surface temperature at hour i and $T_{a,i}$ and S_i are the air temperature and solar radiation at hour i. Taken as a whole, these models test both the suitability of predicting drone-derived mean land surface temperatures based upon meteorological variables, as well as the generalizability of our findings by including data from sites in two different geomorphologic and climatic regions.

3. Results

3.1. Surface Temperature Variability

We evaluated the land surface temperature variability of each flight across common land use types and generally found that green surfaces had a greater degree of variability than gray surfaces, with the exception being the rubber rooftop. As an example, the distribution of surface temperature data (1,986,543 total data points) is shown in Figure 3 for a flight recorded on July 11, 2018. The six gray surfaces recorded a smaller distribution of temperature on average but had more extreme values than green surfaces (Figure 3a). Gray surfaces retain more heat from the sun because of their high emissivity and ATI and therefore typically have higher surface temperatures. Additionally, non-normal behavior was identified for both canopy cover and rubber rooftop (Figure 3b). Canopy cover exhibits a left skew while the rubber rooftop exhibits a right skew. The canopy cover had a variation of tree types and therefore a variation of leaf area indices (LAI), which may be a reason for the skew in the temperature data. The rubber rooftop also exhibited a strong right skew, which may be due to small materials on the roof surface, such as ventilation pipes and drainage grates, that were difficult to detect and may not have been removed from the dataset. Therefore, this caused a distribution of lower temperatures to be recorded.

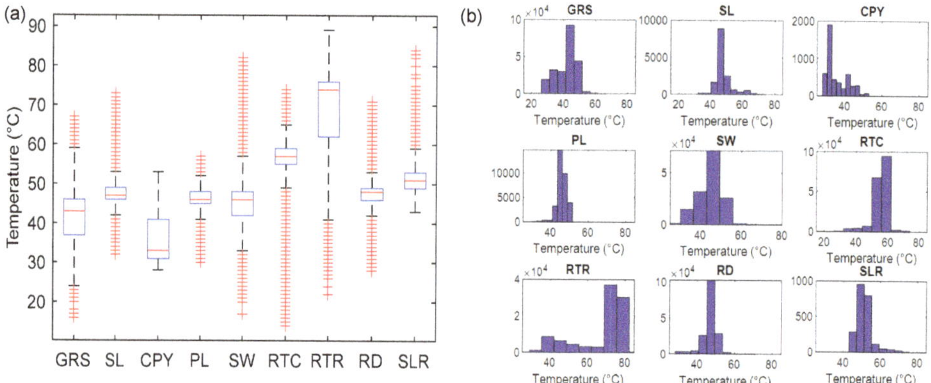

Figure 3. Boxplot distribution of surface temperature (**a**); and histogram of surface temperature (**b**). Data from flight recorded on July 11, 2018. Note GRS = grass; SM = shrub/mulch; CPY = canopy; PL = parking lot; SW = sidewalk; RTC = composite rooftop; RTR = rubber rooftop; RD = road; SLR = solar.

We then summarized the average temperature, standard deviation and coefficient of variation of each surface type for the fourteen recorded flights in Milwaukee, WI and the single flight in El Paso, TX (Table 5). Generally, gray surfaces exhibited higher temperatures throughout the year than green surfaces. In El Paso, the asphalt parking lot exhibited the highest average temperature (51.7 °C) and grass exhibited the lowest (41.6 °C), while in Milwaukee the black rubber rooftop exhibited the highest average temperature (57.4 °C) and canopy cover exhibited the lowest (30.4 °C). In terms of variation, the lowest degrees of variation typically occurred in the parking lots and grass. However, there is a noted difference in the variation between the two locations; the road in Milwaukee had the highest coefficient of variation of 0.32, while the road in El Paso had the lowest at 0.04. This may be due to a difference in traffic on the days that flights were conducted. The location in Milwaukee is located near the city center and is subject to heavy and constant vehicular traffic, while the location in El Paso is in a restricted traffic area and experienced very low vehicle activity on the weekend that the flight was conducted.

Table 5. Average temperature, standard deviation and coefficient of variation of nine surface types From 14 recorded flights on Marquette University campus and from one flight recorded on UTEP's campus.

Location	Surface Type	Temp (°C)	Standard Dev (°C)	Coeff. of Variation
	Grass	34.7	7.9	0.15
	Shrub/mulch	40.7	6.2	0.12
	Canopy	30.4	7.2	0.16
	Parking Lot	38.7	4.2	0.08
MU	Sidewalk	36.3	11.2	0.21
	Rooftop—Composite	47.6	6.2	0.1
	Rooftop—Rubber	57.4	15.8	0.22
	Road	32.5	15.4	0.32
	Solar Panels	47.0	7.2	0.13
	Grass	41.6	6.1	0.1
	Canopy	46.6	6.3	10
	Desert Shrub	46.2	6.7	0.11
UTEP	Parking Lot (asphalt)	51.7	4.9	0.07
	Parking (concrete)	45.3	7.4	0.12
	Sidewalk	43.1	8.2	0.13
	Rooftop—Composite	47.3	7.2	0.11
	Rooftop—Dzong	46.2	6.3	0.1
	Road	48.4	2.8	0.04

The distribution of the average temperature, standard deviation and coefficient of variation for the fourteen Milwaukee, WI flights is further illustrated in Figure 4. The shrub/mulch, composite rooftop and solar panels have the most consistent variability among the land use types as shown in the boxplot distribution of their coefficient of variation, while the greatest spread in variation occurred in the road and sidewalk. This may indicate that areas that are not subject to human traffic (e.g., shrub/mulch flower beds, rooftops and solar panels) have more consistent variability in their temperatures, while other areas that are subject to intermittent human traffic (e.g., roads and sidewalks) have inconsistent temperature variabilities. We also evaluated if the variability in land surface temperature correlated with any meteorological parameters but found no statistically significant predictors.

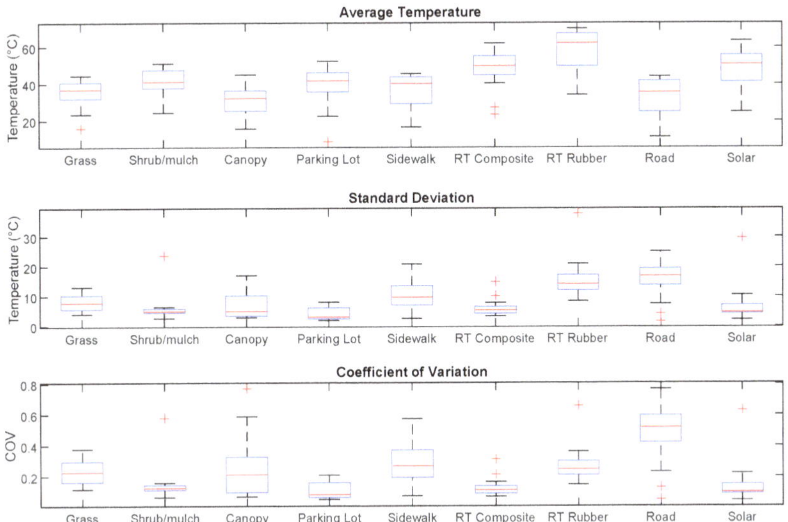

Figure 4. Boxplot distribution of average temperature, standard deviation and coefficient of variation from 14 recorded flights in Milwaukee, WI.

We evaluated the variation in surface temperatures throughout the day and found that the highest degree of variation occurred at noon. This is demonstrated in the Figure 5, which shows box plots of the standard deviation for six land use types: grass, canopy, parking lot, sidewalk, composite roof and road for data from both MU and UTEP. As illustrated, all land use types have the greatest standard deviation in temperatures during 12:00 PM, with lower levels of deviation in the morning and late afternoon. This trend suggests that as surfaces heat up, they do so at different rates, which contributes to more variability during mid-day.

3.2. Impact of the Built Environment

We also evaluated the spatial distribution of surface temperature to locate and identify factors of the built environment that contribute to temperature variability. Figure 5 illustrates the spatial distribution of surface temperatures for a flight on July 8th, 2018. One factor of variability is the reflectance and shaded cover from nearby buildings. For example, sidewalks in close proximity to Engineering Hall exhibited higher temperatures, most likely due to the sun's reflectance off its glass paneling. Two similarly sized sidewalk areas were compared and results show the average temperature was 4.7 °C hotter for the location closer to the building than the one farther away. In comparison to the sidewalk, parking lot land uses had more consistent variability, perhaps because there were fewer nearby buildings or large trees to exacerbate (glass reflectance) or reduce (shaded cover) their temperature. This indicates proximity to nearby buildings or other structures can be a significant factor of uncertainty in predicting surface temperatures.

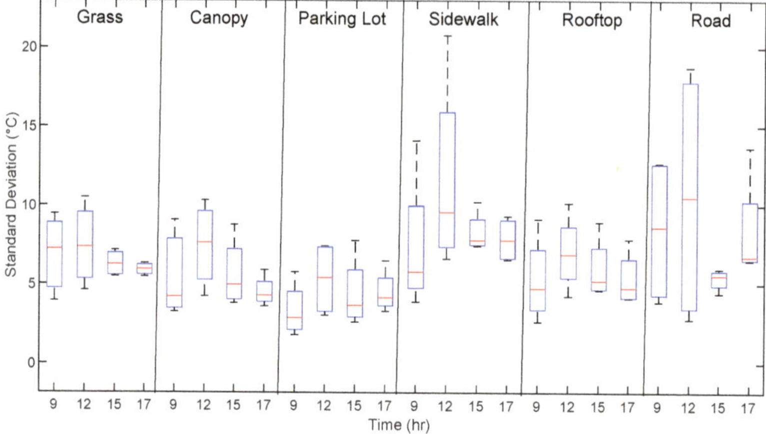

Figure 5. Standard deviation distributions for six land use types at hour 9, 12, 15 and 17 at both the MU and UTEP locations.

Other sources of land surface temperature uncertainty are traffic and parked cars. Traffic flow along a roadway intermittently blocks the suns radiation, thereby impacting the surface temperatures of the roadway pavement below. This creates a concentrated pocket of cooler surface temperatures called a heat shadow, which results in variations in surface temperatures across the pavement. This is especially pronounced in pavement lots with parked cars as illustrated in Figure 6b, which shows the distribution of surface temperatures within a parking lot. In this figure a parked car rooftop, pavement surface and heat shadow recorded temperatures of 69.6 °C, 47.8 °C and 49.0 °C, respectively, all within a space of ~50 m^2.

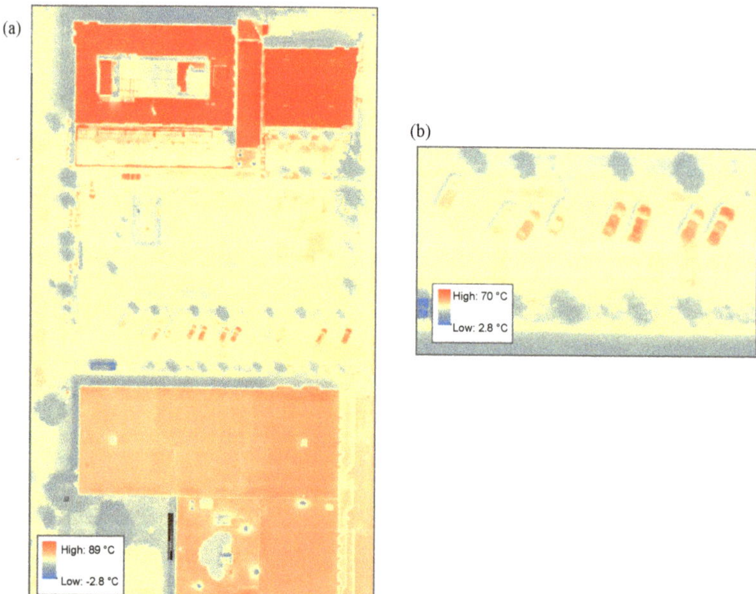

Figure 6. Spatial distribution of temperature from a flight recorded on 11 July 2018 (**a**) and zoomed in spatial distribution of temperature for the concrete parking lot from a flight recorded on 11 July 2018 (**b**). The hotter surfaces (red) in the right image are parked cars and the cooler surfaces (blue) are heat shadows visible after parked cars leave.

3.3. Surface Properties and Land Surface Temperature

Drone data was applied to derive surface properties including albedo, NDVI and ATI, of the surface types in the case study (Table 6). The light concrete parking lot exhibited the highest albedo (0.673) while grass exhibited the lowest (0.317). The spatial distribution of temperature, albedo, NDVI and ATI at the Milwaukee, WI case study location is shown in Figure 7. As illustrated, these surface material properties have a large degree of variation across the case study area.

Table 6. Average albedo, normalized difference vegetation index (NDVI) and apparent thermal inertia (ATI) values for each surface type.

Surface Type	Albedo	NDVI	ATI
Grass	0.317	0.369	0.198
Shrub/mulch	0.502	0.402	0.183
Canopy	0.378	0.490	0.209
Parking Lot	0.673	0.091	0.121
Sidewalk	0.472	0.144	0.195
Rooftop–Composite	0.580	0.101	0.156
Rooftop–Rubber	0.406	0.096	0.219
Road	0.518	0.117	0.179
Solar	0.333	0.143	0.217

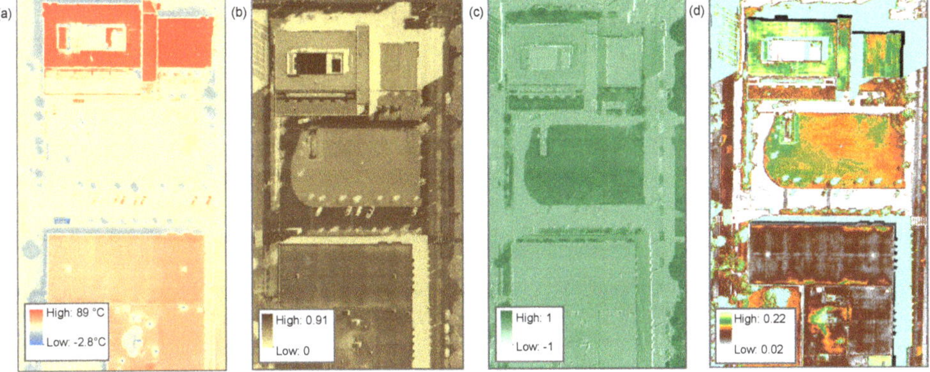

Figure 7. Spatial distribution of tempearture (**a**), albedo (**b**), NDVI (**c**) and ATI (**d**) for a flight recorded on 11 August 2018.

To further explore this variability and assess its impact on surface temperatures, we plotted these surface properties against land surface temperature. Figure 8 illustrates temperature plotted against its respective albedo for the 611,460 total data points captured by the drone imagery and results show clusters that form for different surface types. Some of these clusters exhibit either a (1) low range in albedo and high range in temperature or (2) high range in albedo and low range in temperature. For example, the road exhibits a low range in albedo and high range in temperature, implying the variability in roadway temperatures are more dependent on meteorological (e.g., exposure to solar radiation) and human (e.g., traffic) variables than physical properties (e.g., albedo). On the other hand, the parking lot has a higher but similar range in albedo, yet it has a much lower variability in temperature. This could be due to the fact that the parking lot has a range of materials from asphalt to concrete coupled with a much lower level of traffic as compared to the roadway, which is more homogenous and experiences constant vehicular traffic that intercepts land surface exposure to solar radiation. Therefore, this graphic may support the previous statement that there are anthropogenic variables, such as intermittent human foot or vehicular traffic, that are significant to land surface temperature processes. Overall these results suggest that patterns in the physical properties of urban materials may provide insight into surface temperature variability.

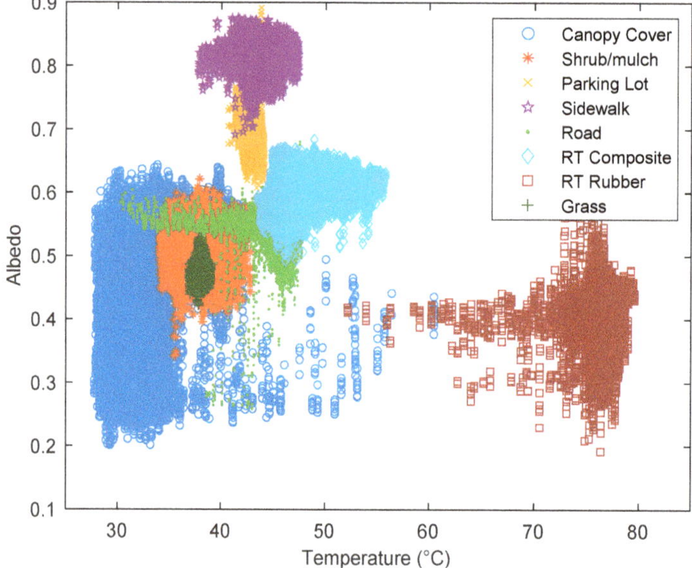

Figure 8. Surface temperature data plotted against albedo from a flight recorded on 11 August 2018.

3.4. Temperature Prediction Models

Drone observations were applied to develop empirical models of land surface temperature. These include (1) a regression model to predict spatially averaged surface temperatures at 12:00 PM based upon environmental variables and (2) a diurnal model to predict surface temperatures throughout a given day.

3.4.1. Spatially Averaged Surface Temperature Regression Model

Multi-variable linear regression models were developed to predict spatially averaged surface temperature and it was found that air temperature and solar radiation are significant predictors (Figure 9). Standard least squares regression was applied to develop models that predict the surface temperature of six land use types: grass, canopy cover, parking lot (concrete), sidewalk, rooftop (composite) and road. The models had an average R^2 of 0.71 with the parking lot having the greatest of (0.89) and the road the lowest (0.37). The parked cars and heat shadows were clipped out as inconsistencies before analysis occurred and therefore the parking lot surface had the most homogenous distribution of temperatures. The grass model had the second greatest R^2 (0.84) and had a similarly homogenous distribution. Contrarily, the roadway surface had a much less homogenous distribution of temperatures and thus the road model had a low predictive power and statistical significance. This may be due in large part due to the difficulty of clipping out inconsistencies related to nonstationary objects (e.g., moving cars) combined with their impact on pavement temperatures.

The data collected in El Paso, TX was evaluated for influence and leverage and it was found that it did not have high influence or leverage in any of the six models. To evaluate influence we used Cook's D and found that the El Paso data points all fell below the threshold of 2.4 (max 0.19) to be considered high-influence points [38]. In addition, we used the hat matrix to evaluate leverage and found that no El Paso data points exhibited high leverage in the model. The agreeability of the data across the two case study areas indicates that the findings in this study may have generalizability beyond the case study locations.

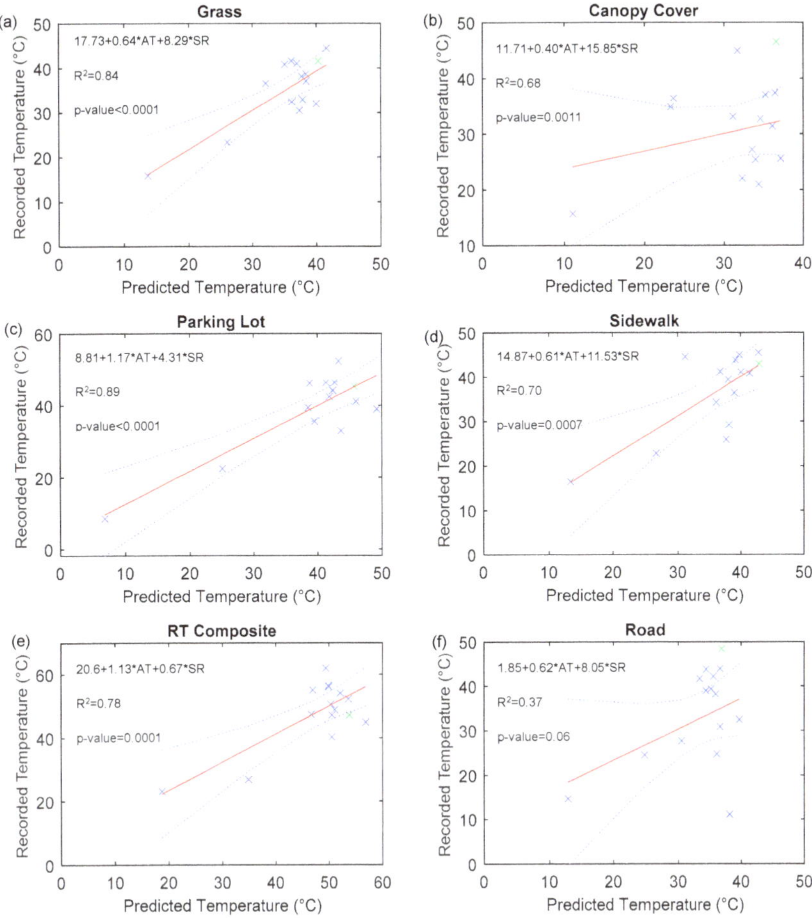

Figure 9. Temperature prediction models of six surface types: grass (**a**), canopy cover (**b**), parking lot (**c**), sidewalk (**d**), composite rooftop (**e**) and road (**f**). UTEP datapoint is fitted in green. Note the 95% confidence intervals are in blue.

3.4.2. Diurnal Prediction Model

Finally, models were developed to predict land surface temperature throughout the day based upon the air temperature and solar radiation (Equations (7)–(9)). The diurnal data was fit with a Gaussian peak distribution and it was found that the parking lot and composite rooftop had the best model fit with an R^2 of 0.83 and 0.78, respectively, while all other models had an R^2 value of 0.53 or below (Figure 10). While this approach is constrained by a limited number of data points from four flights and only four numerical x-axis variables, there are a few insights we can gain from these results. The first is that these models confirm what was found in the previous regression models: it is much easier to predict the land surface temperature of homogenous materials, such as pavements and rooftops, than it is to predict land surfaces that have a greater distribution in texture and material, such as canopy. The second is that anthropogenic variables, such as pedestrians and vehicular traffic that are difficult to quantify, may influence the ability to predict surface temperatures based upon meteorological variables. This was shown by the lower model fit in the high-traffic roadways and sidewalks as compared to the low-traffic parking lot.

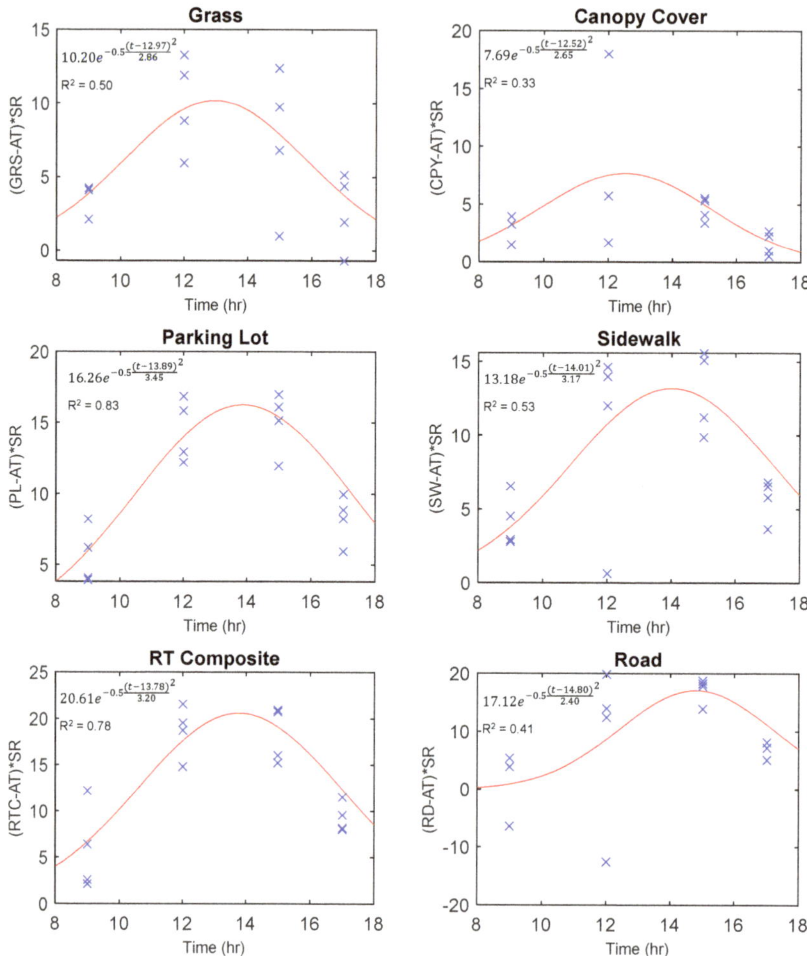

Figure 10. Gaussian peak models of six surface types: grass (**a**), canopy cover (**b**), parking lot (**c**), sidewalk (**d**), composite rooftop (**e**) and road (**f**). Note that GRS = grass; CPY = canopy; PL = parking lot; SW = sidewalk; RTC = composite rooftop; AT = air temperature; SR = solar radiation; t = time.

4. Discussion

We have presented a case study that applied high resolution drone measurements (13 cm) to evaluate urban surface temperatures and results indicate that there is a wide variability in surface temperature behavior across urban land use types. Some of the uncertainty in land surface temperature variability may be attributable to human movement patterns, land surface properties or urban geometry. Results indicate that mean land surface temperatures can be predicted based upon solar radiation and air temperature. By elucidating some of the factors that influence land surface temperature variability, we hope to contribute to the growing body of knowledge centered around land surface temperature in the urban environment.

To this end, our findings suggest that when parameterizing models, it is important to understand the unique relationship between surface material properties, urban geometry, weather and human movement. For example, the results indicate that pedestrian or vehicular traffic may have an impact on land surface temperature variability across sidewalks, parking lots and streets. Depending on the volume of cars, either parked or moving, this can greatly impact the temperature profile of paved surfaces. Parked cars can create heat shadows which cool the surface below and our study demonstrates that when a car moves it can reveal temperatures as low as 8.3 °C cooler than the exposed surface.

In addition, results have identified several factors of urban geometry that affect land surface temperatures. Urban factors such as building reflectivity and surface altitude can impact solar radiation, which then influences surface temperatures in locations impacted by these effects. For example, sidewalks often lie near buildings and depending on a buildings reflectance or shadows this can make sidewalk temperatures more vulnerable to temperature fluctuations. In this study, sidewalk temperatures impacted by glass reflectance were on average 4.7 °C hotter that sidewalks not impacted by reflectance. Therefore, knowledge of the spatial distribution of urban geometry is important for predicting and evaluating land surface temperatures in the built environment. While addressing urban geometry or pedestrian and vehicular traffic within our prediction models is outside of the scope of this project, future work should evaluate how to incorporate these important parameters into land surface temperature predictions.

Results indicate air temperature and solar radiation are significant predictors of mean land surface temperature in both of our models and it was found this relationship holds true in both Milwaukee, WI and El Paso, TX. Because the model holds true across two different climatic regions, the models developed in this project may be generalizable beyond their case study regions. In addition, these models can also be easily applied as air temperature and solar radiation are commonly measured across the world. The generalizability of these findings also has important implications for engineering applications that use predictions of land surface temperatures. Urban land surface temperatures are often used by public health officials to mitigate the impact of the urban heat island effect on human health [2], in developing binders and mixers of pavement in roadway designs [41] or to estimate the impact of land surface temperatures on receiving stream temperatures [42–44].

This study also demonstrates several advantages and disadvantages of using drones as compared to satellite or in-situ imagery. The case studies we evaluated were restricted to the size of a city block around 46,000 m^2 and even though battery life would have allowed us to collect an area ten times this size, we were restricted by United States Federal Aviation Administration UAV pilot rules that restrict the flight of UAVs to within line of sight of the pilot. In an urban environment with tall buildings the line of sight may be the primary constraint on coverage area. Therefore, a disadvantage of UAVs is that flight time and legal restrictions may constrain the flight areas to small portions of a city. However, this could be overcome with fixed-wing drones that are able cover a greater area, in addition to relaxed regulations that allow flights beyond the line of sight [45]. Despite the restriction on the spatial extent of the study area, advantages of UAVs over satellites or in-situ methods are their ability to collect distributed temperature data at spatial resolutions (13 cm) that reflect small scale changes in the urban environment. In addition, satellite data is restricted to daily to weekly observations while drones can be flown on-demand, which allows them to capture temperature changes throughout the day.

Overall, this study highlights the utility of using drone observations to capture the variability of urban land surface temperatures at small spatial scales. Urban environments are spatially complex, making it difficult to capture the spatial distribution of observable phenomena outside of high-resolution remote sensing techniques. Our findings suggest that drones could also be good tools for evaluating the variability of other parameters of the urban environment that are important for environmental studies such as soil moisture, leaf area index or impervious cover. Therefore, it is important for studies such as this one that evaluate the spatial complexities of the urban environment in order to improve the methods that we use to model and understand urban systems.

5. Conclusions

The main objectives of this work were to apply drone imagery to capture land surface temperature variability and develop models to predict mean land surface temperatures. This was done through the application of high-resolution thermal imagery as a parameterizing tool for model development. The results revealed that land surface temperature variability is extensive and influenced by numerous variables related to urban environments and that air temperature and solar radiation are significant predictors of mean land surface temperature. Conclusions from this study hold true in both Milwaukee, WI and El Paso, TX, indicating they could be generalizable to regions beyond these two case study locations.

The key findings from this study were:

- Land surface variability was significant and ranged between (3.9–15.8 °C) for common land use types.
- Areas that experienced pedestrian or vehicular traffic exhibited higher variabilities than comparable surfaces that did not. In Milwaukee, the high-traffic road had a coefficient of variation of 0.32 as compared to 0.08 for the low-traffic parking lot. This indicates that human traffic may impact land surface temperatures due to the heat-shadow effect.
- Urban geometry has an influence on land surface temperatures; shadows and reflectance from buildings showed a significant influence on the temperatures of nearby land surfaces throughout the day. Sidewalk temperatures impacted by glass reflectance were on average 4.7 °C hotter that sidewalks not impacted by reflectance.
- Land surface temperature variability is low in the morning, peaks at noon and goes back down in the evening. This may indicate that as surfaces heat up, they do so at different rates, which contributes to more variability during mid-day.
- Air temperature and solar radiation were significant predictors of spatially averaged surface temperature in both of our models.
- Data were consistent in the models between Milwaukee, WI and El Paso, TX, suggesting that the findings in this study may be generalizable beyond the case study locations.

Overall, our findings suggest that land surface temperature variability in the urban environment can come from several sources including surface material properties, urban geometry, weather and pedestrian and vehicular traffic. This has direct implications for land surface temperature models that are used for urban environmental studies. As climate change and urbanization continue to exacerbate the SUHI, studies such as this are important for gaining a better understanding of the complexities of land surface temperatures. Ultimately this improved understanding will help to develop better methods and procedures to mitigate the impact of land surface temperatures on human and environmental health.

Author Contributions: J.N. provided investigation, data collection, data analysis, and writing of the original draft. W.M. contributed conceptualization, supervision, data collection and draft editing.

Funding: This project was funded by the Marquette OPUS College of Engineering Earl B. and Charlotte Nelson Award.

Acknowledgments: The authors would like to acknowledge and sincerely thank Saurav Kumar and Wissam Atwah at the University of Texas El Paso for their help in collecting data in El Paso, Tx.

Conflicts of Interest: The authors declare no conflict of interest.

References

1. Foley, J.A.; DeFries, R.; Asner, G.P.; Barford, C.; Bonan, G.; Carpenter, S.R.; Chapin, F.S.; Coe, M.T.; Daily, G.C.; Gibbs, H.K.; et al. Global consequences of land use. *Science* **2005**, *309*, 570. [CrossRef] [PubMed]
2. Shahmohamadi, P.; Che-Ani, A.I.; Etessam, I.; Maulud, K.N.A.; Tawil, N.M. Healthy Environment: The Need to Mitigate Urban Heat Island Effects on Human Health. *Procedia Eng.* **2011**, *20*, 61–70. [CrossRef]
3. U.S. Environmental Protection Agency. *The Assessment, Total Maximum Daily Load (TMDL) Tracking and Implementation System (ATTAINS)*; U.S. Environmental Protection Agency: Washington, DC, USA, 2019.
4. White, P.A.; Kalff, J.; Rasmussen, J.B.; Gasol, J.M. The effect of temperature and algal biomass on bacterial production and specific growth rate in freshwater and marine habitats. *Microb. Ecol.* **1991**, *21*, 99–118. [CrossRef] [PubMed]
5. Beitinger, T.L.; Bennett, W.A.; McCauley, R.W. Temperature Tolerances of North American Freshwater Fishes Exposed to Dynamic Changes in Temperature. *Environ. Biol. Fishes* **2000**, *58*, 237–275. [CrossRef]

6. Lazzarini, M.; Marpu, P.R.; Ghedira, H. Temperature-land cover interactions: The inversion of urban heat island phenomenon in desert city areas. *Remote Sens. Environ.* **2013**, *130*, 136–152. [CrossRef]

7. Jiang, Y.; Fu, P.; Weng, Q. Assessing the impacts of urbanization-associated land use/cover change on land surface temperature and surface moisture: A case study in the midwestern united states. *Remote Sens.* **2015**, *7*, 4880–4898. [CrossRef]

8. Liu, L.; Zhang, Y. Urban heat island analysis using the landsat TM data and ASTER Data: A case study in Hong Kong. *Remote Sens.* **2011**, *3*, 1535–1552. [CrossRef]

9. Sekertekin, A.; Kutoglu, S.H.; Kaya, S. Evaluation of spatio-temporal variability in Land Surface Temperature: A case study of Zonguldak, Turkey. *Environ. Monit. Assess.* **2016**, *188*, 1–15. [CrossRef]

10. Sobrino, J.A.; Oltra-carrió, R.; Sòria, G.; Jiménez-muñoz, J.C.; Franch, B.; Hidalgo, V.; Mattar, C.; Julien, Y.; Cuenca, J.; Romaguera, M.; et al. Evaluation of the surface urban heat island effect in the city of Madrid. *Int. J. Remote Sens.* **2013**, *34*, 3177–3192. [CrossRef]

11. Mallick, J.; Kant, Y.; Bharath, B.D. Estimation of land surface temperature over Delhi using Landsat-7 ETM+. *J. Ind. Geophys. Union.* **2008**, *12*, 131–140.

12. Zhou, D.; Xiao, J.; Bonafoni, S.; Berger, C.; Deilami, K.; Zhou, Y.; Frolking, S.; Yao, R.; Qiao, Z.; Sobrino, J.A. Satellite Remote Sensing of Surface Urban Heat Islands: Progress, Challenges, and Perspectives. *Remote Sens.* **2018**, *11*, 48. [CrossRef]

13. Granero-Belinchon, C.; Michel, A.; Lagouarde, J.-P.; Sobrino, J.A.; Briottet, X. Multi-Resolution Study of Thermal Unmixing Techniques over Madrid Urban Area: Case Study of TRISHNA Mission. *Remote Sens.* **2019**, *11*, 1251. [CrossRef]

14. Nguyen, K.A.; Liou, Y.A. Global mapping of eco-environmental vulnerability from human and nature disturbances. *Sci. Total Environ.* **2019**, *664*, 995–1004. [CrossRef] [PubMed]

15. Deng, C.; Wu, C. Estimating very high resolution urban surface temperature using a spectral unmixing and thermal mixing approach. *Int. J. Appl. Earth Obs. Geoinf.* **2013**, *23*, 155–164. [CrossRef]

16. Nichol, J.E.; Wong, M.S. High resolution remote sensing of densely urbanised regions: A case study of Hong Kong. *Sensors* **2009**, *9*, 4695–4708. [CrossRef] [PubMed]

17. Cooper, B.E.; Dymond, R.L.; Shao, Y. Impervious Comparison of NLCD versus a Detailed Dataset Over Time. *Photogramm. Eng. Remote Sens.* **2017**, *83*, 429–437. [CrossRef]

18. Li, J.; Song, C.; Cao, L.; Zhu, F.; Meng, X.; Wu, J. Impacts of landscape structure on surface urban heat islands: A case study of Shanghai, China. *Remote Sens. Environ.* **2011**, *115*, 3249–3263. [CrossRef]

19. Zhou, W.; Huang, G.; Cadenasso, M.L. Does spatial configuration matter? Understanding the effects of land cover pattern on land surface temperature in urban landscapes. *Landsc. Urban Plan.* **2011**, *102*, 54–63. [CrossRef]

20. Chudnovsky, A.; Ben-Dor, E.; Saaroni, H. Diurnal thermal behavior of selected urban objects using remote sensing measurements. *Energy Build.* **2004**, *36*, 1063–1074. [CrossRef]

21. Gaitani, N.; Burud, I.; Thiis, T.; Santamouris, M. High-resolution spectral mapping of urban thermal properties with Unmanned Aerial Vehicles. *Build. Environ.* **2017**, *121*, 215–224. [CrossRef]

22. Kottek, M.; Grieser, J.; Beck, C.; Rudolf, B.; Rubel, F. World map of the Köppen-Geiger climate classification updated. *Meteorol. Z.* **2006**, *15*, 259–263. [CrossRef]

23. National Weather Service Forcast Office. Available online: https://www.weather.gov/mkx/ (accessed on 17 July 2019).

24. Marshall, S.J. We Need To Know More About Infrared Emissivity. *Proc. SPIE* **1982**, *313*, 119–128.

25. Blonquist, J.M.; Norman, J.M.; Bugbee, B. Automated measurement of canopy stomatal conductance based on infrared temperature. *Agric. For. Meteorol.* **2009**, *149*, 1931–1945. [CrossRef]

26. Humes, K.S.; Kustas, W.P.; Moran, M.S.; Nichols, W.D.; Weltz, M.A. Variability of emissivity and surface temperature over a sparsely vegetated surface. *Water Resour. Res.* **1994**, *30*, 1299–1310. [CrossRef]

27. Wittich, K.P. Some simple relationships between land-surface emissivity, greenness and the plant cover fraction for use in satellite remote sensing. *Int. J. Biometeorol.* **1997**, *41*, 58–64. [CrossRef]

28. Ramamurthy, P.; Bou-Zeid, E. Contribution of impervious surfaces to urban evaporation. *Water Resour. Res.* **2014**, *50*, 2889–2902. [CrossRef]

29. Chen, H.; Ooka, R.; Huang, H.; Tsuchiya, T. Study on mitigation measures for outdoor thermal environment on present urban blocks in Tokyo using coupled simulation. *Build. Environ.* **2009**, *44*, 2290–2299. [CrossRef]

30. Larsson, O.; Thelandersson, S. Estimating extreme values of thermal gradients in concrete structures. *Mater. Struct.* **2011**, *44*, 1491–1500. [CrossRef]

31. Salamanca, F.; Krpo, A.; Martilli, A.; Clappier, A. A new building energy model coupled with an urban canopy parameterization for urban climate simulations—Part I. formulation, verification, and sensitivity analysis of the model. *Theor. Appl. Climatol.* **2009**, *99*, 331. [CrossRef]

32. Herb, W.R.; Janke, B.; Mohseni, O.; Stefan, H.G. Thermal pollution of streams by runoff from paved surfaces. *Hydrol. Process.* **2008**, *22*, 987–999. [CrossRef]

33. Spectrolab, Inc. *Spectrolab Photovotaic Products Data Sheet*; Spectrolab, Inc.: Sylmar, CA, USA, 2012.

34. Ban-Weiss, G.A.; Woods, J.; Levinson, R. Using remote sensing to quantify albedo of roofs in seven California cities, Part 1: Methods. *Sol. Energy* **2015**, *115*, 777–790. [CrossRef]

35. Carlson, T.N.; Ripley, D.A. On the relation between NDVI, fractional vegetation cover, and leaf area index. *Remote Sens. Environ.* **1997**, *62*, 241–252. [CrossRef]

36. Sohrabinia, M.; Rack, W.; Zawar-Reza, P. Soil moisture derived using two apparent thermal inertia functions over Canterbury, New Zealand. *J. Appl. Remote Sens.* **2014**, *8*, 1–16.

37. SAS Institute. *JMP®8 User Guide*, 2nd ed.; SAS Institute Inc.: Cary, NC, USA, 2009.

38. Helsel, D.R.; Hirsch, R.M. *Statistical Methods in Water Resources*; Elsevier: Amsterdam, The Netherlands, 2002; Volume 36.

39. Thompson, A.M.; Wilson, T.; Norman, J.M.; Gemechu, A.L.; Roa-Espinosa, A. Modeling the effect of summertime heating on urban runoff temperature. *J. Am. Water Resour. Assoc.* **2008**, *44*, 1548–1563. [CrossRef]

40. Guo, H. A Simple Algorithm for Fitting a Gaussian Function [DSP Tips and Tricks]. *IEEE Signal Process. Mag.* **2011**, *28*, 134–137. [CrossRef]

41. Solaimanian, M.; Kennedy, T.W. *Predicting Maximum Pavement Surface Temperature Using Maximum Air Temperature and Hourly Solar Radiation*; Transportation Research Board: Washington, DC, USA, 1993.

42. Herb, W.R.; Janke, B.; Mohseni, O.; Stefan, H.G. *MINUHET (Minnesota Urban Heat Export Tool): A Software Tool for the Analysis of Stream Thermal Loading by Stormwater Runoff*; Project Report No. 526; St. Anthony Falls Laboratory, The University of Minnesota: Minneapolis, MN, USA, 2009.

43. Roa-Espinosa, A.; Wilson, T.B.; Norman, J.M.; Johnson, K. Predicting the Impact of Urban Development on Stream Temperature Using a Thermal Urban Runoff Model. *Methods* **2003**, 369–389.

44. Hailegeorgis, T.T.; Alfredsen, K. High spatial–temporal resolution and integrated surface and subsurface precipitation-runoff modelling for a small stormwater catchment. *J. Hydrol.* **2018**, *557*, 613–630. [CrossRef]

45. Ravich, T. A Comparative Global Analysis of Drone Laws: Best Practices and Policies. In *The Future of Drone Use: Opportunities and Threats from Ethical and Legal Perspectives*; Custers, B., Ed.; T.M.C. Asser Press: The Hague, The Netherlands, 2016; pp. 301–322. ISBN 978-94-6265-132-6.

 *remote sensing*

Article

Quantifying the Congruence between Air and Land Surface Temperatures for Various Climatic and Elevation Zones of Western Himalaya

Shaktiman Singh [1,*], Anshuman Bhardwaj [1], Atar Singh [2], Lydia Sam [1], Mayank Shekhar [3], F. Javier Martín-Torres [1,4] and María-Paz Zorzano [1,5]

1   Division of Space Technology, Department of Computer Science, Electrical and Space Engineering, Luleå University of Technology, 97187 Luleå, Sweden; anshuman.bhardwaj@ltu.se (A.B.); lydia.sam@ltu.se (L.S.); javier.martin-torres@ltu.se (F.J.M.-T.); zorzanomm@cab.inta-csic.es (M.-P.Z.)
2   Department of Environmental Science, Sharda University, Greater Noida 201310, India; 2015002854atar@dr.sharda.ac.in
3   Birbal Sahni Institute of Palaeosciences, Lucknow 226007, India; mayank_shekhar@bsip.res.in
4   Instituto Andaluz de Ciencias de la Tierra (CSIC-UGR), Armilla, 18100 Granada, Spain
5   Centro de Astrobiología (INTA-CSIC), Torrejón de Ardoz, 28850 Madrid, Spain
*   Correspondence: shaktiman.singh@ltu.se; Tel.: +46-920-492-043

Received: 21 October 2019; Accepted: 2 December 2019; Published: 4 December 2019

Abstract: The surface and near-surface air temperature observations are primary data for glacio-hydro-climatological studies. The in situ air temperature (T_a) observations require intense logistic and financial investments, making it sparse and fragmented particularly in remote and extreme environments. The temperatures in Himalaya are controlled by a complex system driven by topography, seasons, and cryosphere which further makes it difficult to record or predict its spatial heterogeneity. In this regard, finding a way to fill the observational spatiotemporal gaps in data becomes more crucial. Here, we show the comparison of T_a recorded at 11 high altitude stations in Western Himalaya with their respective land surface temperatures (T_s) recorded by Moderate Resolution Imagining Spectroradiometer (MODIS) Aqua and Terra satellites in cloud-free conditions. We found remarkable seasonal and spatial trends in the T_a vs. T_s relationship: (i) T_s are strongly correlated with T_a (R^2 = 0.77, root mean square difference (RMSD) = 5.9 °C, n = 11,101 at daily scale and R^2 = 0.80, RMSD = 5.7 °C, n = 3552 at 8-day scale); (ii) in general, the RMSD is lower for the winter months in comparison to summer months for all the stations, (iii) the RMSD is directly proportional to the elevations; (iv) the RMSD is inversely proportional to the annual precipitation. Our results demonstrate the statistically strong and previously unreported T_a vs. T_s relationship and spatial and seasonal variations in its intensity at daily resolution for the Western Himalaya. We anticipate that our results will provide the scientists in Himalaya or similar data-deficient extreme environments with an option to use freely available remotely observed T_s products in their models to fill-up the spatiotemporal data gaps related to in situ monitoring at daily resolution. Substituting T_a by T_s as input in various geophysical models can even improve the model accuracy as using spatially continuous satellite derived T_s in place of discrete in situ T_a extrapolated to different elevations using a constant lapse rate can provide more realistic estimates.

Keywords: Himalaya; land surface temperature; air temperature; topography; MODIS

1. Introduction

Air temperature (T_a) is an important proxy for energy exchange between land-surface and atmosphere, making T_a one of the most important parameters in climate research [1,2]. T_a is generally observed at a height of about 2 m above the land surface and it is considered as a critical parameter for glacio-hydrological studies because it controls the rate of melting of snow and ice and the proportion of form of precipitation [3,4]. In addition, it also regulates the evolution of flora and fauna in an area, ultimately controlling the evolution of the ecological niche [5]. T_a is also important for determining the atmospheric water vapor saturation point and thus for the formation of fogs and clouds. The gradient between the air and ground temperature is relevant for estimating the sensible heat flux (i.e., the convective heat flux loss from surface to the air) for calculations of the surface energy balance [6]. The surface-to-air temperature difference is particularly important for evapotranspiration [7]. In other regions, such as the Arctic, the T_a difference is taken as a critical parameter to monitor climate change [8]. Therefore, it is imperative to have accurate estimates of T_a for various natural science disciplines including glaciology, hydrology, ecology, and climatology. The measurement of T_a using in situ automatic meteorological stations is cost intensive due to involved instrumentation and maintenance which makes the spatial continuity of data sparse, particularly in remote environments. This spatially discontinuous nature of in situ T_a measurements adds uncertainty in geospatial modelling in mountainous terrain when the T_a representing single data points are extrapolated to a continuous surface based on fixed lapse rates [9–11].

The land surface temperature (T_s) in a remote sensing perspective is the measure of how hot or cold the top canopy skin layer of the Earth at a particular location will feel when touched [12]. The measure of T_s is largely dependent on net solar radiation, sensible heat flux, reflectance property of the surface, aerodynamic resistance, and the density of air [13]. Although the T_s is closely related to T_a, it can be significantly influenced by the surface characteristics, buffering effects of vegetation and the periodicity of the shortwave radiation emitted from the sun [14]. Over the past decade, the remotely-sensed T_s measurements have been used to map permafrost in different parts of the world [14–17]. There have been several attempts to estimate T_a using remotely-sensed T_s in different ecological systems [18–22]. The root mean square difference (RMSD) between T_a from meteorological stations and T_s from Moderate Resolution Imagining Spectroradiometer (MODIS) on Terra [23] and Aqua [24] satellites was estimated to be ±2.20 °C in Indo-Gangetic plain [25], ±1.33 °C in Portugal [18], ±5.48 °C in mountainous regions of Nevada, United states of America [19], ±2.97 to ±7.45 °C in northern Tibetan Plateau, China [26], ±4.09 to ±4.53 °C in a mountainous region of sub-Arctic Canada [27], and ±1.51 to ±3.74 °C over different ecosystems in Africa [22]. A recent study attempted to analyze the temperature trend using the 8-day T_s corrected using the difference between T_s and T_a calculated for 87 meteorological stations in the Chinese part of Himalaya and Tibetan Plateau [28]. Most of these published studies have compared the T_a and T_s at monthly or 8-day scales while several prominently used ecological and glacio-hydrological models in Himalaya that require daily temperature data as input parameter [4,29]. Moreover, such comparative studies for high mountains of Central or Western parts of Himalaya are completely missing.

The observed temperatures in Himalaya are scarce and fragmented in spatiotemporal domain due to difficult terrain, inhospitable weather conditions, and logistic difficulties in setting-up the weather stations [29]. The Himalayan mountains serve as a source of fresh water supply [30,31] and hydropower generation [32] to the densely populated mountainous regions of Indian Subcontinent. The Himalayan rivers mainly consist of the meltwaters coming from snow and glaciers [30] and this runoff is largely dependent on the seasonal temperatures [4,33]. The glaciers in Himalaya are losing mass in general with a few exceptions [29,33]. However, the quantification of the changes evident in glacierized regions in Himalaya with respect to the changing temperatures are largely uncertain due to unavailability of well-distributed and spatiotemporally continuous network of meteorological stations [29]. Furthermore, the lack of a definite and abiding framework for mutual climatological data sharing among various research and academic organizations in Himalayan countries makes regional-scale glacio-hydro-climatological modelling and interpretations more uncertain [29]. In this respect, there are two significant research gaps: (i) the studies comparing T_a with T_s for a large spatial domain are completely missing for the Central and Western Himalaya, and (ii) owing to this research gap, the glacio-hydrological community is further unsure of the significant role that spatiotemporally continuous satellite-derived surface temperatures can play as a substitute for spatially discontinuous T_a observations. The land-surface temperature is more likely proxy of energy exchange between land-surface and atmosphere for phenomena which are more strongly linked to ground processes [27]. The main aim of the present study is to understand and quantify the statistically significant trends in $T_s - T_a$ variation over a large spatiotemporal domain in Western Himalaya. Here, we start with providing a description of the study area, followed by data and used methods, and finally we discuss and conclude the main findings of the analyses.

2. Study Area

In the present paper, we analyze the relationship between daily and 8-day mean T_a from 11 high-altitude weather stations (Table 1) in Western Himalaya (Figure 1) and the respective daily and 8-day mean T_s measured by MODIS. The daily and 8-day night- and day-time T_s observations from Version 6 of Terra MODIS (MOD11A1 and MOD11A2 available from February, 2000) and Aqua MODIS (MYD11A1 and MYD11A2 available from July, 2002) were used to calculate average daily and 8-day T_s, respectively.

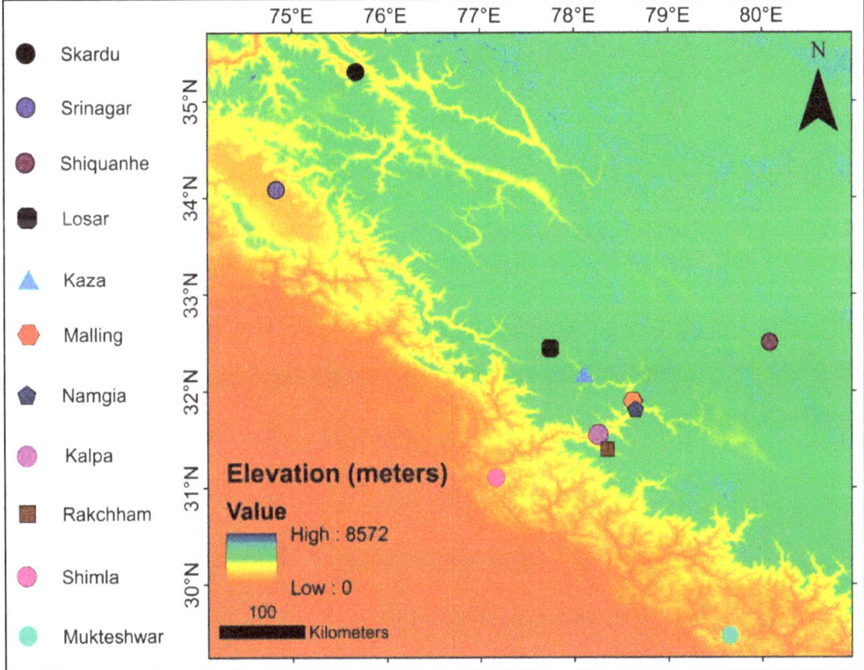

Figure 1. The map showing the location of the stations considered in the study for comparison of land surface temperature (T_s) and air temperature (T_a) with elevation (Aster GDEM v2, 2011) profile of the region.

The stations in the present study are located over a large elevation range above sea level (1587–4280 m) (Figure 1). All the stations used in the analysis are located above 2100 m except for Srinagar which is located at 1587 m. These stations are located in three different Himalayan states of India namely Uttarakhand, Himachal Pradesh, and Jammu and Kashmir. In addition, we have also included one station (Shiquanhe) from the Chinese part of the region in the study (Figure 1). These stations represent various precipitation regimes of the region such as monsoon dominance, westerly dominance, the precipitation-transition zone from monsoon-to-westerly dominance, and orographic precipitation-shadow zone. The topographic variations, i.e., altitudes and orography, among the Himalayan ranges not only govern the temperatures but also the precipitation [34]. Here, we aim to further untangle the degree of control of the altitude and orography in deciding the correlation between T_s and T_a in the Himalaya.

Table 1. Details of meteorological stations used for comparison in the present study.

Sl. No.	Name of the Station	Elevation (m)	Period	Organization	Precipitation Regime
1.	Kalpa	2707	07-Jul-02–31-Dec-09	IMD	Transition
2.	Kaza	3631	07-Jul-02–31-Dec-09	IMD	Shadow
3.	Namgia	2832	07-Jul-02–31-Dec-09	BBMB	Transition
4.	Rakchham	3046	07-Jul-02–31-Dec-09	BBMB	Transition
5.	Malling	3588	07-Jul-02–31-Dec-09	BBMB	Transition
6.	Losar	4122	07-Jul-02–31-Dec-09	BBMB	Shadow
7.	Mukteshwar	2311	13-Dec-15–30-Sept-19	GHCN	Monsoon
8.	Shimla	2202	01-Jan-16–30-Sept-19	GHCN	Monsoon
9.	Shiquanhe	4280	06-Jul-02–30-Sept-19	GHCN	Shadow
10.	Skardu	2181	02-Oct-02–30-Sept-19	GHCN	Shadow
11.	Srinagar	1587	05-Jul-02–30-Sept-19	GHCN	Shadow

3. Materials and Methods

3.1. Air Temperature (T_a)

The T_a is generally observed at a height of about 2 m above the land surface. T_a data observed at 11 different stations was used in the present study (Table 1). The data for five stations from Global Historical Climatology Network (GHCN) which was acquired from National Centre for Environmental Information (NCEI), NOAA web portal (https://www.ncei.noaa.gov/) [35] had observed daily mean T_a estimated using hourly or 6-hourly observations (Version 3). For the other six stations of Bhakra Beas Management Board (BBMB) and India Meteorology Department (IMD), the T_a was calculated using the daily maximum and minimum observations due to unavailability of daily mean T_a. This method of averaging the daily maximum and minimum temperatures for calculation daily mean temperature is widely used due to the instrumentation, logistic, and computational simplicity [36]. Although the method produces bias in the output due to inability to track the diurnal asymmetry [37], it has been used by considerable number of studies to make acceptable estimates requiring T_a [36]. In order to understand the degree of bias for our study area, we compared the given daily mean T_a with mean of daily maximum and minimum T_a for the five stations of GHCN for which all the three parameters were available. The analysis showed that the RMSD was less than 1.62 °C with a very high correlation ($R^2 > 0.96$) for all the stations. In addition to the daily T_a, we also calculated 8-day mean T_a for comparison with the corresponding 8-day T_s explained in next section. The observations were carefully checked for systematic and random errors before using it for further comparison. The stations used in the present study are distributed broadly over four different precipitation zones (Table 1, Figure 1). The precipitation varies significantly in these precipitation zones. The monsoon dominated, transition zone, westerlies dominated, and precipitation shadow zone receive >1500, 200–800, 600–800, and <150 mm total annual precipitation on an average, respectively.

3.2. MODIS Data

The daily, and 8-day night- and day-time T_s from MODIS satellites on Terra (MOD11A1 and MOD11A2 available from February, 2000) and Aqua (MYD11A1 and MYD11A2 available from July, 2002) satellites [23,24] was downloaded from NASA Earthdata portal (https://earthdata.nasa.gov/) [38] and was used to calculate average of daily and 8-day T_s (Table 2). The remotely-sensed T_s from MODIS (version 006) has been observed to have RMSD of less than 0.5 K in comparison to the in situ measurements of the T_s [39] and therefore has been widely used for multiple scientific applications [18,19,22,25–28,40].

Table 2. Details of Moderate Resolution Imagining Spectroradiometer (MODIS) data used in the present study.

Sl. No.	Data Characteristics	Terra		Aqua	
		MOD11A1	MOD11A2	MYD11A1	MYD11A2
1.	Temporal resolution	Daily	8-day	Daily	8-day
2.	Spatial resolution	1 km		1 km	
3.	Available from	February, 2000		July, 2002	
4.	Local day time of observation	10:30–11:30		12:30–13:30	
5.	Local night time of observation	21:30–22:30		00:30–01:30	

The local time for the pass over the study area for Terra is around 10:30 and 22:30 and for Aqua is around 13:30 and 01:30 during day and night, respectively. The 8-day land surface temperature data MOD11A2 and MYD11A2 is a simple pixel wise average of all the respective MOD11A1 and MYD11A2 data collected during the 8-day period. The days with all the four observations, including the day and night-time measurements available from both Terra and Aqua were included in the comparison at both daily and 8-day scale. Equation (1) was used to compute the average of four MODIS observations during a day or 32 MODIS observations during an 8-day period (referred as T_s in °C) from the pixel value corresponding to every station given in Table 1.

$$T_s = \frac{T_d^T + T_n^T + T_d^A + T_n^A}{4} - c,$$

(1)

where,

T_d^T = Terra day-time observation

T_n^T = Terra night-time observation

T_d^A = Aqua day-time observation

T_n^A = Aqua night-time observation

c = Constant for conversion from kelvin to Celsius (273.15)

For every data point of daily and 8-day T_s, two night-time and two day-time satellite observations were used. It moderated the calculated daily T_s for further comparison with T_a. Therefore, every data point of daily T_s is average of four MODIS observations during that day and 8-day T_s is average of 32 MODIS observations during that 8-day period. Since, the satellite observation from MODIS is unavailable in cloud cover condition and the calculated daily T_s for comparison with T_a can have large data gaps, we decided to also include MODIS 8-day T_s in the analyses. For 8-day T_s, the data available is comparatively more continuous due to correction of cloud contamination [39]. Although, the 8-day MODIS observations are more efficient in terms of temporal continuity, it compromises with the temporal resolution. Additionally, the number of data points available for comparison for 8-day average is significantly less than the dataset available for daily average. Thus, the average T_s used in our analyses, and referred to hereafter, is essentially the average of four-times daily and 8-day MODIS T_s observations and all the results should be considered accordingly. Therefore, based on these data limitations, we finally compared the average of four-times daily and 8-day MODIS T_s observations with observed daily mean T_a for five GHCN stations and with the average of observed daily maximum and minimum temperatures for the remaining six stations.

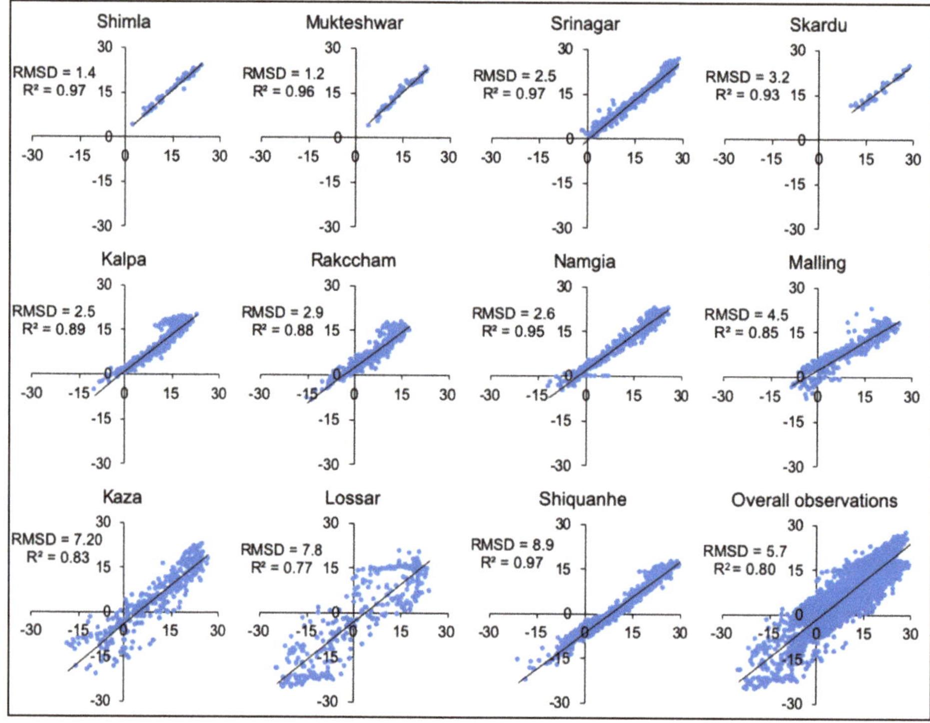

Figure 3. The scatter plot between 8-day T_s (x-axis) and T_a (y-axis) for all the stations and overall observations with respective coefficient of determination (R^2) and RMSD in °C.

The number of data points available for 8-day analysis is significantly less in comparison to the daily analysis (Table 1). Additionally, the use of 8-day data gives spatiotemporal continuity due to correction of cloud contaminated pixels but poses a restriction on the frequency of comparisons. We decided to represent the analysis of the relationship between T_a and T_s at daily scale henceforth because there was very small improvement in the results observed after the use of 8-day data and the daily observation are crucial for different geophysical models as explained in the Introduction section. We noticed certain spatial patterns in the daily differences between T_s and T_a which are discussed in detail in the following paragraphs.

4.2. Altitudinal Relationship

First, we present the relationship between $T_s - T_a$ by considering altitudinal positions of the stations. The variation in altitude affects the T_a due to difference in density of air which causes a reduced green-house effect in the higher reaches. The RMSD between T_s and T_a for stations has a direct correlation with the elevation of the station (Figures 2–4). Although, the RMSD increases systematically with increase in elevation in general, small variation in this trend is observed for northernmost stations (Skardu and Srinagar). The annual mean RMSD is strongly correlated to the elevation ($R^2 = 0.74$) in general (Figure 5a) except for two stations (Skardu and Srinagar), which even when located at comparatively low elevations show higher magnitude of RMSD (Figure 4). The R^2 is stronger for monsoon season (Figure 5b) in comparison to annual (Figure 5a) and summer season values (Figure 5c). The observed T_a was unavailable for Skardu for winter months (Figure 5d). Therefore, the R^2 is highest for winter when compared to monsoon, summer, and annual analysis. Furthermore, the magnitude of $T_s - T_a$ is observed to be higher in summer months in comparison to the winter months for all the stations in general (Figure 4).

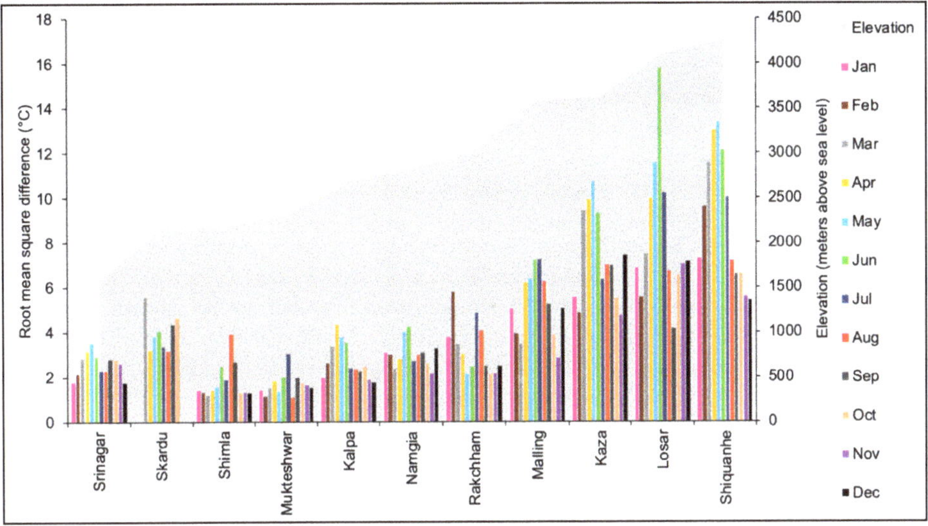

Figure 4. Graph showing the monthly mean of RMSD between daily T_s and T_a for different stations with respective elevation. The RMSD is higher in summer months and increases with increase in elevation.

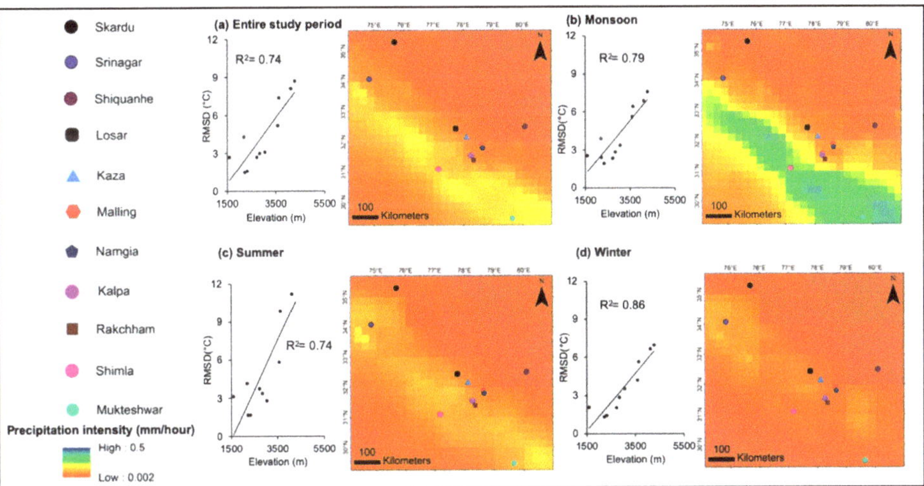

Figure 5. The graph showing the relationship between average of RMSD between daily T_s and T_a for each station and its corresponding elevation for (**a**) the entire study period, (**b**) monsoon (JASO), (**c**) summer (MAMJ), and (**d**) winter (NDJF). The map shows the precipitation intensity (mm/h) data from Tropical Rainfall Measuring Mission (GES DISC, 2016) for the period 1 January 1999 to 31 December 2017 plotted through GIOVAANI (https://giovanni.gsfc.nasa.gov/giovanni/) [41].

4.3. Seasonal Relationship

The difference between mean monthly T_a and T_s (i.e., $T_s - T_a$) for the entire period of study shows high inter-monthly variability for all the stations except for the stations in monsoon-dominated regions (Figures 5 and 6). The mean monthly T_s is lower in comparison to mean monthly T_a for southern slopes (Figure 6a) and increases with increasing latitudes (Figure 6c) except for the stations in westerlies dominated areas (Figure 6d). The magnitude of difference between mean monthly T_s and mean monthly T_a is negative for the stations in monsoon-dominated areas and positive for the stations in precipitation shadow and westerly-dominated regions. In the precipitation-transition zone, the difference is positive for summer months and negative for winter months except for Rakchham, the southernmost station of the transition zone (Figure 5). For Rakchham, the $T_s - T_a$ values are negative throughout the year similar to the stations in monsoon-dominated areas (Figure 6b). This might be a result of the added effect of humidity in the near-surface atmosphere and presence of snow on land surface which moderates the difference between T_s and T_a [42] throughout the year in monsoon-dominated regions. In the precipitation-transition zone, the difference is partly moderated by the presence of snow during winter months and partly humidity during summer months, particularly for the southernmost stations of the zone (Rakchham and Kalpa) (Figures 5 and 6b) which receive enough precipitation through both monsoon and westerlies. $T_s - T_a$ values for the stations in westerly-dominated region are regulated mainly by the presence of snow during winters (Figures 5 and 6d), which tends to cool the surface due to high albedo [42]. The $T_s - T_a$ values for the stations in precipitation-shadow zone are significantly high and positive in magnitude throughout the year in comparison to all the other stations due to the perennial cold-arid atmospheric conditions (Figures 5 and 6c). This confirms the role of water cycle on this gradient and shows that in the absence of soil-atmosphere water cycle (dry conditions) the magnitude of the difference between $T_s - T_a$ increases and is more positive.

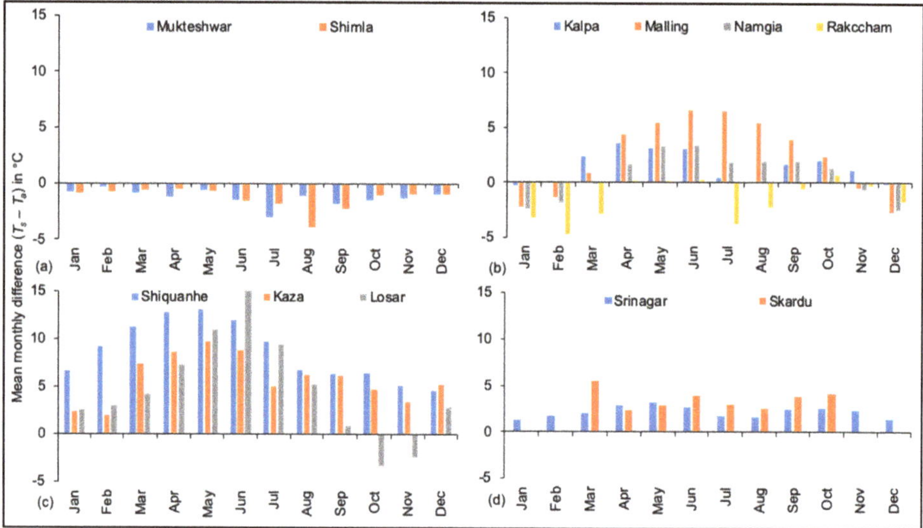

Figure 6. Graph showing the mean monthly difference between daily T_s and T_a for the entire period for which the data is available for stations in (**a**) monsoon-dominated areas, (**b**) transition zone, (**c**) precipitation shadow zone, and (**d**) westerlies-dominated areas. The observed T_a for Skardu for winter months was unavailable. The $T_s - T_a$ values for the stations in the precipitation-shadow zone (**c**) are significantly higher and positive in magnitude, due to the perennial cold-arid atmospheric conditions.

The comparison of T_s and T_a showed high inter-monthly variability throughout the study period. Therefore, we performed an additional analysis where we estimated the seasonal effect of each month on the difference between T_s and T_a (Figure 7) in reference to a base month. For this multiple regression analysis, January was considered as the base month since the $T_s - T_a$ values in January were least for all the stations in general (Figure 6). This analysis further corroborates the above-discussed aspect that the $T_s - T_a$ coefficient values are larger in summer months in comparison to winter months (Figure 7). Additionally, the difference in coefficient and RMSD is high for stations in precipitation shadow regions (Kaza, Shiquanhe and Losar) in comparison to the stations in monsoon-dominated areas (Shimla and Mukteshwar) (Figures 6 and 7).

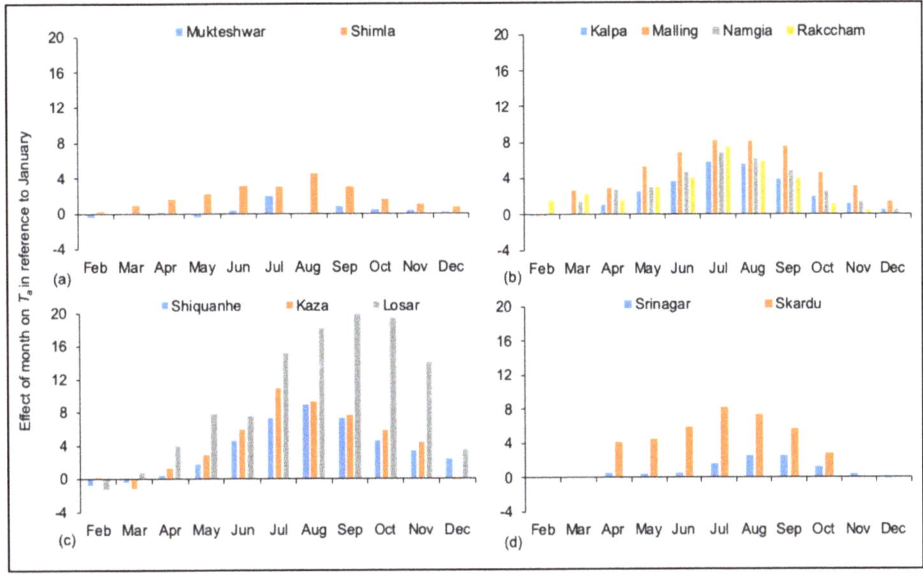

Figure 7. Graph showing the effect of each month on the T_a in reference to the month of January for stations in (**a**) monsoon-dominated areas, (**b**) transition zone, (**c**) precipitation shadow zone, and (**d**) westerlies-dominated areas. The observed T_a for Skardu for winter months was unavailable.

To further corroborate the effect of seasonality and the presence of snow and humidity on the $T_s - T_a$ values, we created monthly box and whisker plots of daily difference between T_s and T_a (Figure 8). The whisker for the stations in precipitation shadow zone and transition zone is longer showing the high monthly variability of the difference value in comparison to the stations in monsoon and westerlies-dominated areas. This is due to the presence of snow during the winter and humidity in the atmosphere in summer regulating the difference between T_s and T_a in the monsoon dominated areas. Additionally, the size of the boxes are smaller for the stations in monsoon and westerlies-dominated areas explaining the presence of maximum data points close to the median representing that throughout the year the difference between T_s and T_a is regulated by presence of snow or atmospheric moisture. On the contrary, the boxes for stations in precipitation shadow zones which receive significantly less precipitation throughout the year, are wider in size representing large variation in the difference between T_s and T_a throughout the year. The boxes for the stations in the southern part of the transition zone are smaller in summer and wider in winter showing the effect of humidity due to some influence of monsoon owing to their spatial closeness to the monsoon-dominated region. Besides, both the boxes and whiskers for the stations in north-eastern part of the transition zone, closer to the precipitation shadow zone, are wider in size showing the variability due to lack of both snow and humidity.

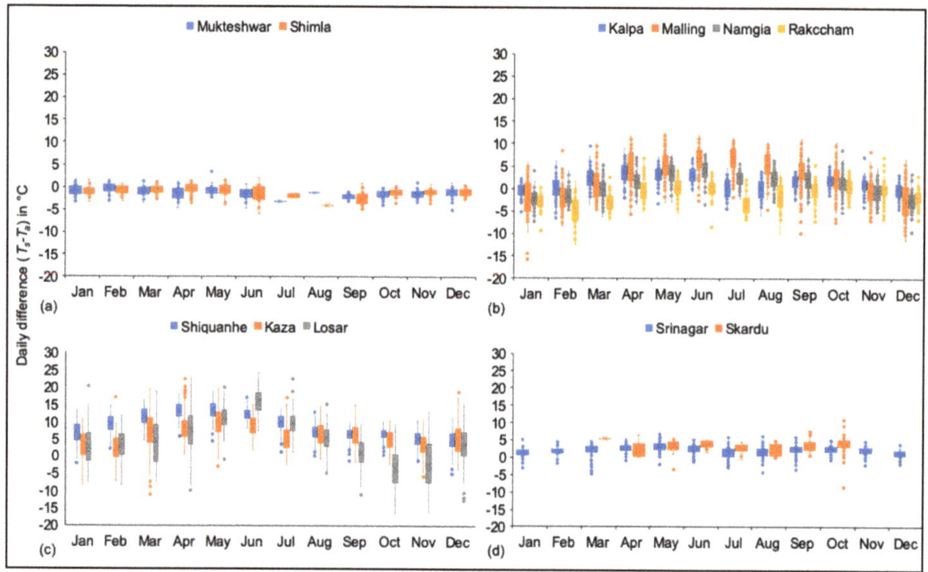

Figure 8. Box and whisker plots showing the monthly variation of daily difference between T_s and T_a for the entire period for which the data is available for stations in (**a**) monsoon-dominated areas, (**b**) transition zone, (**c**) precipitation shadow zone, and (**d**) westerlies-dominated areas.

5. Discussion

The observed near-surface air temperature is one of the most important climate parameters used in different kinds of environmental studies particularly in Himalaya where the interaction between high elevation, climate, and cryosphere is highly significant and complex. It is extremely difficult to capture the spatial heterogeneity of the near-surface temperature [43] which is the primary forcing data for different glacio-hydrological models [3,4,44–46]. It is also used as primary data for climate change assessment [47,48], agro-climatic [40], ecological [49,50], and socio-economic [51,52] studies. Our results present a freely available substitute for station recorded T_a with high temporal and spatial resolution. Conclusively, the T_s is highly correlated with T_a throughout the study area at both daily and 8-day scales. The correlation is highest at the stations located at Southern slope (Shimla and Mukteshwar) with significantly low RMSD in comparison to the stations located in the Eastern part (Losar and Shiquanhe). Although, the degree of congruence between T_s and T_a is slightly higher in the 8-day dataset ($R^2 > 0.77$) in comparison to the daily dataset ($R^2 > 0.69$), the number of data points available for comparison is significantly low. The overall RMSD improved by 0.2 °C on an average by using the 8-day dataset. The largest improvement in RMSD was observed for Skardu (1.1 °C) but the number of data points available for correlation was significantly less than other stations. The overall SE improved by 0.38 °C except for Kalpa and Kaza for which it deteriorated by 0.02 and 0.08 °C, respectively. It is interesting to note that for Shiquanhe which is located in precipitation shadow zone and highest altitude among all the stations, shows largest improvement in SE (by 1.63 °C) and reduction in RMSD (0.2 °C).

The difference between T_s and T_a is primarily controlled by elevation, the land surface cover characteristics, and near-surface humidity. At higher altitudes, the thinner atmosphere shows lesser water holding capacity and the atmosphere saturates faster, thus allowing for lesser evaporation/sublimation in a given pressure-temperature scenario [53]. This puts a constraint on the limit of specific humidity in the high elevations and the comparatively lesser number of available water molecules in the near-surface atmosphere cannot trap the same amount of heat as those at lower elevations. This can provide a basis for the observed high values of $T_s - T_a$ at the higher altitudes. The intercept of the regression between T_s and T_a shows increase for the stations in monsoon, transition, and westerlies-dominated areas. On the contrary, the stations in precipitation shadow zone show a sharp decrease in the intercept of the regression at high elevations (>3600 m). The slope of the regression between T_s and T_a is higher for stations in low elevation and precipitation-dominated areas (0.80–1.03) in comparison to the stations in high elevation and in transition-to-precipitation shadow zones (0.59–0.86). This observation is supported by a study which shows decrease in slope and degree of correlation in high elevation [54]. The high difference between T_s and T_a for the stations in dry atmosphere at high altitude may partially be due to the heat from the Sun and cooling of near-surface atmosphere due to heat exchange from surrounding air and temperature lapse rate [55]. The presence of more humidity moderates the difference between T_s and T_a in precipitation dominant areas. The difference between T_s and T_a is highest with positive magnitude when the land surface is snow-free and the near-surface atmosphere is dry. On the contrary, the T_s and T_a is negative and lower in magnitude when the land surface is covered by snow and/or atmosphere is more saturated with moisture regardless of high altitude. In addition to the elevation and precipitation regime, season was observed to have significant control over the difference between T_s and T_a. The summer months were observed to have a significantly higher effect on T_a in reference to January, in general for all the stations. The inter-monthly variability was observed to be very high for year-round humidity-deficient transition zones and precipitation shadow zones in comparison to the monsoon-dominated and westerlies-dominated regions. It can be interpreted that the energy exchange between the surface and near-surface atmosphere in the precipitation dominant areas is more efficient in comparison to the precipitation deficient areas.

The lower magnitude of RMSD between T_s and T_a represents lower gradient of temperature between the land surface and near-surface air due to the cold bias caused by snow cover which protects the surface from warming because of its high albedo [42]. A possible contributing factor to this seasonal disparity can be the reported perpetually melting seasonal snow in Himalayan mountains [55] under the changing regional climate. The causal mechanism for this relationship deserves a separate detailed investigation. However, a possible cause of such observations can be linked to the fact that the diurnal temperatures even in the mid-winter months often cross the freezing point causing a certain degree of melting to prevail [56]. This can start a cascading event where during the preliminary warming phase, the average snowpack temperature reaches and stays at 0 °C isotherm until the melting typically starts within the snowpack prior to the ripening phase as meltwater is retained within the snowpack [57]. This meltwater may subsequently refreeze owing to the diurnal cycles of temperature and the latent heat released during this process can additionally warm the snow surface and the surrounding air, further minimizing the temperature difference [56]. In addition to the seasonal change in the land cover characteristics, the variation in humidity in the near-surface atmosphere is an important factor controlling the difference between T_s and T_a. It was recently proposed that the amount of moisture content on the land surface has a cooling effect on land surface temperature [40]. Thus, the precipitation regime in which a particular station is located can further provide us several clues regarding the observed variations in $T_s - T_a$ values. These precipitation regimes have been previously characterized [34] and in the following discussion, we take a focused approach towards revisiting the $T_s - T_a$ variations with respect to the respective precipitation scenarios.

All the statistical results and the regression equation between T_s and T_a have specific trends for particular climate setting and elevation which can be used to estimate T_a using T_s (Table 3) for glacio-hydrological and climate change studies in data-deficient Himalaya. For example, the RMSD ranges between 1.2–1.6, 2.6–5., 2.5–4.3, and 7.2–8.9 °C for the stations in monsoon-dominated, transition, westerlies-dominated, and precipitation shadow zones, respectively, for both daily and 8-day products. The slope and intercept of the regression equations between T_s and T_a are also similar for the stations in the same precipitation regime. The paper demonstrates different patterns of variation of $T_s - T_a$ in different climate regimes within the region of study. Due to the inherent limitations of the available data, some of this analysis may be revised in the future by specific dedicated studies, in particular to asses if the relationships hold on daily scales and with what error bar. Some possible error sources for this analysis may come from the scarcity of the data, and the fact that we compared data from different instrumentation accuracies and cadences. T_a is measured by three different organizations and two calculation methods are used for daily mean air temperature during different observation periods. The correlation of the instantaneous observation of T_a in relation to satellite derived T_s can be investigated by analyzing the diurnal variation of T_a in relation to the time of pass of the satellite [58]. There are different parameters like wind speed and fractional vegetation which have additional effects on the difference between T_s and T_a, which have not been investigated in the present study [54,55] and can be interesting research questions for future investigations in the region.

6. Conclusions

Unavailability of reliable temperature observations with spatial continuity along with the extreme weather conditions and difficult terrain in the remoteness of Himalaya hampers our understanding of the cryosphere-climate coupling in these mountains. Here we attempt to compare remotely sensed T_s with respect to in situ T_a observations over different precipitation and altitudinal zones of the Western Himalaya. Although, there are several studies available from different parts of the globe attempting to estimate T_a using T_s or vice-versa using monthly or 8-day MODIS data, we provide an understanding of the spatiotemporal variability of the T_a vs. T_s relationship at diurnal scales. The results show a strong and statistically significant relationship between T_s and T_a in general with a spatiotemporal consistency, thus projecting satellite-derived T_s as a viable alternative to the in situ T_a for glacio-hydro-climatological studies. We also provide regression equations to facilitate modeling of gridded T_a using corresponding T_s for different regions of Western Himalaya. MODIS in combination with Sea and Land Surface Temperature Radiometer (SLSTR) onboard Sentinel-3 can provide better capability to overcome cloud gaps and ensuring spatiotemporal continuity for T_s future studies in this direction.

Author Contributions: S.S. conceptualized and designed the research and wrote the manuscript. S.S., A.B., A.S., L.S., and M.S. performed statistical tests, and wrote the manuscript and methods section with inputs from all the co-authors. S.S., A.B., and A.S., performed raw data generation and analysis. F.J.M.-T. and M.-P.Z. helped in analyzing the results and correlating them with the different variables.

Funding: This research received no external funding.

Acknowledgments: The authors would like to acknowledge National Snow and Ice Data Centre, USA and National Oceanic and Atmospheric Administration, USA for providing freely available MODIS satellite products and Global Historical Climatology Network station data, respectively. The authors are also grateful to India Meteorology Department (IMD), India, Bhakhra Beas Management Board (BBMB), India and Hendrik Wulf, University of Zurich, Switzerland for providing the station data. A.B. acknowledges the Swedish Research Council for supporting his research in Himalaya. M.S. acknowledges Director, Birbal Sahni Institute of Palaeosciences and Birbal Sahni Research Associate fellowship.

Conflicts of Interest: The authors declare no conflict of interest.

References

1. Hansen, J.; Ruedy, R.; Sato, M.; Lo, K. Global surface temperature change. *Rev. Geophys.* **2010**, *48*, RG4004. [CrossRef]
2. Jones, P.D. Hemispheric Surface Air Temperature Variations: A Reanalysis and an Update to 1993. *J. Clim.* **1994**, *7*, 1794–1802. [CrossRef]
3. Hock, R. Temperature index melt modelling in mountain areas. *J. Hydrol.* **2003**, *282*, 104–115. [CrossRef]
4. Kumar, R.; Singh, S.; Kumar, R.; Singh, A.; Bhardwaj, A.; Sam, L.; Randhawa, S.S.; Gupta, A. Development of a glacio-hydrological model for discharge and mass balance reconstruction. *Water Resour. Manag.* **2016**, *30*, 3475–3492. [CrossRef]
5. Hatfield, J.L.; Prueger, J.H. Temperature extremes: Effect on plant growth and development. *Weather Clim. Extrem.* **2015**, *10*, 4–10. [CrossRef]

6. Stoll, M.J.; Brazel, A.J. Surface-air temperature relationships in the urban environment of Phoenix, Arizona. *Phys. Geogr.* **1992**, *13*, 160–179. [CrossRef]

7. Seiler, C.; Moene, A.F. Estimating Actual Evapotranspiration from Satellite and Meteorological Data in Central Bolivia. *Earth Interact.* **2011**, *15*, 1–24. [CrossRef]

8. Screen, J.A.; Simmonds, I. The central role of diminishing sea ice in recent Arctic temperature amplification. *Nature* **2010**, *464*, 1334. [CrossRef]

9. Ishida, T.; Kawashima, S. Use of cokriging to estimate surface air temperature from elevation. *Theor. Appl. Climatol.* **1993**, *47*, 147–157. [CrossRef]

10. Snehmani; Bhardwaj, A.; Singh, M.K.; Gupta, R.D.; Joshi, P.K.; Ganju, A. Modelling the hypsometric seasonal snow cover using meteorological parameters. *J. Spat. Sci.* **2015**, *60*, 51–64. [CrossRef]

11. Willmott, C.J.; Robeson, S.M. Climatologically aided interpolation (CAI) of terrestrial air temperature. *Int. J. Climatol.* **1995**, *15*, 221–229. [CrossRef]

12. Bense, V.F.; Read, T.; Verhoef, A. Using distributed temperature sensing to monitor field scale dynamics of ground surface temperature and related substrate heat flux. *Agric. For. Meteorol.* **2016**, *220*, 207–215. [CrossRef]

13. Sellers, P.J.; Dickinson, R.E.; Randall, D.A.; Betts, A.K.; Hall, F.G.; Berry, J.A.; Collatz, G.J.; Denning, A.S.; Mooney, H.A.; Nobre, C.A.; et al. Modeling the exchanges of energy, water, and carbon between continents and the atmosphere. *Science* **1997**, *275*, 502–509. [CrossRef] [PubMed]

14. Luo, D.; Jina, H.; Marchenkoa, S.S.; Romanovsky, V.E. Difference between near-surface air, land surface and ground surface temperatures and their influences on the frozen ground on the Qinghai-Tibet Plateau. *Geoderma* **2018**, *312*, 74–85. [CrossRef]

15. Hachem, S.; Allard, M.; Duguay, C. Using the MODIS land surface temperature product for mapping permafrost: An application to northern Québec and Labrador, Canada. *Permafr. Periglac. Process.* **2009**, *20*, 407–416. [CrossRef]

16. Hachem, S.; Duguay, C.R.; Allard, M. Comparison of MODIS-derived land surface temperatures with ground surface and air temperature measurements in continuous permafrost terrain. *Cryosphere* **2012**, *6*, 51–69. [CrossRef]

17. Ran, Y.; Li, X.; Jin, R.; Guo, J. Remote sensing of the mean annual surface temperature and surface frost number for mapping permafrost in China. *Arct. Antarct. Alp. Res.* **2015**, *47*, 255–265. [CrossRef]

18. Benali, A.; Carvalho, A.C.; Nunes, J.P.; Carvalhais, N.; Santos, A. Estimating air surface temperature in Portugal using MODIS LST data. *Remote Sens. Environ.* **2012**, *124*, 108–121. [CrossRef]

19. Mutiibwa, D.; Strachan, S.; Albright, T. Land Surface Temperature and Surface Air Temperature in Complex Terrain. *IEEE J. Sel. Top. Appl. Earth Obs. Remote Sens.* **2015**, *8*, 4762–4774. [CrossRef]

20. Prihodko, L.; Goward, S.N. Estimation of air temperature from remotely sensed surface observations. *Remote Sens. Environ.* **1997**, *60*, 335–346. [CrossRef]

21. Stisen, S.; Sandholt, I.; Nørgaard, A.; Fensholt, R.; Eklundh, L. Estimation of diurnal air temperature using MSG SEVIRI data in West Africa. *Remote Sens. Environ.* **2007**, *110*, 262–274. [CrossRef]

22. Vancutsem, C.; Ceccato, P.; Dinku, T.; Connor, S.J. Evaluation of MODIS land surface temperature data to estimate air temperature in different ecosystems over Africa. *Remote Sens. Environ.* **2010**, *114*, 449–465. [CrossRef]

23. Wan, Z.; Hook, S.; Hulley, G. MOD11A1 MODIS/Terra Land Surface Temperature/Emissivity Daily L3 Global 1km SIN Grid V006 [Data set]. *NASA EOSDIS LP DAAC* **2015**. [CrossRef]

24. Wan, Z.; Hook, S.; Hulley, G. MYD11A1 MODIS/Aqua Land Surface Temperature/Emissivity Daily L3 Global 1km SIN Grid V006 [Data set]. *NASA EOSDIS LP DAAC* **2015**. [CrossRef]

25. Shah, D.B.; Pandya, M.R.; Trivedi, H.J.; Jani, A.R. Estimating minimum and maximum air temperature using MODIS data over Indo-Gangetic Plain. *J. Earth Syst. Sci.* **2013**, *122*, 1593–1605. [CrossRef]

26. Zhu, W.; Lü, A.; Jia, S. Estimation of daily maximum and minimum air temperature using MODIS land surface temperature products. *Remote Sens. Environ.* **2013**, *130*, 62–73. [CrossRef]

27. Williamson, S.; Hik, D.; Gamon, J.; Kavanaugh, J.; Flowers, G. Estimating temperature fields from MODIS land surface temperature and air temperature observations in a sub-arctic alpine environment. *Remote Sens.* **2014**, *6*, 946–963. [CrossRef]

28. Pepin, N.; Deng, H.; Zhang, H.; Zhang, F.; Kang, S.; Yao, T. An examination of temperature trends at high elevations across the Tibetan Plateau: The use of MODIS LST to understand patterns of elevation-dependent warming. *J. Geophys. Res. Atmos.* **2019**, *124*, 5738–5756. [CrossRef]

29. Singh, S.; Kumar, R.; Bhardwaj, A.; Sam, L.; Shekhar, M.; Singh, A.; Kumar, R.; Gupta, A. The Indian Himalayan Region: A review of signatures of changing climate and vulnerability. *WIREs Clim. Chang.* **2016**, *7*, 393–410. [CrossRef]

30. Immerzeel, W.W.; Van Beek, L.P.H.; Bierkens, M.F.P. Climate Change Will Affect the Asian Water Towers. *Science* **2010**, *328*, 1382–1385. [CrossRef]

31. Sam, L.; Bhardwaj, A.; Kumar, R.; Buchroithner, M.F.; Martín-Torres, F.J. Heterogeneity in topographic control on velocities of Western Himalayan glaciers. *Sci. Rep.* **2018**, *8*, 12843. [CrossRef] [PubMed]

32. Sam, L.; Bhardwaj, A.; Sinha, V.S.P.; Joshi, P.K.; Kumar, R. Use of Geospatial Tools to Prioritize Zones of Hydro-Energy Potential in Glaciated Himalayan Terrain. *J. Indian Soc. Remote Sens.* **2016**, *44*, 409–420. [CrossRef]

33. Shekhar, M.; Bhardwaj, A.; Singh, S.; Ranhotra, P.S.; Bhattacharyya, A.; Pal, A.K.; Roy, I.; Martín-Torres, F.J.; María-Paz, Z. Himalayan glaciers experienced significant mass loss during later phases of little ice age. *Sci. Rep.* **2017**, *7*, 10305. [CrossRef] [PubMed]

34. Bookhagen, B.; Burbank, D.W. Toward a complete Himalayan hydrological budget: Spatiotemporal distribution of snowmelt and rainfall and their impact on river discharge. *J. Geophys. Res.* **2010**, *115*. [CrossRef]

35. National Centre for Environmental Information (NCEI). Available online: https://www.ncei.noaa.gov/ (accessed on 5 November 2019).

36. Villarini, G.; Khouakhi, A.; Cunningham, E. On the impacts of computing daily temperatures as the average of the daily minimum and maximum temperatures. *Atmos. Res.* **2017**, *198*, 145–150. [CrossRef]

37. Ma, Y.; Guttorp, P. Estimating daily mean temperature from synoptic climate observations. *Int. J. Climatol.* **2013**, *33*, 1264–1269. [CrossRef]

38. NASA Earthdata Portal. Available online: https://earthdata.nasa.gov/ (accessed on 5 November 2019).

39. Wan, Z. New refinements and validation of the MODIS land-surface temperature/emissivity products. *Remote Sens. Environ.* **2008**, *112*, 59–74. [CrossRef]

40. Shah, H.L.; Zhou, T.; Huang, M.; Mishra, V. Strong influence of irrigation on water budget and land surface temperature in Indian sub-continental river basins. *J. Geophys. Res. Atmos.* **2019**, *124*, 1449–1462. [CrossRef]

41. Goddard Earth Sciences Data and Information Services Center. *TRMM (TMPA) Precipitation L3 1 Day 0.25 Degree × 0.25 Degree V7*; Savtchenko, A., Ed.; GES DISC: Greenbelt, MD, USA, 2016. [CrossRef]

42. Williamson, S.N.; Hik, D.S.; Gamon, J.A.; Jarosch, A.H.; Anslow, F.S.; Clarke, G.K.; Rupp, T.S. Spring and summer monthly MODIS LST is inherently biased compared to air temperature in snow covered sub-Arctic mountains. *Remote Sens. Environ.* **2017**, *189*, 14–24. [CrossRef]

43. Immerzeel, W.W.; Petersen, L.; Ragettli, S.; Pellicciotti, F. The importance of observed gradients of air temperature and precipitation for modeling runoff from a glacierized watershed in the Nepalese Himalayas. *Water Resour. Res.* **2014**, *50*, 2212–2226. [CrossRef]

44. Singh, S.; Kumar, R.; Bhardwaj, A.; Kumar, R.; Singh, A. Changing climate and glacio-hydrology: A case study of Shaune Garang basin, Himachal Pradesh. *Int. J. Hydrol. Sci. Technol.* **2018**, *8*, 258–272. [CrossRef]

45. Singh, P.; Haritashya, U.K.; Kumar, N. Modelling and estimation of different components of streamflow for Gangotri Glacier basin, Himalayas. *Hydrol. Sci. J.* **2008**, *53*, 309–322. [CrossRef]

46. Wulf, H.; Bookhagen, B.; Scherler, D. Differentiating between rain, snow, and glacier contributions to river discharge in the western Himalaya using remote-sensing data and distributed hydrological modeling. *Adv. Water Resour.* **2016**, *88*, 152–169. [CrossRef]

47. Kraaijenbrink, P.D.A.; Bierkens, M.F.P.; Lutz, A.F.; Immerzeel, W.W. Impact of a global temperature rise of 1.5 degrees Celsius on Asia's glaciers. *Nature* **2017**, *549*, 257–260. [CrossRef]

48. Shekhar, M.S.; Chand, H.; Kumar, S.; Srinivasan, K.; Ganju, A. Climate-change studies in the western Himalaya. *Ann. Glaciol.* **2010**, *51*, 105–112. [CrossRef]

49. Liang, E.; Dawadi, B.; Pederson, N.; Eckstein, D. Is the growth of birch at the upper timberline in the Himalayas limited by moisture or by temperature? *Ecology* **2014**, *95*, 2453–2465. [CrossRef]

50. Xu, J.; Grumbine, R.E.; Shrestha, A.; Eriksson, M.; Yang, X.; Wang, Y.; Wilkes, A. The Melting Himalaya: Cascading Effects of Climate Change on Water, Biodiversity, and Livelihoods. *Conserv. Biol.* **2009**, *23*, 520–530. [CrossRef]

51. Archer, D.R.; Forsythe, N.; Fowler, H.J.; Shah, S.M. Sustainability of water resources management in the Indus Basin under changing climatic and socio economic conditions. *Hydrol. Earth Syst. Sci.* **2010**, *14*, 1669–1680. [CrossRef]

52. Malla, G. Climate Change and Its Impact on Nepalese Agriculture. *J. Agric. Environ.* **2009**, *9*, 62–71. [CrossRef]

53. Kimball, J.S.; Running, S.W.; Nemani, R. An improved method for estimating surface humidity from daily minimum temperature. *Agric. For. Meteorol.* **1997**, *85*, 87–98. [CrossRef]

54. Good, E.J. An in situ-based analysis of the relationship between land surface "skin" and screen-level air temperatures. *J. Geophys. Res. Atmos.* **2016**, *121*, 8801–8819. [CrossRef]

55. Good, E.J.; Ghent, D.J.; Bulgin, C.E.; Remedios, J.J. A spatiotemporal analysis of the relationship between near-surface air temperature and satellite land surface temperatures using 17 years of data from the ATSR series. *J. Geophys. Res. Atmos.* **2017**, *122*, 9185–9210. [CrossRef]

56. Kulkarni, A.V.; Rathore, B.P.; Singh, S.K. Distribution of seasonal snow cover in central and western Himalaya. *Ann. Glaciol.* **2010**, *51*, 123–128. [CrossRef]

57. Bhardwaj, A.; Singh, S.; Sam, L.; Bhardwaj, A.; Martin-Torres, F.J.; Singh, A.; Kumar, R. MODIS-based estimates of strong snow surface temperature anomaly related to high altitude earthquakes of 2015. *Remote Sens. Environ.* **2017**, *188*, 1–8. [CrossRef]

58. Niclos, R.; Valiente, J.A.; Barberà, M.J.; Caselles, V. Land surface air temperature retrieval from EOS-MODIS images. *IEEE Geosci. Remote Sens. Lett.* **2013**, *11*, 1380–1384. [CrossRef]

Article

Noise-sensitivity Analysis and Improvement of Automatic Retrieval of Temperature and Emissivity Using Spectral Smoothness

Honglan Shao [1,2], **Chengyu Liu** [1,*], **Feng Xie** [1,2], **Chunlai Li** [1,2] and **Jianyu Wang** [1,2,3]

1   Key Lab of Space Active Opto-Electronic Techniques, Shanghai Institute of Technical Physics, Chinese Academy of Sciences, Shanghai 200083, China; shaohonglan10@mails.ucas.ac.cn (H.S.); xf@mail.sitp.ac.cn (F.X.); lichunlai@mail.sitp.ac.cn (C.L.); jywang@mail.sitp.ac.cn (J.W.)

2   University of Chinese Academy of Sciences, Beijing 100049, China

3   Hangzhou Institute for Advanced Study, UCAS, Hangzhou 310024, China

*   Correspondence: liuchengyu@mail.sitp.ac.cn

Received: 5 June 2020; Accepted: 15 July 2020; Published: 17 July 2020

Abstract: There are numerous algorithms that can be used to retrieve land surface temperature (LST) and land surface emissivity (LSE) from hyperspectral thermal infrared (HTIR) data. The algorithms are sensitive to a number of factors, where noise is difficult to handle due to its unpredictability. Although there is a lot of research regarding the influence of noise on retrieval errors, few studies have focused on the mechanism. In this study, we selected the automatic retrieval of temperature and emissivity using spectral smoothness (ARTEMISS) algorithm—the representative of the iterative spectral smoothness temperature-emissivity separation algorithm family—as the research object and proposed an improved algorithm. First, we analyzed the influence mechanism of noise on the retrieval errors of ARTEMISS in theory. Second, we carried out a simulation and inversion experiment and analyzed the relationship between instrument spectral resolution, noise level, the ARTEMISS parameter setting and the retrieval errors separately. Last, we proposed an improved method (resolution-degrade-based spectral smoothness algorithm, RDSS) based on the mechanism and law of the influence of noise on retrieval errors and provided corresponding suggestions on instrument design. The results show that RDSS improves the accuracy of temperature inversion and is more effective for thermal infrared data with a high noise level and high spectral resolution, which can reduce the LST inversion error by up to 0.75 K and the LSE median absolute deviation (MAD) by 31%. In the presence of noise in HTIR data, the RDSS algorithm performs better than the ARTEMISS algorithm in terms of temperature-emissivity separation.

Keywords: hyperspectral thermal infrared; spectral smoothness; temperature-emissivity separation; sensitivity analysis; noise

1. Introduction

Land surface temperature (LST) and land surface emissivity (LSE) are two key physical parameters characterizing the state of the land surface, which are applied in various areas such as mineral mapping [1–3], gas plume detection [4], soil moisture inversion [5] and oil-film thicknesses measurement [6]. The LST is sensitive to the external environment and reflects the energy budget of an object during a period; the LSE, determined by the object itself, reflects its physical and chemical properties. The LST and LSE of an object can be simultaneously derived from thermal infrared (generally 8–14 µm) data through a temperature-emissivity separation (TES) algorithm. TES is a crucial issue in thermal infrared remote sensing, which can be seen as an underdetermined equation problem, i.e., obtaining $N+1$ unknowns (N emissivities at each band and one temperature) by solving

N equations (one sensor output radiance with N bands). Consequently, it needs at least one additional constraint to make the underdetermined equations solvable.

For thermal infrared data with low or medium spectral resolution, LST is the principal retrieval goal compared with LSE. In this case, the primary task of TES algorithms for multiband thermal infrared data is to determine LSE. Many algorithms were developed for the inversion of LST from multispectral thermal infrared data, such as the split-window method [7], grey body emissivity [8], reference channel method [9], normalized emissivity method (NEM) [10,11], temperature independent spectral indices (TISI) [12], spectral ratio method [13], NDVI-based emissivity method (NBEM) [14–17] and advanced spaceborne thermal emission and reflection radiometer temperature-emissivity separation (ASTER TES) [18]. Among them, NBEM, TISI and ASTER TES are widely used methods [19]. Based on the relationship between NDVI and LSE, NBEM determines the LSE at specific bands with the help of visible near-infrared data. The ASTER TES method iteratively adjusts the average emissivity according to the empirical relationship between the change range of emissivity and the average emissivity to obtain the optimal LST. TISI transforms ground-leaving radiance under some approximation of the Planck function to make it reflect the spectral shape of emissivity. The day and night algorithm, based on TISI, has obtained good results on LST inversion of multi-band thermal infrared remote-sensing data [17]. TISI is theoretically suitable for thermal infrared remote sensing data of any spectral resolution; however, it cannot obtain absolute values of the retrieval emissivity spectrum. Therefore, TISI requires other data or theoretical constraints to solve the problem. The improvement of hyperspectral thermal infrared (HTIR) remote sensing technology brings new opportunities for the development of thermal infrared remote sensing. Unlike multi-spectral remote sensing, HTIR remote sensing is applied to retrieve not only LST but also LSE more accurately, which gives full play to the role of LSE on the fields of classification, identification and parameter inversion of the land surface (especially urban land surfaces of various types). As a result, the application scenarios of HTIR remote sensing are different to those of multispectral thermal infrared remote sensing. To adapt to new application scenarios, the goal of TES of HTIR remote sensing is to obtain the LST and LSE with the highest accuracy under the smallest number of universal and reasonable additional constraints. At present, scholars have proposed or improved a number of TES algorithms for HTIR data [20–35]. Unlike multispectral thermal infrared inversion methods, most TES algorithms for HTIR data have far fewer constraints. One well-known assumption of HTIR TES algorithms is that the LSE of most objects [36] is much smoother than the spectral curve of atmospheric downwelling radiance.

Iterative spectrally smooth temperature-emissivity separation (ISSTES) [21] is a representative TES algorithm based on the concept of spectral smoothness for HTIR data. ISSTES uses a smoothness index as a cost function measuring the atmospheric residue in the estimated LSE, and iteratively optimizes the estimated LST until the cost function is minimized and the corresponding LSE is considered optimally smoothed. The automatic retrieval of temperature and emissivity using spectral smoothness (ARTEMISS) improved ISSTES [22] by updating its cost function. ARTEMISS uses the standard deviation of the estimated at-sensor radiance and true at-sensor radiance as the cost function, instead of the absolute deviation of the estimated LSE and true LSE. Based on ARTEMISS, the quick temperature-emissivity separation (QTES) algorithm [23] was modified to retrieve LST and LSE from HTIR data measured near the ground. QTES uses a narrow spectral range (9.5–10 μm) instead of the entire spectral range of data when searching for the optimal temperature. In addition to the three algorithms proposed by Borel mentioned above, researchers have successively proposed many other TES algorithms based on the concept of spectral smoothness. The spectral smoothness method (SpSm) [24] uses the first derivative of estimated LSE spectrum as cost function. The downwelling radiance residual index (DRRI) [25] algorithm uses downwelling radiance residual index to measure roughness of LSE curve on several selected three-channel groups. The correlation-based temperature and emissivity separation (CBTES) method [26] constructs a cost function to measure the correlation between the estimated LSE and the atmospheric downwelling radiation, based on the concept that LSE is not directly related to atmospheric downwelling radiation, but when estimated LST is not accurate, estimated LSE will include residual atmospheric absorption information. The stepwise refining temperature and emissivity separation (SRTES) [27] considers self-emitting radiation of the land surface a function of wavenumber in a narrow spectral range, where LSE can be seen as a constant and the Planck function can be expressed in a linear form. However, atmospheric downwelling radiation still contains unignorable atmospheric absorption features. SRTES utilizes the residual atmospheric downwelling radiation of calculated self-emitting radiation as a criterion and obtains both the LSE at specified bands and the LST by the stepwise refining method. The correlation-wavelet method [28] considers the land surface self-emission curve as a fundamental-frequency signal and atmospheric downwelling radiance as a high-frequency signal, and obtains the land surface self-emission from ground-leaving radiance by filtering out the atmospheric characteristics based on the spectral smoothness concept. The correlation-wavelet method calculates the correlation between the emissivity curves calculated by taking into the atmospheric downwelling radiance and by direct wavelet filtering without considering atmospheric downwelling radiance. At the maximum correlation, the corresponding temperature is seen as the optimal retrieval temperature. Meanwhile, the emissivity curve is synthesized by wavelet signals of different scales whose proportion is calculated based on the correlation. A TES algorithm for low-emissivity materials [29] is based on the bias characteristic of atmospheric downwelling radiance, namely when retrieval LSE deviates from its true value, atmospheric downwelling radiance also deviates from its true value and the deviations are the same at the bands of the atmospheric absorption peak and valley. Some researchers have tried to reduce the unknowns of equations by reducing the dimensions of the emissivity, making the underdetermined equations solvable. One way to reduce the dimensionality is to divide the emissivity curve into several segments, each of which is the linear function of the wavelength. Based on the segmented linear constraints, three TES methods have been proposed, namely, the linear spectral emissivity constraint (LSEC) [30], the improved LSEC (I-LSEC) [31] and the pre-estimate shape LSEC (PES-LSEC) [32], which all take the sum of squared residuals of the estimated and true ground-leaving radiance as cost function. Another way to reduce the dimensionality is using the wavelet transform. The wavelet transform method for separating temperature and emissivity (WTTES) [33], which expresses LSE as a function of low-frequency wavelet coefficients, reduces the number of unknowns from N + 1 to N/2 + 1, and iteratively searches for the optimal wavelet coefficient and LST. The multi-scale wavelet-based TES algorithm (MSWTES) [34] is based on the fact that both high frequencies of ground-leaving radiance and the LSE calculated according to inaccurate LST are closely correlated with the atmospheric downwelling radiance. From the perspective of the spectral curve, the above TES algorithms follow similar basic concepts. For eliminating the influence of the "thorns" of a curve due to atmospheric absorption lines, the TES algorithms first aim to obtain a degraded approximate spectral curve close to the true spectral curve. For example, ISSTES estimates the LSE from equations directly; LSEC and WTTES estimate LSE after modifying its expression. Then, they search for the optimal LST, which minimizes the cost function and calculates its corresponding LSE as the final retrieval LSE. In addition to the idea of spectral smoothness, there is a new idea from the perspective of statistics, which assumes that the LSE spectra of natural and man-made materials can be well represented in a given subspace of the original data space. Based on the new idea, the dictionary subspace based temperature and emissivity separation (D-SBTES) [35,36] uses a singular value decomposition to extract the basis matrix of the subspace from the emissivity spectra dictionary to obtain the retrieval emissivity. However, D-SBTES suffers from several factors, such as noise, the rank of basis matrix adopted to address the emissivity subspace and the true land surface temperature, and it is more suitable for high-emissivity objects.

Overall, spectral smoothness has been a reasonable strategy widely used in the field of TES. Among the TES algorithms mentioned above, the most commonly used cost function is the standard deviation of the simulated at-sensor radiance and true at-sensor radiance, and a classic and common method to approximately process the LSE curve is the boxcar average. Therefore, for this study we researched ARTEMISS, which is the representative algorithm of the spectral smoothness TES family and has first-rate performance.

There are many impact factors to the performance of a TES algorithm for HTIR data, mainly including (1) sensor-related parameters (e.g., sensor altitude, spatial resolution, spectral range, spectral resolution and instrument noise), (2) data pre-processing (before TES) related residual errors (e.g., radiation calibration, spectral calibration and atmospheric correction) and (3) the algorithm's limitations (e.g., the adopted assumptions or constraints) [37–48]. The sensor altitude affects the amount of atmospheric radiation entering the sensor. The higher the sensor altitude, the greater the inversion error [37,44]. If the spatial resolution is low, non-isothermal mixed pixels will appear in the thermal infrared hyperspectral image, which makes it difficult to separate the temperature emissivity accurately [39,41,42]. The spectral range of 7.5–8 μm, containing dense atmospheric absorption lines, is usually used to retrieve atmospheric parameters; however, the inversion requires a sufficient spectral resolution. The spectral range of 8–12.5 μm is usually used to retrieve the LST and LSE; the higher the spectral resolution, the smaller the retrieval errors [42,43]. A sufficient signal-to-noise ratio is necessary to obtain a unique solution for TES, and the retrieval error increases linearly with instrument noise [37,40,42]. The radiation calibration needs to have sufficient accuracy in order to make the simulated and measured radiance match [42]. The spectral calibration error greatly affects the temperature and emissivity errors even if the atmospheric correction has no error [40,45–48]. The influence of atmospheric correction errors (atmospheric upwelling radiance, atmospheric downwelling radiance and atmospheric transmittance) on the retrieval results varies by the TES algorithm. Among the three atmospheric parameters, atmospheric downwelling radiance has most of the influence on retrieval errors, especially for low-emissivity objects [41,43,45,46]. The adopted cost function is a mathematical expression of the adopted assumption for a TES algorithm, which certainly affects algorithm performance, regardless of whether the algorithm solves the underdetermined problem by increasing the constraints (e.g., spectral smoothness assumption) or reducing unknowns (e.g., piecewise linear constraint) [37,38,44,47,48].

The above factors, excluding the algorithm's limitations, can be classified as systematic errors and random errors. The systematic errors can be reduced or removed by some correction methods based on the study of error patterns. For example, besides laboratory calibration, scene-based spectral calibration is able to improve further the spectral calibration accuracy of the field-measured data, which selects some atmospheric absorption channels as the reference channel to perform spectral calibration on the measured data, and the minimum error of the center wavelength shift is within 1 nm [49]. The sensor altitude, spectral range and resolution depend on the equipment development level and the external conditions during data acquisition. They mainly affect errors of the atmospheric parameters (input parameters for separating temperature and emissivity), and the accuracy of atmospheric correction increases with the improvement of technology [40,50,51]. The random errors, however, are much more difficult to deal with. Like noise, it is uncontrollable and limited by the detector technology, and it is difficult to remove the noise non-destructively after data acquisition. Therefore, the influence of random errors on the TES algorithms is also widely researched [25–28,30–32,37,40–44,47,48]. However, most research on this issue has been about how much noise causes the temperature and emissivity inversion errors, and only a few studies have focused on the influence mechanism of noise on the TES algorithms.

Here, we selected the ARTEMISS algorithm—a representative of the iterative spectral smoothness TES algorithm family with a good application effect—as the research object. The influence mechanism of noise on the retrieval errors of ARTEMISS is derived from the cost function. The relationships between the instrument spectral resolution, noise level, the ARTEMISS parameter setting and the retrieval errors are investigated through simulation experiment. On the basis of the mechanism and law of the influence of noise on the retrieval errors obtained from the experiment, the resolution-degrade-based spectral smoothness (RDSS) algorithm, an improved method, is proposed. Corresponding suggestions on the instrument design are also provided. Section 1 reviews TES algorithms and impact factors. Section 2 theoretically analyzes how noise affects the retrieval results of ARTEMISS. Section 3 describes the data simulation experiment, implementation of TES and the evaluation metrics for the retrieval results (Section 3.1) and the results of a sensitivity analysis of ARTEMISS regarding noise and smoothing window size (Section 3.2). Section 4 proposes the RDSS algorithm based on ARTEMISS, and presents the validation results. Sections 5 and 6 present the discussion and conclusions, respectively.

2. Background

The temperature-emissivity separation is based on radiative transfer theory. Ignoring multiple scattering, the radiative transfer equation (RTE) [9,52,53] is as follows:

$$L(\lambda_i, T) \ = \ L_g(\lambda_i, T)\tau(\lambda_i) + L_u(\lambda_i) \tag{1}$$

$$L_g(\lambda_i, T) \ = \ \varepsilon(\lambda_i)B(\lambda_i, T) + [1 - \varepsilon(\lambda_i)]L_d(\lambda_i) \tag{2}$$

$$B(\lambda_i, T) \ = \ \frac{c_1}{\lambda_i{}^5 \cdot exp\left(\frac{c_2}{\lambda_i T} - 1\right)} \tag{3}$$

where $L(\lambda_i, T)$ is the at-sensor radiance at i-th band and LST of T, $i \in [1, N]$; N is the number of bands; $L_g(\lambda_i, T)$ is the ground-leaving radiance, including the object's self-emitting radiance and reflected atmospheric downwelling radiance; $L_u(\lambda_i)$ is the atmospheric upwelling radiance; $\tau(\lambda_i)$ is the atmospheric transmittance; $\varepsilon(\lambda_i)$ is the land surface emissivity; $B(\lambda_i, T)$ is the blackbody radiance at T and $L_d(\lambda_i)$ is the atmospheric downwelling radiance.

The original ARTEMISS algorithm includes two parts: atmospheric correction and temperature-emissivity separation. This study focuses on the influence of instrument noise on the retrieval LST and LSE of ARTEMISS. Therefore, this study only discusses the temperature-emissivity separation process of ARTEMISS, assuming the three atmospheric parameters ($\tau(\lambda_i)$, $L_d(\lambda_i)$ and $L_u(\lambda_i)$) known and without errors. Then, we know $L_g(\lambda_i, T)$ according to Equation (1). Given temperature estimation $\hat{T}$, we can obtain the LSE according to Equation (2). The emissivity is calculated by,

$$\hat{\varepsilon}(\lambda_i) \ = \ \frac{L_g(\lambda_i, T) - L_d(\lambda_i)}{B\left(\lambda_i, \hat{T}\right) - L_d(\lambda_i)} \tag{4}$$

where $\hat{\varepsilon}(\lambda_i)$ is the LSE estimation and $B\left(\lambda_i, \hat{T}\right)$ is the blackbody radiance at $\hat{T}$.

According to the concept of spectral smoothness, the goal of ARTEMISS is to find an optimal LST estimation where the corresponding LSE is the smoothest. ARTEMISS uses the standard deviation of estimated at-sensor radiance and true at-sensor radiance as the cost function. When the cost function reaches the minimum, the corresponding LST estimation is the optimal LST. The process of ARTEMISS temperature-emissivity separation is as follows:

Given an LST estimation, the LSE estimation can be calculated using Equation (4); then, a boxcar average is performed on it and one can obtain the smoothed LSE estimation $\overline{\varepsilon}(\lambda_i)$:

$$\overline{\varepsilon}(\lambda_i) \ = \ \frac{1}{3} \sum_{j\,=\,i-1}^{i+1} \hat{\varepsilon}\left(\lambda_j\right) \tag{5}$$

Then, the estimated at-sensor radiance, $L_{fit}(\lambda, \hat{T}, \overline{\varepsilon})$ is calculated according to the RTE:

$$L_{fit}(\lambda, \hat{T}, \overline{\varepsilon}) \ = \ \overline{\varepsilon}(\lambda)B(\lambda, T)\tau(\lambda) + [1 - \overline{\varepsilon}(\lambda)]L_d(\lambda)\tau(\lambda) + L_u(\lambda) \tag{6}$$

Then, the standard deviation (cost function) of the estimated at-sensor radiance and the true at-sensor radiance is calculated:

$$\sigma(\hat{T}, \overline{\varepsilon}) \ = \ \sigma(L_{fit}(\lambda, \hat{T}, \overline{\varepsilon}) - L(\lambda, \hat{T}, \overline{\varepsilon})) \tag{7}$$

We iteratively cycle through Equations (5)–(7) until the cost function reaches the minimum, and the corresponding temperature is seen as the optimal LST. Then we obtain the retrieval emissivity with the optimal temperature through Equation (4).

The above processes do not consider instrument noise. In fact, the output radiance of the thermal infrared hyperspectral sensor inevitably includes instrument noise. Instrument noise is a kind of random noise. In this study, the instrument noise is represented by additive Gaussian white noise, and the bands are uncorrelated. That is, to the right side of in Equation (1), add a noise term $\eta(\lambda_i)$. Accordingly, to the right side of in Equation (2), also add a noise term $\frac{\eta(\lambda_i)}{\tau(\lambda_i)}$. Then $\hat{\varepsilon}(\lambda_i)$ in Equation (4) becomes,

$$
\begin{aligned}
\hat{\varepsilon}(\lambda_i) &= \frac{[B(\lambda_i,T)-L_d(\lambda_i)]\varepsilon(\lambda_i)+\frac{\eta(\lambda_i)}{\tau(\lambda_i)}}{B(\lambda_i,\hat{T})-L_d(\lambda_i)} \\[2mm]
&= \frac{B(\lambda_i,T)-L_d(\lambda_i)}{B(\lambda_i,\hat{T})-L_d(\lambda_i)}\varepsilon(\lambda_i) + \frac{\frac{\eta(\lambda_i)}{\tau(\lambda_i)}}{B(\lambda_i,\hat{T})-L_d(\lambda_i)}
\end{aligned}
\tag{8}
$$

where $\hat{\varepsilon}(\lambda_i)$ and $\varepsilon(\lambda_i)$ are the estimated emissivity and its true value respectively and $\hat{T}$ and T are the estimated temperature and its true value respectively.

For simplicity, $\hat{\varepsilon}(\lambda_i)$, $\varepsilon(\lambda_i)$, $B(\lambda_i, \hat{T})$, $B(\lambda_i, T)$, $\tau(\lambda_i)$, $L_u(\lambda_i)$, $L_d(\lambda_i)$ and $\eta(\lambda_i)$ will be abbreviated in the following equations as $\hat{\varepsilon}_i$, ε_i, $\hat{B}_i$, B_i, τ_i, $L_i^{\uparrow}$, $L_i^{\downarrow}$ and η_i respectively. We bring Equations (5) and (8) into Equation (7), and the cost function of ARTEMISS, i.e., Equation (7), becomes

$$
\begin{aligned}
\sigma(\hat{T},\overline{\hat{\varepsilon}}) &= \sigma\left(\left[\frac{\varepsilon_{i-1}+\varepsilon_i+\varepsilon_{i+1}}{3}\left(\hat{B}_i - L_i^d\right)\tau_i + L_i^d\tau_i\right] - \left[\left(B_i - L_i^d\right)\varepsilon_i\tau_i + L_i^d\tau_i + \eta_i\right]\right) \\[2mm]
&= \sigma\left(\left(\begin{array}{c}\frac{B_{i-1}-L_{i-1}^d}{\hat{B}_{i-1}-L_{i-1}^d}\frac{\varepsilon_{i-1}}{3} + \frac{B_i-L_i^d}{\hat{B}_i-L_i^d}\frac{\varepsilon_i}{3} + \frac{B_{i+1}-L_{i+1}^d}{\hat{B}_{i+1}-L_{i+1}^d}\frac{\varepsilon_{i+1}}{3} \\[3mm] +\frac{1}{3}\frac{\frac{\eta_{i-1}}{\tau_{i-1}}}{\hat{B}_{i-1}-L_{i-1}^d} + \frac{1}{3}\frac{\frac{\eta_i}{\tau_i}}{\hat{B}_i-L_i^d} + \frac{1}{3}\frac{\frac{\eta_{i+1}}{\tau_{i+1}}}{\hat{B}_{i+1}-L_{i+1}^d}\end{array}\right)\left(\hat{B}_i - L_i^d\right)\tau_i - \left(B_i - L_i^d\right)\varepsilon_i\tau_i - \eta_i\right) \\[2mm]
&= \sigma\left(\begin{array}{c}\frac{1}{3}\frac{B_{i-1}-L_{i-1}^d}{\hat{B}_{i-1}-L_{i-1}^d}\left(\hat{B}_i - L_i^d\right)\varepsilon_{i-1}\tau_i + \frac{1}{3}\frac{B_{i+1}-L_{i+1}^d}{\hat{B}_{i+1}-L_{i+1}^d}\left(\hat{B}_i - L_i^d\right)\varepsilon_{i+1}\tau_i - \frac{2}{3}\left(B_i - L_i^d\right)\varepsilon_i\tau_i \\[3mm] +\frac{1}{3}\frac{\hat{B}_i-L_i^d}{\hat{B}_{i-1}-L_{i-1}^d}\frac{\eta_{i-1}\tau_i}{\tau_{i-1}} + \frac{1}{3}\frac{\hat{B}_i-L_i^d}{\hat{B}_{i+1}-L_{i+1}^d}\frac{\eta_{i+1}\tau_i}{\tau_{i+1}} - \frac{2}{3}\eta_i\end{array}\right)
\end{aligned}
\tag{9}
$$

where $\sigma(\cdot)$ in the right side of Equation (9) refers to $\sqrt{\frac{1}{N}\sum_{i=1}^{N}[(\cdot)^2]}$.

In Equation (9), the first three items are not related to noise and the last three items are noise-related. The items not related to noise are fixed and determined by the temperature. In general, the cost function is smallest when $\hat{T} = T$. However, after noise appears, the cost function may not be the minimum value when $\hat{T} = T$. That is to say, after the coupling of noise, temperature and transmittance, different noise levels may cause different optimal $\hat{T}$ appearances for the same emissivity.

3. Sensitivity Analysis of ARTEMISS

This section analyzed the relationship between instrument spectral resolution, the noise level, the ARTEMISS parameter setting and the retrieval errors of ARTEMISS through simulation data. This section described the simulation experiment in Section 3.1 and the results in Section 3.2.

3.1. Simulation Experiment

A thermal infrared hyperspectral imager is the primary tool used to acquire HTIR remote sensing data. The instrument's settings and performance directly affect data quality, which in turn affects the accuracy of subsequent TES process. On the contrary, the TES algorithms also influence the development of HTIR instruments over time. Based on this background and recent research [48], this study mainly uses the instrument's key parameters, such as spectral range, noise level, spectral resolution and ARTEMISS parameter setting, to explore their influence on the accuracy of the TES algorithm. Note that the scenes simulated in this study were only isothermal homogeneous pixels, the complex radiation transfer process among non-isothermal heterogeneous pixels involves other complex problems, such as pixel decomposition, which may not be conducive to focusing on evaluating the relationship between the accuracy of the TES algorithm and the selected impact factors.

During the simulation of at-sensor radiance, this study considered three impact factors unrelated to the algorithm itself (i.e., spectral range, noise level and spectral resolution) and four input parameters (i.e., atmospheric model, sensor altitude, surface temperature and surface emissivity), as shown in Table 1. They were set as follows:

Table 1. Primary variables setup of simulation experiment.

Variable	Value	Number
Spectral range	7.5–12.5 µm, 8–12.5 µm	2
Random noise	From 0 to 0.5 K with an incremental step of 0.05 K	11
Spectral resolution	50 nm, 35 nm, 10 nm, 5 nm	4
Sensor altitude	2 km, 10 km, 750 km	3
LST	Atmospheric temperature + offset Offset: (1) from −5 K to 20 K with an incremental step of 5 K when atmospheric temperature ≤ 280 K; (2) from −10 K to 15 K with an incremental step of 5 K when atmospheric temperature > 280 K.	6
LSE	All emissivity spectra covering thermal infrared wavelengths from ASTER 2.0	1524
Atmospheric model	Tropical Model, 299.7 K Mid-Latitude Summer Model, 294.2 K Mid-Latitude Winter Model, 272.2 K Sub-Arctic Summer Model, 287.2 K Sub-Arctic Winter Model, 257.2 K	5
Total	—	12,070,080

(1) Spectral range: 7.5–13.5 µm is generally regarded as the atmospheric window of thermal infrared remote sensing. Considering the low detector response efficiency at 12.5–13.5µm and the atmospheric absorption at 7.5–8.0 µm, this study used two spectral ranges: 7.5–12.5 µm and 8–12.5 µm.

(2) Noise level: considering that a HTIR imager is a system, the noise in calibrated data is an accumulation of noise from the radiance entering the imager to have been calibrated, including detector noise, circuit noise, calibration source noise, etc. The calibration errors include both systematic and random errors. This study only considered the random calibration error, which is more difficult to be removed than systematic error. Therefore, the noise here refers to a comprehensive noise equivalent temperature difference, namely the random error in calibrated data, which is simulated by a Gaussian function [37,44]. Taking the current noise equivalent temperature difference (NEDT) of the instrument, which is generally below 0.3 K [54], and other sources of random errors into account, in this study we tested 11 noise levels, and the specific value was from 0 to 0.5 K with an incremental increase of 0.05 K.

(3) Spectral resolution: the HTIR imager splits the thermal infrared radiation spectrum into hundreds and thousands, approximately continuous and narrow bands, and the spectral response of each channel conforms to the Gaussian distribution [55,56]. When dealing with multi-spectral thermal infrared data, the spectral response function is usually obtained by direct measurement [57,58]. However, for the HTIR imager, the spectral response performance at each channel is usually fitted by a Gaussian function after spectral calibration by devices such as a monochromator, and described by the center wavelength and full width at half height (FWHM) obtained by fitting. Practically, the spectral response function of the HTIR imager is obtained by putting the center wavelength and FWHM at each band into a Gaussian function. In general, the spectral resolution of HTIR imagers is higher than 100 nm (one hundredth of a wavelength). The spectral resolution of most current airborne instruments is above 50 nm [54]. Four spectral resolutions (50 nm, 35.2 nm, 10 nm and 5 nm) were tested to analyze the influence of the spectral range and spectral resolution on the retrieval results (Table 1). The two spectral resolutions of 50 nm and 35.2 nm were from the airborne thermal-infrared hyperspectral imager system

(ATHIS) [55] and hyperspectral thermal emission spectrometer (HyTES) [59] respectively. The other two spectral resolutions of 10 nm and 5 nm were set to provide theoretical analysis for the design and development of HTIR instruments with higher spectral resolution in the future.

(4) Atmospheric model: to include as many atmospheric conditions as possible, we chose five atmospheric models from moderate resolution atmospheric transmission (MODTRAN) software developed by Spectral Sciences, Inc., Burlington, United States [60], including tropical atmospheric model, mid-latitude summer (winter) atmospheric model and sub-arctic summer (winter) atmospheric model.

(5) Sensor altitude: the development of HTIR imagers is currently in its infancy. There are no spaceborne HTIR instruments in orbit yet, and airborne HTIR imagers are still the mainstream. Thus, airborne HTIR data are the main source currently available. The sensor altitude directly affects atmospheric transmittance and path radiance. With due consideration of the airborne HTIR remote sensing scene, this study sets three sensor altitudes of 2 km, 10 km and 750 km for low aerial, high aerial and spaceborne instruments, respectively.

(6) Surface temperature: the heat between the land surface and atmosphere is frequently exchanged. LST generally fluctuates more than atmospheric bottom temperature. Therefore, the surface temperature is set around the atmospheric temperature. That is, the surface temperature is equal to the atmospheric temperature plus an offset. In the simulation, when the atmospheric temperature was higher than 280 K, the offset was set from −5 to 20 K with an incremental step of 5 K; when the atmospheric temperature was lower than 280 K, the offset was set from −10 to 15 K with an incremental step of 5 K.

(7) Surface emissivity: the theoretical basis of ARTEMISS is that the smoothness of LSE is higher than the downwelling radiance spectrum of the atmosphere. It is necessary to choose a dataset containing a large number of spectra to test the influence of the fluctuation of spectral curves of different ground objects on inversion accuracy. We chose the advanced spaceborne thermal emission and reflection radiometer (ASTER) 2.0 spectral library [61], which is often adopted in a remote sensing numerical simulation experiment. There are 2445 spectra of different ground objects in the library. Among them, 1524 emissivity curves covering thermal infrared wavelengths were selected, simulating the majority of scenarios in practical applications. The selected dataset involves common ground object types including stony mineral (909), rock (388), manmade (84), soil (75), stony meteorite (60), vegetation (4), water (4), snow (4) and ice (1) [48].

The influence of ARTEMISS's smooth window size on the inversion accuracy was studied during the process of TES in this study. The method used in [48] was adopted in the terms of searching for the optimal temperature; the specific settings were as follows:

(1) Smooth window size setting: here, the algorithm parameter refers to the best window size for smoothing, where the retrieval error is the smallest among the several window sizes for a group of data at a certain noise level. We used the ARTEMISS algorithm with 17 smoothing window sizes (from 3 to 35 with an incremental step of 2) to retrieve the simulated at-sensor radiances to analyze the relationship between the ARTEMISS spectral smoothing window sizes and the retrieval errors.

(2) The strategy of optimal temperature solution: to reduce the computational complexity and avoid the local minimum of the cost function, the strategy of the optimal temperature solution adopted in this study was the same as in [48]. First, a search range of temperature was set, as in [48]. Then, the cost function was calculated for each temperature in the temperature search range. Finally, the temperature corresponding to the minimum cost function value was seen as the optimal temperature. According to the literature [48], the optimal temperature obtained with a search range of true LST ± 100 K is basically same with that obtained with a search range of true LST ± 20 K. Therefore, in order to save calculation time, the true surface temperature ± 20 K was set as the temperature search range in this study.

We used the root mean square error (RMSE) metrics to evaluate the retrieval temperature error. The RMSE of a group of retrieval LSTs with a certain spectral parameter and noise level is defined as:

$$RMSE_T = \sqrt{\frac{1}{N_s} \sum_{j=1}^{N_s} \left(\hat{T}_j - T_j\right)^2} \tag{10}$$

where $j \in [1, N_s]$, N_s is the number of the samples in the group of data to be evaluated; and $\hat{T}_j$ and T_j are the retrieval LST and the real LST of a sample in the group of data to be evaluated respectively.

The RMSE is sensitive to the amplitude of error at some bands (especially large errors), but it is not sensitive to the overall deviation tendency of retrieval LSE from its true LSE. The overall deviation means that the amplitudes of error at all bands are about the same. Therefore, we used both the RMSE and median absolute deviation (MAD) to evaluate the retrieval LSE. This strategy can simultaneously measure the emissivity error in terms of both details and overall morphologies.

The RMSE and MAD of a group of retrieval LSEs with a certain spectral parameter and noise level are defined as:

$$RMSE_\varepsilon = \frac{1}{N_s} \sum_{j=1}^{N_s} \sqrt{\frac{1}{N} \sum_{i=1}^{N} \left(\hat{\varepsilon}_j(\lambda_i) - \varepsilon_j(\lambda_i)\right)^2} \tag{11}$$

$$MAD_\varepsilon = \frac{1}{N_s} \sum_{j=1}^{N_s} median\left(\hat{\varepsilon}_j(\lambda_i) - \varepsilon_j(\lambda_i)\right) \tag{12}$$

where $j \in [1, N_s]$, N_s is the number of samples in a group of data to be evaluated; $i \in [1, N]$, N is the number of bands of a retrieval LSE spectrum; $\sqrt{\frac{1}{N} \sum_{i=1}^{N} \left(\hat{\varepsilon}_j(\lambda_i) - \varepsilon_j(\lambda_i)\right)^2}$ is RMSE of retrieval LSE spectrum; $median\left(\hat{\varepsilon}_j(\lambda_i) - \varepsilon_j(\lambda_i)\right)$ is MAD of retrieval LSE spectrum and $\hat{\varepsilon}_j(\lambda_i)$ and $\varepsilon_j(\lambda_i)$ is the retrieval and true LSE of sample j at the i-th band, respectively.

The data simulation included two processes: the ultra-high-resolution (1 nm) at-sensor radiance simulation and sensor output radiance simulation, as detailed in article [48]. The simulation experiment was based on the interface data language (IDL), Exelis Visual Information Solutions, Inc. MODTRAN was called by an IDL program to simulate the atmospheric parameters. All other experimental processes were implemented by the IDL code. In order to facilitate the drawing and explanation of the results, the simulation results were classified into eight groups according to spectral range and spectral resolution in this article, as shown in Table 2.

Table 2. Groups of simulated hyperspectral thermal infrared (HTIR) data.

Name	Spectral Range/μm	Spectral Resolution/nm	Similar Sensor
Group1	7.5–12.5	50	
Group2	7.5–12.5	35.2	HyTES
Group3	7.5–12.5	10	
Group4	7.5–12.5	5	
Group5	8–12.5	50	ATHIS
Group6	8–12.5	35.2	
Group7	8–12.5	10	
Group8	8–12.5	5	

3.2. Results Analysis

3.2.1. The Relationship of the Retrieval Errors of ARTEMISS vs. the Noise Level and Spectral Parameters

This study retrieved eight groups of simulated HTIR data with two spectral ranges (7.5–12.5 μm and 8–12.5 μm) and four spectral resolution (50 nm, 35.2 nm, 10 nm and 5 nm) by the ARTEMISS algorithm and measured the retrieved LST and LSE errors. Figure 1 shows the retrieved LST errors of the eight groups of simulated HTIR data as a function of the noise level.

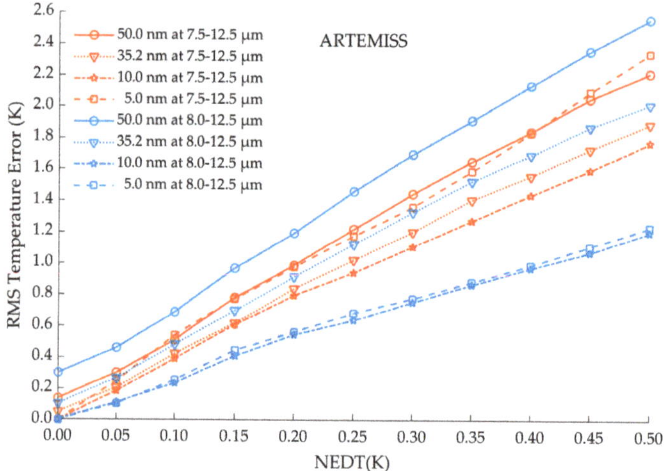

Figure 1. Plot of automatic retrieval of temperature and emissivity using spectral smoothness (ARTEMISS) land surface temperature (LST) inversion errors varying with noise equivalent temperature difference (NEDT), spectral range and spectral resolution.

For all eight groups of simulated HTIR data shown in Figure 1, the retrieved LST errors of ARTEMISS increased linearly, roughly with the noise level. For the four groups of simulated data with high spectral resolution of 10 nm and 5 nm, after the NEDT of 0.2 K, the retrieved LST errors increased with the noise level at a slower growth rate. For the eight groups of simulated data except the two with the spectral resolution of 5 nm, the retrieved LST errors of ARTEMISS decreased as the spectral resolution increased. However, the two groups of data with a spectral resolution of 5 nm show differences to each other. For the spectral range of 8–12.5 μm, the retrieved LST errors of 5 nm at 8–12.5 μm (refers to Group8 with a spectral resolution of 5 nm and in a spectral range of 8–12.5 μm) were smaller than those with a lower spectral resolution (50 nm and 35.2 nm) and were consistent with those of 10 nm and a spectral range of 8–12.5 μm in terms of both amplitude and trend. For the spectral range of 7.5–12.5 μm, the retrieved LST errors of 5 nm at 7.5–12.5 μm were larger than those with a lower spectral resolution (35.2 nm and 10 nm) and were numerically similar with those of 50 nm at 7.5–12.5 μm. For the spectral resolutions of 50 nm and 32.5 nm, the retrieved LST errors with a spectral range of 7.5–12.5 μm were smaller than those in the spectral range of 8–12.5 μm. Among the eight groups of results, the group with 10 nm at 8–12.5 μm had the smallest retrieved LST error (e.g., LST RMSE = 0.11 K, 0.55 K and 1.10 K when NEDT = 0.05 K, 0.20 K and 0.50 K, respectively), 50 nm at 8–12.5 μm had the largest retrieval LST error (e.g., LST RMSE = 0.46 K, 1.20 K and 2.56 K when NEDT = 0.05 K, 0.20 K and 0.50 K, respectively). The results suggest that for temperature inversion, a higher spectral resolution and wider spectral range do not necessarily lead to smaller LST inversion errors. The results of this study show that, considering noise, 10 nm at 8–12.5 μm was the most suitable setting for thermal infrared hyperspectral temperature inversion among the eight spectral settings investigated.

Figure 2a,b shows the RMSE and MAD of ARTEMISS retrieved LSE for eight groups of simulated HTIR data. Figure 2a shows that the retrieved LSE RMSEs of ARTEMISS increased approximately linearly with the noise level for all eight groups of simulated HTIR data, which is consistent with the retrieved LST RMSEs in Figure 1. However, the retrieved LSE RMSEs show a significant difference between the two spectral ranges. The retrieved LSE RMSEs of all four groups of data in the spectral range of 8–12.5 μm were smaller than the retrieved LSE RMSEs of four groups of data with in spectral range of 7.5–12.5 μm. Moreover, the relationship between the retrieved LSE RMSEs and the spectral resolution were different in the two spectral ranges. For the four groups of data in spectral range of

7.5–12.5 μm, the retrieved LSE RMSEs of ARTEMISS increased as the spectral resolution increased; in contrast, for the four groups of data in the spectral range of 8–12.5 μm, the retrieved LSE RMSEs of ARTEMISS decreased as the spectral resolution increased (except the group of 5 nm at 8–12.5 μm). Among the eight groups, 10 nm at 8–12.5 μm had the smallest LSE RMSEs (0.0021, 0.0082 and 0.0203 when NEDT = 0.05 K, 0.20 K and 0.5 K, respectively), while the 5 nm at 7.5–12.5 μm had the largest LSE RMSEs (0.0173, 0.0418 and 0.0796 when NEDT = 0.05 K, 0.20 K and 0.5 K, respectively). Figure 2b shows that the trend of the retrieved LSE MADs of ARTEMISS was close to that of the retrieved LST RMSEs for the eight groups. For the eight groups of data with two spectral ranges, the retrieved LSE MADs of ARTEMISS roughly increased linearly with the noise level. Moreover, the relationship between the retrieved LSE MADs and the spectral resolution were the same for the two spectral ranges. The retrieved LSE MADs decreased as the spectral resolution increased, except for 5 nm at 7.5–12.5 μm. Among the eight groups, 5 nm at 8–12.5 μm had the smallest retrieved LSE MADs and 50 nm at 8–12.5 μm had the largest retrieved LSE MADs.

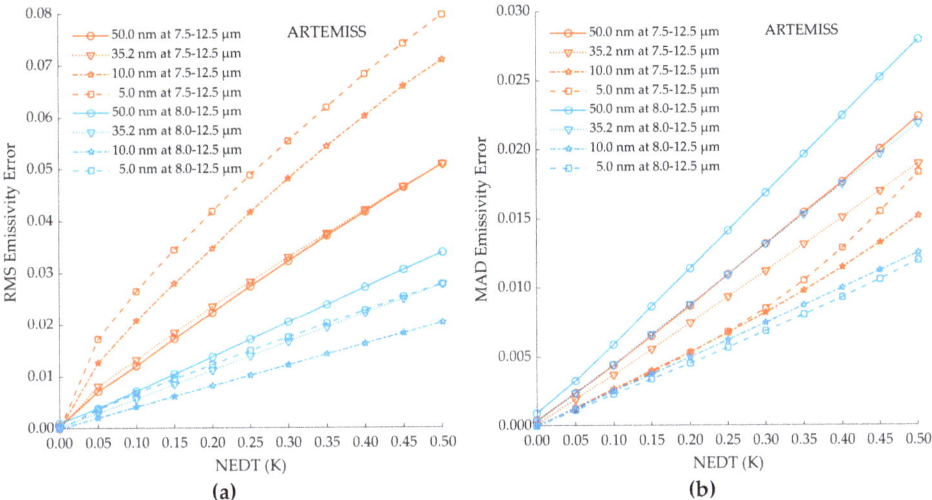

Figure 2. Plots of ARTEMISS retrieved land surface emissivity (LSE) errors varying with NEDT, spectral range and spectral resolution. (**a**) The retrieved LSE root mean square error (RMSE) and (**b**) the retrieved LSE median absolute deviation (MAD).

The LSE RMSE represents the degree to which the retrieved emissivity spectrum deviates from the true value at each band numerically, while LSE MAD represents the degree to which the retrieved emissivity spectrum deviates from the true spectral curve in the overall shape. Figures 1 and 2b show that the smaller the temperature inversion error, the smaller the emissivity MAD, indicating that the more accurate the temperature inversion, the closer the overall trend of the inversion emissivity curve is to the true spectrum curve. Both the retrieved LST errors and the retrieved LSE MADs of the four groups of data with a spectral range of 7.5–12.5 μm are between those of 10 nm at 8–12.5 μm and 50 nm at 8–12.5 μm with a spectral range of 8–12.5 μm. However, the retrieved LSE RMSEs of all four groups of data with a spectral range of 7.5–12.5 μm were larger than those with a spectral range of 8–12.5 μm. The reason for this seemingly contradictory phenomenon is that, compared to 8–12.5 μm, the retrieved LSE spectra with a spectral range of 7.5–12.5 μm (especially 7.5–8.0 μm is where dense atmospheric absorption lines exist) have large outliers in some bands, as shown in Figure 3. The existence of the huge outliers causes the cases that although the retrieved LSE has a large RMSE, the overall shape of the retrieved emissivity is closer to the true curve than some retrieved LSE with a small RMSE. The finding that in terms of the spectral range of 7.5–12.5 μm, the higher the spectral resolution is, the larger the retrieved LSE RMSE, suggests that for the cases with a spectral range of 7.5–12.5 μm, the higher the spectral resolution was, the more outliers of the retrieved LSE RMSE occurred. This also suggests that for thermal infrared hyperspectral remote sensing, the spectral range of 8–12.5 μm is more suitable for temperature inversion, which is similar to the statement in [42].

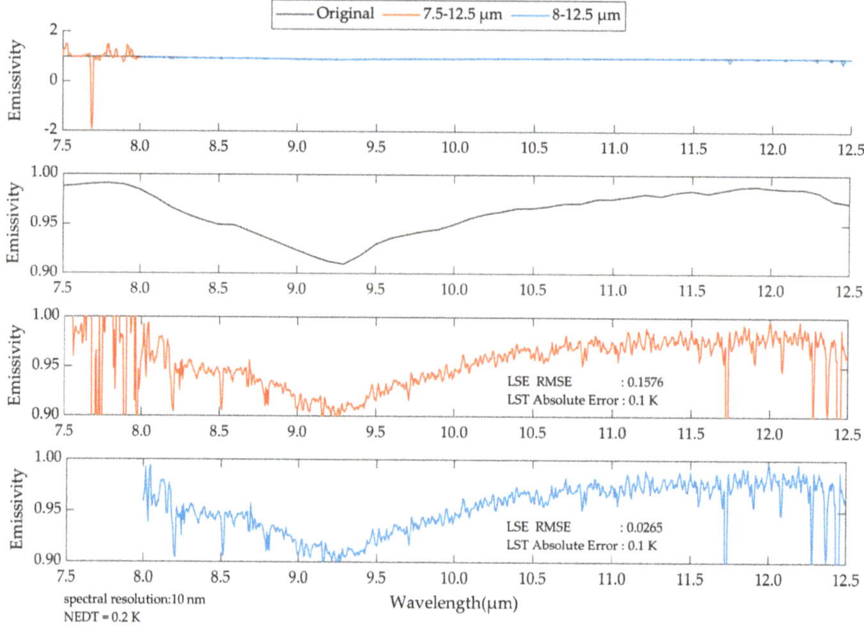

Figure 3. A case of outliers occurring in the retrieved LSE spectrum.

3.2.2. The Relationship of the Retrieval Errors vs. the Optimal Window Size Spectral Smoothing

To investigate the effect of the window size for the spectral smoothing index in ARTEMISS on the retrieved LST and LSE errors, we tested 17 spectral smoothing window sizes for ARTEMISS, going from 3 to 35 in increments of 2 for eight groups of simulated data with two spectral ranges and four spectral resolutions. In the process of a simulated radiance spectrum inversion, we looped the window size for spectral smoothing, calculated the cost functions and found the smallest one among all of the cost functions under 17 window sizes. The temperature at the minimum cost function was considered as the retrieved LST, the corresponding emissivity was the retrieved LSE, and the corresponding window size was the optimal spectral smoothing window size because the retrieved LST error was close to zero in the majority of cases. Figure 4 has eight subplots for the eight groups of simulated data, each of which shows the distribution of the optimal spectral smoothing window size vs. the noise level under a certain spectral setting. The distributions of the optimal spectral smoothing window size vs. the noise level show a consistent trend for the eight spectral settings. For each noise level, the window size with the largest proportion is 3. Generally, the larger the window size, the smaller the proportion (except 35, at the end of the window size range). In the ideal situation without noise, a window size of 3 has the most advantageous proportion in the optimal spectral smoothing window size distribution, especially when the spectral resolution is high (e.g., 10 nm and 5 nm); as the noise level increased, the proportion of the window size 3 gradually decreased for a certain spectral setting. The results suggest that when the hyperspectral thermal infrared data is noisy, it may be beneficial to improve the accuracy of temperature inversion by increasing the spectral smoothing window size in some cases.

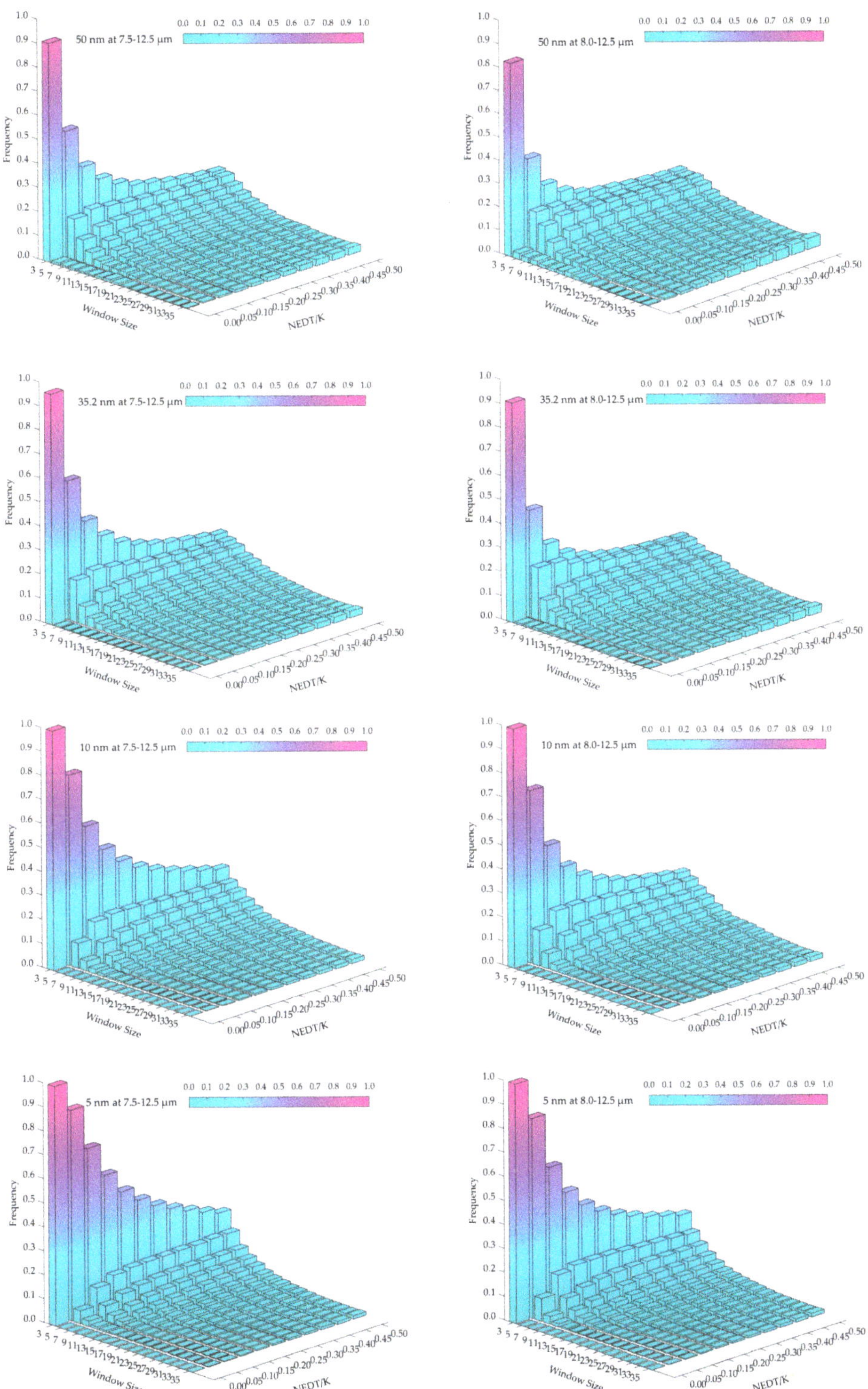

Figure 4. The distribution of the optimal spectral smoothing window size varying with the noise level.

4. The Proposal and Validation of the RDSS Algorithm

4.1. RDSS—An Improved TES Algorithm Based on ARTEMISS

The coupling of temperature and emissivity, as shown in Equation (9), suggests that the temperature is easier to determine than the emissivity in the TES process. Therefore, most algorithms adopt the strategy of first determining the temperature and then calculating the emissivity. Hence, accurately determining the temperature has also become the key to TES. We rewrite Equation (9) by combining the items determined by the temperature estimation and the items determined by both the temperature estimation and noise as follows:

$$\sigma\left(\hat{T}, \overline{\hat{\varepsilon}}\right) = \sigma\left(\left(f_\varepsilon\left(\hat{B}_i\right) + f_\eta\left(\hat{B}_i\right)\right)\left(\hat{B}_i - L_i^d\right)\tau_i - \left(B_i - L_i^d\right)\varepsilon_i\tau_i - \eta_i\right) \tag{13}$$

where $f_\varepsilon\left(\hat{B}_i\right)$ and $f_\eta\left(\hat{B}_i\right)$ are the emissivity estimation after box-average smoothing and the residual containing noise respectively. For the algorithms that use the standard deviation of the simulated at-sensor radiance and the actual at-sensor radiance as the cost function, $f_\varepsilon\left(\hat{B}_i\right)$ and $f_\eta\left(\hat{B}_i\right)$ can also be considered as the approximately processed emissivity and residual noise, respectively. They can be expressed as

$$f_\varepsilon\left(\hat{B}_i\right) = \frac{B_{i-1} - L_{i-1}^d}{\hat{B}_{i-1} - L_{i-1}^d}\frac{\varepsilon_{i-1}}{3} + \frac{B_i - L_i^d}{\hat{B}_i - L_i^d}\frac{\varepsilon_i}{3} + \frac{B_{i+1} - L_{i+1}^d}{\hat{B}_{i+1} - L_{i+1}^d}\frac{\varepsilon_{i+1}}{3} \tag{14}$$

$$f_\eta\left(\hat{B}_i\right) = \frac{1}{3}\frac{\frac{\eta_{i-1}}{\tau_{i-1}}}{\hat{B}_{i-1} - L_{i-1}^d} + \frac{1}{3}\frac{\frac{\eta_i}{\tau_i}}{\hat{B}_i - L_i^d} + \frac{1}{3}\frac{\frac{\eta_{i+1}}{\tau_{i+1}}}{\hat{B}_{i+1} - L_{i+1}^d} \tag{15}$$

When the emissivity approximation processing can remove the "thorns" in a curve caused by atmospheric background radiation and obtain relatively accurate $f_\varepsilon\left(\hat{B}_i\right)$, and the noise residuals $f_\eta\left(\hat{B}_i\right)$ with a noise item η_i of 0 or close to 0, the searched optimal temperature is likely to be equal or much closer to the true temperature. As the magnitude of noise residuals $f_\eta\left(\hat{B}_i\right)$ and noise term η_i increase, it is likely that the signs of the items will cancel each other out, resulting in the searched optimal temperature deviating from the true surface temperature. The results of ARTEMISS sensitivity on the spectral smoothing window size in Section 3.2.2 suggest that increasing the spectral smoothing window size did not effectively reduce the inversion error; different window sizes are needed to find the optimal temperature as close to the true one possible when different noises, emissivity and transmittances are coupled together. However, it will add computation complexity and cost much more computation time. That is to say, it is difficult to remove the influence of noise on the optimal temperature search simply by emissivity approximation processing. Therefore, in this study, from the perspective of reducing the noise of the data, we filtered the variables participating in the calculation of the emissivity estimation with a unified filter. The unified filter ensured the variables were at the same spectral resolution while eliminating the noise in the at-sensor radiances. We named the TES algorithm proposed in this study the resolution-degrade-based spectral smoothness (RDSS) algorithm. The calculation process of RDSS is:

(1) Unified filtering: before calculating the emissivity with Equation (4), we filtered each variable participating in the calculation of the emissivity estimation with mean filtering. The calculation method was as follows:

$$\widetilde{V} = F(V) \tag{16}$$

where F is the filter function, V is the variable to be filtered and $\widetilde{V}$ is the filtered variable. Here, the variables refer to the ground-leaving radiance, the atmospheric downwelling radiance and the blackbody radiance.

Specifically, the calculation method for filtering the variables in Equation (4) with a mean filter was as follows:

$$\left[\widetilde{L}_g(\lambda_i, T), \widetilde{L}_d(\lambda_i), \widetilde{B}(\lambda_i, \hat{T})\right] = \frac{1}{2N_r + 1} \sum_{j=-N_r}^{N_r} \left[L_g(\lambda_{i+j}, T), L_d(\lambda_{i+j}), B(\lambda_{i+j}, \hat{T})\right] \tag{17}$$

where $\widetilde{L}_g(\lambda_i, T)$, $\widetilde{L}_d(\lambda_i)$ and $\widetilde{B}(\lambda_i, \hat{T})$ are the filtered $L_g(\lambda_{i+j}, T)$, $L_d(\lambda_{i+j})$ and $B(\lambda_{i+j}, \hat{T})$, respectively, with the mean filter; and $2N_r + 1$ is the spectral smoothing window size.

(2) Emissivity estimation and approximate processing: to avoid the local minimum value issue, taking advantage of the fact that the object temperature has a finite value range as a physical parameter, this study adopted the strategy of an exhaustive search to find the optimal temperature. We determined a temperature search range $Tsup_{inf}$ and the incremental step δT (also named as the temperature resolution), then calculated the emissivity estimations for all of the temperatures in the search range and approximately processed the emissivity estimations (i.e., boxcar average):

$$\hat{\varepsilon}\left(\lambda_i, T_{inf}() \frac{\widetilde{L}_g(\lambda_i, T) - \widetilde{L}_d(\lambda_i)}{\widetilde{B}(\lambda_i, T\widetilde{=}_{diinf}}\right) \tag{18}$$

$$\widetilde{\varepsilon}\left(\lambda_i, T_{inf}() \frac{1}{3} \sum_{j=-1}^{1\Sigma} \hat{\varepsilon}(\lambda_{i+j}, T_{inf}())\right) \tag{19}$$

where $\hat{\varepsilon}(\lambda_i, T_{inf}())$ is the emissivity estimation at the temperature of T_{inf}, k is the steps, $k = 0, 1, 2, \ldots, (T_{sup} - T_{inf})/\delta T$ and $\widetilde{\varepsilon}(\lambda_i, T_{inf}())$ is the corresponding emissivity estimation of $\hat{\varepsilon}(\lambda_i, T_{inf}())$ after approximate processing.

(3) Cost function calculation and the optimal temperature determination. We calculated the cost functions for all temperatures in the temperature search range according to the approximate processed emissivity. The calculation method was as follows:

$$\sigma(T_{\mathrm{inf}} + \delta T \cdot k) = \sigma\left(\left(\widetilde{B}(\lambda_i, T_{\mathrm{inf}} + \delta T \cdot k) - \widetilde{L}_d(\lambda_i)\right)\widetilde{\varepsilon}(\lambda_i, T_{\mathrm{inf}} + \delta T \cdot k) + \widetilde{L}_d(\lambda_i) - \widetilde{L}_g(\lambda_i, T)\right) \tag{20}$$

$$\underset{k}{\inf\arg}\ min\left(\sigma\left(T_{inf}()\right)()\right)\hat{T}_{opt} = T \tag{21}$$

where $\hat{T}_{opt}$ is the optimal temperature estimation, i.e., the retrieved temperature.

(4) Final emissivity calculation: we substituted the retrieved temperature from (3) and the other three variables before filtering them into Equation (4) to obtain the final emissivity. The calculation method was:

$$\hat{\varepsilon}_{opt}(\lambda_i) = \frac{L_g(\lambda_i, T) - L_d(\lambda_i)}{B(\lambda_i, \hat{T}_{opt}) - L_d(\lambda_i)} \tag{22}$$

where $\hat{\varepsilon}_{opt}(\lambda_i)$ is the final emissivity estimation, i.e., the retrieved emissivity.

4.2. Validation Results Analysis

4.2.1. The Retrieval Errors of the RDSS Algorithm

We retrieved the eight groups of simulated HTIR data with two spectral ranges (7.5–12.5 μm and 8–12.5 μm) and four spectral resolutions (50 nm, 35.2 nm, 10 nm and 5 nm) using the RDSS algorithm and measured the retrieved LST and LSE errors. Figures 5 and 6 show the retrieved LST errors and the retrieved LSE errors of the eight groups of simulated HTIR data as a function of the noise level, respectively. The change trends of the retrieved LST and LSE errors of the RDSS algorithm with noise, spectral resolution and spectral range were basically the same as those of the ARTEMISS algorithm, namely both the LST and LSE errors increase with noise linearly. The difference is that in terms of the magnitude of errors, the LST and LSE errors of RDSS were smaller than those of ARTEMISS except for some LSE RMSEs.

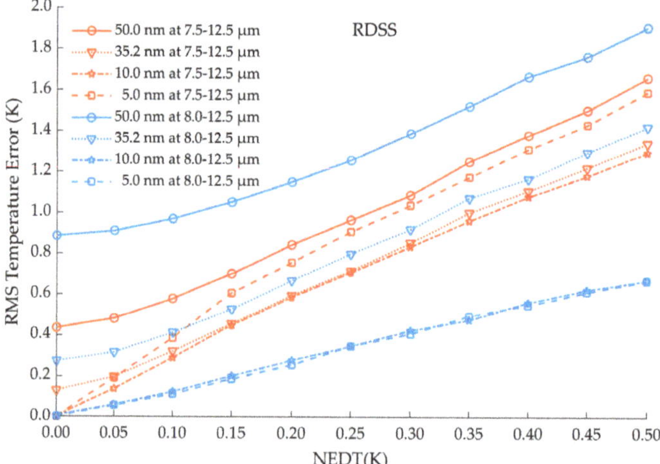

Figure 5. The plot of resolution-degrade-based spectral smoothness (RDSS) algorithm LST inversion errors varying with NEDT, spectral range and spectral resolution.

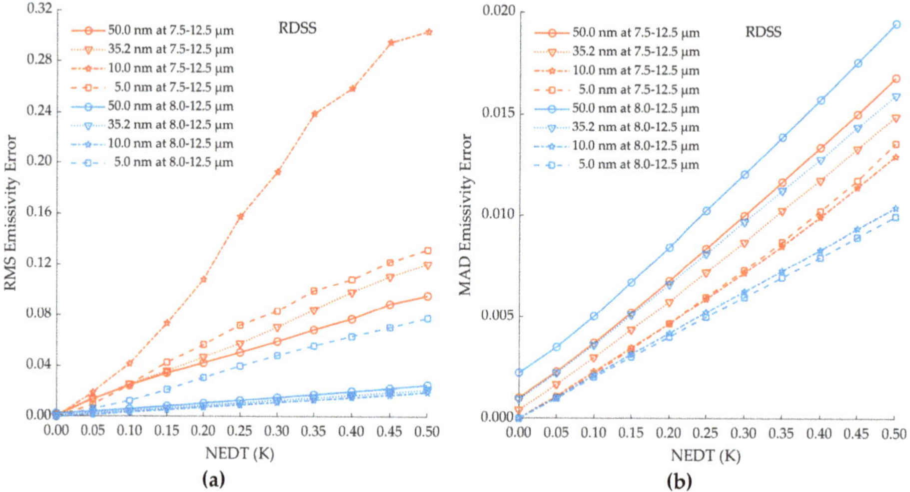

Figure 6. Plots of RDSS retrieved LSE errors varying with NEDT, spectral range and spectral resolution. (**a**) The retrieved LSE RMSE and (**b**) the retrieved LSE MAD.

Figure 7 shows a comparison chart of RDSS and ARTEMISS on the retrieved LST error, and Table 3 shows the corresponding error list. They show that under most noise and spectral settings, the retrieved LST error of RDSS are reduced by varying degrees, with a maximum reduction of 0.75 K (Group4, NDET = 0.5 K). The reduced values of the retrieved LST errors vary for different noise levels and spectral settings. Specifically, the reduced degrees of RDSS LST errors increase with the noise level; however, LST errors of RDSS are larger than ARTEMISS when there are no noise or less noise. The initial noise level at which LST errors of RDSS became smaller than ARTEMISS decreased as the spectral resolution increased. For example, the initial noise level decreased from 0.15 (Group1) and 0.2 K

(Group5) with a spectral resolution of 50 nm to 0.05 K and 0.05 K (Group4 and Group8) with the spectral resolution of 5 nm, respectively. The above trends indicate that in terms of temperature inversion, RDSS was more effective for data with high noise levels and a high spectral resolution. The same temperature inversion accuracy can be obtained by RDSS, while the instrument design specifications or data correction accuracy were reduced.

Table 3. The retrieved LST error list of the RDSS and ARTEMISS and their differences.

Groups\NEDT(K)		0.00	0.05	0.10	0.15	0.20	0.25	0.30	0.35	0.40	0.45	0.50
G1 [1]	A [2]	0.14	0.30	0.51	0.78	0.99	1.22	1.45	1.65	1.84	2.05	2.22
	R [3]	0.44	0.48	0.58	0.70	0.84	0.96	1.09	1.25	1.38	1.50	1.66
	Dif [4]	−0.30	−0.18	−0.07	0.08	0.15	0.26	0.36	0.40	0.46	0.55	0.56
	RDif [5]	—	—	—	10%	15%	21%	25%	24%	25%	27%	25%
G2	A	0.05	0.21	0.42	0.62	0.84	1.02	1.20	1.41	1.56	1.73	1.89
	R	0.13	0.20	0.32	0.46	0.59	0.72	0.85	1.00	1.11	1.22	1.34
	Dif	−0.08	0.01	0.10	0.16	0.24	0.31	0.35	0.41	0.45	0.51	0.55
	RDif	—	5%	24%	26%	29%	30%	29%	29%	29%	29%	29%
G3	A	0.00	0.19	0.39	0.61	0.80	0.94	1.11	1.27	1.44	1.60	1.77
	R	0.01	0.14	0.29	0.45	0.59	0.71	0.83	0.96	1.08	1.18	1.30
	Dif	0.00	0.05	0.10	0.16	0.21	0.23	0.27	0.31	0.36	0.42	0.48
	RDif	—	26%	26%	26%	26%	24%	24%	24%	25%	26%	27%
G4	A	0.00	0.25	0.54	0.77	0.98	1.18	1.36	1.59	1.83	2.10	2.34
	R	0.00	0.19	0.39	0.61	0.76	0.91	1.04	1.18	1.31	1.43	1.59
	Dif	0.00	0.06	0.15	0.17	0.22	0.27	0.33	0.41	0.52	0.66	**0.75**
	RDif	—	24%	28%	22%	22%	23%	24%	26%	28%	31%	32%
G5	A	0.30	0.46	0.69	0.97	1.20	1.46	1.70	1.91	2.14	2.36	2.56
	R	0.89	0.91	0.97	1.05	1.15	1.26	1.39	1.52	1.67	1.76	1.91
	Dif	−0.58	−0.45	−0.28	−0.08	0.04	0.20	0.31	0.39	0.47	0.59	0.65
	RDif	—	—	—	—	3%	14%	18%	20%	22%	25%	25%
G6	A	0.11	0.27	0.48	0.70	0.92	1.13	1.33	1.53	1.69	1.87	2.02
	R	0.27	0.32	0.41	0.53	0.67	0.80	0.92	1.07	1.17	1.30	1.42
	Dif	−0.17	−0.05	0.07	0.17	0.25	0.33	0.41	0.45	0.53	0.57	0.59
	RDif	—	—	15%	24%	27%	29%	31%	29%	31%	30%	29%
G7	A	0.00	0.11	0.24	0.41	0.55	0.64	0.75	0.87	0.97	1.07	1.20
	R	0.01	0.06	0.13	0.20	0.28	0.35	0.43	0.48	0.56	0.62	0.67
	Dif	0.00	0.05	0.11	0.21	0.27	0.29	0.33	0.39	0.41	0.45	0.53
	RDif	—	45%	46%	51%	49%	45%	44%	45%	42%	42%	44%
G8	A	0.00	0.11	0.26	0.45	0.57	0.69	0.78	0.88	0.99	1.11	1.23
	R	0.00	0.06	0.11	0.19	0.26	0.35	0.41	0.50	0.55	0.61	0.67
	Dif	0.00	0.05	0.14	0.26	0.31	0.34	0.37	0.39	0.44	0.50	0.57
	RDif	—	45%	54%	**58%**	54%	49%	47%	44%	44%	45%	46%

[1] Group 1. [2] ARTEMISS. [3] RDSS. [4] Difference between RDSS and ARTEMISS (i.e., RDSS − ARTEMISS). [5] Relative difference between RDSS and ARTEMISS (RDSS − ARTEMISS)/ARTEMISS). [6] The gray values show how much RDSS performs better than ARTEMISS and the bold values are the maximum reduction in error. Figure 8 shows a comparison of RDSS and ARTEMISS on the retrieved LSE RMSEs, and Table 4 is a list of the RMSE corresponding to Figure 8. Figure 9 shows a comparison chart of RDSS and ARTEMISS on the retrieval LSE MADs, and Table 5 is a list of the MAD corresponding to Figure 9. The results of LSE RMSE show that unlike the retrieval LST RMSE, for a few of the eleven noise levels and eight spectral settings (31 of 88 cases), the retrieved LSE RMSEs from the RDSS are smaller than those of the ARTEMISS algorithm. From the perspective of LSE RMSE, it seems that the RDSS had limited improvement on the emissivity inversion accuracy, and was not as good as ARTEMISS in terms of the overall effect. However, the results for LSE MAD tell another story.

Table 4. The retrieved LSE RMSE for RDSS and ARTEMISS and their differences.

Groups\NEDT(K)		0.00	0.05	0.10	0.15	0.20	0.25	0.30	0.35	0.40	0.45	0.50
G1	A	0.0005	0.0071	0.0121	0.0173	0.0224	0.0274	0.0322	0.0371	0.0417	0.0464	0.0509
	R	0.0016	0.0140	0.0248	0.0343	0.0423	0.0506	0.0592	0.0685	0.0772	0.0888	0.0954
	Dif	−0.0011	−0.0068	−0.0127	−0.0171	−0.0199	−0.0232	−0.0270	−0.0314	−0.0355	−0.0425	−0.0445
	RDif	—	—	—	—	—	—	—	—	—	—	—
G2	A	0.0002	0.0082	0.0133	0.0185	0.0235	0.0283	0.0330	0.0376	0.0420	0.0465	0.0508
	R	0.0006	0.0138	0.0254	0.0358	0.0468	0.0576	0.0705	0.0843	0.0982	0.1107	0.1203
	Dif	−0.0004	−0.0056	−0.0121	−0.0174	−0.0233	−0.0293	−0.0374	−0.0467	−0.0562	−0.0642	−0.0695
	RDif	—	—	—	—	—	—	—	—	—	—	—
G3	A	0.0000	0.0127	0.0208	0.0280	0.0348	0.0417	0.0481	0.0544	0.0602	0.0660	0.0710
	R	0.0000	0.0192	0.0416	0.0735	0.1081	0.1576	0.1927	0.2392	0.2591	0.2950	0.3037
	Dif	0.0000	−0.0065	−0.0208	−0.0455	−0.0733	−0.1159	−0.1446	−0.1848	−0.1989	−0.2290	−0.2327
	RDif	—	—	—	—	—	—	—	—	—	—	—
G4	A	0.0000	0.0172	0.0265	0.0345	0.0418	0.0488	0.0554	0.0619	0.0683	0.0741	0.0796
	R	0.0000	0.0095	0.0252	0.0427	0.0565	0.0720	0.0831	0.0994	0.1082	0.1221	0.1318
	Dif	0.0000	0.0078	0.0013	−0.0082	−0.0147	−0.0232	−0.0277	−0.0375	−0.0399	−0.0480	−0.0522
	RDif	—	45%	5%	—	—	—	—	—	—	—	—
G5	A	0.0010	0.0039	0.0071	0.0104	0.0138	0.0171	0.0204	0.0238	0.0272	0.0306	0.0339
	R	0.0026	0.0042	0.0063	0.0085	0.0108	0.0131	0.0155	0.0179	0.0203	0.0226	0.0251
	Dif	−0.0015	−0.0003	0.0008	0.0019	0.0030	0.0040	0.0050	0.0059	0.0069	0.0079	**0.0088**
	RDif	—	—	11%	18%	22%	23%	25%	25%	25%	26%	26%
G6	A	0.0004	0.0030	0.0057	0.0084	0.0112	0.0139	0.0167	0.0195	0.0222	0.0250	0.0278
	R	0.0011	0.0028	0.0048	0.0068	0.0089	0.0110	0.0132	0.0153	0.0174	0.0196	0.0217
	Dif	−0.0007	0.0001	0.0009	0.0016	0.0023	0.0029	0.0036	0.0042	0.0049	0.0054	0.0061
	RDif	—	3%	16%	19%	21%	21%	22%	22%	22%	22%	22%
G7	A	0.0000	0.0021	0.0041	0.0061	0.0082	0.0102	0.0122	0.0143	0.0163	0.0183	0.0203
	R	0.0000	0.0018	0.0038	0.0057	0.0076	0.0095	0.0115	0.0134	0.0156	0.0175	0.0198
	Dif	0.0000	0.0002	0.0004	0.0005	0.0006	0.0007	0.0007	0.0008	0.0007	0.0008	0.0005
	RDif	—	10%	10%	8%	7%	7%	6%	6%	4%	4%	2%
G8	A	0.0000	0.0037	0.0067	0.0095	0.0123	0.0150	0.0175	0.0202	0.0227	0.0254	0.0278
	R	0.0000	0.0061	0.0128	0.0216	0.0308	0.0399	0.0483	0.0560	0.0635	0.0706	0.0778
	Dif	0.0000	−0.0024	−0.0061	−0.0120	−0.0185	−0.0249	−0.0307	−0.0359	−0.0407	−0.0453	−0.0500
	RDif	—	—	—	—	—	—	—	—	—	—	—

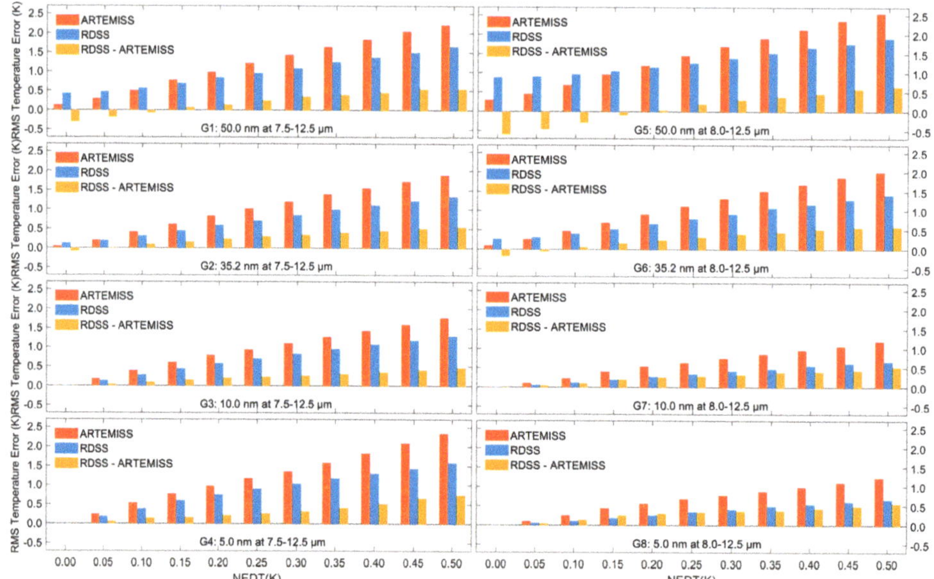

Figure 7. The comparison of the retrieved LST errors of RDSS and ARTEMISS.

Table 5. The retrieved LSE MAD for the RDSS and ARTEMISS algorithms and their differences.

Groups\NEDT(K)		0.00	0.05	0.10	0.15	0.20	0.25	0.30	0.35	0.40	0.45	0.50
G1	A	0.0004	0.0024	0.0044	0.0065	0.0087	0.0109	0.0131	0.0154	0.0176	0.0200	0.0223
	R	0.0010	0.0023	0.0037	0.0052	0.0068	0.0084	0.0100	0.0116	0.0134	0.0150	0.0168
	Dif	−0.0006	0.0001	0.0007	0.0013	0.0019	0.0026	0.0032	0.0038	0.0043	0.0050	0.0055
	RDif	—	4%	16%	20%	22%	24%	24%	25%	24%	25%	25%
G2	A	0.0001	0.0020	0.0037	0.0056	0.0075	0.0093	0.0112	0.0131	0.0151	0.0170	0.0190
	R	0.0004	0.0017	0.0030	0.0044	0.0058	0.0072	0.0087	0.0102	0.0117	0.0133	0.0149
	Dif	−0.0003	0.0003	0.0007	0.0012	0.0017	0.0021	0.0025	0.0029	0.0033	0.0037	0.0041
	RDif	—	15%	19%	21%	23%	23%	22%	22%	22%	22%	22%
G3	A	0.0000	0.0013	0.0027	0.0040	0.0054	0.0067	0.0082	0.0098	0.0115	0.0132	0.0152
	R	0.0000	0.0011	0.0023	0.0035	0.0047	0.0059	0.0072	0.0085	0.0099	0.0114	0.0129
	Dif	0.0000	0.0002	0.0004	0.0005	0.0007	0.0009	0.0010	0.0013	0.0015	0.0019	0.0022
	RDif	—	15%	15%	13%	13%	13%	12%	13%	13%	14%	14%
G4	A	0.0000	0.0013	0.0026	0.0039	0.0053	0.0068	0.0085	0.0105	0.0128	0.0155	0.0183
	R	0.0000	0.0011	0.0022	0.0034	0.0047	0.0060	0.0073	0.0087	0.0102	0.0117	0.0136
	Dif	0.0000	0.0002	0.0003	0.0005	0.0006	0.0008	0.0012	0.0018	0.0026	0.0038	0.0047
	RDif	—	15%	12%	13%	11%	12%	14%	17%	20%	25%	26%
G5	A	0.0009	0.0033	0.0059	0.0086	0.0114	0.0141	0.0169	0.0196	0.0224	0.0252	0.0279
	R	0.0022	0.0035	0.0050	0.0067	0.0084	0.0102	0.0120	0.0139	0.0157	0.0175	0.0195
	Dif	−0.0013	−0.0002	0.0009	0.0019	0.0029	0.0039	0.0048	0.0058	0.0067	0.0077	0.0085
	RDif	—	—	15%	22%	25%	28%	28%	30%	30%	**31%**	30%
G6	A	0.0003	0.0024	0.0045	0.0066	0.0088	0.0109	0.0132	0.0153	0.0175	0.0197	0.0218
	R	0.0010	0.0022	0.0036	0.0051	0.0066	0.0081	0.0097	0.0112	0.0128	0.0144	0.0160
	Dif	−0.0006	0.0002	0.0009	0.0015	0.0022	0.0028	0.0035	0.0041	0.0047	0.0053	0.0059
	RDif	—	8%	20%	23%	25%	26%	27%	27%	27%	27%	27%
G7	A	0.0000	0.0013	0.0025	0.0038	0.0050	0.0062	0.0075	0.0087	0.0100	0.0112	0.0125
	R	0.0000	0.0010	0.0021	0.0032	0.0042	0.0052	0.0063	0.0073	0.0083	0.0094	0.0104
	Dif	0.0000	0.0002	0.0004	0.0006	0.0008	0.0010	0.0012	0.0014	0.0017	0.0019	0.0021
	RDif	—	15%	16%	16%	16%	16%	16%	16%	17%	17%	17%
G8	A	0.0000	0.0011	0.0023	0.0034	0.0045	0.0057	0.0069	0.0081	0.0093	0.0106	0.0120
	R	0.0000	0.0010	0.0020	0.0030	0.0040	0.0050	0.0060	0.0070	0.0080	0.0090	0.0100
	Dif	0.0000	0.0001	0.0003	0.0004	0.0005	0.0007	0.0009	0.0011	0.0013	0.0016	0.0020
	RDif	—	9%	13%	12%	11%	12%	13%	14%	14%	15%	17%

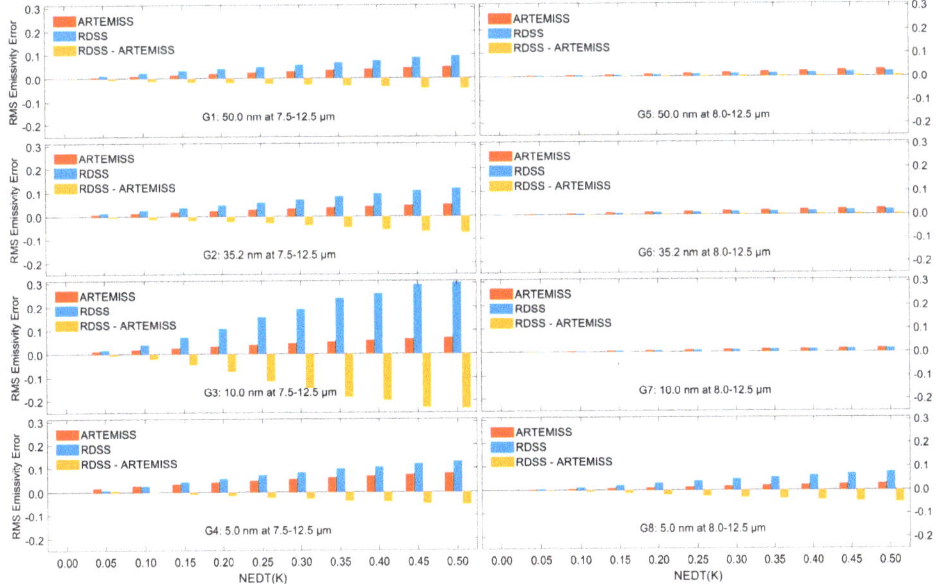

Figure 8. The comparison of the retrieved LSE RMSEs of RDSS and ARTEMISS.

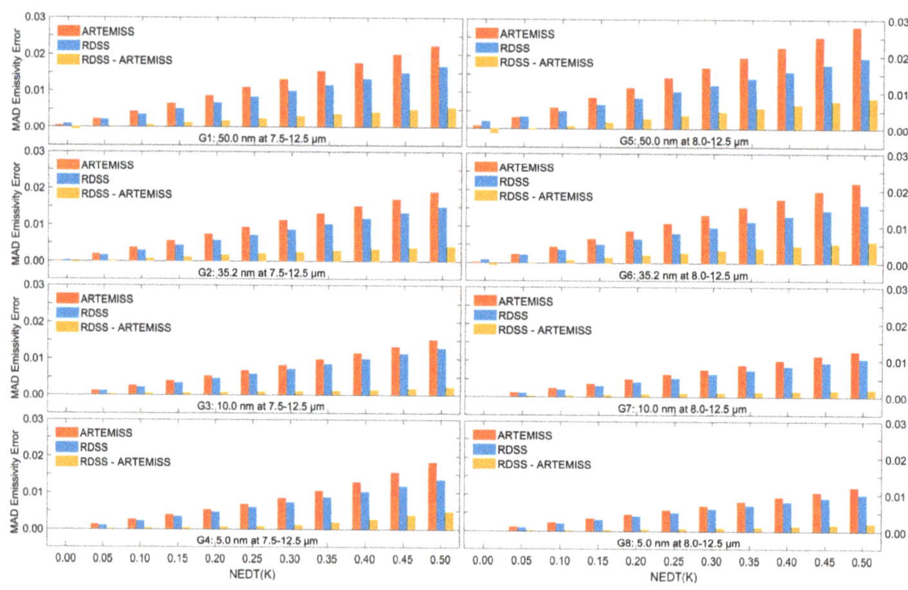

Figure 9. A comparison of the retrieved LSE MADs using the RDSS and ARTEMISS algorithms.

Figure 8 shows a comparison of RDSS and ARTEMISS on the retrieved LSE RMSEs, and Table 4 is a list of the RMSE corresponding to Figure 8. Figure 9 shows a comparison chart of RDSS and ARTEMISS on the retrieval LSE MADs, and Table 5 is a list of the MAD corresponding to Figure 9. The results of LSE RMSE show that unlike the retrieval LST RMSE, for a few of the eleven noise levels and eight spectral settings (31 of 88 cases), the retrieved LSE RMSEs from the RDSS are smaller than those of the ARTEMISS algorithm. From the perspective of LSE RMSE, it seems that the RDSS had limited improvement on the emissivity inversion accuracy, and was not as good as ARTEMISS in terms of the overall effect. However, the results for LSE MAD tell another story.

Figure 9 and Table 5 show that the retrieved LSE MADs of RDSS have similar distribution with the retrieved LST RMSE for the eight groups of data. For the majority of cases (79 out of 88 cases), the retrieved LSE MADs of RDSS were smaller than those of ARTEMISS. The results of LSE RMSE and LSE MAD behaved differently, probably due to the RMSE itself. The RMSE of an emissivity will be much larger because of some large abnormal values, although the retrieved temperature was accurate. As shown in Figure 10, there were some cases where the retrieved LSE of RDSS with a small LST error and LSE MAD was much closer to the true value from the perspective of the trend of the emissivity curve; however, the RMSE of the emissivity calculated using the accurate temperature was relatively large due to the influence of individual outliers. The phenomenon that the RMSE of an emissivity did not decrease but increased when the retrieved temperature was more accurate was also described in another paper [34].

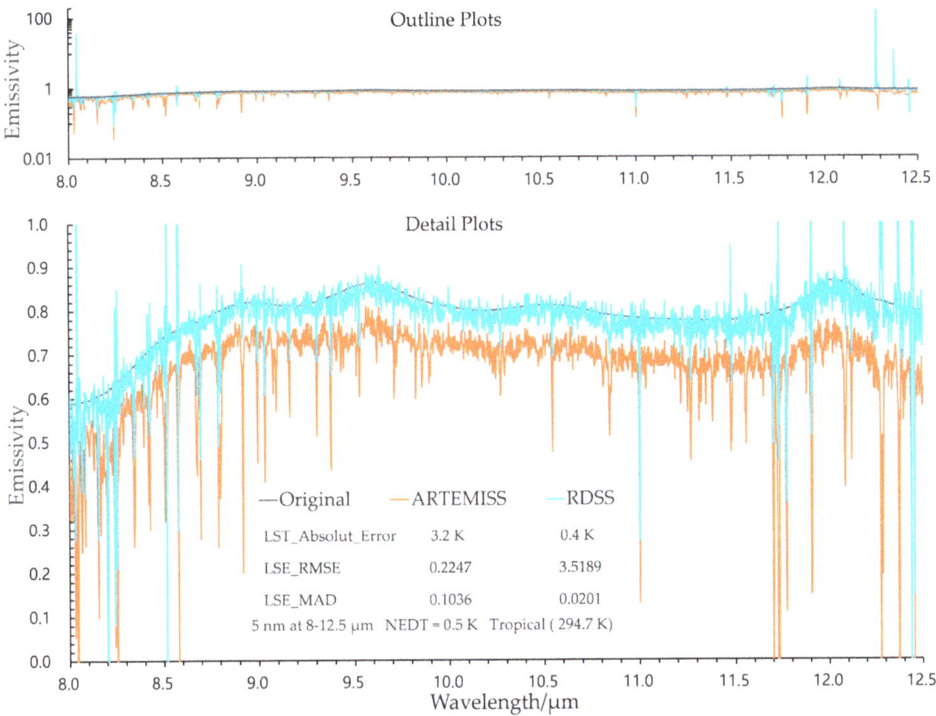

Figure 10. A case with a small LST error and LSE MAD but large LSE RMSE.

This study used MAD to evaluate the retrieved LSE error to measure the quality of the retrieved LSE curve from the perspective of the overall degree of deviation of emissivity from the true value. However, MAD cannot be compared with RMSE in terms of values, because MAD focuses on the deviation degree of the overall trend, which is different from the commonly used RMSE in terms of magnitude. We used the percentage of MAD to measure the reduction of the retrieved LSE error, as an indicator reflecting how close the trend of the emissivity curve retrieved by RDSS was to the true emissivity curve. Table 5 shows that the maximum relative decline in the percentage of LSE MAD can reach 31% (G5, 0.45 K). As the noise increased, the decline in the percentage of the LSE MAD of RDSS increased accordingly. When there was no noise or less noise, however, the MAD of RDSS instead increased. This is similar to the retrieved LST error in terms of change characteristics. However, the decline in the percentage of MAD of RDSS did not increase with the spectral resolution. In the range of 7.5–12.5 μm (G1–G4), the decline in the percentage of retrieval LSE MAD firstly decreased, and then increased along with the spectral resolution (it reached a minimum when the spectral resolution was 10 nm); in the range of 8.5–12.5 μm (G5–G8), the decline in the percentage of retrieved LSE MAD gradually decreased as spectral resolution increased. Overall, when the noise level was greater than 0.1 K, the decline in the percentage of the retrieved LSE MAD of RDSS was greater than 10%, which proved the effectiveness of the RDSS algorithm.

4.2.2. The Relationship of RDSS and the Window Setting

During spectral degradation, the window setting ($2N_r + 1$ in Equation (17)) is a key parameter. Different window settings may lead to different results, and temperature inversion is the key of the iterative spectral smoothness TES algorithms. The results of this study, at least the LSE MAD results, suggest that the improvement of temperature inversion accuracy is beneficial to the improvement of emissivity inversion accuracy. Therefore, we set 17 spectral smoothing window sizes, ranging from 3 increasing incrementally by 2–35, and separated the temperature and emissivity of the eight sets of simulated data to study the effect of window size on the performance of RDSS. First, we performed a mean filter cyclically with a window size on a radiance curve to obtain spectral degraded radiance curves under different window settings; second, we retrieved the temperatures under different window settings by RDSS, and looped the two steps for the eight groups of data; finally, we measured the retrieved temperature errors under different window settings.

Figure 11 shows the retrieved LST errors for the RDSS algorithm varying with 11 noise levels under 17 window settings for the eight groups of data. For the four groups of data with lower spectral resolutions (G1–2 and G5–6), 3 was the optimal window setting using RDSS, where the retrieved LST errors were the lowest among the 17 window settings under all or most noise levels. For the four groups of data with higher spectral resolutions (G3–4 and G7–8), the optimal window setting increased rapidly with the noise level. For the two groups of data with a spectral range of 7.5–12.5 μm (G3 and G4), the optimal window setting tended to be stable when the noise level was above 0.15 K, i.e., 27 for G3 and 31 for G4. For the group of data in the spectral range of 8–12.5 μm and the spectral resolution of 10 nm (G7), the optimal window setting tended to be stable at a noise level above 0.25 K—5 for G7. For the group of data in a spectral range of 8–12.5 μm and with a spectral resolution of 5 nm (G8), the optimal window setting was not stable but fluctuated in the range 5–21 when the noise level was above 0.05 K. It is worth noting that for the cases in G8 (5 nm at 8–12.5 μm), as the noise increased, the retrieved LST errors corresponding to the optimal window setting and the second optimal one, or even some other window settings, were very close to each other, and their difference was within 0.01 K. Therefore, it can be considered that when separating the LST and LSE with RDSS for HTIR data, we obtained the minimum LST error with a window setting of 3 for HTIR data with a spectral resolution in the tens of nanometers, and need to increase the window setting according to the spectral range and spectral resolution for HTIR data with a spectral resolution of 10 nm.

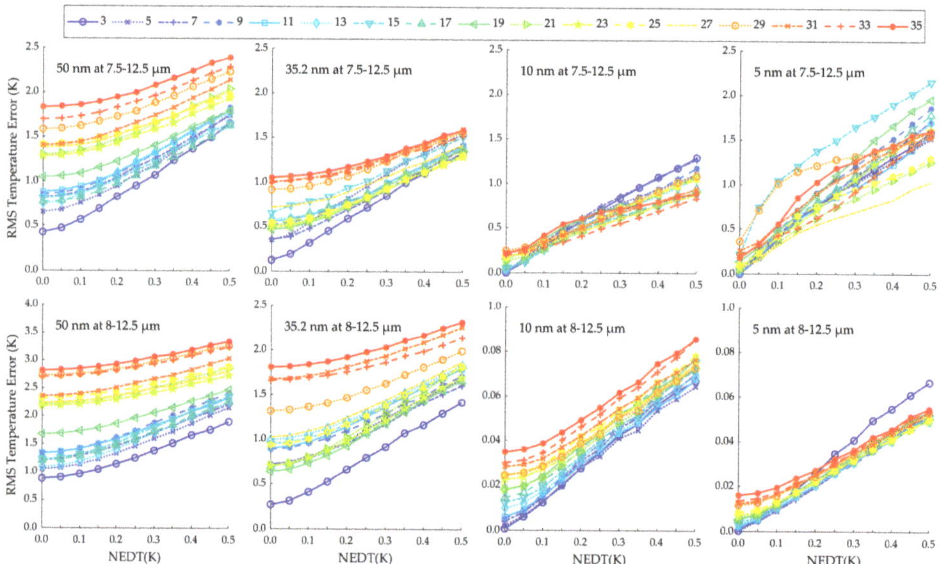

Figure 11. The retrieved LST errors of RDSS under different window size settings.

5. Discussion

This study focused on the sensitivity of an iterative spectral smoothness TES algorithm to the spectral range, spectral resolution, noise and smooth window size. We analyzed the characteristics of the influence of the four variables on temperature and emissivity inversion accuracy, and proposed an improved algorithm called RDSS. Compared with previous studies, this study analyzed the influence of the key spectrometer indices on the inversion error more comprehensively, especially the spectral resolution and the smooth window size.

The improved method, RDSS, performs spectral degradation on the input data, which can eliminate the effect of random noise on the search for the optimal temperature and improve the accuracy of temperature and emissivity inversion. In terms of method validation, the effect of random noise on RDSS was mainly studied in this study, while the effect of atmospheric error on RDSS needs further study. However, RDSS can also further improve the inversion accuracy by effectively removing random noise in atmospheric correction errors. For the systematic errors in atmospheric correction, the performance of the RDSS algorithm is similar to the ARTEMISS algorithm.

The results of this study also provide a valuable reference for the performance improvement of hyperspectral thermal infrared instruments and data processing. The improvement in spectral resolution is beneficial to reducing the error of temperature inversion. However, for a HTIR imager, the spectral resolution is dependent on the spatial resolution. In other words, the spatial resolution may decrease while the spectral resolution is improved, which will lead to the increase of mixed pixels. Without a good pixel decomposition algorithm, the accuracy of TES may decrease. Therefore, it is not a case of "the higher the spectral resolution, the better". The choice of spectral resolution needs to consider the application requirements, inversion algorithm, instrument design and other factors, and seek the optimal solution. This may be a very meaningful research topic.

From the perspective of the coupling of temperature and emissivity, accurately determining the temperature is an extremely critical step in the TES process. It is naturally believed that the improvement on the accuracy of temperature inversion will bring corresponding improvements in the accuracy of emissivity inversion. However, there are some anomalies that when the retrieval LST error is reduced to a certain value, the RMSE of retrieved LSE increases instead of decreasing. The emissivity of an object is a spectral curve composed of a set of digits, and the evaluation of the retrieval LSE error is equivalent to the measurement of the relationship between the two sets of digits, i.e., the retrieved LSE and the true value. The temperature of an object, however, is a unique value. Therefore, there is a certain difference between the retrieved LSE errors and the retrieved LST errors on the evaluation method. Moreover, the emissivity retrieved from the HTIR data is mostly used for target identification or surface parameters inversion. We adopted MAD alongside RMSE by combining the evaluation index and the application target in the evaluation of the retrieved LSE. In addition, the retrieved LSEs were not denoised before accuracy evaluation in the experiment, resulting in the retention of many outliers. In the practical application of HTIR data, however, the retrieved LSE curves can be denoised to eliminate outliers; at this moment, the evaluation of the trend of LSE becomes particularly important. Furthermore, the highest spectral resolution in the simulation experiment of the study was set to 5 nm, which is mainly considering that the spectral resolution of the emissivity of the ground objects in the ASTER spectral library is 10 nm or lower. Therefore, cases with a higher spectral resolution than 5 nm need further research when there are emissivity spectra of ground objects with sufficiently high spectral resolution.

6. Conclusions

We focused on the study of instrument noise, a widely existing factor that is difficult to eliminate losslessly in hyperspectral thermal infrared data, and focused on optimizing instrument design. Taking the ARTEMISS algorithm—the representative of the iterative spectral smoothness TES algorithm family with a good application effect—as the object of study. Supplemented by atmospheric radiation transfer simulation software and emissivity spectral library, we carried out simulation and inversion experiments; studied the relationship between the spectral resolution of the instrument, noise level, the ARTEMISS parameter setting and the inversion error; and proposed an improved method—RDSS—based on the mechanism and law of the influence of noise on the inversion error.

The law of the spectral response range, spectral resolution and noise on the inversion error obtained in this study provide a reference for the future development of airborne and spaceborne thermal infrared hyperspectral imagers and optimize the instrument design. The improved TES algorithm can be used for temperature and emissivity inversion of various thermal infrared hyperspectral remote sensing data. The resistance of this method also provides a certain surplus space for the design of the instrument in NEDT.

Author Contributions: Conceptualization, H.S., C.L. (Chengyu Liu), F.X. and J.W.; methodology, H.S. and C.L. (Chengyu Liu); software, H.S.; validation, H.S. and C.L. (Chengyu Liu); formal analysis, H.S.; investigation, H.S.; resources, H.S.; data curation, H.S. and C.L. (Chengyu Liu); writing—original draft preparation, H.S.; writing—review and editing, H.S. and C.L. (Chengyu Liu); visualization, H.S.; supervision, C.L. (Chengyu Liu), F.X., C.L. (Chunlai Liu) and J.W.; project administration, F.X., C.L. (Chunlai Li) and J.W.; funding acquisition, F.X., C.L. (Chunlai Li) and J.W. All authors have read and agreed to the published version of the manuscript.

Funding: This research was funded by the Project of Science and Technology Commission of Shanghai Municipality (grant number 18511102202) and the Project of civil space technology pre-research of the 13th five-year plan (grant number D040104).

Acknowledgments: The authors would like to sincerely thank Jet Propulsion Laboratory for the ASTER emissivity library. The authors would also like to sincerely thank the editors and anonymous reviewers for their careful reading and constructive comments.

Conflicts of Interest: The authors declare no conflict of interest.

References

1. Aslett, Z.; Taranik, J.V.; Riley, D.N. Mapping Rock-forming Minerals at Daylight Pass, Death Valley National Park, California, Using SEBASS Thermal-infrared Hyperspectral Image Data. In Proceedings of the 2008 IEEE International Geoscience and Remote Sensing Symposium, Boston, MA, USA, 7–11 July 2008; IEEE Press: Piscataway, NJ, USA, 2009; pp. 366–369.

2. Cui, J.; Yan, B.; Dong, X.; Zhang, S.; Zhang, J.; Tian, F.; Wang, R. Temperature and emissivity separation and mineral mapping based on airborne TASI hyperspectral thermal infrared data. *Int. J. Appl. Earth Obs. Geoinf.* **2015**, *40*, 19–28. [CrossRef]

3. Vaughan, R.G.; Calvin, W.M.; Taranik, J.V. SEBASS hyperspectral thermal infrared data: Surface emissivity measurement and mineral mapping. *Remote Sens. Environ.* **2003**, *85*, 48–63. [CrossRef]

4. Johnson, W.; Hulley, G.; Hook, S. Remote gas plume sensing and imaging with NASA's Hyperspectral Thermal Emission Spectrometer (HyTES). In Proceedings of the SPIE Sensing Technology + Applications, Baltimore, MA, USA, 28 May 2014; pp. 91010V-1–91010V-7.

5. Palombo, A.; Pascucci, S.; Loperte, A.; Lettino, A.; Castaldi, F.; Muolo, M.R.; Santini, F. Soil Moisture Retrieval by Integrating TASI-600 Airborne Thermal Data, WorldView 2 Satellite Data and Field Measurements: Petacciato Case Study. *Sensors* **2019**, *19*, 1515. [CrossRef]

6. Guo, G.; Liu, B.; Liu, C. Thermal Infrared Spectral Characteristics of Bunker Fuel Oil to Determine Oil-Film Thickness and API. *J. Mar. Sci. Eng.* **2020**, *8*, 135. [CrossRef]

7. Wan, Z.; Dozier, J. A generalized split-window algorithm for retrieving land-surface temperature from space. *IEEE Trans. Geosci. Remote Sens.* **1996**, *34*, 892–905. [CrossRef]

8. Barducci, A.; Pippi, I. Temperature and Emissivity Retrieval from remotely sensed images using the "grey body emissivity" method. *IEEE Trans. Geosci. Remote Sens.* **1996**, *34*, 681–695. [CrossRef]

9. Kahle, A.B.; Madura, D.P.; Soha, J.M. Middle infrared multispectral aircraft scanner data: Analysis for geological applications. *Appl. Opt.* **1980**, *19*, 2279–2290. [CrossRef]

10. Gillespie, A.R. Lithologic mapping of silicate rocks using TIMS. In Proceedings of the TIMS Data Users' Workshop, Pasadena, CA, USA, 18–19 June 1985; pp. 29–44.

11. Realmuto, V.J. Separating the effects of temperature and emissivity: Emissivity spectrum normalization. In Proceedings of the Second Thermal Infrared Multispectral Scanner (TIMS) Workshop, Pasadena, CA, USA, 22 May 1991; pp. 31–35.

12. Becker, F.; Li, Z.-L. Temperature-independent spectral indices in thermal infrared bands. *Remote Sens. Environ.* **1990**, *32*, 17–33. [CrossRef]

13. Watson, K. Spectral ratio method for measuring emissivity. *Remote Sens. Environ.* **1992**, *42*, 113–116. [CrossRef]

14. De Griend, A.A.V.; Owe, M. On the relationship between thermal emissivity and the normalized difference vegetation index for natural surfaces. *Int. J. Remote Sens.* **1993**, *14*, 1119–1131. [CrossRef]

15. Valor, E.; Caselles, V. Mapping land surface emissivity from NDVI: Application to European, African, and South American areas. *Remote Sens. Environ.* **1996**, *57*, 167–184. [CrossRef]

16. Sobrino, J.A.; Raissouni, N. Toward remote sensing methods for land cover dynamic monitoring: Application to Morocco. *Int. J. Remote Sens.* **2000**, *21*, 353–366. [CrossRef]

17. Sobrino, J.A.; Raissouni, N.; Li, Z.-L. A Comparative Study of Land Surface Emissivity Retrieval from NOAA Data. *Remote Sens. Environ.* **2001**, *75*, 256–266. [CrossRef]

18. Abrams, M. The Advanced Spaceborne Thermal Emission and Reflection Radiometer (ASTER): Data products for the high spatial resolution imager on NASA's Terra platform. *Int. J. Remote Sens.* **2000**, *21*, 847–859. [CrossRef]

19. Li, Z.-L.; Wu, H.; Wang, N.; Qiu, S.; Sobrino, J.A.; Wan, Z.; Tang, B.-H.; Yan, G. Land surface emissivity retrieval from satellite data. *Int. J. Remote Sens.* **2013**, *34*, 3084–3127. [CrossRef]

20. Salisbury, J.W.; D'Aria, D.M. Emissivity of terrestrial materials in the 8–14 μm atmospheric window. *Remote Sens. Environ.* **1992**, *42*, 83–106. [CrossRef]

21. Borel, C. Iterative Retrieval of Surface Emissivity and Temperature for a Hyperspectral Sensor. In Proceedings of the First JPL Workshop on Remote Sensing of Land Surface Emissivity, Pasadena, CA, USA, 6–8 May 1997.

22. Borel, C.C. ARTEMISS–an algorithm to retrieve temperature and emissivity from hyper-spectral thermal image data. In Proceedings of the 28th Annual GOMACTech Conference, Hyperspectral Imaging Session, Tampa, FL, USA, 31 March–3 April 2003; pp. 1–4.

23. Borel, C.; Rosario, D.; Romano, J. Data processing and temperature-emissivity separation for tower-based imaging Fourier transform spectrometer data. *Int. J. Remote Sens.* **2015**, *36*, 4779–4792. [CrossRef]

24. Kanani, K.; Poutier, L.; Nerry, F.; Stoll, M.-P. Directional Effects Consideration to Improve out-doors Emissivity Retrieval in the 3–13 μm Domain. *Opt. Express* **2007**, *15*, 12464–12482. [CrossRef]

25. Wang, X.; OuYang, X.; Tang, B.; Li, Z.; Zhang, R. A New Method for Temperature/Emissivity Separation from Hyperspectral Thermal Infrared Data. In Proceedings of the IGARSS 2008—2008 IEEE International Geoscience and Remote Sensing Symposium, Boston, MA, USA, 7–11 July 2008; pp. 286–289.

26. Cheng, J.; Liu, Q.; Li, X.; Xiao, Q.; Liu, Q.; Du, Y. Correlation-based temperature and emissivity separation algorithm. *Sci. China Ser. D Earth Sci.* **2008**, *51*, 357–369. [CrossRef]

27. Cheng, J.; Liang, S.; Wang, J.; Li, X. A Stepwise Refining Algorithm of Temperature and Emissivity Separation for Hyperspectral Thermal Infrared Data. *IEEE Trans. Geosci. Remote Sens.* **2010**, *48*, 1588–1597. [CrossRef]

28. Zhao, H.; Li, J.; Jia, G.; Qiu, X. Correlation-wavelet method for separation of hyperspectral thermal infrared temperature and emissivity. *Opt. Precis. Eng.* **2019**, *27*, 1737–1744. [CrossRef]

29. Chen, M.; Jiang, X.; Wu, H.; Qian, Y.; Wang, N. Temperature and emissivity retrieval from low-emissivity materials using hyperspectral thermal infrared data. *Int. J. Remote Sens.* **2019**, *40*, 1655–1671. [CrossRef]

30. Wang, N.; Wu, H.; Nerry, F.; Li, C.; Li, Z. Temperature and Emissivity Retrievals From Hyperspectral Thermal Infrared Data Using Linear Spectral Emissivity Constraint. *IEEE Trans. Geosci. Remote Sens.* **2011**, *49*, 1291–1303. [CrossRef]

31. Ni, L.; Wu, H.; Zhang, B.; Zhang, W.; Gao, L. Improvement of linear spectral emissivity constraint method for temperature and emissivity separation of hyperspectral thermal infrared data. In Proceedings of the 2015 7th Workshop on Hyperspectral Image and Signal Processing: Evolution in Remote Sensing (WHISPERS), Tokyo, Japan, 2–5 June 2015; pp. 1–4.

32. Lan, X.; Zhao, E.; Li, Z.-L.; Labed, J.; Nerry, F. An Improved Linear Spectral Emissivity Constraint Method for Temperature and Emissivity Separation Using Hyperspectral Thermal Infrared Data. *Sensors* **2019**, *19*, 5552. [CrossRef]

33. Zhang, Y.-Z.; Wu, H.; Jiang, X.-G.; Jiang, Y.-Z.; Liu, Z.-X.; Nerry, F. Land Surface Temperature and Emissivity Retrieval from Field-Measured Hyperspectral Thermal Infrared Data Using Wavelet Transform. *Remote Sens.* **2017**, *9*, 454. [CrossRef]

34. Zhou, S.; Cheng, J. A multi-scale wavelet-based temperature and emissivity separation algorithm for hyperspectral thermal infrared data. *Int. J. Remote Sens.* **2018**, *39*, 8092–8112. [CrossRef]

35. Acito, N.; Diani, M.; Corsini, G. Dictionary Based Temperature and Emissivity Separation Algorithm in LWIR Hyperspectral Data. In Proceedings of the 2018 9th Workshop on Hyperspectral Image and Signal Processing: Evolution in Remote Sensing (WHISPERS), Amsterdam, Netherlands, 23–26 September 2018; pp. 1–5.

36. Acito, N.; Diani, M.; Corsini, G. Subspace-Based Temperature and Emissivity Separation Algorithms in LWIR Hyperspectral Data. *IEEE Trans. Geosci. Remote Sens.* **2019**, *57*, 1523–1537. [CrossRef]

37. Ingram, P.M.; Muse, A.H. Sensitivity of iterative spectrally smooth temperature/emissivity separation to algorithmic assumptions and measurement noise. *IEEE Trans. Geosci. Remote Sens.* **2001**, *39*, 2158–2167. [CrossRef]

38. Cheng, J.; Xiao, Q.; Liu, Q.; Li, X. Effects of Smoothness Index on the Accuracy of Iterative Spectrally Smooth Temperature/Emissivity Separation Algorithm. *J. Atmos. Environ. Opt.* **2007**, *2*, 376–380.

39. Cheng, J.; Du, Y.; Liu, Q.; Li, X.; Xiao, Q.; Liu, Q. Analysis of the Non-isothermal Effects on the Temperature/Emissvity Separation of Flat Mixed Pixel. *J. Atmos. Environ. Opt.* **2008**, *3*, 57–64.

40. Borel, C. Error analysis for a temperature and emissivity retrieval algorithm for hyperspectral imaging data. *Int. J. Remote Sens.* **2008**, *29*, 5029–5045. [CrossRef]

41. OuYang, X.; Wang, N.; Wu, H.; Li, Z.-L. Errors analysis on temperature and emissivity determination from hyperspectral thermal infrared data. *Opt. Express* **2010**, *18*, 544–550. [CrossRef]

42. Borel, C.C. Temperature-Emissivity Separation Algorithms for Hyperspectral Imaging Spectrometers. In Proceedings of the Fourier Transform Spectroscopy and Hyperspectral Imaging and Sounding of the Environment, Lake Arrowhead, CA, USA, 1–4 March 2015.

43. Qian, Y.; Wang, N.; Ma, L.; Mengshuo, C.; Wu, H.; Liu, L.; Han, Q.; Gao, C.; Yuanyuan, J.; Tang, L.; et al. Evaluation of Temperature and Emissivity Retrieval using Spectral Smoothness Method for Low-Emissivity Materials. *IEEE J. Sel. Top. Appl. Earth Obs. Remote Sens.* **2016**, *9*, 4307–4315. [CrossRef]

44. Pieper, M.; Manolakis, D.; Truslow, E.; Cooley, T.; Brueggeman, M.; Jacobson, J.; Weisner, A. Performance limitations of temperature–emissivity separation techniques in long-wave infrared hyperspectral imaging applications. *Opt. Eng.* **2017**, *56*, 081804. [CrossRef]

45. Pieper, M.; Manolakis, D.; Truslow, E.; Jacobson, J.; Weisner, A.; Cooley, T.; Ingle, V. *Sensitivity of Temperature and Emissivity Separation to Atmospheric Errors in LWIR Hyperspectral Imagery*; SPIE: Bellingham, WA, USA, 2018; Volume 10644.

46. Pieper, M.; Manolakis, D.G.; Truslow, E.; Cooley, T.; Brueggeman, M.; Jacobson, J.; Weisner, A. Effects of Wavelength Calibration Mismatch on Temperature-Emissivity Separation Techniques. *IEEE J. Sel. Top. Appl. Earth Obs. Remote Sens.* **2018**, *11*, 1315–1324. [CrossRef]

47. Wang, N.; Qian, Y.; Wu, H.; Ma, L.; Tang, L.; Li, C. Evaluation and comparison of hyperspectral temperature and emissivity separation methods influenced by sensor spectral properties. *Int. J. Remote Sens.* **2019**, *40*, 1693–1708. [CrossRef]

48. Shao, H.; Liu, C.; Li, C.; Wang, J.; Xie, F. Temperature and Emissivity Inversion Accuracy of Spectral Parameter Changes and Noise of Hyperspectral Thermal Infrared Imaging Spectrometers. *Sensors* **2020**, *20*, 2109. [CrossRef] [PubMed]

49. Xie, F.; Liu, C.; Shao, H.; Zhang, C.; Yang, G.; Wang, J. Scene-based spectral calibration for thermal infrared hyperspectral data. *Infrared Laser Eng.* **2017**, *46*. [CrossRef]

50. Gu, D.; Gillespie, A.R.; Kahle, A.B.; Palluconi, F.D. Autonomous atmospheric compensation (AAC) of high resolution hyperspectral thermal infrared remote-sensing imagery. *IEEE Trans. Geosci. Remote Sens.* **2000**, *38*, 2557–2570. [CrossRef]

51. Young, S.J. An in-scene method for atmospheric compensation of thermal hyperspectral data. *J. Geophys. Res.* **2002**, *107*, 20. [CrossRef]

52. Chandrasekhar, S. *Radiative Transfer*; Dover: New York, NY, USA, 1960.

53. Lenoble, J. *Radiative Transfer in Scattering and Absorbing Atmospheres: Standard Computational Procedures*; A. Deepak: Hampton, VA, USA, 1985.

54. Hulley, G.C.; Duren, R.M.; Hopkins, F.M.; Hook, S.J.; Vance, N.; Guillevic, P.C.; Johnson, W.R.; Eng, B.T.; Mihaly, J.M.; Jovanovic, V.M.; et al. High spatial resolution imaging of methane and other trace gases with the airborne Hyperspectral Thermal Emission Spectrometer (HyTES). *Atmos. Meas. Tech.* **2016**, *9*, 2393–2408. [CrossRef]

55. Yuan, L.; He, Z.; Lv, G.; Wang, Y.; Li, C.; Xie, J.n.; Wang, J. Optical design, laboratory test, and calibration of airborne long wave infrared imaging spectrometer. *Opt. Express* **2017**, *25*, 22440–22454. [CrossRef]

56. Hook, S.J.; Eng, B.T.; Gunapala, S.D.; Hill, C.J.; Johnson, W.R.; Lamborn, A.U.; Mouroulis, P.; Mumolo, J.M.; Realmuto, V.J.; Paine, C.G. Qwest and HyTES: Two New Hyperspectral Thermal Infrared Imaging Spectrometers for Earth Science. In Proceedings of the SET-151 Specialist Meeting on Thermal Hyperspectral Imagery, Brussels, Belgium, 26–27 October 2009; NATORTO: Brussels, Belgium, 2009; pp. 1–8.

57. Jiménez-Muñoz, J.C.; Sobrino, J.A.; Skoković, D.; Mattar, C.; Cristóbal, J. Land Surface Temperature Retrieval Methods From Landsat-8 Thermal Infrared Sensor Data. *IEEE Geosci. Remote Sens. Lett.* **2014**, *11*, 1840–1843. [CrossRef]

58. Lahraoua, M.; Raissouni, N.; Chahboun, A.; Azyat, A.; Achhab, N.B. Split-Window LST Algorithms Estimation From AVHRR/NOAA Satellites (7, 9, 11, 12, 14, 15, 16, 17, 18, 19) Using Gaussian Filter Function. *Int. J. Inf. Netw. Secur.* **2012**, *2*, 68–77. [CrossRef]

59. Johnson, W.R.; Hook, S.J.; Mouroulis, P.; Wilson, D.W.; Gunapala, S.D.; Realmuto, V.; Lamborn, A.; Paine, C.; Mumolo, J.M.; Eng, B.T. HyTES: Thermal imaging spectrometer development. In Proceedings of the 2011 Aerospace Conference, Big Sky, MT, USA, 5–12 March 2011; pp. 1–8.

60. MODTRAN. Available online: http://modtran.spectral.com/ (accessed on 7 July 2020).

61. Baldridge, A.M.; Hook, S.J.; Grove, C.I.; Rivera, G. The ASTER spectral library version 2.0. *Remote Sens. Environ.* **2009**, *113*, 711–715. [CrossRef]

 *remote sensing* 

Article

Inter-Comparison of Field- and Laboratory-Derived Surface Emissivities of Natural and Manmade Materials in Support of Land Surface Temperature (LST) Remote Sensing

Mary F. Langsdale [1,*], Thomas P. F. Dowling [1,2], Martin Wooster [1], James Johnson [1], Mark J. Grosvenor [1], Mark C. de Jong [1], William R. Johnson [3], Simon J. Hook [3] and Gerardo Rivera [3]

[1] NERC National Centre for Earth Observation (NCEO), c/o Department of Geography, King's College London, London WC2B 4BG, UK; thomas.dowling@unep-wcmc.org (T.P.F.D.); martin.wooster@kcl.ac.uk (M.W.); james.johnson@kcl.ac.uk (J.J.); mark.grosvenor@kcl.ac.uk (M.J.G.); mark.dejong@kcl.ac.uk (M.C.d.J.)
[2] United Nations Environment Programme World Conservation Monitoring Centre, 219 Huntingdon Road, Cambridge CB3 0DL, UK
[3] National Aeronautics and Space Administration-Jet Propulsion Laboratory (NASA-JPL), 4800 Oak Grove Drive, Pasadena, CA 91109, USA; william.r.johnson@jpl.nasa.gov (W.R.J.); simon.j.hook@jpl.nasa.gov (S.J.H.); gerardo.rivera@jpl.nasa.gov (G.R.)
*   Correspondence: mary.langsdale@kcl.ac.uk

Received: 31 October 2020; Accepted: 11 December 2020; Published: 17 December 2020

Abstract: Correct specification of a target's longwave infrared (LWIR) surface emissivity has been identified as one of the greatest sources of uncertainty in the remote sensing of land surface temperature (LST). Field and laboratory emissivity measurements are essential for improving and validating LST retrievals, but there are differing approaches to making such measurements and the conditions that they are made under can affect their performance. To better understand these impacts we made measurements of fourteen manmade and natural samples under different environmental conditions, both in situ and in the laboratory. We used Fourier transform infrared (FTIR) spectrometers to deliver spectral emissivities and an emissivity box to deliver broadband emissivities. Field- and laboratory-measured spectral emissivities were generally within 1–2% in the key 8–12 micron region of the LWIR atmospheric window for most samples, though greater variability was observed for vegetation and inhomogeneous samples. Differences between laboratory and field spectral measurements highlighted the importance of field methods for these samples, with the laboratory setup unable to capture sample structure or inhomogeneity. The emissivity box delivered broadband emissivities with a consistent negative bias compared to the FTIR-based approaches, with differences of up to 5%. The emissivities retrieved using the different approaches result in LST retrieval differences of between 1 and 4 °C, stressing the importance of correct emissivity specification.

Keywords: land surface temperature; land surface emissivity; measurement uncertainties; emissivity box method; Fourier transform infrared spectrometer; portable spectrometer

---

1. Introduction

Emissivity is a spectrally varying property of a material, describing at any particular wavelength the efficiency at which an object emits electromagnetic radiation as a function of its temperature. It is mathematically defined as the ratio between the electromagnetic radiation actually emitted by the object at the wavelength in question, and that emitted by a black body at the object's thermodynamic (or kinetic) temperature [1]. Kirchhoff's law of thermal radiation furthermore states that at any particular wavelength, the absorptivity of a surface is equal to the emissivity of the surface if it is in thermal equilibrium with its surroundings, meaning for example that a perfect blackbody absorbs all the arriving electromagnetic radiation and re-emits the absorbed energy according to Planck's radiation law [2]. However, natural materials are not perfect blackbodies, and most are selective radiators, which may emit electromagnetic radiation according to Planck's radiation law at certain wavelengths, but not others. It is therefore important to understand their spectrally varying emissivity across the electromagnetic spectrum, including within the longwave infrared (LWIR) spectral atmospheric window (8–13 μm) where most remote sensing of land surface temperature (LST) is conducted. This is particularly the case when estimating LST remotely, where knowledge of the target's surface emissivity in the LWIR is essential when converting infrared brightness temperature (BT) measurements into accurate estimates of LST [3].

Emissivity depends on the chemical makeup of a material, and its geometry, surface roughness, and moisture content and as such can show strong seasonally varying cycles and land use/land cover variability [4]. Most soils and vegetation emissivities vary between a minimum of around 0.6 to a maximum of at or close to 1, while pure metals for example can have far lower values [2]. Unfortunately, relatively small errors in the assumed emissivity of a surface can induce quite large impacts on the finally estimated LST. For typical earth surface conditions, Jiménez-Muñoz and Sobrino [5] calculated that emissivity uncertainties of 0.01 typically result in LST uncertainties of around 0.6 K. Given the many applications of LST–such as deriving evapotranspiration and monitoring droughts [6]–recent years have seen an increase in interest in improving the accuracy of LST retrieval as evidenced by the development of new thermal infrared (TIR) sensors capable of LST retrieval such as the ECOsystem Spaceborne Thermal Radiometer Experiment on Space Station (ECOSTRESS) [7] and the classification of LST as an essential climate variable (ECV) by the World Meteorological Organisation's Global Climate Observing System (GCOS) [8]. The correct specification of surface spectral emissivity has been identified as the greatest source of error in current satellite-based measurements of LST [9] and it therefore is essential to try to minimise emissivity uncertainties in order to maximise the accuracy of remotely sensed LST estimates.

Multiple field and laboratory techniques for measuring emissivity have been developed, enabling both spectral emissivity measurement and broadband emissivity retrieval [10–14]. Although unable to perfectly capture field conditions known to impact surface spectral emissivity, such as soil moisture [15] or canopy structure [16], laboratory-based emissivity measurements are often preferred to field-based measurements (for samples that can be transported without modifying the sample and its emissivity). This is because, unlike field measurements, laboratory measurements can be collected under highly controlled conditions, thus reducing errors that might result from changing atmospheric or thermal conditions in the field for example [17]. Online spectral emissivity libraries consist predominately of such laboratory-derived emissivity spectra [18–22]. Data from these spectral libraries have been used extensively to "ground-truth" airborne and satellite LST and emissivity outputs [14,23–25], for the derivation LST algorithm coefficients [26–28] and in the calibration of LWIR satellite and airborne sensors [29]. However, a recent inter-comparison of laboratory emissivity measurements of the same samples reported some quite significant differences in emissivity values from different laboratory measurement setups [30]. For example, they found standard deviations of ±2.52% (0.024) in the emissivities derived for distilled water within the LWIR atmospheric window (8–14 μm). These uncertainties are much larger than those previously reported for laboratory setups [15] and larger than those typically reported with field measurements [11,31,32], thus highlighting the continuing importance of field-based surface emissivity measurements. This is particularly true given that such in situ measurement approaches allow measurements of emissivity under "natural conditions"—for example for samples such as vegetation that is difficult to preserve while transported.

Given that the correct specification of surface spectral emissivity is the greatest source of error in current satellite-based measurements of LST [9], and the discrepancies that have been found both within laboratory and between field and laboratory measurements detailed above, there is a need for further rigorous examination of the degree of agreement between current approaches to emissivity measurement. With this in mind, we conducted a study to compare different field and laboratory spectral emissivity measurement approaches, using the same targets to better understand the emissivity differences that can result from use of different measurement approaches and/or different measurement conditions. We have focused on Fourier transform infrared (FTIR) spectrometer-based emissivity measurement systems since these are the most common type used to provide spectral emissivity measurements, applying to the measured spectra a variety of different post-processing approaches to derive the surface emissivity information. We also include a comparison of these spectrally resolved data to the broadband emissivities produced using an "emissivity box", a popular low-cost method of broadband field emissivity determination that uses a sequence of LWIR radiometer measurements and a specially constructed box [33]. The impact of the emissivity measurement uncertainties from these methods on calculation of in situ LSTs is assessed as the last stage of our investigation.

2. Emissivity Measurement Techniques

Summarised in Table 1 are different field emissivity measurement techniques deployed in previous studies. The most utilised are variants of the emissivity box method, detailed in Rubio et al. [33,34], which provide broadband LWIR emissivity estimates, and approaches based on spectral radiance measurements made by field portable FTIR spectrometers, which provide spectrally resolved LWIR emissivity data [31]. As detailed by Rubio et al. [33,34], the two primary variants of the box method are the two-lid approach [35] and the one lid approach [36]. Both involve a bottomless box with highly reflective (for example polished aluminium) inner walls and a LWIR radiometer to make the broadband measurements. During each measurement, the box is covered by a lid with a small central hole through which the radiometric measurements are made, with the lid having either high reflectance (the "cold lid") or high emissivity (the "hot lid"). A sequence of four radiometer measurements with the box and lids in different configurations provide the data to estimate the broadband emissivity of the surface over which the box is placed [33,34].

Table 1. Overview of various different field emissivity measurement techniques, with variants of the first approaches considered in this study.

Method	Overview	References
Emissivity Box Method	One- and two-lid variants of the emissivity box method, used to determine the LWIR broadband emissivity of a surface	[33–39]
Portable FTIR Spectrometer Approach	Use of field-portable Fourier Transform Infrared (FTIR) spectrometer to estimate LWIR surface spectral emissivity	[31,32,40,41]
Temperature and Emissivity (TES) retrieval algorithm applied to a multi-band radiometer	Application of the Advanced Spaceborne Thermal Emission and Reflection Radiometer (ASTER) Temperature and Emissivity Separation (TES) algorithm with in situ radiance measurements obtained using multi-band radiometers	[42,43]
Novel Emissiometer	Novel instrument combining an oscillating TIR radiance source with digital signal processing to determine the band-effective emissivity of a radiometer	[12,44]
Sun Shadow Method	Similar approach to day/night LST retrieval algorithm adapted to in situ measurements in sun and sun-shadow with spectroradiometer to derive spectral emissivities	[45,46]

The field spectrometer approach to the emissivity measurement is detailed by Salvaggio and Miller [32], and involves the spectrometer measuring the emitted LWIR signal from the surface and using this, along with a measurement of the downwelling LWIR atmospheric radiation, to derive the surface's spectral emissivity. The downwelling component is most commonly assessed using a downward looking measurement of a gold Lambertian panel, which reflects almost all of the LWIR atmospheric radiation irradiating it.

In addition to field emissivity approaches, there exist a number of laboratory-based methods to assess surface emissivity, generally based on FTIR spectroscopic techniques, which provide surface spectral emissivity values. The spectrometers measure either LWIR sample emission or directional hemispherical reflectance (DHR) [10]. In the emission mode the emissivity estimate is derived through comparison of the spectral radiance emitted by the sample to that emitted by a blackbody at the same temperature (for example [19]). Being lab-based, this approach generally means the sample must be heated to temperatures significantly above the laboratory such that any low emissivity features in the resulting emissivity spectra are not simply "filled in" by reflected LWIR radiation coming from the surroundings at the same temperature as the sample. Consequently, the method is unsuited to samples such as vegetation [47]. To avoid this, FTIR spectrometers operating in the DHR mode are used, generally with a source of intense LWIR radiation that is used to illuminate the sample and assess its LWIR reflectance via consecutive measurements of the sample and a highly reflective reference standard such as Infragold [48]. Emissivity is then calculated from the LWIR reflectance spectra using Kirchhoff's law [49].

Many field emissivity and LST validation studies have used the box method since the equipment is relatively simple, inexpensive, easily portable, and with minimal power requirements (e.g., [11,14,50–54]). Multiple studies have assessed the quality of emissivities derived using the approach, typically by comparing them to full spectral emissivity data coming from laboratory measurements convolved to the waveband of the LWIR radiometer used in the box [15]. The conclusions of these studies generally indicate that the quality of the box-derived field emissivity data is highly dependent on the measurement conditions, particularly for the one-lid variant [11,15,33]. Under favourable measurement conditions, a strong degree of agreement is seen between the data derived by the box method and that of the various laboratory spectral measurement approaches applied. Mira et al. [15] and Nerry et al. [38] for example both observed that the two-lid box method produced broadband LWIR emissivity estimates with a mean error of ±0.5% under stable field conditions (low winds and constant cloud conditions that help keep the sample surface temperature consistent while the measurements are made). Göttsche and Hulley [11] reported less than 1% difference for sand samples in Gobabeb (Namibia) where clear, cold skies with low winds made measurement conditions optimum. However, under less suitable conditions (e.g., high winds and variable cloud cover), the sensitivity of the derived surface emissivity value to changes in the sample temperature during the measurement can result in large errors. A change of 3 K over the measurement period results in emissivity errors of up to 2% in the one-lid method for example [34]. While such percentage errors seem small, due to coupling of LST and emissivity, a 1% error in specified emissivity will generally result in about a 0.5 K error in the derived LST [9]. Hence the accuracy of surface emissivity data is key to accurate LST derivation.

Compared to the box method, fewer studies exist comparing field- and lab-derived emissivity data based on FTIR spectrometer measurements [31,32,47,55,56]. However, as with the box method, the studies that have been conducted found that the accuracy of the field-derived data is highly dependent on the environmental conditions that existed during the measurement. For example, Salvaggio and Miller [32] assessed the field spectral emissivity data coming from measurements made with the Designs and Prototypes (D&P) μFTIR system, specifically designed for field emissivity measurement. Under ideal measurement conditions (stable, low winds and clear skies), the mean absolute emissivity error was less than 1% for most surface samples, with the D&P spectral measurements processed to spectral emissivity using Horton et al.'s [57] spectral smoothness approach. However, more problems were observed with measurements made under conditions of high humidity and air temperature, and/or more variable conditions [31,55]. Horton et al. [57] found that a 0.5 K change in sample temperature during the measurement procedure resulted in errors in the final calculated emissivity of 2.5%. As a result, samples with relatively low thermal inertia (such as dry soils) or samples that experience rapid evaporative cooling in the near-surface layer (such as water, damp soils, or dewy vegetation) can show higher errors under changing environmental conditions, such as high winds [17].

As well as these meteorological factors, observed differences between laboratory and field emissivity measurements (whether the latter be from the box- or FTIR-based approaches) are often attributed to physical changes in the sample, which may occur between the field and the laboratory, for example in terms of its structure and surface moisture [15]. Such possibilities for error further highlight the importance of field emissivity measurements. However, since the accuracy assessment of the field methodologies is often performed through comparison with laboratory-derived measurements, any differences between the laboratory and field sample conditions can affect the evaluation. Studies that intercompare the emissivities of the same samples derived by different field measurement approaches may help to redress this issue, but few such studies exist. Those that have been conducted considered are restricted to few sample types (e.g., soils or sands) or have been based on rather limited comparisons, for example due to differing instrument spectral responses [12]. A critical finding is that of Mira et al. [15], who observed emissivity differences between 2% and 7% in the 8–9 µm LWIR band between the values derived using the two-lid emissivity box and the TES-retrieved radiometer approach (see Table 1), corresponding to a 0.7–2.6 K error in derived LST.

3. Methods

Measurements were made of multiple manmade and natural samples with varying physical structures during two field campaigns in the UK and Italy using four methods: (i) a laboratory FTIR spectrometer setup at King's College London operating in DHR mode, (ii–iii) two portable field FTIR spectrometers with different processing approaches, and (iv) a two-lid emissivity box constructed at King's College London.

3.1. Instrumentation, Measurements, and Post-Processing

3.1.1. Emissivity Determination Using the Laboratory FTIR Spectrometer

In the laboratory, high spectral resolution (4 cm^{-1}) surface emissivity spectra of the target samples were derived from directional hemispherical reflectance LWIR spectral measurements made by a Bruker Vertex 70 FTIR spectrometer with an external highly reflective gold integrating sphere and an external thermal infrared source, as shown in Figure 1. The full measurement setup is detailed in Langsdale et al. [30], and the measurements covered the spectral range 2.5–16 µm, extending beyond the normal LWIR atmospheric window (8–13 µm).

The data coming from the laboratory system shown in Figure 1 can be processed to surface spectral emissivity using either the substitution or comparative methods, which are detailed in Hecker at al. [58]. The authors of [30] found that surface spectral emissivities derived using the comparative method, which uses the internal wall of the diffusely coated gold integrating sphere as the reference target, on the same laboratory system were within 1.5% of the mean of those derived by a wide range of international laboratories' measurements (spectral range 8–14 µm). To measure emissivity using this comparative method, the target surface was placed directly underneath the sample port of the external integrating sphere and illuminated with the LWIR beam coming from the external source. The reflected spectral radiance ($V_s(\lambda)$) was then measured and compared to a subsequent measurement of the reflected radiance from the internal wall of the integrating sphere ($V_r(\lambda)$), enabling the calculation of sample reflectance ($\rho_s(\lambda)$):

$$\rho_s(\lambda) = \frac{V_s(\lambda) - V_o(\lambda)}{V_r(\lambda) - V_o(\lambda)} \rho_r(\lambda) \tag{1}$$

where $V_o(\lambda)$ is an open port measurement as detailed in Hecker et al. [58] and $\rho_r(\lambda)$ is the absolute reflectance of the internal gold wall of the integrating sphere ($\rho_r(\lambda) \approx 0.97$ across 2.5–14 µm) used as the reference target. An internal rotating mirror was used to move the infrared beam illumination between the sample and the reference position. Sample spectral emissivity ($\varepsilon_s(\lambda)$) was then calculated from reflectance using Kirchhoff's law [59]:

$$\varepsilon_s(\lambda) = 1 - \rho_s(\lambda) \tag{2}$$

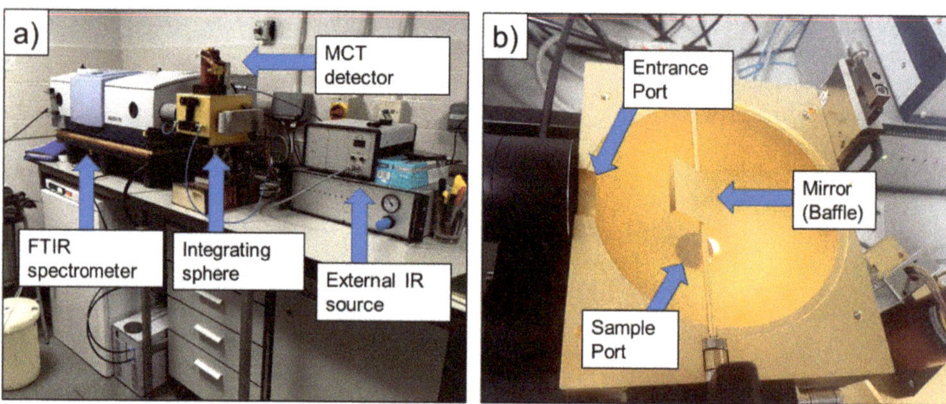

Figure 1. (**a**) Laboratory setup for surface spectral emissivity determination based on measurements made by a Bruker Vertex 70 FTIR spectrometer installed at King's College London along with an external water-cooled longwave infrared (LWIR) radiation source and a gold-coated integrating sphere. (**b**) Details of the inside of the integrating sphere, showing the gold coating used as the reference target in the comparative method and rotating mirror to direct the measurement beam from entrance port to sample port/internal wall.

For each sample, a minimum of the three emissivity measurements was collected to enable consideration of measurement variability. More measurements were made for low reflectance samples (card, grass and water) and for inhomogeneous samples (gravel and grass). Each individual spectral measurement consisted of either 500 or 1000 coadded scans, with the higher number of scans used for samples with low LWIR reflectances.

3.1.2. Emissivity Determination Using the Field Portable FTIR Spectrometers

Two portable FTIR spectrometers were deployed in this study (Table 2). The first was the aforementioned D&P µFTIR spectrometer [31,32,40], specifically designed for surface emissivity measurement in the field (Figure 2). The instrument operates in passive emission mode to measure emitted LWIR radiation, and uses the two-temperature blackbody approach for its calibration. It has a 45° mirror (rotating to allow angled measurements) within an enclosed tube. The main improvement on the µFTIR design described originally in Korb et al. [31] and Hook and Kahle [40] is the inclusion of a Stirling cycle cooling for the detector in place of liquid nitrogen. The second FTIR deployed was a Bruker EM27, also with a Stirling cycle cooled detector and an internal blackbody target that can be rapidly heated and cooled to provide the necessary two point calibration [30]. Though this system is designed primarily for atmospheric remote sensing, it is easily adapted to assess surface emitted LWIR radiation via attachment of a 45° flat high IR reflectance gold mirror as shown in Figure 2. This mirror can then be used to reflect the surface target emitted and gold-panel reflected LWIR radiation into the spectrometer. The system, its 12 V battery/inverter and a controlling laptop were mounted on a rugged trolley for relatively easy transport around a field site. Spectral resolutions used were the maximum for the two systems, namely 0.5 cm^{-1} for the EM27 and 4 cm^{-1} for the D&P μFTIR, with spectral sampling intervals in practice of 0.25 cm^{-1} and 3 cm^{-1}. The D&P was available for the measurements in Italy only.

Table 2. Instrument specifications for the portable FTIR spectrometers deployed herein, namely a Bruker EM27 FTIR spectrometer and the Designs and Prototypes μFTIR [31,40,60].

Parameter	Bruker EM27	Designs and Prototypes μFTIR
Spectral Range	4.5–14.3 μm (700–2200 cm^{-1})	2.0–14.0 μm
Spectral Resolution	0.5 cm^{-1}	4 cm^{-1}
Sampling Rate	0.25 cm^{-1}	3 cm^{-1}
Interferometer	Michelson	Michelson
Detector	HgCdTe	HgCdTe
Dimensions (cm)	40 × 36 × 27	33 × 46 × 32
Weight	18 kg	12.4 kg
Power	40 W (Average), 80 W (Max)	18 W
FOV (@ 1 m)	1.7° (60 mm)	4.8° (78 mm)

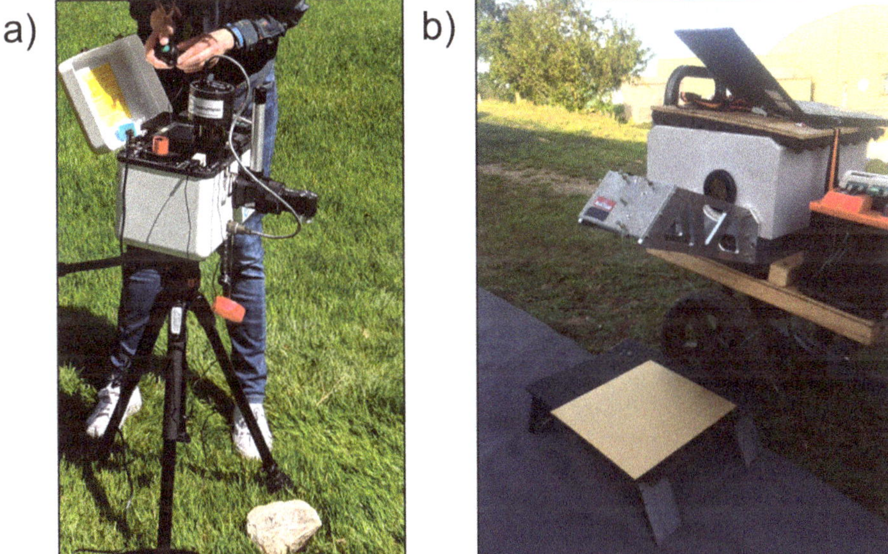

Figure 2. (**a**) Field-deployed Designs and Prototypes μFTIR spectrometer and (**b**) field-deployed Bruker EM27 FTIR spectrometer with a 45° mirror attachment fitted to view upwelling radiation from the surface, here shown assessing downwelling LWIR atmospheric radiation via observations of an Infragold panel.

To retrieve surface spectral emissivities from the passive LWIR spectra collected by either of the FTIR instruments, the calibration blackbody temperatures were first set to appropriately bracket the sample temperature [17]. As recommended by Salvaggio and Miller [41], the hot and cold blackbody temperatures used for the calibrations were set to approximately 10 K above and below the estimated sample temperature to reduce extrapolation error, although very high ambient air temperatures encountered at the Italian field site required the cool blackbody to be elevated above this limit. Sample temperature was itself estimated using a FLIR i7 handheld LWIR thermal imaging camera. Consecutive spectral measurements were then made of the sample (L) and of a 13 cm × 13 cm Labsphere Infragold panel (L_{panel}), with the panel measurement used as a proxy for the downwelling

LWIR atmospheric signal. The panel has a known and spectrally flat emissivity (ε_{panel}), provided by the manufacturer as 0.03 ± 0.01 across 2.5–14 μm range. The panel was placed in the same configuration as the sample, positioned just above the sample location.

The D&P μFTIR spectrometer comes with its own software to estimate sample emissivity across 7.5–12.0 μm from these measurements. Sample temperatures are estimated using the "Maximum Spectral Temperature" method detailed in Salvaggio and Miller [32] and developed by Korb et al. [31] and Hook and Kahle [40]. Emissivity uncertainties were taken as the standard deviation of multiple measurements. The measurement procedure takes around 25 min for a complete set of measurements. For the EM27 we developed our own emissivity measurement approach and software, with the full measurement sequence (three consecutive and repeated measurements of the sample and Infragold panel along with spectral calibration) typically taking around 20 min. From the measurements of the gold panel, downwelling radiance ($L^{\downarrow}$) spectra were first estimated as:

$$L^{\downarrow}(\lambda) = \frac{L_{panel}(\lambda) - \varepsilon_{panel}(\lambda)L_{BB}\left(T_{panel}, \lambda\right)}{1 - \varepsilon_{panel}(\lambda)} \tag{3}$$

where T_{panel} is the kinetic temperature of the Infragold panel (K), measured with a contact k-type thermocouple (manufacturer-stated accuracy ±0.1 K) and $L_{BB}\left(T_{panel}, \lambda\right)$ is the blackbody spectral radiance at temperature T_{panel} calculated using the Planck function such that:

$$L_{BB}\left(T_{panel}, \lambda\right) = \frac{2hc^2}{\lambda^5\left(e^{\frac{hc}{\lambda k T_{panel}}} - 1\right)} \tag{4}$$

where h is the Planck constant ($6.62606957 \times 10^{-34}$ Js), c is the speed of light ($299,792,458$ ms^{-1}), and k is the Boltzmann constant ($1.3806488 \times 10^{-23}$ JK^{-1}).

If the sample temperature (T_s) is accurately known, the surface spectral emissivity ($\varepsilon(\lambda)$) of the sample can be retrieved through use of a rearranged radiative transfer equation appropriate to a surface-viewing sensor positioned close to the target [32]:

$$\varepsilon(\lambda) = \frac{L(\lambda) - L^{\downarrow}(\lambda)}{L_{BB}(T_s, \lambda) - L^{\downarrow}(\lambda)} \tag{5}$$

where $L_{BB}(T_s)$ is the blackbody spectral radiance at temperature (T_s). However, sample temperature can vary even over short timescales (e.g., due to wind), and can be hard to assess accurately in the field for certain targets (e.g., vegetation) even under good measurement conditions. We therefore avoided having to specify T_s by using the "spectral smoothness" approach [57], an approach determined as optimal for emissivity derivation based on field-portable FTIR measurements by Salvaggio and Miller [32]. We implemented this by identifying a realistic sample temperature range as in Salvaggio and Miller [32] and calculating emissivity using Equation (5) for all temperatures within this range (in increments of 0.01 K). The sample temperature was taken to be the temperature, which minimised residuals in the resulting emissivity spectra that were clearly associated with atmospheric absorption and emission features in the 8.12–8.60 μm range of the short wavelength lobe of the silicate doublet. Emissivity uncertainties for the EM27 were calculated through propagation of the uncertainties in the input parameters. Uncertainties in the gold panel temperature and emissivity were taken as the manufacturer-stated accuracy (0.01 K and 0.01 respectively), which in the sample temperature as the 0.01 K precision, and that in the sample and gold panel spectra as the standard deviation of the measurement coadds (minimum 12 per spectra). Due to its extremely low emissivity, no adjustment was made for self-emission of the 45° mirror used to direct LWIR radiation from the sample to the spectrometer, nor for errors related to the fact that the cold sky temperatures are far lower than the minimum temperature of the two-point blackbody calibration. The typically low values of the downwelling radiation from the cold (clear) sky compared to the LWIR emission from the samples, along with the high emissivities (and thus low reflectances) of the samples, mean that uncertainties in the assessment of the downwelling sky radiation do not have much impact on the final emissivity uncertainty [31].

3.1.3. Emissivity Determination Using the Emissivity Box

We constructed a two-lid emissivity box at King's College London (Figure 3), based on the design of Rubio et al. [33,34] with (i) an improved outer thermal insulation layer surrounded by a robust outer case, (ii) a 3D-printed angled port to hold the radiometer at a constant view zenith angle of 5° to reduce the "Narcissus" effect, as in Göttsche and Olesen [37], (iii) continuous 1 Hz sampling of the radiometer measurements as in Göttsche and Olesen [37] to enable identification and rejection of erroneous readings (e.g., when conditions were not stable) during post-processing readings, and (iv) the addition of a "heating tray" to help the hot lid more quickly achieve its optimal temperature, while also reducing heat loss in cooler conditions. While our design could also be used for the one-lid approach, the two-lid method is considered more robust in windy or otherwise variable field conditions [34], some of which we encountered during our study.

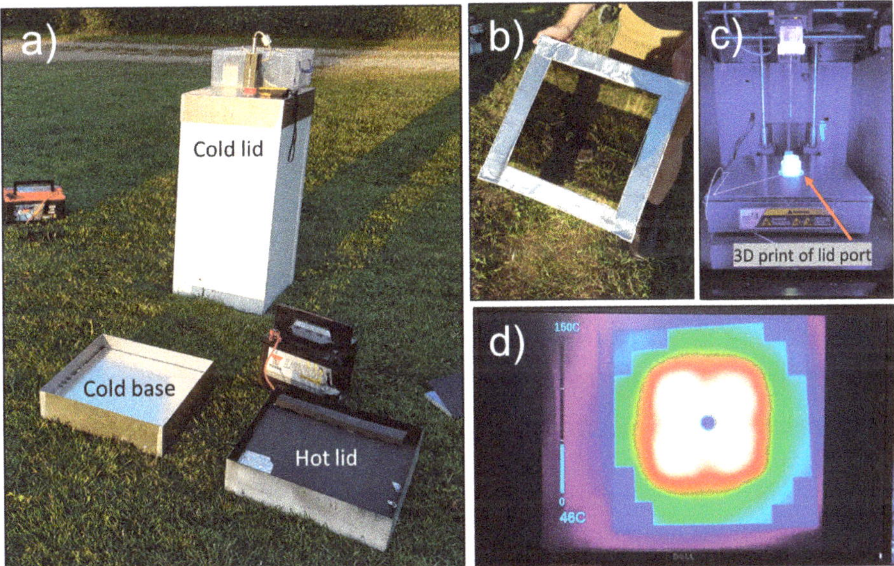

Figure 3. (**a**) Insulated emissivity box with the cold base and battery-powered hot lid. Details include (**b**) internal walls of highly polished aluminium, (**c**) a 3D printed 5° radiometer port for consistent off-centre angled sampling to avoid the "Narcissus" effect, and (**d**) evenly distributed electronic heating pads on the hot lid enabling heating up to at least 60 °C when combined with the heating tray.

Broadband surface emissivity was determined using our emissivity box via a sequence of BT measurements made with a Heitronics KT15.85 IIP radiometer fitted in the angled radiometer port to sequentially view the target surface and the base, as described in [31]. This radiometer is the same model as that used at the four permanent LST validation stations described in Göttsche et al. [50] and operates over the spectral range 9.6–11.5 μm, which is located well within the LWIR atmospheric window (Figure 4). BT measurements from the radiometer (T, kelvin) were converted into spectral radiances (L, Wm^{-2}sr^{-1}μm^{-1}) using Planck's radiation law (Equation (4)) evaluated at the effective radiometer central wavelength (approximately 10.55 μm; the exact value depending on the target temperature). Laboratory calibration tests confirmed the radiometer to have an absolute accuracy of

±0.5 K plus 0.7% of the difference between the BT of the target and the radiometer body temperature (taken as the ambient temperature). For example, if the ambient temperature was 300 K and the target BT was recorded as 295 K, the absolute accuracy was determined to be ±(0.5 + [0.7/100 × 5]) = ±0.535 K. When fitted into the angled port, the observed surface area was 170 cm^2.

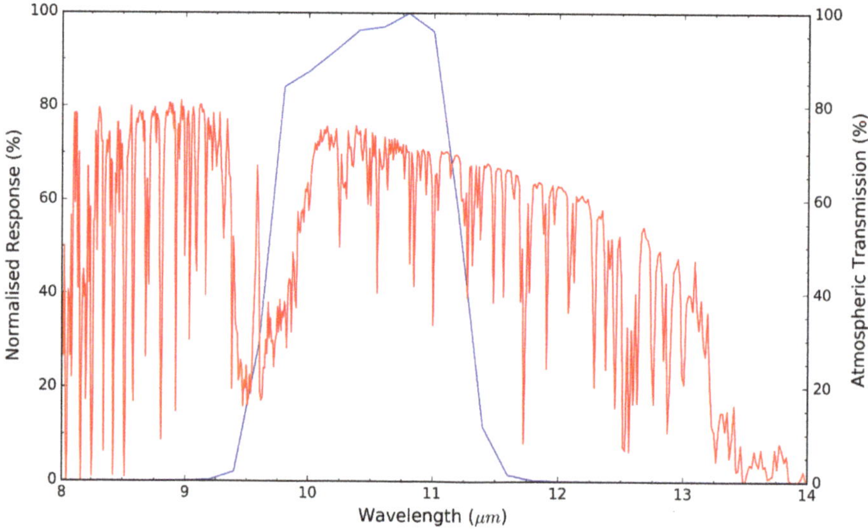

Figure 4. Spectral response function of the Heitronics KT15.85 radiometer (blue, left axis) overplotted on the atmospheric transmission of a standard mid-latitude summer atmosphere (red, right axis) calculated using MODTRAN 5.0 [61].

To make the measurements necessary to estimate the targets surface emissivity, the box was first placed on the target surface sample and left for two minutes to ensure stabilised temperatures. Measurements then proceed as in Figure 5, with the rationale for this sequence explained in Rubio et al. [33]. Using the same nomenclature as in Rubio et al. [33] and Figure 5 the broadband emissivity (ε_0) of the target surface if the box were ideal was calculated as:

$$\varepsilon_0 = \frac{L^3 - L^1}{L^3 - L^2} \tag{6}$$

where (in order of measurement), L^2 is the sample radiance measured when the box is over the ground sample with the cold lid in use, L^1 the sample radiance made with the hot lid in use instead of the cold lid and L^3 the radiance obtained when putting the box with the hot lid on over a cold base with the same emissivity as the cold lid.

However, the box departs from non-ideal behaviour (because the emissivity of the hot lid cannot be 1 and the emissivity of the cold lid cannot be 0) as detailed by Rubio et al. [34], who developed a correction factor ($\delta\varepsilon$) for these effects equal to:

$$\delta\varepsilon = (1 - \varepsilon_0)\left(1 - \frac{\left(L^3 - L^2\right)(1 - \varepsilon_c)}{(L^3 - L^2) - (L^3 - L^1)P + (L^2 - B_c)Q}\right) \tag{7}$$

where ε_c is the emissivity of the polished aluminium, the term B_c the radiance measured through the box with the cold lid on top and cold base below (effectively equal to the blackbody spectral radiance at the temperature of the aluminium), and P and Q are box-specific parameters between 0 and 1 defined by the box geometry and material properties such that $P = f(\varepsilon_c, \varepsilon_h)$ and $Q = g(\varepsilon_c)$ (see [34] for exact expressions of P and Q).

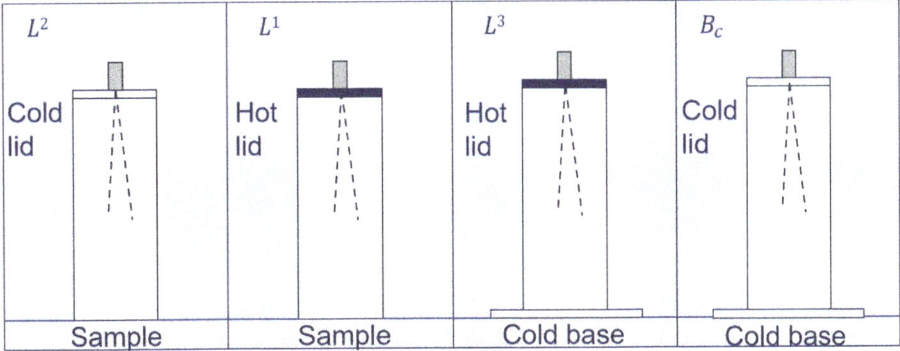

Figure 5. Radiance measurement sequence for the two-lid variant of the emissivity box method, using nomenclature from Rubio et al. [33]. Measurements proceed from left (L^2) to right (B_c).

The final broadband emissivity estimate of the target sample surface is then given by the sum of the outputs from Equations (6) and (7):

$$\varepsilon = \varepsilon_0 + \delta\varepsilon \tag{8}$$

The emissivity of the hot lid is that of the high emissivity paint it was covered in (provided by the manufacturer as $\varepsilon_h = 0.98$). Prior to field deployment, measurements were conducted to determine the emissivity of the cold lid as derived in Appendix 1 of Rubio et al. [34]:

$$\varepsilon = \frac{B(T_{rad})}{\sigma T_{kin}^4} \tag{9}$$

where $B(T)$ is the spectral radiance derived from Planck's radiation law at temperature T (kelvin) as in Equation (4) and σ is the Stefan–Boltzmann constant $\left(5.67 \times 10^{-8}\,\mathrm{Js}^{-1}\mathrm{m}^{-2}\mathrm{K}^{-4}\right)$. Through this method, the emissivity of the polished aluminium (ε_c) of the cold lid was determined as 0.05, resulting in P and Q as 0.0123 and 0.4223 respectively for the box deployed herein. These values were derived again based on measurements made at the end of each field campaign to identify any changes, associated for example with oxidisation of the box interior aluminium surface or damage to the paint of the hot lid, but no such changes were found.

A minimum of five repeated measurements were collected per sample, with averages and standard deviations of the multiple measurements calculated. Uncertainties for each sample were taken as the standard deviation of the multiple measurements. Example values for the two-lid box method developed at King's College London are shown in Table 3.

Table 3. Example values for the two-lid box measurements, with gravel, grass, and sand shown. Measured temperatures are expressed in kelvin and equivalent radiances in brackets in $\mathrm{Wm}^{-2}\mathrm{sr}^{-1}\mu\mathrm{m}^{-1}$.

Sample	L^2	L^1	L^3	B_c	ε_0	$\delta\varepsilon$	ε
Gravel	303.25 (10.27)	304.02 (10.39)	318.36 (12.77)	304.59 (10.48)	0.952	0.001	0.953
Grass	299.23 (9.65)	299.81 (9.74)	317.98 (12.71)	304.02 (10.39)	0.971	−0.002	0.969
Sand	304.02 (10.39)	305.74 (10.66)	322.38 (13.49)	303.44 (10.30)	0.913	0.004	0.917

3.2. Sites and Experimental Samples

Field measurements for the study were made in three locations: a disused airfield at Alconbury (UK), a farm in Grosseto (Italy), and Duxford Aerodrome (UK; Figure 6). In total, fourteen different surface samples were considered, detailed in Table 4 and shown in Figure 7. These included manmade samples such as a large tarpaulin used as calibration targets in an accompanying airborne campaign, and natural samples such as homogeneous areas of grass, sand, and water.

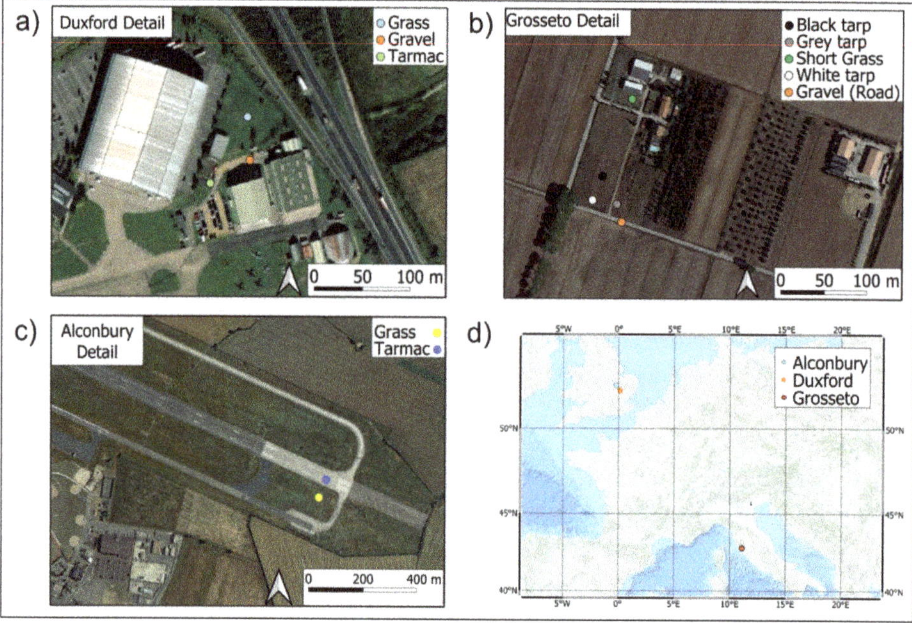

Figure 6. Sites in the UK and Italy where the study measurements took place, showing (**a**) detail of Duxford where measurements of grass, gravel, tarmac, and the tarpaulins were collected; (**b**) detail of Grosseto indicating locations of the same tarpaulin measurements there; (**c**) detail of the runway area at Alconbury, showing the grass and the tarmac where all non-grass samples were placed for field measurement, and (**d**) the relative locations of the three field sites.

Field measurements were collected in the late afternoon at all sites to try to maximise thermal contrast between target radiance and downwelling radiance, as recommended in Salvaggio and Miller [41]. All data were collected under stable and generally low wind, clear sky conditions, with these considered suitable for application of the two-lid box method. Air temperatures at the UK sites were between 19 and 25 °C, with wind speeds recorded in Alconbury and Duxford of 3.6 ± 0.9 ms^{-1} and 4.3 ± 0.3 ms^{-1} respectively. Air temperatures in Italy were higher (between 30 and 32 °C) with a recorded wind speed of 2.3 ± 0.5 ms^{-1}. Relative humidity throughout the measurement period was $44\% \pm 4\%$, $56\% \pm 6\%$, and $60\% \pm 7\%$ in Alconbury, Duxford and Grosseto respectively.

Figure 7. All samples considered in this study other than distilled water. Samples (**a**) to (**d**) were measured for the Alconbury comparison and show (**a**) black hardboard card on the tarmac, (**b**) construction sand, (**c**) green grass in Alconbury, and (**d**) polystyrene. Samples (**e**) to (**l**) were measured in Italy, with (**e**) the beach near Grosseto from where sand was collected, (**f**) the short green grass in Grosseto, (**g**) the gravel drive in Grosseto with emissivity box pictured, (**h**) a close-up the grey tarpaulin in Grosseto, (**i**) the white, grey and black tarpaulin photographed from the plane in Grosseto, Italy while in use as calibration targets, (**j**) gravel in Duxford during measurement, (**k**) green grass in Duxford, and (**l**) tarmac in Duxford.

Table 4. Samples considered for the field and laboratory emissivity inter-comparison, with the number of measurements made using each indicated in brackets in the final column. Note that the three tarpaulins were measured by the EM27 in both Duxford (UK) and Grosseto (Italy).

Sample Description	Sample ID	Location (Field)	Date	Instruments	
				Field	**Lab**
Black hardboard card (5 mm × 240 mm × 303 mm)	Card	Alconbury, UK	May-18	EM27 (3)	Vertex (5)
Polystyrene (40 mm × 160 mm × 150 mm)	Polystyrene	Alconbury, UK	May-18	EM27 (3)	Vertex (4)
Green grass (max. 200 mm height)	Grass_Alc	Alconbury, UK	May-18	EM27 (3)	Vertex (5)
Construction sand	Sand_Alc	Alconbury, UK	May-18	EM27 (3)	Vertex (3)
Distilled Water	DistilledWater	Alconbury, UK	May-18	EM27 (3)	Vertex (4)
Sandy gravel drive	Gravel_Gro	Grosseto, Italy	Jun-19	Box (5), EM27 (1), D&P (3)	-
Short dry grass	Grass_Gro	Grosseto, Italy	Jun-19	Box (5), EM27 (1), D&P (3)	-

Table 4. *Cont.*

Sample Description	Sample ID	Location (Field)	Date	Instruments	
				Field	Lab
Beach sand	Sand_Gro	Grosseto, Italy	Jun-19	Box (5), EM27 (1–Grosseto, 3–Duxford)	Vertex (3)
White PVC tarpaulin (630 gsm)	WhiteTarp	Duxford, UK (Box, EM27); Grosseto, Italy (EM27)	Jun-19	Box (5), EM27 (1–Grosseto, 3–Duxford)	Vertex (3)
Grey polyester tarpaulin (matte finish)	GreyTarp	Duxford, UK (Box, EM27); Grosseto, Italy (EM27; D&P)	Jun-19	Box (5), EM27 (1–Grosseto, 3–Duxford), D&P (3)	Vertex (3)
Black polyester tarpaulin (matte finish)	BlackTarp	Duxford, UK (Box, EM27); Grosseto, Italy (EM27)	Jun-19	Box (5), EM27 (1–Grosseto, 3–Duxford)	Vertex (3)
Gravel driveway (gravel pieces 10–40 mm)	Gravel_Dux	Duxford, UK	Jun-19	Box (5), EM27 (3)	Vertex (6)
Short green grass mixed with clover	Grass_Dux	Duxford, UK	Jun-19	Box (5), EM27 (3)	-
Homogeneous road tarmac	Tarmac	Duxford, UK	Jun-19	Box (5), EM27 (3)	-

3.2.1. Sample Preparation

For the EM27-based field emissivity measurements conducted in Alconbury in May 2018, the black card, construction sand, and polystyrene (Figure 7a,b,d) were placed onto the runway tarmac shown in Figure 6c for measurement. For measurement of the sand, an area greater than the instrument FOV and with a depth of at least 3 cm was prepared on the tarmac. Distilled water was poured into a plastic tray to a depth of 15 mm for measurement, while the grass measurement (Figure 7c) was conducted on the vegetated area neighbouring the runway as indicated in Figure 6c.

For the other field-based emissivity measurements, all samples aside from the beach sand (Sand_Gro) were measured as found and as shown in Figures 6 and 7. A sample of beach sand (Sand_Gro) was collected from the beach shown in Figure 7e for measurement the following day at the same time as the other targets. The emissivity measurements of the tarpaulins, which were being used as calibration targets for airborne remote sensing measurements, were collected when the tarpaulins were laid out for the overhead flights as shown in Figure 6b.

For the laboratory emissivity measurements, flat samples such as the tarps, card, and polystyrene were placed under the sample port of the integrating sphere shown in Figure 1a, with no gap between port and sample. To preserve the structure and moisture content of the Alconbury grass sample, a section of turf was extracted (Figure 8a–b) and measured the following morning in a foil container of high reflectivity, with the grass pressed underneath the sample port while ensuring no blades went inside the integrating sphere. This method was chosen over the method detailed in Salisbury and D'Aria [48] as it better mimicked field conditions. Distilled water, construction sand, and gravel were all placed into petri dishes such as that in Figure 8c, and measured as close to the sphere port as possible without risk of contaminating the inside of the sphere. Due to the uneven shapes and surfaces of the gravel, some gaps were observed between the sample and sphere (distances < 10 mm) as shown in Figure 8d. However, as with the grass, it was determined as preferable to measure the sample unaltered rather than change the structural composition so as to best mimic field conditions.

Figure 8. Laboratory preparation of (**a**) the grass sample from Alconbury from the side and (**b**) from above, (**c**) the gravel sample from Duxford from above, and (**d**) underneath the sample port of the external integrating sphere.

3.3. Emissivity Measurement Comparison

The surface spectral emissivities derived from measurements made by the two portable FTIR spectrometers (EM27 and D&P) along with those from the laboratory setup (Vertex 70) were compared to determine the absolute emissivity differences and the degree of agreement of the identified spectral features. The comparison was limited to the spectral range 8–13 μm, since this covers the wavelength range commonly employed in LST retrieval algorithms [62]. Measurements of the tarpaulins made using the EM27 in both Italy and the UK were compared to enable assessment of the performance of the same method in different environments.

Each FTIR-derived emissivity spectrum was then convolved with the Heitronics K15.85 radiometer spectral response function (Figure 4) to obtain broadband emissivity values comparable with those derived from the emissivity box.

3.4. Evaluation of Impact on LST

To understand the impact of any noted emissivity differences on LST estimation, a scenario was simulated for a near-surface Heitronics KT15.85 radiometer observing six samples measured in Grosseto and Duxford. Atmospheric transmissivity and path radiance effects were negligible due to the near-surface nature of the simulated observations, and sample-specific input values for land surface BT and sky BT were taken from in-situ LWIR radiometer measurements collected during the campaign (Table 5). Input emissivities (ε) used were the broadband emissivities derived for each of the measurement methods used to assess the emissivity of that sample. LSTs corresponding to the radiometer were calculated for each sample input emissivity as in Guillevic et al. [3]:

$$LST = B^{-1}\left[\frac{1}{\varepsilon}\left(L_{surf} - (1-\varepsilon)L_{sky}^{\downarrow}\right)\right] \tag{10}$$

where $B^{-1}(L)$ is the inverse Planck function describing the blackbody equivalent temperature (T, kelvin) of spectral radiance (L, $W.m^{-2}.sr^{-1}.\mu m^{-1}$), L_{surf} the spectral radiance ($W.m^{-2}.sr^{-1}.\mu m^{-1}$) corresponding to the input surface viewing BT, and $L_{sky}^{\downarrow}$ the downwelling atmospheric LWIR spectral radiance ($W.m^{-2}.sr^{-1}.\mu m^{-1}$) corresponding to the sky viewed BT. Uncertainties were calculated and propagated as in Ghent et al. [63] and detailed in Appendix A.

Table 5. Input surface viewing and sky viewing brightness temperatures used to simulate land surface temperature with the measured emissivities.

Sample	Location	Surface Viewing BT (K)	Sky Viewing BT (K)
White Tarpaulin	Grosseto, IT	300	250
Black Tarpaulin	Grosseto, IT	330	250
Grey Tarpaulin	Grosseto, IT	330	250
Grass	Duxford, UK	300	240
Gravel	Duxford, UK	310	240
Tarmac	Duxford, UK	320	240

4. Results

4.1. Emissivity Measurement Inter-comparison

4.1.1. Spectral Emissivities

Emissivity spectra (8–13 μm) for the Alconbury surface samples are shown in Figure 9, as measured in the field using the EM27 and in the laboratory using the Vertex 70. Figure 10 shows the spectral emissivities of the white, grey, and black tarpaulin as measured in the laboratory using the Vertex 70 and in Grosseto and Duxford using the EM27 and D&P spectrometers. Field- and laboratory-derived emissivity spectra of the other samples from Grosseto and Duxford are shown in Figure 11.

Alconbury

Results from Alconbury (Figure 9) enable a direct comparison of the laboratory (Vertex 70) and field-measured (EM27) spectral emissivities. The closest absolute agreement is found for the graybody samples (grass and water), with high and spectrally flat emissivities within 1% of each other between 8 and 12 μm. The laboratory and field measurements of polystyrene and card are generally within 2% across 8–12 μm, but differences of up to 0.04 are observed at points for the polystyrene (for example around 9.56 μm). For the sand sample, the laboratory and field measurements are within 2% between 9.8 μm and 12 μm but there are differences of 10–15% in the restrahlen bands over 8.0–9.5 μm, with the laboratory measured emissivity higher than the EM27 field measured emissivity. For the other samples, between 8 and 12 μm the field-derived emissivities tend generally to be slightly higher than the laboratory-derived values. Beyond 12 μm, there is increased noise in the field-measured (EM27) emissivity spectra, which could be due to increased atmospheric effects at these longer wavelengths (Figure 4).

Some non-physical spectral emissivities (>1) are observed in the laboratory measurements of the graybody samples and in the field measurements of the grass shown in Figure 9. Increased uncertainties are also observed for both field and laboratory measurements of the graybody samples compared to the other samples. Given that the surface temperatures of water and grass are sensitive to even low winds [17], increased field uncertainties and noise for these sample measurements are probably due to varying sample temperatures during the measurement. The emissivities greater than unity found in the laboratory measured data appear also to be largely due to noise. Increased noise for these samples is expected due to the limitation of measuring samples of high spectral emissivity (low spectral reflectance) using a laboratory setup operating in directional hemispherical reflectance mode. An alternative explanation of the non-physical emissivities of the grass sample for both field and laboratory measurements could be canopy scattering, with increased emissivities due to the cavity effect [16].

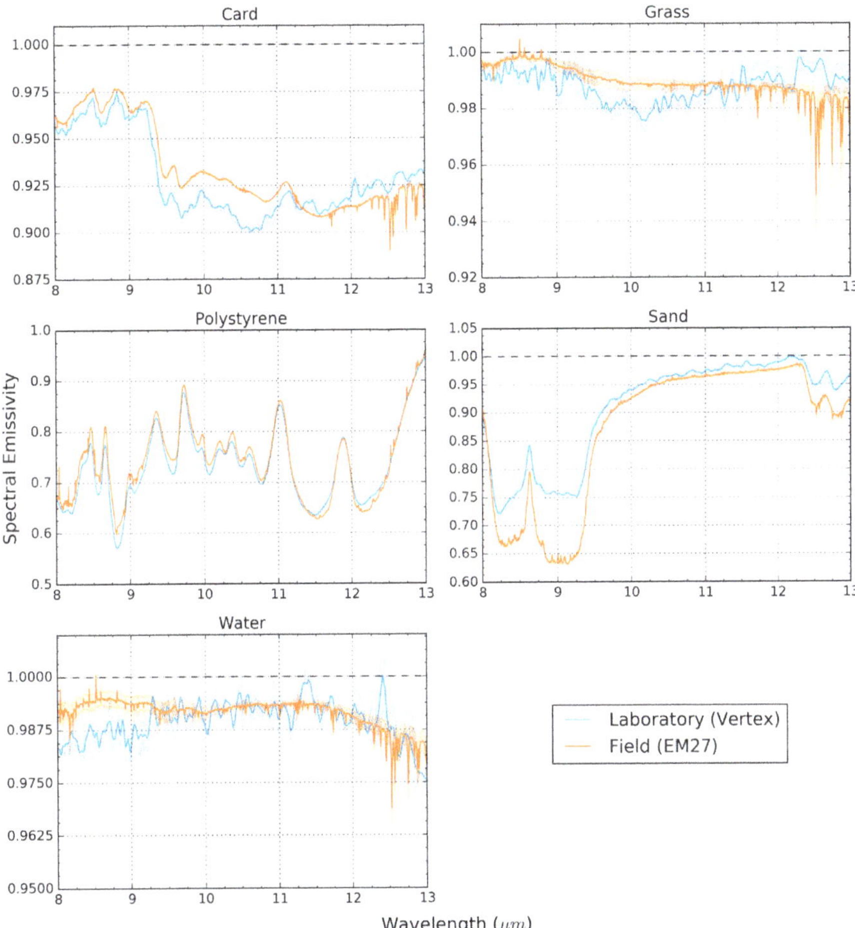

Figure 9. Laboratory-measured (Vertex) and field-measured (EM27) surface spectral emissivities of five different samples, as measured in Alconbury (UK) in May 2018. Values are the mean of all measurements, with the surrounding shaded area indicating the corresponding uncertainty as detailed in Section 3.1. The numbers of measurements made of each sample were listed in Table 4. Grey shaded area indicates the spectral range of the Heitronics KT15.85 IIP radiometer used for the emissivity box measurements.

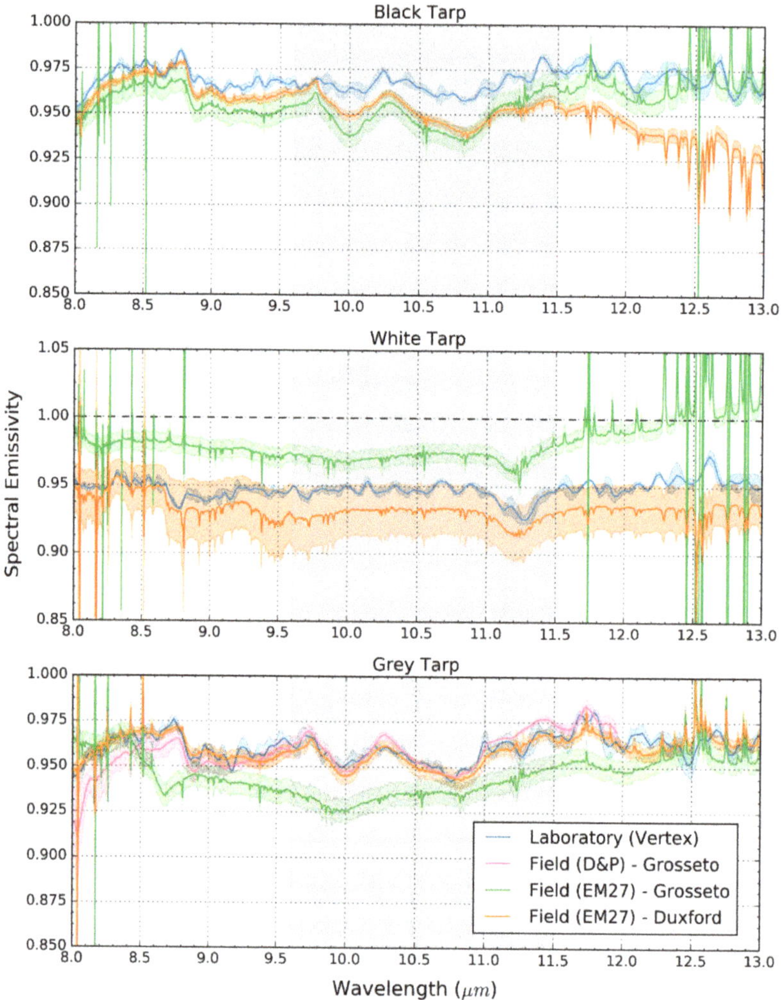

Figure 10. Spectral emissivities of (top to bottom) the black tarpaulin, white tarpaulin, and grey tarpaulin based on data collected in the laboratory (Vertex) and field using (i) the Bruker EM27 FTIR spectrometer (EM27) in both Grosseto, Italy and Duxford, UK and (ii) the Designs and Prototypes µFTIR spectrometer (D&P, grey tarpaulin only). Values are the mean of all measurements, with the surrounding shaded area indicating the corresponding uncertainty as detailed in Section 3.1. The numbers of measurements made of each sample were listed in Table 3. Grey shaded area indicates the spectral range of the Heitronics KT15.85 IIP radiometer used for the emissivity box measurements.

Considerable spectral variability is observed in the three non-graybody surfaces measured at Alconbury, with emissivities going down to about 0.6 (polystyrene, ~8.8 µm). The restrahlen bands (8–9.5 µm) and the Christiansen peak near 12.3 µm are clearly evident in both field and laboratory spectra of sand, although the minima in the restrahlen bands are weaker in the laboratory measurement. Despite absolute differences, the wavelengths at which specific spectral features are observed at correspond very well between the field (EM27) and laboratory (Vertex 70) measurements of the non-graybody samples, particularly for the polystyrene.

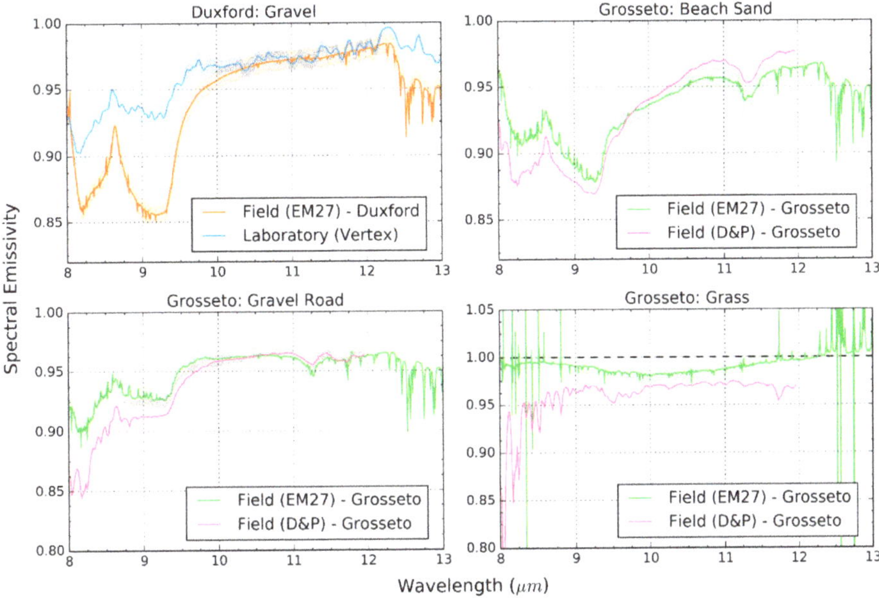

Figure 11. From top left clockwise, spectral emissivity measurements of (i) gravel from Duxford, (ii) beach sand in Grosseto, (iii) the sandy gravel drive in Grosseto, and (iv) short grass in Grosseto. Measurements were made using a Bruker Vertex 70 laboratory setup, a Designs and Prototypes µFTIR spectrometer (D&P) operated in the field, and a Bruker EM27 also operated in the field. Values are the mean of all measurements, with the surrounding shaded area indicating the corresponding uncertainty as detailed in Section 3.1. The numbers of measurements made of each sample were listed in Table 4. Grey shaded area indicates the spectral range of the Heitronics KT15.85 IIP radiometer used for the emissivity box measurements.

Grosseto and Duxford

Considering the spectral measurements of the samples from Grosseto and Duxford, the measurements of the tarpaulin made in the laboratory and those collected using the D&P µFTIR in Grosseto and the EM27 in the field at Duxford are all within 2% of each other between 8.0 and 12.0 µm (Figure 10). These differences are in line with the Alconbury measurements, and comparable with other studies that have compared emissivity measurement approaches [11,15]. However, agreement between the laboratory and field measurements of the gravel sample from Duxford is poor by comparison, with a difference of up to 8% observed between the EM27 (field) and Vertex (lab) measurements in the restrahlen bands between 8.0 and 9.5 µm (Figure 11). Furthermore, as with the sand measurement from Alconbury shown in Figure 9, while the restrahlen bands are clearly evident in the EM27 measurements of the gravel in Duxford (Gravel_Dux), these minima are weaker in the laboratory measurements.

The increase in noise in the derived spectral emissivity data beyond 12 µm for field measurements made using the EM27 in Alconbury (Figure 9) is again observed for all EM27 field measurements in Grosseto and Duxford (Figures 10 and 11 respectively). The EM27 measurements of the white tarpaulin and grass in Grosseto additionally show non-physical emissivities (>1) above 12 µm, which appeared systematic and not attributable solely to noise. Conversely, a decrease in spectral emissivity above 12 µm is observed in the EM27 measurements of the gravel in Grosseto, beach sand in Grosseto, and gravel in Duxford (Figure 11).

Spectral emissivity data of the same tarpaulins measured in the field in Duxford and in Grosseto using the same EM27 setup show larger differences than anticipated, in both spectral shape and magnitude (Figure 10). The Duxford EM27 measurements were in better agreement with the laboratory- and D&P-measured spectra than the Grosseto EM27 measurements, particularly so for the grey tarpaulin where the EM27 measurements collected in Grosseto failed to identify certain spectral features. The EM27 measurements in Duxford by contrast were in close agreement (<1%) to those of the D&P (from Grosseto) and the laboratory measurements. Despite the Duxford measurements appearing to perform relatively better than those from Grosseto, high uncertainties are also observed in the Duxford spectral emissivity measurements of the white tarpaulin. The PVC coating of this particular target had slightly specular characteristics, which may have made the emissivity more variable between measurements as the EM27's retrieval method is intended for samples with Lambertian behaviour surfaces.

The sand, grass, and gravel samples from Grosseto shown in Figure 11 enable direct comparison of the data from the two portable FTIR spectrometers, with measurements collected almost simultaneously and under identical field conditions. The spectral emissivities of sand derived with the EM27 and the D&P instruments were within 1% of each other between 8.5 and 12.0 μm, and the gravel emissivities were within 2% of one another over the same range, with the increased differences for the gravel likely attributable to the increased variability within this material. There was also strong agreement seen between the spectral features for the beach sand and gravel road. These results promote confidence in the emissivity data derived from both FTIR instruments over the 8.5–12.0 μm range. While the measured emissivities of the grass from the two FTIR spectrometers were also within 2% of each, non-physical (>1) noisy emissivities are observed in the EM27 data of the grass sample in Grosseto, as was also observed in this instrument's measurement of the Alconbury grass sample (Figure 9).

Below 8.5 μm, spectral emissivities retrieved using the D&P μFTIR spectrometer seem to be consistently lower than those derived using the EM27, particularly for the grass sample (Grass_Gro) in Figure 11. This could indicate insufficient correction of atmospheric features in the post-processing of the D&P data, since this region has increased absorbance from atmospheric water vapour [17]. Outside this spectral region, EM27-derived emissivity spectra appeared consistently noisier than those from the D&P, as can be observed again in the grass measurements from Grosseto (Figure 11).

4.1.2. Broadband Emissivities

The derived broadband emissivities for all samples measured in Grosseto and Duxford with the EM27 and D&P systems are shown in Figure 11, alongside those derived using the two-lid emissivity box. Agreement between the FTIR-derived values and those of the emissivity box is excellent for some samples, such as the gravel road in Grosseto (Gravel_Gro) where the EM27 and D&P measurements were within 0.1% (0.001) of those of the emissivity box (Figure 12). However, for other samples the emissivity box provides broadband emissivities consistently lower than those of the FTIR systems, with differences of over 5% (0.05) for the grey tarpaulin for example (whereas for this sample the EM27 in Duxford, D&P in Grosseto and the laboratory Vertex 70 deliver broadband emissivities within 0.5% of each other).

Considering the different sample types, we observe that there were consistent discrepancies between the measurements of the two grass samples collected in Grosseto and Duxford using the EM27 and the emissivity box. For both, the emissivities of the vegetation as measured by the box method had a negative bias compared to the EM27 (Figure 12), with the EM27-derived values more in line with vegetation measurements reported elsewhere [39,51]. However, the measurement of the grass sample in Grosseto collected using the D&P was similar to the box-derived emissivity of that sample. Given this unclear performance and the non-physical emissivities observed in the EM27-derived emissivities of the grass samples from Alconbury (Grass_Alc, Figure 9) and Grosseto (Grass_Gro, Figure 11), further work is recommended to understand the performance of the EM27-based system over such targets.

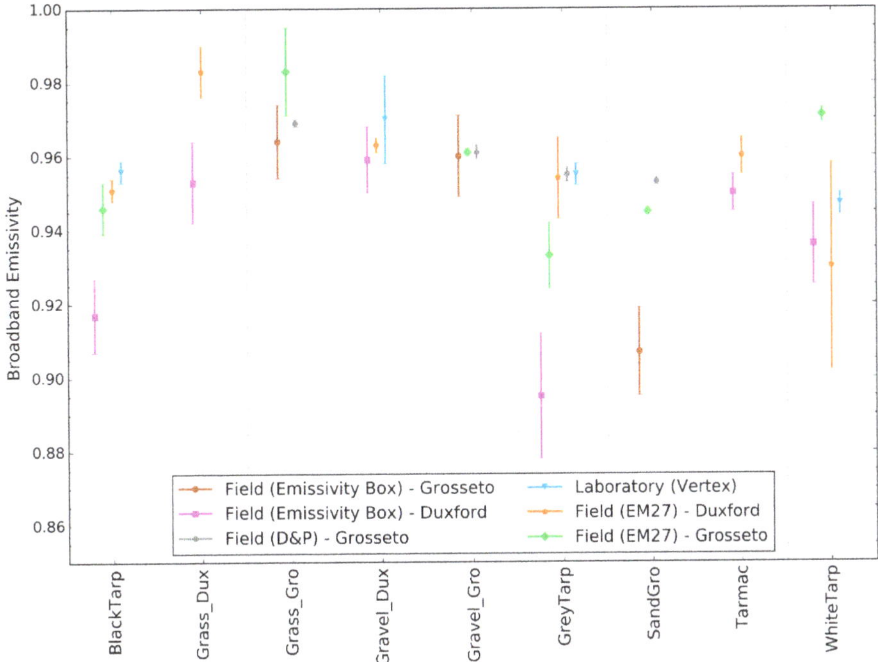

Figure 12. Broadband LWIR emissivities of target samples measured in Grosseto and Duxford by the EM27 and D&P FTIR systems, derived via the convolution of these surface spectral emissivity data with the spectral response function of the Heitronics KT15.85 radiometer shown in Figure 4. Matching broadband emissivity values derived using a two-lid emissivity box and that same Heitronics radiometer are shown alongside. Error bars show the uncertainties of each estimate.

Uncertainties were consistently around 0.010 for the box emissivity measurements, which is comparable with other studies making use of the emissivity box [15,37,52]. However, these were consistently higher than uncertainties from the other methods (with the exception of the white tarpaulin measurement collected using the EM27 in Duxford, which had a band-averaged uncertainty of 0.028). The increased uncertainty, and the lack of consistency found here in the relative emissivity values in comparison to the range of other sensors deployed, lead us to question the ability to reliably use the box method—and thus its suitability for calibration and validation studies.

4.2. Impact of Measurement Differences on LST Estimation

Table 6 shows the LST values and uncertainties calculated as detailed in Section 3.4 for each sample and input emissivity, with maximum and minimum derived LSTs highlighted in red and blue respectively. LSTs derived using emissivities from the two-lid emissivity box were the highest for all samples other than for the white tarpaulin, reflecting the consistent negatively biased emissivities delivered by this method relative to the others. In the case of the grey tarpaulin, the derived LSTs using the box-derived emissivities were 3.92 °C higher than those based on emissivities from the D&P μFTIR, EM27, or Vertex FTIR spectrometer as inputs. The magnitude of this bias again questions the reliability of the emissivity box approach.

Table 6. Calculated LSTs and LST uncertainties (°C) for the six samples, each calculated using the emissivities derived with the various field and laboratory emissivity measurement methods considered herein, with the median and interquartile range (IQR) for each sample.

Sample	Calculated LST (°C)						
	TL-Box	D&P	EM27 (Grosseto)	EM27 (Duxford)	Lab	Median	IQR
WhiteTarp	29.48 ± 0.38		28.00 ± 0.36	29.74 ± 0.45	29.00 ± 0.37	29.24	0.80
GreyTarp	63.44 ± 0.39	59.52 ± 0.37	60.91 ± 0.39	59.59 ± 0.38	59.52 ± 0.37	59.59	1.39
BlackTarp	61.95 ± 0.38		60.09 ± 0.38	59.77 ± 0.37	59.46 ± 0.37	59.93	0.86
Gravel_Dux	28.72 ± 0.37			28.52 ± 0.36	28.20 ± 0.37	28.53	0.26
Grass_Dux	39.23 ± 0.37			37.70 ± 0.36		38.49	0.79
Tarmac	49.75 ± 0.37			49.15 ± 0.37		49.45	0.30

Differences between LSTs calculated using the broadband emissivities derived from the FTIR-based methods deployed herein are smaller than those resulting from use of the box-derived emissivities. However, we still observe LSTs differing by up to 1 °C (white tarpaulin), which, given that the GCOS target accuracy and currently achievable requirements for LST as an ECV are 1 °C and 2–3 °C respectively [28], highlights the continuing importance of reducing uncertainties on emissivity retrieval.

5. Discussion

Comparison of the spectral emissivities found the majority of emissivities from field and laboratory spectrometers to be within 1–2% of each other for most of the spectral range 8.5–12.0 μm. These levels are broadly in line with other studies [11,12] and give confidence in the measurements and methods for selected samples. Outside of 8.5–12.0 μm, differences between emissivity measurements increased and increased noise levels were observed in the EM27 spectra. A potential cause for the reduced performance of the EM27 beyond 12 μm could be extrapolation error from calibration to the cold sky temperatures, which Korb et al. [31] observed to create spectral artefacts between 11 and 13 μm due to nonlinearity of the MCT detector responses over wide signal ranges. However, if this were the case, this decrease in emissivities beyond 12 μm would likely be observed in the EM27 measured spectra of all samples, which was not the case. Furthermore, no systematic distortion in the EM27 spectra was apparent at wavelengths below 12 μm as would likely be if this were the cause. While emissivities from 8.5 to 12.0 μm were satisfactory for most applications from field to satellite scale, with the majority of satellite thermal bands used for LST calculation located in this region [2], further investigation is required if field emissivities accurate outside of this range are required.

The differences between the laboratory (Vertex 70) and field (EM27) measurements in the restrahlen bands of the sand sample from Alconbury (Figure 9) and the gravel sample from Duxford (Figure 11) raise questions about the relative performances of these laboratory and field setups in this region for samples with strong restrahlen features. Further investigation in particular is recommended for the laboratory setup, since the EM27 measurements of the sand and gravel drive samples in Grosseto compared well with the D&P measurements of the same samples (Figure 11). In the case of the gravel sample from Duxford, a likely cause of this lack of agreement is the inhomogeneity of the sample (Figure 8c) together with the different field-of-regards between the two measurement techniques, with the diameter of the sample port in the laboratory setup half that of the field-of-regard for the EM27 in the setup deployed in Duxford (diameters of 25 mm and 50 mm respectively). A contributing factor to the higher emissivity in the laboratory measurement could also be that the gap between the gravel sample and sample port in the laboratory (discussed in Section 3.2.1) decreased the measured sample reflectance and increased the derived emissivity. Although neither of these interpretations are applicable to the sand measurements, these issues highlight the impact that the different scales and designs of laboratory and field instrumentation can have on the retrieved emissivity. This is particularly important when using emissivity from in situ measurements for calibration/validation activities over targets such as gravel that are apparently homogeneous in satellite/airborne sensor pixels but heterogeneous at the sub-pixel level [64]. In this study, field measurements were found to be vital for such samples as they observe a larger area than laboratory measurements and can more easily cope with the target sample structure.

A key limitation of this study was that only one sample was measured by all four methods (GreyTarp, Figure 9). With this sample however, strong agreement was seen between the laboratory, D&P and EM27 (Duxford) while the broadband emissivity from the emissivity box measurement of the sample was considerably lower. The negative bias in emissivity box measurements of this sample and other samples compared to the other methods as shown in Figure 12 questioned the reliability of the emissivity box approach for calibration and validation activities, particularly since use of the emissivities from the box in this study were found to result in positive biases of up to 4 °C when used to simulate in situ LSTs from radiometer observations.

The differences between the EM27 measurements of the tarpaulins made in Grosseto and Duxford indicate field conditions have a strong impact on output emissivities. Interestingly, the EM27 measurements of the tarpaulin from Grosseto were found to be in poorer agreement with the laboratory and D&P measurements from Grosseto than the EM27 measurements of the tarpaulin from Duxford, despite being made at the same time and location as the D&P measurements. However, as shown in Figure 11, other measurements made with the EM27 and D&P in Grosseto were in closer agreement, with the D&P and EM27 measurements of gravel, grass, and sand made in Grosseto within 2% across 8.5–12 µm. More measurements with the D&P (e.g., in Duxford) would have assisted here in understanding the relative performance of these two field spectrometers. Increased noise in the EM27 measurements of the grass sample compared to the D&P measurements was likely an artefact of the different instrument spectral resolutions (EM27 at 0.5 cm^{-1} and D&P at 4 cm^{-1}). However it could also indicate reduced sensitivity in the EM27 instrument compared to D&P. The latter interpretation was identified as the calculated sample temperature of the grass that was five degrees lower than the measured ambient air temperature (30.5 °C), thus reducing signal to noise [41]. This should be investigated further since if this is the case the performance of the EM27 will be limited for measurement of cold samples in hot environments due to the reduced signal to noise ratio for these samples unless modifications are made to improve the sensitivity of the instrument.

Two potential causes were identified for the poor agreement in the EM27 tarpaulin measurements from Grosseto compared to those from Duxford. Firstly, the calculated temperatures of the white and grey tarpaulin were within just 2 °C of the cold blackbody temperature used in the EM27's two temperature calibration, with Hook and Kahle [40] finding that absolute errors in field emissivity measurements increased where the sample temperature was close to the temperature of the calibration blackbody. Care should therefore be taken to ensure the blackbody temperatures are at least 5 °C either side of the estimated sample temperature, but the high ambient temperatures in Grosseto meant the power required to cool the blackbody to the necessary temperature was insufficient—an issue now resolved by the use of a more powerful inverter in the EM27 setup. This could also have been the cause of the high uncertainties in the EM27 measurement of the white tarpaulin from Duxford (Figure 10), since the calculated sample temperature was also only 0.5 °C above the cold blackbody temperature. A second potential cause of the differences in the tarpaulin measurements made in Grosseto and Duxford using the EM27 was identified through comparison of the measured downwelling radiances and humidity in Grosseto and Duxford (not shown). Increased downwelling radiances were observed throughout the Grosseto campaign compared to Duxford, which corresponded with increased humidity (Section 3.2). Furthermore, greater variability between consecutive measurements of the gold panel in Grosseto than in Duxford indicates downwelling radiances were changing more rapidly with time in Grosseto—providing increased scope for changes between the sequential measurements of the panel and the sample being collected. Environmental conditions therefore indicate reduced stability and increased humidity, both factors known to impact the accuracy of retrieved emissivities [17,31]. This interpretation was supported by the increased atmospheric emission lines between 8 and 9 µm apparent in the EM27 spectra collected in Grosseto compared to Duxford (Figure 10). This may have impacted the EM27 measurements more than it did the D&P measurements due to (i) the higher spectral resolution and (ii) the spectral smoothness retrieval method used to derive emissivity from the EM27, which relies on minimizing atmospheric features across 8.12–8.60 µm and therefore is optimal for stable atmospheres. Consideration of the tarpaulin emissivity measurements derived from the two different locations (Grossetto and Duxford) therefore highlights that measurement accuracies and uncertainties were highly sensitive to environmental conditions, and that care should be taken to ensure the blackbody calibration temperatures properly bracket the sample temperature.

The measurement of vegetation was shown to prove challenging for all methods, with non-physical emissivities and high noise levels observed and differences between the broadband emissivities. In the case of the EM27, non-physical emissivities and high noise were attributed to increased variability in sample temperatures during measurement [57]. However, we also observed that the calculated sample temperature of the Alconbury grass (Grass_Alc, Figure 9) was just 1 K hotter than the cold blackbody temperature. As discussed above with the tarpaulin measurements, this could have led to increased errors. With respect to the laboratory measurements, while non-physical emissivities and high noise levels were observed in the measurements of the Alconbury grass (Figure 9), it is difficult to determine if there is a systematic problem with measurement of vegetation or if this was an exception as only one sample was measured in the laboratory. Measurements of additional vegetation samples in the laboratory would have enabled further analysis but no samples were collected in Grosseto and Duxford, as they would have deteriorated before measurement in the laboratory due to the gap between the collection date and measurement date. An increased number of scans would improve the signal-to-noise ratio and is therefore recommended for future measurement of low reflectance samples. However it should also be considered whether a setup operating in the DHR mode could be used to make measurements of vegetation, since vegetation tends to have non-isothermal properties (with different temperatures in different parts of the sample) but Kirchhoff's law theoretically requires samples to be isothermal [47]. Salisbury and D'Aria [48] avoided this by cutting vegetation samples and arranging them in a continuous monolayer on an adhesive tape substrate. However, this is also known to impact the emissivity by changing the structural composition and does not take into account any exposed soil components [65]. The non-isothermal properties of vegetation samples could also be the cause of the non-physical emissivities of the grass sample for the EM27-derived field measurements (which assumes a uniform sample temperature to calculate emissivity with the spectral smoothness method). This supports Ribeiro de Luz and Crowley's [47] argument for development of radiative transfer models that account for non-isothermal structures. Given that one of the major applications of LST from satellite and airborne sensors is monitoring evapotranspiration and crop health [66], further work on measurement of vegetation samples in both the laboratory and the field is therefore recommended. This is particularly important since the vegetation samples considered in both Duxford and Grosseto were limited to homogeneous short cropped grass, while in reality more complex samples containing exposed soil and more complex canopy structures are likely to also need assessing.

This study considered the impact these emissivity differences would have on LST algorithm validation activities through simulating in situ LSTs from field radiometers. However, in situ emissivity values from the laboratory or field instrumentation are also important for the development of LST and land surface emissivity (LSE) retrieval algorithms from satellite or airborne sensors, despite the development of new hyperspectral and multispectral thermal sensors and new physical retrieval algorithms (e.g., [26,67]) capable of simultaneous LST/LSE retrieval without the need for input emissivity estimates from land cover maps or other sources [62]. An example of such an application is in derivation of the coefficients for the Maximum–Minimum Difference (MMD) module in the TES algorithm [26] used to produce the operational Moderate Resolution Imaging Spectroradiometer (MODIS), ASTER, and ECOSTRESS LST/LSE products [53,68]. In this case, a negative bias in emissivity inputs would cause reduced maximum emissivities for the same min–max difference, thus shifting the regression curve, changing the coefficients in the MMD module and impacting the retrieved LSTs and LSEs. It is crucial therefore for LST and LSE retrieval algorithm development and validation activities that work continues on improving and understanding uncertainties surrounding in situ emissivity measurement methods in the field and laboratory.

6. Summary and Conclusions

We conducted an inter-comparison of four different methods of LWIR surface emissivity retrieval, encompassing methods that derived full spectral emissivity data and broadband emissivities, and which operate in the field and in the laboratory. The methods considered are based on field measurements made with two portable FTIR spectrometers (a Bruker EM27 and a D&P µFTIR) operating in the emission mode, a laboratory FTIR spectrometer (Vertex 70) operating in directional hemispherical reflectance mode, and a two-lid emissivity box based on the design of Rubio et al. [33] also deployed in the field. Fourteen target samples were considered across four field sites covering both the UK and Italy, and these include man-made materials such as tarpaulins and natural materials such as sand, grass, and water.

The majority of the derived spectral emissivities were within 1–2% of each other between the major part of the LWIR atmospheric window (8.5–12.0 µm), with identification of spectral features also in agreement between the different field and laboratory approaches. This degree of agreement is consistent with that found by other studies comparing field and laboratory methods of spectral emissivity determination. Differences of up to 15% were observed between the laboratory and field measurements for samples with strong restrahlen features, suggesting a need for further investigation into the laboratory setup's performance when measuring samples with these features. Consideration of the gravel sample from Duxford suggests that field instrumentation can be more suitable than laboratory directional hemispherical reflectance setups for non-homogeneous samples and samples with complex structures. Beyond 12 µm, significant noise and an unexplained drop off in spectral emissivity was observed in certain of the EM27 retrieved emissivities. As a result, we recommend use of EM27 emissivity spectra should be limited to within the 8.0–12.0 µm region. Similarly, although fewer measurements were made using the D&P, increased noise and a decrease in emissivity below 8.5 µm indicates that the D&P-system may deliver emissivities not fully to be trusted below this wavelength, at least in the configuration used herein.

Differences between field measurements made of the same samples using the EM27 but in different locations under different environmental conditions identified some issues. In particular the power supply was inadequate to cool the internal blackbody to the ideal temperature when ambient conditions were particularly warm, leading to the cold blackbody temperature being probably too similar to the target sample temperature to give well calibrated data. This has now been resolved through installation of a higher power inverter. Some increased noise was also evident in certain EM27 measurements, and we recommend that for comparatively cool samples such as vegetation data collection should be done at times to maximise thermal contrast with the surroundings. The time taken to collect each spectral measurement should also be minimised under conditions of potentially changing atmospheric humidity, for example by reducing the number of scans or lowering the measurement spectral resolution (Salisbury [17] advise that 8 cm^{-1} is generally adequate for spectral emissivity determination).

Measurement of vegetation samples was found to be challenging for all methods due to reduced signal-to-noise, canopy scattering, varying sample temperature during the measurement and non-isothermal properties. Using the measured emissivities to simulate near-surface LST observations of grass found differences of 1.5 °C depending on which method of emissivity determination was used. Given that a major application of LSTs is for agriculture and use in evapotranspiration models [6], accurate measurement of the emissivity of vegetation at the field and laboratory scale is crucial, so further work towards understanding the uncertainties at both the field and laboratory scale is recommended.

We derived broadband emissivities from the spectral emissivity measurements and compared these with those calculated using the two-lid emissivity box method. We found a lack of consistency in the emissivity values measured with the box and increased uncertainties compared to the other methods. This indicates that its performance was inferior to that of the FTIR-based approaches, albeit it is based on far cheaper and more available technology.

Author Contributions: Conceptualization, M.F.L., T.P.F.D. and M.W.; Data curation, M.F.L., T.P.F.D., M.W., M.J.G., M.C.d.J. and W.R.J.; Formal analysis, M.F.L.; Funding acquisition, T.P.F.D. and M.W.; Investigation, M.F.L., T.P.F.D., M.W., M.J.G., M.C.d.J. and W.R.J.; Methodology, M.F.L., T.P.F.D. and M.W.; Project administration, M.F.L., T.P.F.D. and M.W.; Resources, M.F.L., T.P.F.D., M.W., J.J., W.R.J., S.J.H. and G.R.; Software, M.F.L.; Supervision, M.F.L., T.P.F.D. and M.W.; Validation, M.F.L.; Visualization, M.F.L. and T.P.F.D.; Writing—original draft, M.F.L., T.P.F.D. and M.W.; Writing—review and editing, M.F.L., T.P.F.D., M.W., M.J.G., S.J.H. and G.R. All authors have read and agreed to the published version of the manuscript.

Funding: Aspects of this work were part of the joint NASA ESA Temperature Sensing Experiment (NET-Sense) conducted under a programme of, and funded by, the European Space Agency (Contract Number 4000131017/20/NL/FF/ab) and the National Aeronautics and Space Administration (NASA), with part of the research described in this paper carried out in part at the Jet Propulsion Laboratory, California Institute of Technology, under contracts with NASA. In addition, aspects were funded through PRISE (Pest Risk Information Service), a project funded by the UK Space Agency as part of the Global Challenge Research Fund. Support for this research also came partly from NERC National Capability funding to the National Centre for Earth Observation (NE/Ro16518/1).

Acknowledgments: We thank Hannah Nyugen, Bruce Main and Francis O'Shea from King's College London for their assistance in development and testing of the emissivity box on multiple field campaigns. The views in this publication can in no way be taken to reflect the official opinion of the European Space Agency or any other funding body.

Conflicts of Interest: The authors declare no conflict of interest.

Appendix A

This appendix presents the calculation of uncertainties associated with the evaluation of the impact on LST presented in Section 3.4. The error sources on LST were identified as that of the surface radiation ($L_\uparrow$), downwelling radiation ($L_\downarrow$), and emissivity (ε). All terms are wavelength (λ) dependent but the wavelength terms were omitted for clarity.

To calculate the uncertainty on the derived LST observations, the equivalent uncertainties in radiance units $\left(U_{L\uparrow/\downarrow}\right)$ for both surface and sky viewing radiometer observations were first determined from the manufacturer stated uncertainty of the radiometer in temperature units ($U_{T\uparrow/\downarrow}$) through the differential of the Planck function with respect to temperature (T) such that:

$$U_{L} = \left|\frac{\partial B}{\partial T}\right| U_{T} = \frac{c_1 c_2 e^{\frac{c_2}{\lambda T}}}{\lambda^6 T^2 \left(e^{\frac{c_2}{\lambda T}} - 1\right)^2} U_{T} \tag{A1}$$

where c_1 and c_2 are constants such that $c_1 = 2hc^2$ and $c_2 = \frac{hc}{k}$ (with h, c and k as defined in Section 3.1.2).

The uncertainty of the land surface radiance ($U_{L_{surf}}$) was then calculated using Equation (8) in Ghent et al. [63] such that:

$$U_{L_{surf}} = L_{surf} \sqrt{\frac{U_{L\uparrow}^2 + \left((1-\varepsilon)L_\downarrow \sqrt{\frac{U_\varepsilon^2}{(1-\varepsilon)^2} + \frac{U_{L\downarrow}^2}{L_\downarrow^2}}\right)^2}{\left(L_\uparrow - L_\downarrow(1-\varepsilon)\right)^2 + \frac{U_\varepsilon^2}{\varepsilon^2}}} \tag{A2}$$

where U_ε is the uncertainty on the emissivity observation. Using the uncertainty of the surface radiance, we then calculated the absolute uncertainty of a given LST observation (U_{LST}) using Equation (9) in Ghent et al. [63]:

$$U_{LST} = C_2 \left(\frac{c_1\left(\frac{U_{L_{surf}}}{\lambda^5 L_{surf}^2}\right)}{\left(\frac{c_1}{L_{surf}\lambda^5} + 1\right)\lambda\left(\ln\frac{c_1}{L_{surf}\lambda^5} + 1\right)^2}\right) \tag{A3}$$

References

1. Norman, J.M.; Becker, F. Terminology in thermal infrared remote sensing of natural surfaces. *Remote. Sens. Rev.* **1995**, *12*, 159–173. [CrossRef]
2. Kuenzer, C.; Dech, S. Theoretical Background of Thermal Infrared Remote Sensing. In *Land Remote Sensing and Global Environmental Change*; Springer Science and Business Media LLC: Berlin/Heidelberg, Germany, 2013; Volume 17, pp. 1–26.
3. Guillevic, P.; Göttsche, F.; Nickeson, J.; Hulley, G.; Ghent, D.; Yu, Y.; Trigo, I.; Hook, S.; Sobrino, J.A.; Remedios, J.; et al. *Land Surface Temperature Product Validation Best Practice Protocol, Version 1.1.*; National Aeronautics and Space Administration: Washington, DC, USA, 2018; p. 58. [CrossRef]
4. Li, Z.-L.; Tang, B.-H.; Wu, H.; Ren, H.; Yan, G.; Wan, Z.; Trigo, I.F.; Sobrino, J.A. Satellite-derived land surface temperature: Current status and perspectives. *Remote. Sens. Environ.* **2013**, *131*, 14–37. [CrossRef]
5. Jiménez-Muñoz, J.C.; Sobrino, J.A. A generalized single-channel method for retrieving land surface temperature from remote sensing data. *J. Geophys. Res. Space Phys.* **2003**, *108*. [CrossRef]
6. Anderson, M.; Kustas, W. Thermal Remote Sensing of Drought and Evapotranspiration. *Eos* **2008**, *89*, 233–234. [CrossRef]
7. Hulley, G.; Hook, S.; Fisher, J.; Lee, C. ECOSTRESS, A NASA Earth-Ventures Instrument for studying links between the water cycle and plant health over the diurnal cycle. In Proceedings of the 2017 IEEE International Geoscience and Remote Sensing Symposium (IGARSS), Fort Worth, TX, USA, 23–28 July 2017; pp. 5494–5496.
8. World Meteorological Organization Essential Climate Variables. Available online: https://public.wmo.int/en/programmes/global-climate-observing-system/essential-climate-variables (accessed on 16 November 2020).
9. Li, Z.-L.; Wu, H.; Wang, N.; Qiu, S.; Sobrino, J.A.; Wan, Z.; Tang, B.-H.; Yan, G. Land surface emissivity retrieval from satellite data. *Int. J. Remote. Sens.* **2013**, *34*, 3084–3127. [CrossRef]
10. Hecker, C.; Smith, T.E.L.; da Luz, B.R.; Wooster, M.J. Thermal Infrared Spectroscopy in the Laboratory and Field in Support of Land Surface Remote Sensing. In *Land Remote Sensing and Global Environmental Change*; Springer Science and Business Media LLC: Berlin/Heidelberg, Germany, 2013; Volume 17, pp. 43–67.
11. Göttsche, F.-M.; Hulley, G.C. Validation of six satellite-retrieved land surface emissivity products over two land cover types in a hyper-arid region. *Remote. Sens. Environ.* **2012**, *124*, 149–158. [CrossRef]
12. Göttsche, F.-M.; Olesen, F.; Poutier, L.; Langlois, S.; Wimmer, W.; Santos, V.G.; Coll, C.; Niclos, R.; Arbelo, M.; Monchau, J.-P. *Report from the Field Inter-Comparison Experiment (FICE) for Land Surface Temperature (OFE-D130-LSR-FICE-Report-V1-Iss-1-Ver-1*; ESA: Noordwijk, The Netherlands, 2018.
13. Silvestri, M.; Musacchio, M.; Cammarano, D.; Fabrizia Buongiorno, M.; Amici, S.; Piscini, A. Comparison of in-situ measurements and satellite-derived surface emissivity over Italian volcanic areas. *EGUGA* **2016**, *18*, 1751.
14. Sobrino, J.A.; Jimenez, J.C.; Labed-Nachbrand, J.; Nerry, F. Surface emissivity retrieval from Digital Airborne Imaging Spectrometer data. *J. Geophys. Res. Space Phys.* **2002**, *107*, 24. [CrossRef]
15. Mira, M.; Schmugge, T.; Valor, E.; Caselles, V.; Coll, C. Comparison of Thermal Infrared Emissivities Retrieved with the Two-Lid Box and the TES Methods with Laboratory Spectra. *IEEE Trans. Geosci. Remote. Sens.* **2009**, *47*, 1012–1021. [CrossRef]
16. Sobrino, J.A.; Jimenezmunoz, J.; Verhoef, W. Canopy directional emissivity: Comparison between models. *Remote. Sens. Environ.* **2005**, *99*, 304–314. [CrossRef]
17. Salisbury, J.W. *Spectral Measurements Field Guide*; Earth Satellite Corporation: Washington, DC, USA, 1998.
18. Strackerjan, K.-E.; Kerekes, J.P.; Salvaggio, C. Spectral reflectance and emissivity of man-made surfaces contaminated with environmental effects. *Opt. Eng.* **2008**, *47*, 106201. [CrossRef]
19. Kotthaus, S.; Smith, T.E.; Wooster, M.J.; Grimmond, C. Derivation of an urban materials spectral library through emittance and reflectance spectroscopy. *ISPRS J. Photogramm. Remote. Sens.* **2014**, *94*, 194–212. [CrossRef]
20. Meerdink, S.K.; Hook, S.J.; Roberts, D.A.; Abbott, E.A. The ECOSTRESS spectral library version 1.0. *Remote. Sens. Environ.* **2019**, *230*, 111196. [CrossRef]
21. Snyder, W. Thermal Infrared (3–14 μm) bidirectional reflectance measurements of sands and soils. *Remote. Sens. Environ.* **1997**, *60*, 101–109. [CrossRef]
22. Laukamp, C.; Lau, I.; Mason, P.; Warren, P.; Huntington, J.; Green, A.; Whitbourn, L.; Wright, W.; Connor, P.; Lau, I.C. *CSIRO Thermal Infrared Spectral Library—Part 1: Evaluation and Status Report*; CSIRO: Canberra, Australia, 2015.

23. Hulley, G.C.; Hook, S.J. Intercomparison of versions 4, 4.1 and 5 of the MODIS Land Surface Temperature and Emissivity products and validation with laboratory measurements of sand samples from the Namib desert, Namibia. *Remote. Sens. Environ.* **2009**, *113*, 1313–1318. [CrossRef]
24. Sabol, J.D.E.; Gillespie, A.R.; Abbott, E.; Yamada, G. Field validation of the ASTER Temperature–Emissivity Separation algorithm. *Remote. Sens. Environ.* **2009**, *113*, 2328–2344. [CrossRef]
25. Schmugge, T.; Ogawa, K. Validation of Emissivity Estimates from ASTER and MODIS Data. In Proceedings of the 2006 IEEE International Symposium on Geoscience and Remote Sensing, Sydney, Australia, 9–13 July 2001; Institute of Electrical and Electronics Engineers (IEEE): Piscataway, NJ, USA, 2006; pp. 260–262.
26. Gillespie, A.; Rokugawa, S.; Matsunaga, T.; Cothern, J.S.; Hook, S.; Kahle, A.B. A temperature and emissivity separation algorithm for Advanced Spaceborne Thermal Emission and Reflection Radiometer (ASTER) images. *IEEE Trans. Geosci. Remote. Sens.* **1998**, *36*, 1113–1126. [CrossRef]
27. Trigo, I.F.; Peres, L.F.; Dacamara, C.C.; Freitas, S.C. Thermal Land Surface Emissivity Retrieved From SEVIRI/Meteosat. *IEEE Trans. Geosci. Remote. Sens.* **2008**, *46*, 307–315. [CrossRef]
28. Martins, J.P.; Coelho e Freitas, S.; Trigo, I.F.; Barroso, C.; Macedo, J. *Copernicus Global Land Operations-Lot I "Vegetation and Energy" Algorithm Theoretical Basis Document, Land Surface Temperature—LST V1:2, Issue I1.41*; Instituto Português do Mar e da Atmosfera: Lisbon, Portugal, 2019.
29. Sobrino, J.A.; Jimenez, J.C.; Soria, G.; Gomez, M.; Barella-Ortiz, A.; Romaguera, M.; Zaragoza, M.; Julien, Y.; Cuenca, J.; Atitar, M.; et al. Thermal remote sensing in the framework of the SEN2FLEX project: Field measurements, airborne data and applications. *Int. J. Remote. Sens.* **2008**, *29*, 4961–4991. [CrossRef]
30. Langsdale, M.F.; Wooster, M.J.; Harrison, J.J.; Koehl, M.; Hecker, C.A.; Hook, S.J.; Abbott, E.A.; Johnson, W.R.; Maturilli, A.; Poutier, L.; et al. Spectral emissivity (SE) measurement uncertainties across 2.5–14 µm derived from a round-robin study made across international laboratories. *Remote Sens..* submitted.
31. Korb, A.R.; Dybwad, P.; Wadsworth, W.; Salisbury, J.W. Portable Fourier transform infrared spectroradiometer for field measurements of radiance and emissivity. *Appl. Opt.* **1996**, *35*, 1679. [CrossRef] [PubMed]
32. Salvaggio, C.; Miller, C.J. Comparison of field- and laboratory-collected midwave and longwave infrared emissivity spectra/data reduction techniques. In *Aerospace/Defense Sensing, Simulation, and Controls*; Shen, S.S., Descour, M.R., Eds.; SPIE: London, UK, 2001; Volume 4381, pp. 549–558.
33. Rubio, E.; Caselles, V.; Coll, C.; Valour, E.; Sospedra, F. Thermal–infrared emissivities of natural surfaces: Improvements on the experimental set-up and new measurements. *Int. J. Remote. Sens.* **2003**, *24*, 5379–5390. [CrossRef]
34. Rubio, E. Emissivity measurements of several soils and vegetation types in the 8–14, µm Wave band: Analysis of two field methods. *Remote. Sens. Environ.* **1997**, *59*, 490–521. [CrossRef]
35. Buettner, K.J.K.; Kern, C.D. The determination of infrared emissivities of terrestrial surfaces. *J. Geophys. Res. Space Phys.* **1965**, *70*, 1329–1337. [CrossRef]
36. Combs, A.C.; Weickmann, H.K.; Mader, C.; Tebo, A. Application of Infrared Radiometers to Meteorology. *J. Appl. Meteorol.* **1965**, *4*, 253–262. [CrossRef]
37. Göttsche, F.-M.; Olesen, F.-S. Improved field method for determining land surface emissivity. *Recent Adv. Quant. Remote Sens.* **2017**, *55*, 4743–4756.
38. Nerry, F.; Labed, J.; Stoll, M.P. Spectral properties of land surfaces in the thermal infrared: 2. Field method for spectrally averaged emissivity measurements. *J. Geophys. Res. Space Phys.* **1990**, *95*, 7045. [CrossRef]
39. Sobrino, J.A.; Caselles, V. A field method for measuring the thermal infrared emissivity. *ISPRS J. Photogramm. Remote. Sens.* **1993**, *48*, 24–31. [CrossRef]
40. Hook, S.J.; Kahle, A.B. The micro fourier transform interferometer (µFTIR)—A new field spectrometer for acquisition of infrared data of natural surfaces. *Remote. Sens. Environ.* **1996**, *56*, 172–181. [CrossRef]
41. Salvaggio, C.; Miller, C.J. Methodologies and protocols for the collection of midwave and longwave infrared emissivity spectra using a portable field spectrometer. In *Aerospace/Defense Sensing, Simulation, and Controls*; Shen, S.S., Descour, M.R., Eds.; SPIE: London, UK, 2001; Volume 4381, pp. 539–549.
42. Jiménez-Muñoz, J.C.; Sobrino, J.A. Error sources on the land surface temperature retrieved from thermal infrared single channel remote sensing data. *Int. J. Remote. Sens.* **2006**, *27*, 999–1014. [CrossRef]
43. Pérez-Planells, L.; Valor, E.; Coll, C.; Niclòs, R. Comparison and Evaluation of the TES and ANEM Algorithms for Land Surface Temperature and Emissivity Separation over the Area of Valencia, Spain. *Remote. Sens.* **2017**, *9*, 1251. [CrossRef]

44. Monchau, J.-P.; Marchetti, M.; Ibos, L.; Dumoulin, J.; Feuillet, V.; Candau, Y. Infrared Emissivity Measurements of Building and Civil Engineering Materials: A New Device for Measuring Emissivity. *Int. J. Thermophys.* **2013**, *35*, 1817–1831. [CrossRef]

45. Wan, Z. New refinements and validation of the MODIS Land-Surface Temperature/Emissivity products. *Remote. Sens. Environ.* **2008**, *112*, 59–74. [CrossRef]

46. Wan, Z.; Li, Z.-L. MODIS Land Surface Temperature and Emissivity Products. In *Land Remote Sensing and Global Environmental Changes*; Ramachandran, B., Justice, C.O., Abrams, M.J., Eds.; Springer International Publishing: London, UK, 2010; pp. 563–577.

47. da Luz, B.R.; Crowley, J.K. Spectral reflectance and emissivity features of broad leaf plants: Prospects for remote sensing in the thermal infrared (8.0–14.0 μm). *Remote. Sens. Environ.* **2007**, *109*, 393–405. [CrossRef]

48. Salisbury, J.W.; D'Aria, D.M. Emissivity of terrestrial materials in the 8–14 μm atmospheric window. *Remote. Sens. Environ.* **1992**, *42*, 83–106. [CrossRef]

49. Salisbury, J.W.; Wald, A.; D'Aria, D.M. Thermal-infrared remote sensing and Kirchhoff's law: 1. Laboratory measurements. *J. Geophys. Res. Space Phys.* **1994**, *99*, 11897–11911. [CrossRef]

50. Göttsche, F.-M.; Olesen, F.-S.; Trigo, I.F.; Bork-Unkelbach, A.; Martin, M.A. Long Term Validation of Land Surface Temperature Retrieved from MSG/SEVIRI with Continuous in-Situ Measurements in Africa. *Remote. Sens.* **2016**, *8*, 410. [CrossRef]

51. Kant, Y.; Badarinath, K.V.S. Ground-based method for measuring thermal infrared effective emissivities: Implications and perspectives on the measurement of land surface temperature from satellite data. *Int. J. Remote. Sens.* **2002**, *23*, 2179–2191. [CrossRef]

52. Mallick, J.; Singh, C.K.; Shashtri, S.; Rahman, A.; Mukherjee, S. Land surface emissivity retrieval based on moisture index from LANDSAT TM satellite data over heterogeneous surfaces of Delhi city. *Int. J. Appl. Earth Obs. Geoinformat.* **2012**, *19*, 348–358. [CrossRef]

53. Coll, C.; Santos, V.G.; Niclòs, R.; Caselles, V. Test of the MODIS Land Surface Temperature and Emissivity Separation Algorithm with Ground Measurements Over a Rice Paddy. *IEEE Trans. Geosci. Remote. Sens.* **2016**, *54*, 3061–3069. [CrossRef]

54. Niclòs, R.; Galve, J.M.; Valiente, J.A.; Estrela, M.J.; Coll, C. Accuracy assessment of land surface temperature retrievals from MSG2-SEVIRI data. *Remote. Sens. Environ.* **2011**, *115*, 2126–2140. [CrossRef]

55. Kanani, K.; Poutier, L.; Nerry, F.; Stoll, M.-P. Directional effects consideration to improve out-doors emissivity retrieval in the 3–13 m domain. *Opt. Express* **2007**, *15*, 12464. [CrossRef] [PubMed]

56. Ninomiya, Y.; Matsunaga, T.; Yamaguchi, Y.; Ogawa, K.; Rokugawa, S.; Uchida, K.; Muraoka, H.; Kaku, M. A comparison of thermal infrared emissivity spectra measured in situ, in the laboratory, and derived from thermal infrared multispectral scanner (TIMS) data in Cuprite, Nevada, U.S.A. *Int. J. Remote. Sens.* **1997**, *18*, 1571–1581. [CrossRef]

57. Horton, K.A.; Johnson, J.R.; Lucey, P.G. Infrared Measurements of Pristine and Disturbed Soils 2. Environmental Effects and Field Data Reduction. *Remote. Sens. Environ.* **1998**, *64*, 47–52. [CrossRef]

58. Hecker, C.; Hook, S.J.; van der Meijde, M.; Bakker, W.; van Der Werff, H.; Wilbrink, H.; van Ruitenbeek, F.J.; de Smeth, B.; van der Meer, F.D. Thermal Infrared Spectrometer for Earth Science Remote Sensing Applications—Instrument Modifications and Measurement Procedures. *Sensors* **2011**, *11*, 10981–10999. [CrossRef]

59. Nicodemus, F.E. Directional Reflectance and Emissivity of an Opaque Surface. *Appl. Opt.* **1965**, *4*, 767–775. [CrossRef]

60. Bruker Corporation. *Opus RS/E Analysis and Control Sorftware for the Remote Sensing System EM27: User Manual*; Bruker Corporation: Billerica, MA, USA, 2010.

61. Berk, A.; Anderson, G.P.; Acharya, P.K.; Bernstein, L.S.; Muratov, L.; Lee, J.; Fox, M.; Adler-Golden, S.M.; Chetwynd, J.H.; Hoke, M.L.; et al. MODTRAN 5: A reformulated atmospheric band model with auxiliary species and practical multiple scattering options: Update. *Def. Security* **2005**, *5806*, 662–667. [CrossRef]

62. Hulley, G.C.; Ghent, D.; Göttsche, F.M.; Guillevic, P.C.; Mildrexler, D.J.; Coll, C. Land Surface Temperature. In *Taking the Temperature of the Earth*; Elsevier BV: Amsterdam, The Netherlands, 2019; pp. 57–127.

63. Ghent, D.; Dodd, E.; Lerebourg, C. Ground-Based Observations for Validation (GBOV) of Copernicus Global Land Products, Algorithm Theoretical Basis Document Land Surface Temperature products. *Geophysical Res. Abstr.* **2018**, *21*, 1773.

64. Guillevic, P.C.; Privette, J.L.; Coudert, B.; Palecki, M.A.; Demarty, J.; Ottlé, C.; Augustine, J.A. Land Surface Temperature product validation using NOAA's surface climate observation networks—Scaling methodology for the Visible Infrared Imager Radiometer Suite (VIIRS). *Remote. Sens. Environ.* **2012**, *124*, 282–298. [CrossRef]

65. Meerdink, S.K.; Roberts, D.; Hulley, G.; Gader, P.; Pisek, J.; Adamson, K.; King, J.; Hook, S.J. Plant species' spectral emissivity and temperature using the hyperspectral thermal emission spectrometer (HyTES) sensor. *Remote. Sens. Environ.* **2019**, *224*, 421–435. [CrossRef]

66. Hu, X.; Ren, H.; Tansey, K.; Zheng, Y.; Ghent, D.; Liu, X.; Yan, L. Agricultural drought monitoring using European Space Agency Sentinel 3A land surface temperature and normalized difference vegetation index imageries. *Agric. For. Meteorol.* **2019**, *279*, 107707. [CrossRef]

67. Masiello, G.; di Serio, C.; de Feis, I.; Amoroso, M.; Venafra, S.; Trigo, I.F.; Watts, P.D. Kalman filter physical retrieval of surface emissivity and temperature from geostationary infrared radiances. *Atmos. Meas. Tech.* **2013**, *6*, 3613–3634. [CrossRef]

68. Silvestri, M.; Romaniello, V.; Hook, S.; Musacchio, M.; Teggi, S.; Buongiorno, M.F. First Comparisons of Surface Temperature Estimations between ECOSTRESS, ASTER and Landsat 8 over Italian Volcanic and Geothermal Areas. *Remote. Sens.* **2020**, *12*, 184. [CrossRef]

Publisher's Note: MDPI stays neutral with regard to jurisdictional claims in published maps and institutional affiliations.

MDPI

St. Alban-Anlage 66

4052 Basel

Switzerland

Tel. +41 61 683 77 34

Fax +41 61 302 89 18

www.mdpi.com

Remote Sensing Editorial Office

E-mail: remotesensing@mdpi.com

www.mdpi.com/journal/remotesensing